The Soviet Sp

Florida A&M University, Tallahassee
Florida Atlantic University, Boca Raton
Florida Gulf Coast University, Ft. Myers
Florida International University, Miami
Florida State University, Tallahassee
University of Central Florida, Orlando
University of Florida, Gainesville
University of North Florida, Jacksonville
University of South Florida, Tampa
University of West Florida, Pensacola

The Soviet Space Race with Apollo

Asif A. Siddiqi

University Press of Florida

Gainesville · Tallahassee · Tampa · Boca Raton
Pensacola · Orlando · Miami · Jacksonville · Ft. Myers

First cloth printing 2000, as second half of *Challenge to Apollo: The Soviet Union and the Space Race, 1945–1974,* by National Aeronautics and Space Administration, History Division, NASA Sp-2000–4408
First paperback printing 2003, by University Press of Florida
Printed in the United States of America on acid-free paper
All rights reserved for the 2003 edition by University Press of Florida

08 07 06 05 04 03   6 5 4 3 2 1

Library of Congress Cataloging-in-Publication Data
Siddiqi, Asif A., 1966–
The Soviet space race with Apollo / Asif A. Siddiqi.
p. cm.
Originally the second half of Challenge to Apollo: The Soviet Union
    and the Space Race, 1945–1974, a single volume published in 2000.
Includes bibliographical references and index.
ISBN 0-8130-2628-8 (alk. paper)
1. Astronautics–Soviet Union–History. 2. Space race–History.
I. Siddiqi, Asif A., 1966–. II. Title.
TL789.8.S65 S49   2003
629.4'0947–dc21   2002032144

The University Press of Florida is the scholarly publishing agency for the State University System of Florida, comprising Florida A&M University, Florida Atlantic University, Florida Gulf Coast University, Florida International University, Florida State University, University of Central Florida, University of Florida, University of North Florida, University of South Florida, and University of West Florida.

University Press of Florida
15 Northwest 15th Street
Gainesville, FL 32611–2079
http://www.upf.com

To my mother and my father,

who taught me the value of knowledge

Contents

Appendices

The Soviet Space Race with Apollo was originally the second half of Challenge to Apollo: The Soviet Union and the Space Race, 1945–1974, a single volume published in 2000 by the National Aeronautics and Space Administration as part of the NASA History Series. The present volume includes chapters 12–20 accompanied by tables and appendices (pages 517–977 of that book). Its companion volume, Sputnik and the Soviet Space Challenge, includes the first half of the original publication.

History is always written wrong, and so always needs to be rewritten.

—*George Santayana*

You can't cross the sea merely by standing and staring at the water.

—*Rabindranath Tagore*

ACKNOWLEDGMENTS

This book is, in essence, sixteen years in the making. I first attempted to compile a history of the Soviet space program in 1982 when, encouraged by my mother, I put together a rough chronology of the main events. A decade later, while living on a couch in a college friend's apartment, I began writing what I thought would be a short history of the Soviet lunar landing program. The first draft was sixty-nine pages long. Late the following year, I decided to expand the topic to handle all early Soviet piloted exploration programs. That work eventually grew into what you are holding in your hand now. I wrote most of it from 1994 to 1997 in Northampton and Amherst, Massachusetts, and in Philadelphia, Pennsylvania, and completed the manuscript in December 1998.

It would have been difficult, if not impossible, to write this book without the generous assistance of numerous individuals who have spent years and in some cases decades trying to understand and analyze the arcana of the former Soviet space program. Writing this book was as much an exploration into research as it was a long journey making new acquaintances and friends from all over the world.

I am indebted to the staff at the NASA History Office—in particular to Lee Saegesser, Steve Garber, and Nadine Andreassen—for assisting me in my research despite their own busy schedules. Louise Alstork, Jonathan Friedman, Patricia Talbert, and John Edison Betts, Jr., were especially patient in proofreading the entire manuscript, making changes where necessary, and helping with the layout and final production. A most special note of gratitude to NASA Chief Historian Roger Launius, whose comments, encouragement, and guidance were indispensable to the completion of this book. This book would not have existed without his untiring support.

Of all the individuals who helped me with this book, I would particularly like to acknowledge the contributions of Peter Gorin, a recent Guggenheim Fellow at the Smithsonian's National Air and Space Museum. He generously shared notes, books, articles, comments, and his encyclopedic knowledge on any and every topic related to the Soviet space program. He also provided a lion's share of the illustrations and drawings used in this book.

I would also mention U.S. Air Force Lt. Colonel William P. Barry. Our countless exchanges via e-mail or letters were invaluable to coming to some understanding of the political and institutional processes that drove the early Soviet space program. I would have found it virtually impossible to write this book without the use of his remarkable analyses of the institutional history of the Soviet space program.

Dwayne Day at the Space Policy Institute at The George Washington University was an invaluable source for recently declassified U.S. government documents relating to Soviet programs. His insightful comments allowed me to frame many of my arguments in a more cogent manner.

My exchanges over e-mail with Bart Hendrickx and Mark Hillyer were essential to reexamining old events in a new light and eliminating many errors. Bart was very generous in sharing any materials that passed his way—a rare quality that set him apart from many other scholars in the field. Mark's vast knowledge about the sometimes mind-boggling and confusing histories of Soviet-era design bureaus helped me with my own studies on this arcane topic.

I want to give special thanks to T.A. Heppenheimer for his support throughout the project. His exhaustive knowledge of the history of space technology was useful in framing many of the main arguments of this work.

Igor Lissov of VideoCosmos in Moscow warrants special mention. He graciously answered many of my questions.

A very special thank you goes to Dennis Newkirk, the Russian aerospace analyst, for his never-ending efforts to provide me with materials useful for my work. He was one of the first Soviet "space watchers" with whom I communicated, and it was partially because of his encouragement that I began writing this book in the first place.

I would like to thank Glen Swanson, the founding editor of *Quest: The History of Spaceflight Magazine*, for his constant material and moral support during the writing of this book. Glen also graciously shared his complete database of remarkable photographs, many of which I used in the book.

I would like to acknowledge the tireless contributions of the late Maxim Tarasenko for kindly facilitating my contacts with the RKK Energiya archivist Georgiy Vetrov. Maxim was one of the most brilliant space historians of his generation, and he will be sorely missed.

A very special note of thanks goes to Sergey Voevodin of Kostroma, Russia, who was an excellent source of information on the Soviet cosmonaut team.

Special thanks go out to Dr. Sergey Khrushchev, the late Dr. Georgiy Vetrov, and Dr. Gerbert Yefremov, all of whom graciously agreed to answer questions relating to the history of the Soviet space program and their own role in its early years. Their comments were invaluable in ferreting out details of previously hidden events in Soviet space history.

In addition to those named above, I would like to acknowledge the following for not only providing me with Russian-language material and photographs but also in many cases for answering detailed questions on various topics. They are: Igor Afanasyev, Vladimir Agapov, Michael Cassutt, Andrew Chaikin, Phillip Clark, James Harford, Christian Lardier, Jonathan McDowell, Peter Pesavento, and Joel Powell. I would also like to thank the following for providing many of the diagrams or photographs for the work: R. F. Gibbons, Dietrich Haeseler, Don Pealer, Charles Vick, Mark Wade, David Woods, and Steven Zaloga.

On a personal note, I would like to thank my mentor in Bangladesh, Fazlul Huq, and also Paul Thompson, Fred Ruppel, Inji, Sabina, Munmun, Karin Bell, Nadeem, Rahila, Bill Sparks, Rika, Dave Parnell, Pat, Jacqueline, the late musician Greg Black, Becky (who inspires), Heather, Karen Barth, Jill, and Danny. Thanks to the amazing Mandar Jayawant in the hope that he will one day forgive me for not sharing my lunch with him. Thanks also to Paul, Iggy, Kurt, and Maynard.

A very special thank you to Anoo for being her wonderful self and for being so interested.

A very special note of love to my mother, who always told me that I could achieve anything as long as I worked hard enough for it; to my father, who bought me my first encyclopedia on spaceflight when I was twelve years old; to my sister Rochona, who endured years of my uncontrollable excitement over Soyuzes and Salyuts as an adolescent; and to her husband David, who graciously offered me an abode in which I could finish the book. Finally, a special thank you to Karen for being my best friend through all of this. I could not have written this book without her love and friendship.

PREFACE

On the Internet one day, I came upon a discussion of the space dog Layka, launched into orbit by the Soviet Union in 1957. Some people believed that the dog had died when oxygen finally ran out in her cabin. Others had heard that an automatic injection of poison had put her to sleep. Still others had read somewhere that poor Layka had literally burned up in the atmosphere when her capsule gradually fell to Earth. No one could point to a single source with any reasonable claim to authority on Layka's ultimate fate. The same news group carried a spirited discussion of U.S. space policy. The topic of choice was the heady period after the first Apollo Moon landing in 1969, in particular the political maneuvering behind the Nixon administration's approval of the Space Shuttle in 1972. Instead of quibbling over historical events, the emphasis was clearly on interpretation—a problem that had more to do with analysis than simply verifying the facts. The contrast between these two threads of conversation perfectly illustrated both the challenges and the differences in writing histories of the Soviet and American space programs. In one case, we are still disputing elementary facts and sources. In the other, we are disputing interpretations of facts and sources.

As astonishing as it may seem, the story of the Soviet space program, the world's first, has never been told in full. That is not to say that much has not been written on the topic. Western researchers during the 1970s and 1980s were able to interpret official exhortations in the Soviet press and discern some logic of the inner workings of the Soviet space program. All of these works had one major drawback: they were written at a time when the Soviets maintained very strict control over information, especially any that portrayed the space effort in a negative light. Many "facts"—that is, the raw skeleton of the story—were missing. All we had were accounts from the official Soviet media and rumor or speculation from unconfirmed sources—or a combination of both. Thus the range of issues that Western or even Soviet historians could address was severely limited.[1]

Within Russian-language works, there are two relatively clear divisions in the historical record: those published before 1988, when the Soviet censorship apparatus consistently prevented an impartial representation of their efforts to explore space, and those published after, when the doors of the archives finally started opening up. The rupture was so great, it was as if everything written about the Soviet space program—and indeed almost every area of Soviet history—suddenly became obsolete by the turn of the 1990s. Entire programs, personalities, and even space missions of which we never knew all of a sudden came into focus, filling huge gaps in our understanding of the Soviet space effort during the Cold War. But it was not just a matter of filling in the blanks. The revisions and reassessments have been so pervasive that we could point to almost any event in the thirty-year span of the Soviet space and missile

---

1.   For the best Western accounts of the Soviet space program, see F. J. Krieger, *Behind the Sputniks: A Survey of Soviet Space Science* (Washington, DC: Public Affairs Press, 1958); William Shelton, *Soviet Space Exploration: The First Decade* (New York: Washington Square Press, 1968); Nicholas Daniloff, *The Kremlin and the Cosmos* (New York: Alfred A. Knopf, 1972); Peter Smolders, *Soviets in Space* (New York: Taplinger Publishing Co., 1973); Nicholas L. Johnson, *Handbook of Soviet Manned Spaceflight* (San Diego: Univelt, 1980); James E. Oberg, *Red Star in Orbit* (New York: Random House, 1981); Phillip Clark, *The Soviet Manned Space Program: An Illustrated History of the Men, the Missions, and the Spacecraft* (New York: Orion, 1988); Dennis Newkirk, *Almanac of Soviet Manned Space Flight* (Houston: Gulf Publishing Co., 1990); Steven J. Zaloga, *Target America: The Soviet Union and the Strategic Arms Race, 1945–1964* (Novato, CA: Presidio, 1993); James Harford, *Korolev: How One Man Masterminded the Soviet Drive to Beat America to the Moon* (New York: John Wiley & Sons, 1997).

programs after World War II and find that our understanding of that singular occurrence has changed irrevocably.[2]

The recent disclosures have relevance far beyond the limited purview of Soviet space history. In the 1950s and 1960s, U.S. space policy to a large degree was a series of responses to what the Soviets were doing—or at least what policymakers *thought* the Soviets were doing. But despite its key role in shaping American space policy, there continues to be an abundance of ignorance or misinformation on the Soviet program. Many erroneous conjectures on Soviet space motives advanced by Western analysts during the Cold War have remained unquestioned by more recent scholars. Ultimately, any effort to make sense of the dynamics of space exploration during the Cold War, no matter how well-intentioned, will fall short without taking account of the recent revelations from the Russian side. What may be possible now is to take a second look not only at the Soviet space program, but also the U.S. space program—that is, to reconsider again humanity's first attempts to take leave of this planet.

Writing on a topic that has two dynamic parallel histories—one from the Cold War era and one from the post–Cold War era—is, for obvious reasons, a difficult problem. First, there is the challenge of creating context. One could easily lose the main thread of the story by annotating every episode with interpretations from two different time periods—that is, this is how the event was reported in the 1960s, and this is what *really* happened. I have tried as much as possible to avoid the pitfalls of such an approach, but at the same time, I have also not tried to shirk from the opportunity to contrast these two voices when they have served to embellish my story.

A second problem is one of identifying the right sources for the story. As much as possible, I have relied exclusively on Russian-language archival sources available in the post–1988 era. There are, however, several episodes in the narrative that warrant a wider historiographical context. Because of the dual nature of the history of the Soviet space program, different players in the effort have continued to promote contradictory accounts of the same event. For instance, Russian historians have never adequately addressed the use of German expertise in the immediate postwar period. They have generally minimized the German role as extremely peripheral. On the other side, the popular press in the West has had a tradition of dismissing early Soviet successes as merely an extension of German work. Can these two positions be reconciled in a scholarly treatise? In this case, the writing of history as an exercise in impartiality is caught between what is a somewhat dubiously established paradigm of history in the West and what is at best a history with missing chapters on the Soviet side. What I have tried to do is to use recently declassified information to provide a newer perspective, but one that is not necessarily divorced from the existing paradigms of yesteryears.

There are many such other cases in which Soviet space history has been artificially constrained between propaganda and speculation. This is one reason, I believe, that Soviet space achievements have generally been marginalized in the West and mythologized at home. For American historians, there is little debate on the holy grail of space history: it is the first landing of American astronauts on the surface of the Moon in 1969. On that July night in 1969, two men represented humanity's thirst for exploration, serving as ambassadors of the human

2.   For the best recent Russian-language works, see Yu. A. Mozzhorin, *et al.*, eds., *Dorogi v kosmos: I* (Moscow: MAI, 1992); Yu. A. Mozzhorin, *et al.*, eds., *Dorogi v kosmos: II* (Moscow: MAI, 1992); Yaroslav Golovanov, *Korolev: fakty i mify* (Moscow: Nauka, 1994); Yu. A. Mozzhorin, *et al.*, eds., *Nachalo kosmicheskoy ery: vospominaniya veteranov raketno-kosmicheskoy tekhniki i kosmonavtiki: vypusk vtoroy* (Moscow: RNITsKD, 1994); B. Ye. Chertok, *Rakety i lyudi* (Moscow: Mashinostroyeniye, 1994); B. Ye. Chertok, *Rakety i lyudi: Fili Podlipki Tyuratam* (Moscow: Mashinostroyeniye, 1996); Yu. P. Semenov, ed., *Raketno-kosmicheskaya korporatsiya "energiya" imeni S. P. Koroleva* (Korolev: RKK Energiya, named after S. P. Korolev, 1996); V. V. Favorskiy and I. V. Mescheryakov, eds., *Voyenno-kosmicheskiye sily (voenno-istoricheskiy trud): kniga I: kosmonavtika i vooruzhennyye sily* (Moscow: VKS, 1997); B. Ye. Chertok, *Rakety i lyudi: goryachiye dni kholodnoy voyny* (Moscow: Mashinostroyeniye, 1997).

race in our first visit to another celestial body. For most American historians, everything before was simply a prelude and everything after has been a disappointment. Historians in Russia see things much differently. It was, after all, the Soviet Union that launched the first handiwork of humankind into orbit around Earth in 1957—Sputnik, or "fellow traveler." Only four years later came the second big milestone: the Soviet Union sent the first human into space, Yuriy Gagarin. Here was another huge moment, like that of the Apollo landing eight years later: for the first time since human life emerged on this planet, one of us had broken through the atmosphere that surrounds us and sped into the cosmos. But history has remembered Gagarin's short flight much differently. With the race to the Moon won, the American view of the Soviet space program changed dramatically; American historians remembered Sputnik and Gagarin not for their importance in human history, but only as catalysts for the decision to send humans to the Moon. There are works, too numerous to mention, on the repercussions of both Sputnik and Gagarin in the United States, but few on the historical meaning of these events divorced from geopolitics—as there was on Apollo. It is not surprising that this is so. With little film footage, paranoid secrecy, and no advance warning, the Soviets themselves were mostly responsible for consigning these events into that blurry historical limbo between propaganda and speculation. They eventually lost any claim to resonance that they might have had otherwise.

The Soviet space program was, of course, not simply propaganda nor speculation. It emerged from the ashes of World War II, when with Stalin's blessing, a group of ambitious engineers began testing old German missiles from the desert near the Aral Sea. With the onset of the Cold War and the explosion of the first Soviet atomic bomb in 1949, these experiments with rockets gained a new urgency. Many considered rockets, especially long-range ballistic missiles, an ideal way to deliver deadly atomic bombs across continents. Throughout the 1950s, as missile designers made vast advances in rocket design, it became possible to consider options that had little direct military utility—ideas such as space travel. Spurred by a small handful of visionary engineers devoted to the cause of space exploration, the Soviets diverted a strand of their military rocketry program into a single project to launch a satellite into orbit. Conceived as an exercise in scientific research, Sputnik was meant to be a modest contribution to an international effort to study Earth and its surroundings. While its scientific dividends might have been anticipated, no one could have predicted its political repercussions. After the launch of Sputnik on October 4, 1957, in the public image, the Soviet Union moved from being a nation of obsolete agricultural machines to a great technological superpower. Gagarin's flight less than four years later eliminated any remaining doubts about Soviet prowess in space exploration. In both cases, the Americans had lagged behind badly. These two pivotal achievements led eventually to the race to the Moon—a race of epic proportions that culminated in the Apollo landing in 1969. A span of only eight years separated the resounding victory of Gagarin and the crushing humiliation by Apollo. So what happened? What kind of effort did the Soviets mount to compete with Apollo? And why did it fail? I have tried to answer these questions by weaving together a record of the technical, political, and personal histories behind these three endeavors: the launch of Sputnik, the flight of Gagarin, and the challenge to Apollo.

My goal was not to write a history simply because it had never been written before. Certainly, recording the facts is an important exercise, but that would limit the job to a simple chronology. There are several major questions of interpretation that still have to be answered. I have only tackled a few of these.

The first major question has to do with discerning the institutional underpinnings of the Soviet space program. Given the new evidence, can we identify the primary constituencies that drove the effort? What kind of patterns of decision-making did they display? What interests were they serving? The record seems to indicate the importance of both individuals and institutions, all of whom emerged to power not because of the space program, but because of its antecedent ballistic missile development effort.

The second question addresses Soviet space technology. Our conventional understanding of Soviet space technology is generally framed in terms of obsolete products pushed through production-line processes that discourage major innovation. In the evolution of their early missile and space programs, did the Soviets adhere to the idea of incremental advances, or were there technological leaps? Did the Soviets benefit from foreign expertise during these early years? More often than not, the answers to these questions do not conform to our entrenched notion of how the Soviets managed technology in the Cold War era.[3]

Finally, why did the Soviets manage to beat the Americans in launching the first intercontinental ballistic missile, the first satellite, and the first human into space, but fail to beat the United States in landing the first person on the Moon? Was it simply because the last goal was significantly more challenging than the previous three? Or was it because, as was conventionally thought for many years in the West, that the Soviets simply did not want to race the Americans to the Moon? The answers to these questions are not simple: personal, institutional, political, and technological issues intersected in the complex schema of the Soviet Moon program, leading it to its final ignominious failure in 1969.

For this work, I have specifically focused on piloted space programs. In the first four chapters of this book, however, I delve into the origins of the Soviet long-range ballistic missile program and the events leading up to the launch of Sputnik. The following seven chapters cover the rise of the Soviet piloted space program under the tutelage of its founder, Sergey Pavlovich Korolev, ending with his premature death in 1966. The next seven chapters take the story up to 1974, covering the Soviet loss in the Moon race under the direction of Korolev's successor, Vasiliy Pavlovich Mishin. Finally, in the remaining two chapters, I briefly tell the story after 1974.

Note on Transliteration

I have used a modified version of the standard used by the U.S. Board on Geographic Names preferred by the University of Chicago Press. The drawback of this system is that it is often phonetically inappropriate. For example, the letter "ë" is pronounced as "yo" in Russian. Thus "Korolev" should actually be pronounced as "Korolyov." There was one major exception to the Board of Geographic Names system: I have omitted the use of inverted commas (the "soft" and "hard" signs) within Russian words to reduce clutter in the text for those not familiar with the Russian language.

One other note is that NASA's normal convention has been to spell the Soviet cosmodrome "Baikonur," with an "i" instead of a "y." In this book, to be consistent with the rest of the transliteration, it is spelled "Baykonur." The reader will also find a difference in the spelling of some common first names, such as Sergei as Sergey and Yuri as Yuriy.

3. For Western works on the history of Soviet technology, see, for example, Ronald Amann, Julian Cooper, and R. W. Davies, eds., *The Technological Level of Soviet Industry* (New Haven, CT: Yale University Press, 1977); Kendall E. Bailes, *Technology and Society Under Stalin: Origins of the Soviet Intelligentsia, 1917–1941* (Princeton, NJ: Princeton University Press, 1978); Ronald Amann and Julian Cooper, eds., *Industrial Innovation in the Soviet Union* (New Haven, CT: Yale University Press, 1982); Bruce Parrot, *Politics and Technology in the Soviet Union* (Cambridge, MA: MIT Press, 1983); Matthew Evangelista, *How the United States and the Soviet Union Develop New Military Technologies* (Ithaca, NY: Cornell University Press, 1988); Loren R. Graham, *What Have We Learned About Science and Technology from the Russian Experience?* (Stanford, CA: Stanford University Press, 1998).

GLOSSARY

AN SSSR	USSR Academy of Sciences
BOR	Unpiloted Orbital Rocket-Glider
BS	lateral stabilization
CIA	(U.S.) Central Intelligence Agency
CSAGI	Special Committee for the International Geophysical Year
DLB	Long-Duration Lunar Base
DOK	Engine Orientation Complex
DOS	Long-Duration Orbital Station
EOR	Earth-orbit rendezvous
EKR	Experimental Winged Missile
EPAS	Apollo-Soyuz Experimental Flight (Soviet name for the Apollo-Soyuz Test Project)
EPOS	Experimental Piloted Orbital Aircraft
EVA	extravehicular activity
FIAN	(P. N. Lebedev) Physical Institute of the USSR Academy of Sciences
FOBS	Fractional Orbital Bombardment System
GAU	Chief Artillery Directorate
GDL	Gas Dynamics Laboratory
GeoFIAN	Geophysical Institute of the USSR Academy of Sciences
GIPKh	State Institute of Applied Chemistry
GIRD	Group for Study of Reactive Motion
GKAT	State Committee for Aviation Technology
GKB	Lead Design Bureau
GKNPTs	State Space Scientific-Production Center
GKO	State Committee for Defense
GKOT	State Committee for Defense Technology
GKRE	State Committee for Radio Electronics
GMT	Greenwich Mean Time
GNII AiKM	State Scientific-Research Institute for Aviation and Space Medicine
GOGU	Chief Operations and Control Group
Gosplan	State Planning Organ
GR	global missile
GSKB	State Union Design Bureau
GSMZ	State Union Machine Building Plant
GTsP	State Central Range
GU	Chief Directorate
GUKOS	Chief Directorate of Space Assets
GULag	Chief Directorate of Camps
GURVO	Chief Directorate of Reactive Armaments
IAE	Institute of Atomic Energy
IAKM	Institute of Aviation and Space Medicine
ICBM	intercontinental ballistic missiles
IGY	International Geophysical Year
IKI	Institute of Space Research
IMBP	Institute for Biomedical Problems

IP	Measurement Point
IS	Satellite Destroyer
KB	Design Bureau
KB OM	Design Bureau of General Machine Building
KGB	Committee for State Security
KIK	Command-Measurement Complex
KORD	Engine Operation Control (system)
KPSS	Communist Party of the Soviet Union
KSU	command and signal instrument
KVTs	Coordination-Computation Center
LH_2	liquid hydrogen
LII	Flight-Research Institute
LK	Lunar Ship
LKS	Light Space Aircraft
LOK	Lunar Orbital Ship
LOR	lunar-orbit rendezvous
LOX	liquid oxygen
MAI	Moscow Aviation Institute
MAP	Ministry of Aviation Industry
MEK	Martian Expeditionary Complex
MEP	Ministry of Electronics Industry
MIAN	Mathematics Institute of the Academy of Sciences
MIK	Assembly-Testing Building
MKB	Machine-Building Design Bureau
MKBS	Multirole Space Base-Station
MMZ	Moscow Machine-Building Plant
MNII	Moscow Scientific-Research Institute
MNII RS	Moscow Scientific-Research Institute for Radio Communications
MO	Ministry of Defense
MOK	Multirole Orbital Complex or Martian Orbital Complex
MOL	Manned Orbiting Laboratory
MOM	Ministry of General Machine Building
MOP	Ministry of Defense Industry
MP	maneuvering, piloted
MPK	Martian Landing Complex
MRP	Ministry of Radio Industry
MSM	Ministry of Medium Machine Building
MTKVP	Reusable Vertical-Landing Transport Craft
MVTU	Moscow Higher Technical School
MZ	Machine Building Plant
NASA	National Aeronautics and Space Administration
NATO	North Atlantic Treaty Organization
NEK	Scientific-Experimental Complex
NIEI PDS	Scientific-Research and Experimental Institute of the Parachute Landing Service
NII	Scientific-Research Institute
NII AP	Scientific-Research Institute for Automation and Instrument Building
NII AU	Scientific-Research Institute for Aviation Equipment
NII IT	Scientific-Research Institute for Measurement Technology
NII P	Scientific-Research Institute for Instrument Building
NII PM	Scientific-Research Institute for Applied Mechanics

NII RP	Scientific-Research Institute for Rubber Industry
NII TP	Scientific-Research Institute for Precision Instruments *or* Scientific-Research Institute for Thermal Processes
NIIP	Scientific-Research and Test Range
NIP	Scientific-Measurement Point
NITI	Scientific-Research and Technical Institute
NKVD	People's Commissariat for Internal Affairs
NORAD	North American Air Defense Command
NPO	Scientific-Production Association
NS	normal stabilization
NTK	Scientific-Technical Committee
NTS	Scientific-Technical Council
OK	orbital ship
OKB	Special or Experimental Design Bureau
OKB MEI	Experimental Design Bureau of the Moscow Power Institute
OPM	Department of Applied Mathematics
OPS	Orbital Piloted Station
OTK	Special Technical Commission
PKA	Gliding Space Apparatus
PVO	Air Defense Forces
RKK	Rocket-Space Corporation
RKS	Adjustment of Apparent Velocity
RNII	Reactive Scientific-Research Institute
RVSN	Missile Forces of Strategic Designation (Strategic Missile Forces)
SALT	Strategic Arms Limitation Talks
SKB	Special Design Bureau *or* Serial Design Bureau
SKG	Special Design Group
SOBIS	simultaneous emptying of tanks
SOKB	Union Experimental Design Bureau
SOUD	System of Orientation and Motion Control
SOZ	System for Ensuring Firing
SP	Special Publication (NASA)
SR	Suborbital Rocket-Glider
TASS	Telegraph Agency of the Soviet Union (Soviet news agency)
TDU	Braking Engine Unit
TGU	Third Chief Directorate
TKA	Transport-Supply Ship
TLI	translunar-injection
TMK	Heavy Interplanetary Ship
TMKB	Turayevo Machine-Building Design Bureau
TMZ	Tushino Machine Building Plant
TNT	trinitrotoluene
TOS	Heavy Orbital Station
TsAGI	Central Aerohydrodynamics Institute
TsIAM	Central Institute of Aviation Motor Building
TsK	Central Committee
TsKB	Central Design Bureau
TsKBEM	Central Design Bureau of Experimental Machine Building
TsKBM	Central Design Bureau of Machine Building
TsKIK	Central Command-Measurement Complex

TsNII	Central Scientific-Research Institute
TsNIIMash	Central Scientific-Research Institute for Machine Building
TsPK	Cosmonaut Training Center
TsSKB	Central Specialized Design Bureau
TsUKOS	Central Directorate of Space Assets
TsUP	Flight Control Center
UDMH	unsymmetrical dimethyl hydrazine
UHF	ultrahigh frequency
UNKS	Directorate of the Commander of Space Assets
UNRV	Directorate of the Commander of Reactive Armaments
UPMK	Cosmonaut Maneuvering and Motion Unit
UR	universal missile
US	Controlled Satellite
USSR	Union of Soviet Socialist Republics
UZKA	Directorate of the Deputy Commander of Artillery
VHF	very high frequency
VIAM	All-Union Institute for Aviation Materials
VMF	Soviet Navy
VNII	All-Union Scientific-Research Institute
VNII EM	All-Union Scientific-Research Institute for Electro-Mechanics
VNII IT	All-Union Scientific-Research Institute for Current Sources
VPK	Military-Industrial Commission
VSNKh	All-Russian Council of the National Economy
VTs	Computation Center
VVS	Soviet Air Force
YaERD	nuclear-electric rocket engine
ZIKh	M. V. Khrunichev Machine Building Plant

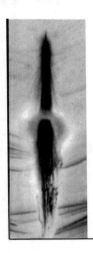

CHAPTER TWELVE
A NEW BEGINNING

Sergey Pavlovich Korolev's death ended one man's unprecedented twenty-year reign over Soviet missile and space programs. He bequeathed to his associates and aides the daunting task of managing an empire whose intricacies had only been clear to him. While many of his deputies were certainly adept in various areas of directing the works of OKB-1, no single person had expertise in managing the design bureau, dealing with Soviet politicians, brokering deals with other chief designers, *and* instilling a vision of space exploration among the thousands who worked at the firm.

Unlike no other chief designer before or since, Korolev dominated the Soviet space program. His informal title in the official Soviet press before his death was not "chief designer of OKB-1," but rather "chief designer of rocket-space systems"—a far-more melodramatic moniker than simply the head of a design bureau. His vast array of roles in the space program did not, for the most part, come from his official appointments (which were many), but rather from his larger-than-life personality. Thus, when he died, there was an unprecedented vacuum. While his successor would inherit the title of chief designer of OKB-1, he would not have Korolev's informal powers accrued through twenty years of making history. In some ways, the post-Korolev period was characterized by an equal playing ground, with the leading chief designers no longer following a single voice. This also meant that there was no single ardent supporter to push projects. The lobbying from the bottom up as a consequence became more diffuse and less imposing in contrast to the Korolev years.

Mishin

The first order of business for a demoralized Soviet space program was to choose a successor to Korolev. The normal procedure for selecting a new chief designer would have been for Minister of General Machine Building Afanasyev to discuss the names of candidates with Secretary of the Central Committee Ustinov. The proposal would be presented to the Central Committee, whose members would pass it on to the Politburo. In the case of OKB-1, Korolev's senior staff did not want to risk having an unwanted individual appointed chief designer by higher-ups, and they tried to take the matter into their own hands. The night after Korolev's death, one of his most beloved former protégés, Chief Designer Viktor P. Makeyev of SKB-385, flew into Moscow from his home base at Miass to try and bring some order into the succession. Makeyev assembled all the senior staff at OKB-1 and asked them for opinions. Some suggested that Makeyev himself take over the design bureau, but he was firmly against doing so; he had as many as sixteen submarine-launched ballistic missile projects ongoing at Miass, far too much work to be suddenly moving to another organization.[1]

---

1.   B. Ye. Chertok, *Rakety i lyudi: goryachiye dni kholodnoy voyny* (Moscow: Mashinostroyeniye, 1997), pp. 368–69.

It took a long time to come to a decision. One who was there, Deputy Chief Designer Yevgeniy V. Shabarov, recalled many years later that:

> . . . through the night we wrote a letter addressed to the Secretary of the [Central Committee], the Chairman of the Military-Industrial Commission, and to [our] minister. In the letter we proposed that in our opinion, Vasiliy Pavlovich Mishin be appointed the successor to Sergey Pavlovich Korolev since he had been [Korolev's] First Deputy. We also offered various other reasons for the choice. At five in the morning the letter was ready and we all signed it.[2]

Bushuyev and Chertok had originally proffered Mishin's name. Only one deputy, Sergey S. Kryukov, had opposed Mishin's candidacy. All other senior staff agreed that Mishin would be the best person for the job. The prompt action by OKB-1 senior staff seems to have surprised government officials, who were not too happy with this internal recommendation. Mishin remembered that:

> my appointment . . . encountered some opposition from Ustinov who at the time was a Secretary of the Central Committee . . . overseeing defense matters. He wanted to use the occasion to limit the authority and jurisdiction of the Chief Designer and put him under an administrative head of OKB-1. Ustinov had made such attempts during Korolev's lifetime but they had run up against Korolev's well-argued objections.[3]

By the time that the senior staff at OKB-1 officially proposed Mishin's name, Communist Party officials had already decided on an alternative person to head the design bureau: Georgiy A. Tyulin, then the First Deputy Minister of General Machine Building. Ustinov believed that by appointing Tyulin as "administrative head" of OKB-1, he would be able to curb some of the undeniable powers of the chief designer of such an important design bureau. The papers for Tyulin's appointment were drawn up, but there were long drawn-out negotiations on the issue, and it took an astonishing five months before the Central Committee agreed to ratify the original proposal from the OKB-1 senior staff. On May 5, 1966, Soviet leader Brezhnev summoned Mishin to the Kremlin and informed him of his promotion, and six days later, Minister of General Machine Building Afanasyev signed an order officially appointing Mishin as the new chief designer of the organization.

Mishin was clearly the most likely choice as a successor, having been groomed by the late Korolev for almost a decade for this position. But he did not have his predecessor's stature or clout. In fact, Mishin had somewhat of a reputation for being blunt and tactless and was not known for his diplomatic skills. He was, however, respected for his engineering skills. One military officer who closely worked with Mishin recalled that he was:

> An excellent mathematician, a fast thinking engineer. He knew the business and, most important, could screen options as fast as a computer. . . . Mishin possessed very specific information. He was always ready to come up with a strong rebuke at the Council of Chief Designers where he was invited. He deferred to no authority as long as the authority in question came up with solutions that defied logic and common sense to serve a hidden agenda. That is why he was not popular.[4]

---

2.   Yu. A. Mozzhorin, et al., eds., *Dorogi v kosmos: I* (Moscow: MAI, 1992), p. 182.
3.   *Ibid.*, p. 121.
4.   Mikhail Rebrov, "The Secrets of Rocket Codes" (English title), *Krasnaya zvezda*, June 3, 1995, p. 6.

*Vasiliy Mishin succeeded Korolev as OKB-1 Chief Designer after Korolev's death in January 1966.
This photo probably dates from early 1968. In the background are Maj. General Aleksandr Kurushin (left),
commander of the Tyura-Tam range at the time, and Maj. General Anatoliy Kirillov (right), Kurushin's deputy.
(copyright Christian Lardier)*

This is an important distinction from Korolev, who, perhaps because he better understood the workings of the political machinery of the Communist Party, was more willing to work out problematic issues than let them languish in deadlock. Mishin, stubborn to the end, refused to budge if his instincts told him so, sticking to his beliefs until the bitter end. Lacking the political instincts of say a Wernher von Braun or a Sergey Korolev, he suffered dearly. Some would argue that so did the Soviet space program in the coming years.

Mishin's appointment as chief designer was only one of several different honors bestowed upon him. He replaced Korolev's vacant position as the head of the somewhat amorphous Council of Chief Designers for programs in which his design bureau had the leading role. Thus, at least during the meetings of the council, he outranked much more senior designers such as Glushko, Pilyugin, and Isayev. In March 1966, Mishin was inducted into the Presidium of the Interdepartmental Scientific-Technical Council on Space Research, headed by Academy of Sciences President Keldysh. That council continued its critical advisory role of implementing the Soviet space program by serving as "expert commissions" for a plethora of projects.[5] Finally on July 1, 1966, Mishin was promoted to the rank of full Academician of the USSR Academy of Sciences. Along with Mishin, three other major space designers—Barmin, Pilyugin, and

5.   Interview, Georgiy Stepanovich Vetrov with the author, November 15, 1996.

Yangel—also became Academicians the same day, joining the select group of Glushko and Chelomey.[6]

These six designers—Barmin, Chelomey, Glushko, Mishin, Pilyugin, and Yangel—all Academicians, commanded great respect among the upper echelons of the space industry, but their ascendance was also evidence of a great diffusion of power. For example, of the six, only one (Yangel) was allowed to become a Candidate Member of the Central Committee of the Communist Party—an unprecedented honor that even Korolev did not enjoy. It was in fact Yangel's new appointment as a Candidate Member that prompted many Western analysts to come to the conclusion that Yangel had "succeeded" Korolev as the "chief" of the Soviet space program, as if the entire effort was run by a single monolithic organization. This was an error in analysis that would not be dispelled until well into the 1970s, when the concept of "design bureaus" filtered out through the curtain of censorship. What was equally unknown at the time was that Yangel's honorary promotion as a Candidate Member of the Central Committee probably stemmed not from his achievements in space, but rather from his clearly notable contributions to the development of strategic ballistic missiles. More evidence of the diffusion of power was the choice of Korolev's replacement as a member of the Presidium of the Academy of Sciences, the highest arbiters of scientific research in the Soviet Union. Neither Glushko, nor Yangel, nor Mishin, nor Chelomey filled the position in May 1967—rather, it was Chief Designer Pilyugin, responsible for guidance systems.[7]

Soon after the changeover in leadership at the design bureau, the Ministry of General Machine Building enacted a ministry-wide change in naming of institutions, which effectively replaced the "OKB-plus-number" system with an even more bewildering designation system. Almost every design bureau involved in the missile and space industry would have the dreary phrase "machine building" attached to its name, perhaps as a somewhat comical way to disguise the true roles of these organizations. Thus on March 6, 1966, OKB-1 became the new Central Design Bureau of Experimental Machine Building, or "TsKBEM" in its Russian abbreviation. Chelomey's OKB-52 meanwhile became the almost identical Central Design Bureau of Machine Building, or "TsKBM," distinguished only by the omission of an "E" in its abbreviation.[8] At the same time, Mishin enacted a large-scale restructuring of his design bureau in

---

6.   Chertok, *Rakety i lyudi: goryachiye dni kholodnoy voyny*, pp. 516–17; Christian Lardier, "Soviet Space Designers When They Were Secrets," presented at the 47th International Astronautical Federation, IAA-96-IAA.2.2.09, Beijing, China, October 7–11, 1996. V. P. Glushko had become an Academician on June 20, 1958, while V. N. Chelomey had become one on June 29, 1962. Note that there were also a number of *scientists* who were Academicians who were involved in the ballistic missile and/or space programs. These included A. A. Blagonravov (became an Academician in September 1943), A. Yu. Ishlinskiy (in June 1960), M. V. Keldysh (in November 1946), S. A. Khristianovich (in September 1943), V. A. Kotelnikov (in October 1953), B. N. Petrov (in June 1960), G. I. Petrov (in June 1958), and L. I. Sedov (in October 1953). In addition, there were several other designers and/or scientists in the space program who were Corresponding Members of the Academy of Sciences—that is, junior to full Academicians. These included A. F. Bogomolov (in July 1966), K. D. Bushuyev (in June 1960), V. I. Kuznetsov (in June 1958), N. S. Lidorenko (in July 1966), A. M. Lyulka (in June 1960), D. Ye. Okhotsimskiy (in June 1960), B. V. Raushenbakh (in July 1966), M. S. Ryazanskiy (in June 1958), S. S. Lavrov (in July 1966), and S. K. Tumanskiy (in June 1964).

7.   Lt. Gen. G. Tyulin, "Look Forward" (English title), *Krasnaya zvezda*, May 18, 1988, p. 4.

8.   Yu. P. Semenov, ed., *Raketno-Kosmicheskaya Korporatsiya "Energiya" imeni S. P. Koroleva* (Korolev: RKK Energiya, named after S. P. Korolev, 1996), p. 158; Mikhail Rudenko, "Designer Chelomey's Rocket Planes" (English title), *Vozdushniy transport* 52 (1995): 8–9; Chertok, *Rakety i lyudi: goryachiye dni kholodnoy voyny*, p. 403. Some of the other organizations whose names were changed included: Glushko's OKB-456, which in January 1966 was renamed the Design Bureau of Power Machine Building (KB EnergoMash); Mozzhorin's NII-88, which in January 1967 was renamed the Central Scientific-Research Institute of Machine Building (TsNIIMash); Barmin's GSKB SpetsMash, which in January 1967 was renamed the Design Bureau of General Machine Building (KB OM); and Yangel's OKB-586, which in October 1966 was renamed the Yuzhnoye Design Bureau (KB Yuzhnoye).

November 1966, creating ten subdivisions, each designated a "complex," dedicated to a specific mission profile. His two First Deputy Chief Designers were Sergey O. Okhapkin and Dmitriy I. Kozlov, both of whom had worked under Korolev since 1946.[9]

Okhapkin, a prematurely gray-haired man full of verve and energy, had served his apprenticeship under such famous Soviet aviation Myasishchev designers as Tupolev, and Ilyushin before joining Korolev's team in 1948 as an expert on dynamics and precision. In December 1952, he became a deputy chief designer, eventually directing planning work on the N1 booster. Upon Mishin's appointment, Okhapkin headed OKB-1's Complex 1, dedicated to rocket systems, which included the N1.[10] Kozlov, on the other hand, had headed the old Branch No. 3 at Kuybyshev since its establishment in 1960. After Korolev's death, the branch remained subordinate to the main center at Kaliningrad, although Kozlov's primary work was related not to piloted systems but rather the development of high-security military reconnaissance satellites. Apart from Okhapkin and Kozlov, there were five remaining deputy chief designers for spacecraft, guidance systems, rocket engines, ground equipment, and testing.[11]

OKB-1 Deputy Chief Designer Sergey Okhapkin was one of the principal forces behind the creation of the N1 rocket. After Mishin's appointment as OKB-1 Chief Designer, Okhapkin served as First Deputy of the organization, primarily responsible for the quickly accelerating work on the N1. (files of Peter Gorin)

As with all notable figures in the space program, the identities of Mishin, Okhapkin, and Kozlov were kept state secrets, and the Soviet press completely refrained from commenting on the nature of the succession to Korolev. Eventually, by the late 1960s and early 1970s, they were allowed to use pseudonyms when writing articles in the popular media.[12] Unlike Korolev, Soviet journalists did not refer to Mishin as the "chief designer of rocket-space systems," but rather the less encompassing "chief designer of piloted spaceships." It was a small, but telling indication that Korolev's old design bureau had reached its zenith of power and that glory days were no longer ahead but consigned to the history pages.

9.   For Okhapkin, see Semenov, ed., *Raketno-Kosmicheskaya Korporatsiya*, p. 158. For Kozlov, see V. Drebkova, "Anniversaries: General Designer D. I. Kozlov—75 Years" (English title), *Novosti kosmonavtiki* 20 (September 24–October 7, 1994): 56. Note that Kozlov assumed his post in 1967, not 1966.

10.   Yaroslav Golovanov, *Korolev: fakty i mify* (Moscow: Nauka, 1994), pp. 478–80.

11.   They were K. D. Bushuyev (spacecraft, Complex 2), B. Ye. Chertok (guidance systems, Complex 3), M. V. Melnikov (rocket engines, Complex 5), A. P. Abramov (ground equipment, Complex 6), and Ya. I. Tregub (testing, Complex 7). Complex 4 was for the Experimental Machine Building Plant (ZEM) attached to the design bureau. It was headed by TsKBEM First Deputy Chief (but not Deputy Chief Designer) V. M. Klyucharev. See Semenov, ed., *Raketno-Kosmicheskaya Korporatsiya*, pp. 158–59. Note that Klyucharev was appointed to his position on September 8, 1967.

12.   Their pseudonyms were M. P. Vasilyev (Mishin), S. O. Osipov (Okhapkin), and D. Ilichev (Kozlov). See Lardier, "Soviet Space Designers When They Were Secrets."

The End of Voskhod

Mishin's first job as Chief Designer of TsKBEM was to assess the state of the Voskhod program. At the time of Korolev's death, there were immediate plans for three to four Voskhod and five Soyuz missions in 1966. The first one, Voskhod 3, was the long-duration mission with cosmonauts Volynov and Shonin, planned for almost a year. The spectacular success of the fourteen-day Gemini VII flight in December 1965 had given the Soviet mission even more of an impetus to get off the ground. There seems to have been some effort from ministerial leaders to substitute the all-woman EVA flight in place of Voskhod 3, but this attempt did not bear fruit.[13] The subsequent Voskhod 4 would be a scientific flight, including artificial gravity experiments with test pilot Beregovoy and scientist Katys, while Voskhod 5 would be a military mission with cosmonauts Shatalov and Demin. An extra mission, with only dogs, would precede Voskhod 3 to test the extended life support systems on the near-obsolete 3KV spacecraft.

The Voskhod 3 mission was timed to coincide with the opening of the 23rd Congress of the Communist Party in early March 1966, as a "gift" to the doctrinal keepers of the Soviet Union. This flight, and the additional two or three Voskhod missions, would also serve to bridge the gap to the inaugural jaunt of the Soyuz spaceship, then slated for sometime in late 1966. From a public relations perspective, the remaining Voskhod expeditions would no doubt deflect worldwide attention from NASA's successful Gemini program. Certainly, the Voskhod 3 mission, dedicated to regaining the mission duration record claimed by Gemini VII, would be an outstanding publicity victory for the Soviet space program.

On January 27, about two weeks after Korolev's death, Mishin hosted the first technical meeting at OKB-1 under his management to discuss the future Voskhod missions. The attendees decided to prepare Voskhod spacecraft 3KV no. 5 for launch with two dogs in the first half of February. Some from the military, particularly Air Force Lt. General Kamanin, opposed such a thirty-day biomedical precursor mission, apparently because he believed that it would unnecessarily delay the Voskhod 3 mission, which was very important to future military operations in space. Cosmonauts had extensively trained to use an infrared optical instrument named *Svinets* ("Lead"), which would allow them to observe plumes from the launches of four Soviet ballistic missiles. At the same time, officials decided to launch spacecraft 3KV no. 6 (Voskhod 3) on an eighteen-day mission during March 10–20, 1966—that is, after the landing of the precursor mission. The primary limiting factor for the extended mission seems to have been the poor performance of the Voskhod spacecraft's life support system, in particular its air regeneration capabilities, which most believed would not guarantee safety for two cosmonauts for a period of eighteen days in space.[14] A second similar meeting on February 10 confirmed the general state of readiness to carry out the two flights.[15]

---

13. See, for example, N. P. Kamanin, *Skrytiy kosmos: kniga vtoraya, 1964–1966gg* (Moscow: Infortekst IF, 1997), pp. 284, 286, 288.

14. *Ibid.*, pp. 293–94. Among those present for this meeting were V. P. Mishin (OKB-1), G. A. Tyulin (MOM), P. V..Tsybin (OKB-1), K. D. Bushuyev (OKB-1), A. I. Burnazyan (Ministry of Health), A. G. Karas (TsUKOS), K. A. Kerimov (MOM), G. I. Voronin (OKB-124), S. G. Darevskiy (SOKB LII), N. P. Kamanin (VVS), N. F. Kuznetsov (TsPK), Yu. A. Gagarin (TsPK), V. M. Komarov (TsPK), Ye. A. Karpov (GKNII AiKM), A. M. Genin (GKNII AiKM), A N. Babiychuk (VVS), S. G. Frolov (VVS), and V. A. Smirnov (VVS).

15. *Ibid.*, pp. 300–01. The meeting was also the forum to formally approve the membership of the first post-Korolev State Commission. This State Commission for Voskhod would now include G. A. Tyulin (Chairman from MOM), M. V. Keldysh (AN SSSR), S. I. Rudenko (VVS), V. P. Mishin (OKB-1), N. N. Smirnitskiy (GURVO), V. A. Kasatanov (affiliation unknown), V. A. Kazakov (MAP), A. G. Karas (TsUKOS), G. P. Melnikov (NII-4), N. P. Kamanin (VVS), A. A. Kurushin (NIIP-5), P. V. Tsybin (OKB-1), I. D. Spitsa (TsKIK), Ye. V. Shabarov (OKB-1), V. N. Pravetskiy (Ministry of Health), and I. T. Bulychev (MO).

There were no major delays in the preparation of the precursor mission, and Tyulin gave the final approval for the launch at a State Commission meeting on February 17. Two dogs, selected for the flight after a rigorous selection process at the Institute for Biomedical Problems in Moscow, would fly a twenty-five-day mission. The 3KV-type Voskhod vehicle, spacecraft no. 5, was launched at 2310 hours Moscow Time on February 22, 1966, and named *Kosmos-110* upon entering orbit. The craft carried dogs named Veterok and Ugolek into a highly elliptical initial orbit of 187 by 904 kilometers at a 51.9-degree inclination. The high apogee of the orbit was evidently an attempt by Soviet scientists to examine the effects of the Van Allen radiation belts on the dogs. It was an element of the flight that had originally emerged as early as 1963 during planning for the Vostok program. The State Commission hoped to launch the subsequent Voskhod 3 craft into a similar orbit not only to study radiation effects, but also to claim the absolute altitude record for a piloted space vehicle. With the launch of Kosmos-110, for the first time in the Soviet space program, a piloted spacecraft used the fifty-one-degree inclination for the orbit—a practice that would be adopted for almost all the remaining crewed space missions in the Soviet era. This inclination not only allowed the 11A57 launch vehicle to lift the heaviest payload into orbit without having to land in China in case of an abort, but it also would provide optimal flight conditions for future missions to the Moon. The total mass of the vehicle was 5,600 kilograms, with 3,000 kilograms of that mass for the spherical descent apparatus that contained the two dogs.[16]

While the primary goal of the flight was to test the life support system in preparation for Voskhod 3, Kosmos-110 also had a number of supplementary scientific goals. Apart from the dogs themselves, there were various types of yeast preparations, samples of blood serums, protein growths, chlorella, and lysogenic bacteria aboard the spacecraft.[17] Throughout the mission, the two dogs were fed anti-radiation drugs and food delivered by means of tubes into their stomachs. Veterok served as the experimental specimen, while Ugolek was the control animal. By March 4, things seemed to be proceeding normally. The only minor problem was a deployment malfunction in one of the communications antennas. On March 14, about twenty days into the flight, the State Commission met to discuss the progress of Kosmos-110. Although the condition of the dogs and cabin atmosphere parameters, such as pressure, temperature, humidity, and carbon dioxide content, were within normal range, there had been "a steady tendency of gradual deterioration of the composition of air in the cabin."[18] Some recommended immediately terminating the flight and recovering the dogs, while others, notably life support system Chief Designer Voronin, expressed confidence in a full twenty-five-day flight. A special landing commission consisting of twenty-five officials discussed the issue in detail throughout the night. By the next morning, all agreed that the flight should be curtailed and the dogs brought down. At 1400 hours Moscow Time on March 16, ground controllers began operations necessary for reentry. Three hours and fifteen minutes later the dogs landed safely 210 kilometers southeast of Saratov, approximately sixty kilometers from the intended landing spot. About thirty to forty minutes later, rescue teams were able to report that the dogs were in safe hands. The flight had lasted nearly twenty-two days.

The physicians who examined the dogs upon their return did not anticipate the poor conditions of the animals. In an official report published two months after the landing, the doctors reported that the animals suffered from muscular reduction, dehydration, calcium loss, and confusion in readjusting to walking. Their motor systems did not return to normal until eight to ten days after the end of the mission, while full restoration of blood circulation system did

16.   G. A. Kustova, *Ot pervogo Sputnika do "Energii"-"Burana" i "Mira"* (Kaliningrad: RKK Energiya, 1994), p. 57.

17.   V. P. Glushko, ed., *Kosmonavtika entsiklopediya* (Moscow: Sovetskaya entsiklopediya, 1985), p. 203.

18.   Kamanin, *Skrytiy kosmos: 1964–1966*, p. 314.

not occur until five days after landing. The doctors added dramatically, "Prolonged space flight and the development of methods to combat unfavorable effects of such flights have raised new problems for space medicine."[19]

The Kosmos-110 mission was to have cleared the way for the piloted Voskhod 3 mission, but during the flight itself, events on the ground had necessitated a second look at safety issues in connection with the 3KV Voskhod spacecraft. As early as February 2, Chief Designer Fedor D. Tkachev of the Scientific-Research and Experimental Institute of the Parachute Landing Service reported that during the past three simulated landing tests of the heavy Voskhod space-craft, the parachutes had ruptured. A fourth consecutive failure soon after did not prevent the launch of Kosmos-110 but raised serious questions about the system as a whole. Continuing problems with the life support system prompted both OKB-124 and the Ministry of Health's Institute for Biomedical Problems to initiate long-duration ground simulations to assess the fea-sibility of carrying out a twenty-day mission in the Voskhod spacecraft. A third technical prob-lem was the bothersome failure of the Blok I third stage of the 11A57 launch vehicle during a ground test in December 1965, apparently because of high-frequency oscillations in the stage. Although the stage had not failed in flight, engineers at OKB-154 in Voronezh had still not identified the reasons for the explosion.[20]

Throughout the Kosmos-110 mission, there were rumors from Moscow that a piloted mis-sion was imminent. On March 9, the United Press International reported that the Soviet Union would launch a multicrewed spacecraft before the end of March 1966, in time for the 23rd Congress of the Communist Party.[21] The rumors were relatively precise and reported that the craft would fly through the Van Allen radiation belts. There was less confidence behind the scenes. The long-duration ground test runs of the life support system did not produce encour-aging results. After fourteen days, the Institute for Biomedical Problems had to terminate its exercise because of a worsening of the atmosphere in the cabin. OKB-124's similar experiment was shut down after sixteen days. Parachute failures meanwhile continued to accumulate throughout March. About the only positive news was on February 28, when the Air Force declared the four cosmonauts training for the flight—Beregovoy, Shatalov, Shonin, and Volynov—ready for launch.[22] Coincidentally, Dr. Norair M. Sisakyan, the Academic Secretary of the Department of Biological Sciences of the Academy of Sciences, died in mid-March amid the intense discussions prior to Voskhod 3.[23] He had played a key role in biomedical aspects of all Soviet piloted space missions beginning with the early suborbital flights of dogs in the early 1950s, and his death must have been a severe blow to Soviet space medicine. By the time of his death, well before the landing of Kosmos-110, the Voskhod 3 mission was quietly moved back to late April at the earliest.

On March 22, Mishin's engineers held a meeting to discuss the problems and assess the results of the Kosmos-110 mission. The only anomalies during the flight had been the failure of the Zarya antenna, a malfunction in the ion sensor, and a problem with the Signal high-frequency transmitter. Biomedicine specialists were already in the midst of two renewed long-duration ground tests of the life support system. If the results from the tests were satisfactory,

19. Raymond H. Anderson, "Soviet Dogs Lost Muscular Control in Space," New York Times, May 17, 1966; Raymond H. Anderson, "Gagarin Hints the Soviet Is Near Orbiting Manned Space Station," New York Times, April 9, 1967, p. 31.

20. Kamanin, Skrytiy kosmos: 1964–1966, pp. 296, 300, 302, 305.

21. "Soviet is Said to Be Preparing Manned Test of Van Allen Belts," New York Times, March 10, 1966, p. 15.

22. Kamanin, Skrytiy kosmos: 1964–1966, pp. 309–11.

23. "Dr. Norair Sisakyan, A Soviet Biochemist," New York Times, March 13, 1966, p. 86.

Voskhod 3 would be launched around April 20–22, 1966. The engineers' perhaps overtly optimistic hopes on carrying out the mission on time were thrown into disarray within days. On March 27, 1966, a Molniya-1 communications satellite lifted off from Tyura-Tam on its 8K78 booster. Unfortunately, the Blok I third stage exploded during the active portion of the trajectory, destroying the payload and the launch vehicle.[24] Because an almost identical variant of Blok I was set for use on the 11A57 booster for Voskhod 3, the failure raised alarms across the board. Several leading State Commission members rightly opposed an early Voskhod launch until investigators had conclusively ascertained the cause of the failure.

All through April, engineers focused on the problem with the Blok I stage, delaying the launch of Voskhod 3 week by week. The tests with the life support system had also proved to be unsatisfactory. Tentatively, officials were hoping for a piloted launch around May 20–27, 1966, but already there was a growing lobby against the flight of Voskhod 3 and in fact the Voskhod program as a whole. The conflict bubbled up to the surface on May 10, 1966, during a meeting of the Military-Industrial Commission. Mishin, Tyulin, Kamanin, and Deputy Minister of Health Burnazyan reported that all resources were ready to support the launch of Voskhod 3 on May 25–28. Military-Industrial Chairman Smirnov, however, stunned everyone by proposing to completely cancel the Voskhod 3 mission, invoking the following reasons:

- An eighteen-day flight would not provide anything new.
- The accomplishment of the Voskhod 3 mission would delay the Soyuz program, which should be the primary focal point for all activities in 1966.
- "[A] flight without maneuvers in orbit and without docking would display [the Soviets'] lag behind the U.S.A. and would be perceived by the public as proof of the superiority of the Americans."[25]

Smirnov clearly had some cogent arguments. NASA was flying Gemini missions at the time that were much more demonstrative of American superiority in piloted spaceflight than anything Voskhod 3 could do. The chairman had the support of a number of other key industrialists, but a whole row of powerful chief designers, academicians, ministers, and military officials strongly resisted Smirnov's suggestion. Smirnov agreed to back down and asked the Voskhod State Commission to look into the matter of terminating the program *as a whole*.

On May 12, the day after Mishin's formal appointment as chief designer of the old Korolev design bureau, the State Commission heard status reports on the various problematic bottlenecks in the Voskhod 3 plans. A designer from OKB-154 assured commission members that the high-frequency oscillations that had caused the Molniya-1 accident would not occur again, but most members remained unconvinced. Despite Chief Designer Voronin's report that the life support system was finally ready, Smirnov's abrupt speech about canceling the project had evidently made a big impression. The numerous technical glitches, combined with Smirnov's well-argued position on the pointlessness of the mission, ground the preparations for the mission into permanent inertia. As engineers argued back and forth throughout May on the reliability of the Blok I stage, State Commission Chairman Tyulin delayed the launch first to June and then to mid-July 1966.[26] The frustrated cosmonauts were sent off on a short holiday; it became increasingly clear that there might never be a Voskhod 3 mission. Despite the occasional

---

24.   Posting to FPSPACE list-server on the Internet by Igor Lissov, December 11, 1996; S. Valyayev, "Russia Cancelled Launch of 'Molniya-M'" (English title), *Novosti kosmonavtiki* 1 (January 1–12, 1997): 29–34. The Molniya-1 satellite in question was 11F67 no. 5.

25.   Kamanin, *Skrytiy kosmos: 1964–1966*, p. 337.

26.   *Ibid.*, pp. 338–39, 343.

murmurs of resuming preparations for the launch as late as November 1966, the Voskhod program was irrevocably over by June.[27]

Smirnov was clearly the instigator in the decision, but it seems that Mishin had played a major role in its termination. Having just assumed the role of chief designer of the most prestigious organization in the Soviet space program, he was no doubt reluctant to start off his tenure with an obsolete spacecraft that would guarantee only marginal safety to its crew. As one Russian journalist later wrote, Mishin "managed to convince the leaders that the 'old junk' couldn't take the country far and would only increase the lag between the United States and Russia."[28] In much the same vein, another source suggests that Mishin was concerned about the obsolete design of the Voskhod spacecraft and persuaded the leaders of the Soviet space program to permit him to terminate the fruitless effort in favor of moving ahead with the much more versatile and advanced Soyuz spacecraft.[29]

In retrospect, Smirnov and Mishin's decision to terminate the Voskhod project was a pragmatic one. Originally planned as a modest extension of the capabilities of the Vostok spacecraft in 1962 and 1963, engineers at OKB-1 continued to formulate plans for the vehicle well into 1966. The spacecraft had extremely poor characteristics and capabilities, and it was only by "cutting corners" that the engineers had managed to establish a manifest that included EVA, long-duration, and high-altitude missions. Voskhod had no capability to change orbits and, therefore, to conduct rendezvous and docking operations, placing it clearly in the first generation of space vehicles rather than the second. To spend the remaining months of 1966 preparing an obsolete spacecraft for flight would have undoubtedly delayed even further any attempts to bring the much more capable Soyuz to quick operational status. It is, however, tempting to consider the effects on public opinion and the U.S. space program if any or all of the projected Voskhod missions had been conducted on time. Many of the same objectives fulfilled in NASA's Gemini program were also planned for Voskhod. Voskhod's EVA mission was flown as Gemini IV (in June 1965), and the two-week-long mission was flown as Gemini VII (in December 1965). Then, the astronaut maneuvering unit was flown on Gemini IX (in June 1966, although the test of the unit never took place because of astronaut Eugene Cernan's troubled spacewalk), and the artificial gravity experiment and high-apogee flight was conducted on Gemini XI (in September 1966).

Some of the remnants of the Voskhod program were incorporated into Soyuz, while some were postponed indefinitely. The eighteen-day long-duration mission fell into the former group, becoming part of planning at an early stage.[30] The female EVA mission lost much of its support when Voskhod was canceled. The four unflown women once again found themselves without a program for which to train, and they were ordered back into their theoretical studies in pursuit of graduate degrees. The extensive medical experiments program, which included surgery on a mammal in orbit, was dropped from any further consideration; science in the Soviet piloted space program continued to be play second fiddle to military or political exigencies. The physicians selected for the Voskhod program never formally entered the cosmonaut team and returned to their jobs with little hope of ever flying into space. The autonomous EVA maneu-

---

27.   On July 23, 1966, Kamanin ordered the Voskhod 3 crews to immediately resume training for the mission so as to be ready by September 15. Then on October 12, Kamanin wrote that Mishin ordered the resumption of preparations for the mission. Finally, on November 25, Kamanin proposed carrying out Voskhod 3 in January 1967. None of these plans were evidently very serious. See *ibid.*, pp. 354, 360, 382, 409.

28.   Leonard Nikishin, "Soviet Space Disaster on the Revolution's Anniversary: How and Why Soviet Cosmonaut Komarov Died," *Moscow News* 9 (March 1–8, 1992): 16.

29.   S. Shamsutdinov and I. Marinin, "Flights Which Never Happened" (English title), *Aviatsiya i kosmonavtika* no. 1 (January 1993): 44–45.

30.   This is hinted at in V. P. Mishin, "Why Didn't We Fly to the Moon?" (English title), *Znaniye: tekhnike: seriya kosmonavtika, astronomiya* no. 12 (December 1990): 3–43.

vering unit named the UPMK, set for use on a later Voskhod mission, was the subject of many delays. Engineers at KB Zvezda did not complete the design of the unit until 1968, two years after Voskhod was canceled. By that time, anticipating little use in the near future, the built units were put in storage for a future time. Soviet cosmonauts would not use a similar contraption until 1990, during a mission to the *Mir* space station. That unit, also developed by Zvezda, was designed on the basis of experience creating the UPMK. There had also been much talk of military missions in the Voskhod program. These lost all justification once the Soyuz came along, particularly the military 7K-VI variant. Finally, the artificial gravity system was found to be too complex. Even before Voskhod's cancellation, in February 1966, Mishin had proposed to Minister Afanasyev to postpone the use of the IT system to a Soyuz mission. Although crews did indeed train with the system, other priorities in the Soyuz program meant that the system was never flown in space.

The Lunar Flotilla

Korolev had adopted the lunar-orbit rendezvous profile for the mission of landing Soviet cosmonauts on the Moon. Through the mid-1960s, engineers continued to fine-tune the plan, motivated by considerations of safety. By 1967, in fact, the single-launch N1-L3 mission plan had grown into a dauntingly complicated flight plan, involving several launch vehicles and spacecraft. Mishin's engineers were most concerned over the conditions at landing. What if the LK lander was damaged upon landing on the surface of the Moon? Could the lone cosmonaut have any way of knowing this before exiting the craft to set foot on the surface? To preclude a premature disembarkation, the engineers decided to launch a separate small lunar rover to inspect the exterior of the lander. Then another question arose: what if the LK was indeed damaged and could not take off? In such a case, TsKBEM engineers proposed having a backup lander launched separately, which would land near the primary one. There were more questions: what if the primary and backup landers were too far from each other for the cosmonaut to walk from one to the other? The pilot would have to travel from site to site via the lunar rover. These complex operations on the surface of the Moon also significantly raised the requirements for precision landing. The engineers introduced two additional lunar orbiters to map the potential landing sites prior to the piloted mission. Finally, there would be supplemental lunar orbital communications satellites to act as relays during landing and surface operations. All of this was motivated because of the tight mass constraints that precluded redundancy of many of the crucial systems on the LK.

The adoption of the more complex plan meant that the piloted lunar program was inextricably linked with the vigorous robotic lunar probe program. The latter had begun in early 1958, when Korolev had proposed a series of probes—the Ye-1, Ye-2, Ye-3, and Ye-4—for initial exploration of the Moon. Of the nine launches of the first generation of probes, only three achieved any modicum of success, but these were some of the most significant firsts in the early years of the "space race." The first was the first probe to achieve escape velocity and enter solar orbit (the *Cosmic Rocket* in January 1959). The second was the first probe to impact on another celestial body (the *Second Cosmic Rocket* in September 1959). The third was the first probe to take photographs of the far side of the Moon (the *Automatic Interplanetary Station* in October 1959).[31] Retroactively called *Luna 1*, *Luna 2*, and *Luna 3*, respectively, these modest spacecraft inaugurated a glorious era of robotic space exploration. By 1959, Korolev was already planning for a more

---

31.   For a Western summary of the early Object Ye lunar missions, see Asif A. Siddiqi, "First to the Moon," *Journal of the British Interplanetary Society* 51 (1998): 231–38.

ambitious series of spacecraft: the Ye-6 lunar soft-lander and the Ye-7 lunar orbiter. In January 1960, the Soviet government approved preliminary work on these two classes of probes.[32]

The Ye-6 lunar lander fared extremely poorly in the ensuing years, hampered partly by the lack of redundant systems on any of the probes because of mass constraints. There were eleven launches of Ye-6 probes between January 1963 and December 1965. Of these, four were orbital launch failures, two failed to leave Earth orbit because of failures in the Blok L acceleration stage, two missed the Moon, and three crashed onto the surface of the Moon.[33] It was a dismal record of missions that no doubt demoralized thousands of engineers. By this time, Korolev had transferred all automated lunar and interplanetary programs to the design bureau of the S. A. Lavochkin State Union Machine Building Plant, led by Chief Designer Georgiy N. Babakin. The first lunar soft-lander type flown under Babakin's command was the Ye-6M, identical to the Ye-6 except for the use of modified shock absorbers and an independent guidance system.[34]

It was seventeen days after Korolev's death, on January 31, 1966, that the first Ye-6M probe, vehicle no. 202, lifted off from Tyura-Tam and headed for the Moon. Once it was dispatched toward the Moon, it was named Luna 9 by the Soviet press. By all standards, Luna 9 and its predecessors designed by Korolev's engineers were ingeniously constructed probes. On its way to the Moon, the probe was about 2.7 meters high and consisted of three sections. At the rear was the S5.5A engine powered by an amine-based fuel and nitric acid with a thrust of 4.64 tons. The main purpose of this engine was to reduce velocity upon the approach to the Moon to facilitate a soft-landing. In addition, there were four arm-mounted thrusters that would be used for the vehicle's stabilization during landing. The central cylindrical section controlled the whole craft and carried telecommunications and command units. Strapped to the central section were two jettisonable units that had a total mass of 312 kilograms. The first of these carried a radar altimeter, which would trigger the final retroburn based on the altitude from the surface of the Moon. This unit also carried attitude control thrusters for mid-course corrections on the way to the Moon. The second unit carried Sun and Moon sensors for attitude reference. The top section of the vehicle was the landing capsule of the probe.[35]

At an altitude of 8,300 kilometers from the surface of the Moon on February 2, the attitude control jets "froze" any rolling motion in the craft and aligned it to a vertical trajectory. The radar then triggered the terminal descent sequence, and the two compartments on the side were ejected. The S5.5A engine then ignited, and five meters from the surface, a deployed sensor made contact with the ground and ordered engine shutdown. At this point, the landing capsule was thrown away from the main bus and bounced separately on the lunar surface not far from the main craft. The exact time of impact was 2145 hours, 4.25 seconds Moscow Time on February 3. Exactly 258 seconds after landing, an automatic timer activated radio transmissions from the fifty-eight-centimeter-diameter spheroid capsule. The Soviets had finally accomplished the first soft-landing of a probe on another heavenly body, nineteen days after the death of Chief Designer Korolev.

---

32.   The earliest published mention of these two variants is in a letter dated March 26, 1960, to Military-Industrial Commission Chairman D. F. Ustinov, published as S. P. Korolev, "On Expediting Operations Concerning Automated Lunar Stations (1960)" (English title) in M. V. Keldysh, ed., *Tvorcheskoye nasprediye Akademika Sergeya Pavlovicha Koroleva: izbrannyye trudy i dokumenty* (Moscow: Nauka, 1980), pp. 414–15. See also Semenov, ed., *Raketno-Kosmicheskaya Korporatsiya*, p. 146.

33.   Semenov, ed., *Raketno-Kosmicheskaya Korporatsiya*, p. 148.

34.   K. Lantratov, "Anniversaries: 25 Years for Lunokhod-1" (English title), *Novosti kosmonavtiki* 23 (November 5–18, 1995): 79–83. Lunar and interplanetary programs were transferred to Babakin in April–May 1965. The Ye-6M program was approved by a decree (no. 055-263) of the Central Committee of the Communist Party and the USSR Council of Ministers on August 3, 1964. What seems to be a technical prospectus for the Ye-6M has been published with disguised designations as S. P. Korolev, et al., "Automatic Stations for the First Landing on the Moon (1964)" (in Russian), in M. V. Keldysh, ed., *Tvorcheskoye nasprediye Akademika*, pp. 515–19.

35.   Andrew Wilson, *Solar System Log* (London: Jane's Publishing Co., 1987), p. 33.

The 105-kilogram probe's internal equipment was protected by shock absorbers and was installed in a pressurized compartment loaded toward the bottom. Four spring-loaded petals opened on top of the lander, and the TV system was activated, returning the first panoramic pictures of the lunar surface. Ironically, the first pictures published from Luna 9 were in the British press, from transmissions intercepted by the famous Jodrell Bank radio telescope. The Soviet bureaucracy's customary inefficiencies prevented *Pravda* from getting the scoop. About nine full or partial scans of the surface were received by the Soviets over the following four days, by which time the batteries were exhausted. The only other experiment on board was a radiation detector measuring the interaction of cosmic rays with the lunar soil.[36] Luna 9 was the first of two such spacecraft to land on the Moon. An almost identical vehicle, *Luna 13*, successfully landed on the Moon in December 1966.

By the time that Luna 9 landed on the Moon, Korolev's design bureau had already spent more than five years developing another robotic lunar probe that figured significantly in the Soviet piloted space program. In early 1960, Mikhail K. Tikhonravov's department at OKB-1 began exploring the possibility of designing and creating a mobile research station to travel the surface of the Moon. Unlike the earlier Ye-6 lunar probes, which were launched by the four-stage 8K78 booster, the new heavier probes would be launched by a variant of the N1 booster.[37] These studies may have had a link to even earlier research from the mid-1950s, which was publicized widely in the Soviet press. In November 1955, Yu. S. Khlebtsevich authored a detailed article in a popular journal on the technical aspects of a mobile "tankette laboratory" for traveling on the surface of the Moon. Bearing a remarkable likeness to early conceptions of such vehicles at OKB-1, Khlebtsevich's design was yet another 1950s-vintage forerunner of Soviet space achievements of the 1960s.[38]

After a slow start exploring various options, such as wheels, tank tracks, and so on, in 1963, Korolev transferred the development of the mobile probe's chassis to the Leningrad-based All-Union Scientific-Research Institute No. 100 (VNII-100) led by Chief Designer Aleksandr L. Kemurdzhian. VNII-100's primary expertise was building tanks for the Soviet Army, but Kemurdzhian had developed a personal interest in remote-controlled space probes.[39] Based on research in 1963 and 1964, Korolev and Kemurdzhian emerged in July 1964 with a conception for a 900-kilogram rover as part of the L2 theme that could support piloted lunar operations. The rover's link with the piloted space program was fortified by the famous August 1964 Soviet Union decree commitment to a human lunar landing program. The rover's primary goal would be detailed photography and research for proposed landing sites for crews on the Moon. By early 1965, engineers at OKB-1 had finished a draft plan for the L2 rover, but at this point, Korolev decided to transfer all robotic exploration probes to the Lavochkin design bureau.[40]

---

36.   *Ibid.*, pp. 33–35.

37.   The specific variant was evidently the N11, with a launch mass of 700 tons and a lifting capacity of twenty tons. See Lantratov, "Anniversaries: 25 Years for Lunokhod-1."

38.   Yu. S. Khlebtsevich, "The Road into Space" (English title), *Nauka i zhizn* no. 11 (November 1955): 33–37. An English translation of this article is included in F. J. Krieger, *Behind the Sputniks: A Survey of Soviet Space Science* (Washington, DC: Public Affairs Press, 1958), pp. 178–88. Note that Khlebtsevich had first spoken of his "tankette laboratory" in an article in *Tekhnika molodezhi* in July 1954. The earliest known description of a mobile probe on the Moon in an *official* Soviet document within the space program is in a letter from USSR Academy of Sciences President M. V. Keldysh to the Soviet government, dated December 22, 1962. This document has been published in an edited version as M. V. Keldysh, "On a Plan for Scientific Research into Outer Space in 1963–1963" (English title), in V. S. Avduyevskiy and T. M. Eneyev, eds., *M. V. Keldysh: izbrannyye trudy: raketnaya tekhnika i kosmonavtika* (Moscow: Nauka, 1988), pp. 460–62.

39.   N. G. Babakin, A. N. Banketov, and V. N. Smorkalov, *G. N. Babakin: zhizn i deyatelnost* (Moscow: Adamant, 1996). p. 56; Lantratov, "Anniversaries: 25 Years for Lunokhod-1."

40.   Lantratov, "Anniversaries: 25 Years for Lunokhod-1." There is an account of a visit in 1965 by Korolev and other engineers from OKB-1 to the premises of VNII-100 in Leningrad to view the test of an early version of the lunar rover on a simulated lunar surface. See Yu. A. Mozzhorin, *et al.*, eds., *Nachalo kosmicheskoy ery: vospominaniya veteranov raketno-kosmicheskoy tekhniki i kosmonavtiki: vypusk vtoroy* (Moscow: RNITsKD, 1994). pp. 48–49. See also Kamanin, *Skrytiy kosmos: 1964–1966*, p. 222, in which the author refers to Korolev's visit to the Likhachev Plant to see a model of the lunar rover.

530

Thus, in May 1965, all documentation and research on the rover ended up in Chief Designer Babakin's lap.

Babakin had had an interesting career. A completely self-taught engineer who received his college degree at the age of forty-three, he was an unusually gifted researcher who held a particular disdain for formal educational learning. He briefly worked at the famous NII-88 from 1949 to 1951, where he first met Korolev. He spent the next fifteen years designing high-priority missiles, including the infamous Burya intercontinental cruise missile at OKB-301 in Khimki under Chief

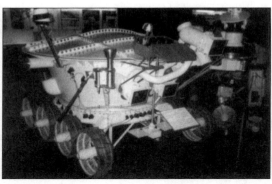

The Ye-8 rover appears here in its final design incarnation in 1971, by which time it was publicly known as Lunokhod. The two square objects in front are cameras, while the container at the top front with the lid open is a laser reflector built by the French.
(copyright Quest *magazine)*

Designer Semyon A. Lavochkin. By 1960, he had risen to the post of deputy chief designer for guidance systems.[41] For a few years in the early 1960s, Babakin worked for Chelomey, when the Lavochkin firm came under the Chelomey's control. When Chelomey lost control of his empire, Babakin rose to the top of the Lavochkin design bureau, at the exact same time that Korolev transferred all automated deep space probes to the organization. He was fifty years old at the time.

Babakin and Kemurdzhian opted to start from scratch on the rover design. By this time, the rover had been renamed Ye-8. To a certain extent, the redesign was dictated by the switch in launch vehicles to Chelomey's UR-500K booster in late 1965. Just like the L1 circumlunar project, the latter would use the Blok D translunar-injection stage to boost the rover to the Moon. More modifications came from data on the lunar soil received from the Luna 9 soft-lander probe. The firmness of the soil as well as the thinness of the dust layer led designers to drop the caterpillar track in favor of eight small wheels for movement. Babakin finished and signed the draft plan for the Ye-8 in the fall of 1966.[42] One of the lead designers of this first mobile probe on the Moon was Oleg G. Ivanovskiy, a veteran from the Korolev days. He had served as the "lead designer" of the Vostok spacecraft and early lunar probes until June 1961, when he left engineering to become the space department head at the Military-Industrial Commission. There for five years, he was responsible for a variety of important tasks, including preparing long-range space goals. In November 1965, he returned to designing as a deputy chief designer responsible for lunar probes at the Lavochkin design bureau.[43]

---

41.   Babakin, Banketov, and Smorkalov, *G. N. Babakin*, pp. 25–29; B. Ye. Chertok, *Rakety i lyudi* (Moscow: Mashinostroyeniye, 1994), pp. 272–73; O. G. Ivanovskiy and M. B. Faynshteyn, "On the Life and Scientific Activities of G. N. Babakin" (English title), in B. V. Raushenbakh, ed., *Issledovaniye tvorchestva osnovopolozhnikov kosmonavtiki i yeye sovremennyye problemy* (Moscow: Nauka, 1989), pp. 29–37.

42.   Konstantin Lantratov, "Anniversaries: 25 Years From Lunokhod-1: Part II" (English title), *Novosti kosmonavtiki* 24 (November 19–December 2, 1995): 70–79. Curiously, one otherwise reliable source states that the first meeting at Babakin's organization to discuss the Ye-8 rover was on June 14, 1967. See Babakin, Banketov, and Smorkalov, *G. N. Babakin*, p. 53.

43.   Yu. A. Mozzhorin, *et al.*, eds., *Dorogi v kosmos: II* (Moscow: MAI, 1992), pp. 13–14; O. G. Ivanovskiy, *Naperekor zemnomy prityazhenyu* (Moscow: Politicheskoy literatury, 1988), p. 265. There were two other deputy chief designers under Babakin: V. G. Perminov (deep space probes) and A. G. Chesnokov (applied themes and satellites).

Although the engineers finished the draft plan for the Ye-8 in 1966, it would be late 1967 before all the design documentation was finished, allowing for the construction of flight models. The complete Ye-8 vehicle had a mass of about 5,700 kilograms and consisted of a lander stage (the KT) and the actual rover (the 8YeL). The latter was designed to operate for three months on the lunar surface. The central components of the KT stage were four eighty-eight-centimeter-diameter spherical propellant tanks arranged in a square-shape linked by cylindrical connections. Two additional pairs of large cylindrical propellant tanks were attached vertically at the opposing sides of the central frame. These detachable tanks had mountings for antennas. One tank also had a nitrogen attitude control system, and another had attitude control sensors for the entire mission to landing. All the tanks contained the same unsymmetrical dimethyl hydrazine and nitric acid propellants, although the cylindrical ones were used only for lunar-orbit insertion and maneuvers in lunar orbit. Four short compressible landing legs were attached to the main spherical tanks, providing a maximum base of approximately four meters in diameter. Attitude control thrusters were positioned at various places, including two on a boom. A radar altimeter similar to the one on the piloted LK lander was installed between the tanks. All eight tanks fed a single engine designed by OKB-2 at Kaliningrad, designated the 11D417, with a variable thrust of 0.75 to 1.92 tons. The engine had a main exhaust supported by two verniers on each side, for use close to the surface so as not to disturb the sampling site. Four additional verniers around the periphery of the base provided stability during flight.

The KT stage was completed by the main pressurized toroidal compartment, which served not only as the primary location for all communications, data processing, and command electronics systems, but also as a platform on which the rover would be placed. The compartment also included gyroscopes for attitude reference and a set of chemical batteries for power. In addition, the stage included two sets of ramps, which would be lowered on each side of the KT following landing. Once the entire vehicle had landed, the ramps would be lowered, and the rover would track down the ramps to start its journey on the lunar surface.[44]

The 8YeL rover, with a total mass of 756 kilograms, was placed on top the KT lander stage. It was a pressurized magnesium alloy lightweight container on wheels, with a height of 1.35 meters and a diameter of 2.15 meters across the top of the compartment. As with most Soviet deep space probes, the majority of the instrumentation was installed within a pressurized compartment (at one atmosphere pressure), which contained communications and control systems. The main chassis had a large hinged convex lid, which opened up to reveal a radiator for daylight exposure. The inside of the lid also contained solar cells for furnishing one kilowatt for the internal batteries of the rover. An additional 350 to 660 watts of power would be furnished by eleven kilograms of radioactive Polonium-210 kept at the rear of the 8YeL to ensure heat for the long lunar nights. To provide information on the rover's movement, the probe used internal gyroscopes; other sensors would cut off power in case the rover attempted to overcome dangerous slopes.

Each of the eight wheels was fifty-one centimeters in diameter and equipped with independent suspension and direct-current electric motors in the hubs, the latter developed by the Krzhizhanovich Power Institute in Moscow under A. I. Moskvitin. The width from left to right at the wheel level was 1.6 meters. The wheels were made out of fine wire mesh and had titanium blades to grip the lunar surface. The 8YeL would be capable of two forward and one fixed reverse speeds, while changes in direction would be achieved by driving the wheels on either side at different speeds or in reverse. In addition, the rover was designed so as to be able to move even if only two wheels on each side were operational. If a particular wheel got stuck, a command from Earth would release a powder charge to burst the shaft, thus making the wheel a passive component. The 8YeL rover was designed from the beginning so as to be controlled from the

44.   Wilson, *Solar System Log*, p. 61.

ground. A five-person team (commander, driver, engineer, navigator, and radio operator) would guide the vehicle together while sitting in front of TV consoles showing views from the lunar surface. Nominal velocity on the surface of the Moon was limited to 100 to 200 meters per hour.

The rover carried four TV slow-image transmission facsimile cameras of the type developed earlier for the Ye-6-class of probes. These would be equipped to return 6,000-line images, which could be assembled into panoramas of the lunar surface. The cameras would be able to scan 360 degrees in the vertical and 180 degrees in the horizontal, thus providing side, down, and rear views. In addition, there were two TV cameras positioned at the forward end of the rover for providing stereo photographs with a 50-degree field of view. Communications for all surface operations would be via two antennas: one steerable high-gain and the other conical low-gain. All cameras were dual purpose—that is, for controlling the vehicle as well as for research on topography. Controllers would determine initial direction by using the panoramic cameras and would negotiate more precisely by the two frontally placed remaining cameras.

Among the scientific instruments eventually included on the 8YeL models built in the late 1960s was a penetrometer to test the soil's mechanical characteristics. The *Rifma* x-ray fluorescence spectrometer was for irradiating the soil and recording the induced radiation to identify elemental quantities of iron, calcium, silicon, magnesium, titanium, aluminum, and other substances. The x-ray device could also be used for measuring extragalactic x-rays.[45]

The Ye-8 lunar rover probe began to figure into the N1-L3 piloted lunar mission profile as early as March 1966; it would select a suitable landing site for the Lunar Ship lander and serve as a radar beacon to allow the LK to make a precision landing at a safe landing site. In December, the Interdepartmental Scientific-Technical Council for Space Research met under Keldysh's supervision to discuss requirements for the rover mission as it related to the L3 piloted landing expedition. The council discussed two different scenarios: a "realistic" one, with the rover having a lifetime of two to three months and a limited radius of operation, and an "unrealistic" one, with the rover having a lifetime of a year and a radius of operation extending to 500 kilometers. Discussions also centered around formulating a specific sequence of launches for the rover in conjunction with the N1-L3. Curiously, the Soviet press was uncharacteristically forthcoming about the rover project. On August 20, 1966, a commentator on Radio Moscow told his listeners, "Soviet experts are designing an automatic mobile station to place on the Moon."[46]

By early 1967, the N1-L3 profile had expanded into a highly complex plan with a flotilla of support missions, most designed to compensate for the poor capabilities of the L3 complex. The first lunar landing mission would be preceded by the launches of two Ye-8LS robot lunar orbiters, which would take detailed high-resolution photographs of the proposed landing sites. The photographs would allow scientists to select two landing sites: a primary one and a reserve one. Once the landing sites were determined, the Soviets would launch two separate Ye-8 rovers within a week of each other on top of UR-500K–Blok D boosters from Tyura-Tam. The rovers would land at the primary and reserve landing sites, respectively, making sure that the specific areas of landing would not pose hazards to the piloted lander. Teams on Earth would control both rovers by remote control.

---

45.   *Ibid.*, pp. 63–64; Christian Lardier, *L'Astronautique Soviétique* (Paris: Armand Colin, 1992), p. 269; Lantratov, "Anniversaries: 25 Years From Lunokhod-1: Part II"; Kenneth Gatland, *Robot Explorers* (London: MacMillan, 1972), pp. 153–57. Note that the data refer to the final version of the 8YeL in 1967–68. In 1966, the mass of the rover was limited to only 650 kilograms.

46.   *Soviet Space Programs, 1966–70: Goals and Purposes, Organization, Resources, Facilities and Hardware, Manned and Unmanned Flight Programs, Bioastronautics, Civil and Military Applications, Projections of Future Plans, Attitudes Toward International Cooperation and Space Law*, prepared for the Committee on Aeronautical and Space Sciences, U.S. Senate, 92d Cong., 1st sess. (Washington, DC: U.S. Government Printing Office, December 1971), p. 363.

A month or two later, the N1 would be launched with a working L3 complex, the latter including a LOK orbiter and the Reserve Lunar Ship (LK$_R$). The LK$_R$ would land automatically at the site of the reserve Ye-8 rover using radio beacons to guide it to a precision landing, thus saving the lander's precious propellant supply. The automated LOK would photograph the landing site from lunar orbit and return to Earth. The Ye-8 rover would then reconnoiter around the LK$_R$, taking photographs of all sides of its exterior and relaying back TV pictures, thus making sure that there had been no damage during landing. Only after an analysis that the LK$_R$ was indeed in working condition would preparations begin for launching the actual L3 complex for the piloted landing. This launch would take place during the following lunar launch window after the landing of the LK$_R$—that is, after about a month. The second N1-L3 would carry out its flight as per the nominal mission profile, with the flight engineer remaining in orbit in the LOK and the commander landing on the Moon in the LK. The actual landing would be effected by using radio beacons from the Ye-8 rovers on the surface of the Moon. The landing was to take place as close as possible to the LK$_R$. The rovers would once again examine the primary LK to ascertain whether the lander was in good external condition for takeoff. If there was no damage, the lone cosmonaut would be allowed to disembark and step onto the surface of the Moon. A nominal EVA would last about two hours, while the total stay on the Moon would be limited to six hours.

In case the primary LK was damaged, the cosmonaut would have to get to the LK$_R$ and lift off in that spacecraft. Because the Soviets were less than confident that the two landers could be landed within walking distance of each other, the Ye-8 rovers would serve as transport vehicles if the connecting distance was too far. The rovers would be equipped with reserve oxygen, allowing the cosmonaut to connect the *Krechet-94* suit to the rover's internal supply. In addition, there would be a small platform for securing the cosmonaut in a standing position for travel from one lander to the other. The cosmonaut could control the movement of the rover via a control panel, allowing a top speed on the surface of 1.2 kilometers per hour. After arrival at the LK$_R$, the cosmonaut would board it and take off to enter lunar orbit. The remainder of the mission would be identical to the standard N1-L3 lunar profile.[47]

There were two more support programs to the N1-L3 landing mission. The first involved mapping mass concentrations on the Moon that profoundly affected lunar-orbital trajectories, and the second was to support reliable communications at lunar distances. Both objectives could be achieved with the use of robotic lunar satellites. Even before these requirements had surfaced, Babakin's team had already begun developing a series of small probing lunar satellites. The first model, the Ye-6S, was built almost accidentally. When the Voskhod 3 mission was postponed, the Communist Party was left without a spectacular space mission to celebrate the 23rd Congress of the Communist Party in Moscow in March 1966. Babakin proposed that he could launch a modest satellite to the Moon if given a month. His engineers used the basic Ye-6 bus to create the Ye-6S probe, which was designed, developed, and built in less than thirty days and launched on March 1. A failure in the guidance system of the Blok L stage prevented the mission's completion, but an identical probe was launched a month later on March 31 and named *Luna 10*. On April 3, Luna 10 became the first artificial satellite of the Moon. Immediately after, the *Internationale*, the anthem of the Communist Party, was played aboard the probe and beamed back directly to the Kremlin Palace of Congresses where the 23rd Congress of the Party was in session. Assembled delegates stood at attention as the anthem was played.[48]

---

47. K. Lantratov, "Anniversaries: The 'Deceased' Lunar Plan" (English title), *Novosti kosmonavtiki* 14 (July 2–15, 1994): 60–61.

48. A. Tarasov, "Missions in Dreams and Reality" (English title), *Pravda*, October 20, 1989, p. 4; *Soviet Space Programs, 1966–70*, p. 17; Babakin, Banketov, and Smorkalov, *G. N. Babakin*, pp. 42–43. Although only two Ye-6S spacecraft were launched, there were apparently a total of five ordered for manufacture by MOM Minister S. A. Afanasyev on February 11, 1966.

The Luna 10 craft was shaped similar to Luna 9, except the lander was replaced by a 245-kilogram orbiter. Although the orbiter had no imaging capability, it relayed micrometeoroid, gamma-ray, infrared, and radiation data from near-Moon space for fifty-six days. Scientists also gathered important information on the pattern of the Moon's gravitational field based on orbital tracking. Radiation detectors revealed that the Moon had no trapped radiation belts comparable to those around Earth.[49] The success of Luna 10 allowed Babakin's engineers to design a dedicated probe primarily to take photographs of the surface of the Moon, the Ye-6LF, two of which were launched in August and October 1966 as *Luna 11* and *Luna 12*, respectively. Both carried cameras for surface photography, although the first failed to return any usable images because of malfunctions in the spacecraft's stabilization engines, which sent the spacecraft into a spin.[50] They also carried the R-1 unit for checking the action in vacuum of motors similar to the ones designed to turn the wheels of the Ye-8 rover.

Tracking during the Luna 10 mission had proved that the Moon had a very heterogeneous gravitational field. For Luna 12, ballistics experts on the ground had predictions for its orbit around the Moon for a six-month period based on prior information. But during the course of the mission, its perilune reduced by three to four kilometers per day, contrary to predictions.[51] A failure in one of the attitude control engines of the probe prevented changing the perilune of the spacecraft. The data gathered during the mission, however, served as a starting point to design and develop a new model of a lunar satellite, one of whose mission goals was to study the Moon's gravitational field to make precise determinations of trajectories for the various elements of the N1-L3 lunar landing plan. Babakin's team began development of the Ye-6LS in late 1966, which also had the dual purpose of testing the Soviet deep space communications network.

Tracking for the Moon

The Soviet tracking and telemetry network, known officially as the Command-Measurement Complex, had grown in steps and bounds since its early days in the late 1950s. Approximately fifteen stations, referred to as Scientific-Measurement Points (NIP), were located throughout the contiguous USSR, serving as stations for use during Earth-orbit and deep space missions, both piloted and automated. The ground stations were augmented in the mid-1960s by the third generation of Soviet tracking ships. In 1965 and 1966, the new *Ristna* and *Bezhitsa* replaced the older *Ilichevsk* and *Krasnodar*. Later in 1967, four new ships were introduced—the *Kegostrov*, the *Nevel*, the *Morzhovets*, and the *Borovichi*—each with a displacement of 6,100 tons and a crew of thirty-six.[52] The same year, all the ships were officially turned

49. Wilson, *Solar System Log*, pp. 35–36.

50. The Ye-6LF probes were designed to take photographs in two regimes: (1) photographing in a stabilized mode from the perilune immediately after the satellite braked into lunar orbit and (2) in slow rotation conditions when oriented relative to the Sun. The failure on Luna 10 was due to "uncompensated parasitic moments" in the stabilization engine system. See Babakin, Banketov, and Smorkalov, *G. N. Babakin*, pp. 45–46. A document dated circa 1965, from Keldysh and Babakin to the government, proposing the Ye-6LF program has been published as M. V. Keldysh and G. N. Babakin, "On Photographing the Lunar Surface with Artificial Satellites of the Moon" (English title), in Avduyevskiy and Eneyev, eds., *M. V. Keldysh*, pp. 480–81.

51. Babakin, Banketov, and Smorkalov, *G. N. Babakin*, pp. 46–48. Both the Luna 11 and Luna 12 spacecraft also had extensive supplementary scientific instruments aboard. The latter carried a gamma-ray detector, a magnetometer, radiation detectors, an infrared radiometer, and meteoroid detectors. Luna 12 detected x-ray emissions from the Moon's surface as a secondary effect of fluorescence under solar x-ray emission. The Soviets later claimed that this was the "birth of x-ray astronomy." See G. V. Petrovich, ed., *The Soviet Encyclopaedia of Space Flight* (Moscow: Mir Publishers, 1969), p. 45. Contact was lost with Luna 11 on October 1, 1966, and with Luna 12 on January 19, 1967.

52. Jacques Villain, ed., *Baïkonour: la porte des étoiles* (Paris: Armand Colin, 1994), p. 92; B. A. Pokrovskiy, *"Zarya"—pozyunoye zemni* (Moscow: Moskovskiy rabochiy, 1987), p. 254.

over to the Department of Naval Expeditionary Work of the Academy of Sciences, although it seems that the "civilian" tag was somewhat of a misnomer because much of the on-board personnel were military servicepersons.[53] The Soviets depended to a great extent on these ships, partly because overflying satellites were in direct visibility of ground stations only nine out of twenty-four hours on the average. In addition, unlike NASA, the Soviets had less luck placing stations in foreign countries, although stations were established in Chad, Cuba, Guinea, Mali, and the United Arab Emirates in 1967–70.[54] The locations in Africa were evidently built specifically for piloted lunar programs because they would be on the ground track for return trajectories from lunar distances.

All the ground stations of the Command-Measurement Complex were under the direct control of the Strategic Missile Forces via military unit no. 32103. This unit, commanded by Maj. General Ivan I. Spitsa since March 1965, had emerged from the auspices of the military NII-4, located in Bolshevo outside Moscow. Since the early days of the ICBM program, NII-4, which was subordinated to the Strategic Missile Forces, was responsible for coordinating tracking and communications with space satellites via its numerous tracking stations across the Soviet Union. In December 1957, NII-4 moved its control center from Bolshevo to Moscow, and in January 1963, this control center was removed from NII-4's jurisdiction and subordinated directly to the General Staff of the Strategic Missile Forces as military unit No. 32103.[55] The Moscow location was the central control node for the early Soviet space program, supporting all communications with robotic and piloted satellites in space.

Unlike NASA, however, the Soviets did not have a dedicated mission control facility for piloted missions until well into the early 1970s. Instead, each mission had its own customized chief operations and control group (GOGU), somewhat analogous to the Western concept of a flight control team, which maintained control over all flight operations, such as docking, EVA, reentry, and so forth. The GOGU was staffed by approximately ten representatives from the design bureaus, the military, the production plants, and the Academy of Sciences.[56] By the time of the early Soyuz missions, the GOGU oversaw up to about 500 individuals, who worked around the clock in three shifts. If there were specific technical issues or problems, specialists from the relevant design bureaus were invited to participate in the operations of the GOGU. Up until 1966, Colonel Amos A. Bolshoy, an officer in the Missile Strategic Forces, led the GOGU for all piloted missions. For a particular flight, the GOGU was given access to the military Command-Measurement Complex, and depending on the circumstances surrounding the mission, the GOGU could be based at one of several locations, including NII-4's Moscow branch (for Vostok missions) or the Ministry of Defense's General Staff control center, also in Moscow (for Voskhod missions). Because the Vostok and Voskhod missions were relatively short, State Commission members usually never departed the launch site at Tyura-Tam after

53.   The commander of the Department of Naval Expeditionary Work (OMEP) was Rear Admiral I. D. Papanin, who served in that capacity from 1951 until his death in 1986.

54.   *Soviet Space Programs: 1976–80: Supporting Vehicles and Launch Vehicles, Political Goals and Purposes, International Cooperation in Space, Administration, Resource Burden, Future Outlook*, prepared for the Committee on Commerce, Science, and Transportation, U.S. Senate, 97th Congress, 2d sess. (Washington, DC: U.S. Government Printing Office, December 1982), p. 124.

55.   K. V. Gerchik, ed., *Nezabyvayemyy Baykonur* (Moscow: Interregional Council of Veterans of the Baykonur Cosmodrome, 1998), p. 379. See also B. A. Pokrovskiy, *Kosmos nachinayetsya na zemlye* (Moscow: Patriot, 1996), p. 272. For military unit No. 32103, see Semenov, ed., *Raketno-Kosmicheskaya Korporatsiya*, p. 351.

56.   For the early Soyuz missions, the GOGU included nine men: P. A. Agadzhanov (TsKIK), S. N. Anokhin (TsKBEM), B. Ye. Chertok (TsKBEM), K. P. Feoktistov (TsKBEM), G. I. Levin (NII-4), Pavlov (affiliation unknown), B. V. Raushenbakh (TsKBEM), M. S. Ryazanskiy (NII Priborostroyeniya), and Ya. I. Tregub (TsKBEM). See Chertok, *Rakety i lyudi: goryachiye dni kholodnoy voyny*, p. 422.

liftoff. Thus, for these early flights, senior officials such as Korolev, Keldysh, or Tyulin would remain at Tyura-Tam and maintain a constant communications link with the Moscow center, which itself maintained contact with the Command-Measurement Complex. The nerve center at Tyura-Tam was usually at site 2 on the second floor of the giant Assembly-Testing Building in the offices of Maj. General Anatoliy S. Kirillov, the famous chief of the First Directorate at the launch range during the early 1960s.[57]

With the commencement of the Soyuz program, officers of the Strategic Missile Forces proposed moving the main control center for piloted missions to a dedicated facility, the Scientific-Measurement Point No. 16 (NIP-16) at Yevpatoriya in Crimea. NIP-16 thus became the second-generation Soviet flight control center, at which the GOGU controlled almost every single Soviet piloted mission from 1966 to 1975. By 1966, the first-generation flight control centers, at NII-4 and the General Staff, were, for the most part, turned over to control automated military satellites.

NIP-16 had originally been built in the late 1950s as a modest station for receiving telemetry from overflying satellites, but its central role in the Soviet space program grew dramatically during the early 1960s. In 1959, when OKB-1 first began developing interplanetary spacecraft to fly to Mars and Venus, Korolev and Keldysh proposed a dedicated site to build a deep space tracking station. The designers had a deadline of just eight months. A special commission quickly selected Yevpatoriya on the shore of the Black Sea. The future facility was named "Object MV" to denote its role in tracking spaceships to Mars and Venus, although it was rumored that the "MV" also stood for Mstislav Vsevolodovich, the first two names of Academician Keldysh. Korolev had initially invited Chief Designer Ryazanskiy of NII-885 to design the radio tracking systems for the facility, but he had declined, believing that it would be impossible to develop antennas capable of tracking signals from a distance of 100 million kilometers. Chief Designer Yevgeniy S. Gubenko of SKB-567 took on the job and proposed that instead of one 100-meter parabolic dish, eight sixteen-meter" bowls," designated ADU-1000, be erected at the site, providing a capability to communicate to distances of 300 million kilometers.[58]

Korolev came up with an ingenious idea to mount the dishes using leftover parts from the Soviet Navy. Construction workers dug a huge crater out of the rocky ground, poured in a foundation, took the revolving gun turret of a former seafaring battleship consigned to the junkyard, and placed it on the foundation. Then the open framework of a railroad bridge was placed over the turret. The bridge itself was covered by the solid hull of a scrapped submarine. The eight antennas were fixed to this hull.[59] Eventually, the Object MV station at NIP-16 consisted of three complexes separated by several kilometers: one designed to send commands and the two others to receive incoming information. Each complex had eight antennas with a diameter of sixteen meters and a surface area of 1,000 square meters. The transmission power was rated at 120 kilowatts, and the maximum range was 300 million kilometers. The sensitivity was sufficient to detect a match struck on the surface of the Moon. The facility came on

57.   *Ibid.*, pp. 413–14; Semenov, ed., *Raketno-Kosmicheskaya Korporatsiya*, pp. 351–53.

58.   Pokrovskiy, *Kosmos nachinayetsya na zemlye*, pp. 309–12; B. Ye. Chertok, *Rakety i lyudi: Fili Podlipki Tyuratam* (Moscow: Mashinostroyeniye, 1996), pp. 301–02. Chief Designer Ye. S. Gubenko died unexpectedly in 1959, and this work was continued by his successor, A. V. Belousev. Other enterprises involved in building the dishes included TsNII-173 (mechanical drives) and MNII-1 (systems for aiming the antennas). Note that Chertok says that the diameter of the dishes was twelve meters. Most other sources suggest sixteen meters. See, for example, Pokrovskiy, *"Zarya"—pozyunoye zemni*, p. 228.

59.   B. Sopelnyak, "The Secret of Facility MV" (English title), *Krasnaya zvezda*, March 22, 1990, p. 4.

line on September 26, 1960, on a provisional standing, and it was fully operational by December 30.[60]

The Yevpatoriya station was supported by several "ballistics centers." These were located at NII-4, at the Institute of Applied Mathematics of the Academy of Sciences in Moscow, at the Central Scientific-Research Institute for Machine Building in Kaliningrad, and at Yevpatoriya itself, for computing all trajectories, orbits, flight parameters, and so forth. The facilities at Yevpatoriya were relatively primitive. Mission controllers had no real-time visual depictions of mission parameters, such as at NASA's much more modern Manned Spacecraft Center in Houston, Texas. The primary mode of communications between the centers and spacecraft were, in fact, old-style telephone and telegraph systems, scrambled to maintain secrecy.

In 1966, Maj. General Pavel A. Agadzhanov, a deputy commander of military unit No. 32103, began his tenure as the head of the GOGU—that is, the "flight director" of Soviet piloted space missions. An amateur radio enthusiast in his youth, he joined NII-4 in 1948 and contributed to the development of tracking networks at Kapustin Yar, Tyura-Tam, and eventually the space Command-Measurement Complex. Based on this work, Agadzhanov earned his Ph.D. in the late 1950s, and he moved into ballistics computation work for the Soviet ground communications segment.[61] For the top-secret piloted lunar

Pavel Agadzhanov was the "flight director" for Soviet piloted space missions during the late 1960s. His early career had been at the military NII-4. Later, he served as a deputy chief of the Command-Measurement Complex, the Soviet tracking network. (files of Peter Gorin)

flights—the UR-500K-LI circumlunar and NI-L3 landing missions—Colonel Nikolay G. Fadeyev, yet another accomplished military officer, headed flight operations in the late 1960s.[62]

The GOGU controlled the missions via the military officers of the Command-Measurement Complex, but the GOGU itself was subordinated to the temporary State Commission, which would receive recommendations from the GOGU, make decisions based on these communications, and then recommend courses of actions. The GOGU would also maintain constant contact with "backup" centers: Group T at Tyura-Tam and Group M at NII-4. TsKBEM played a major role in the operation of the GOGU, because its "technical leader" (the "deputy flight director") was usually a civilian deputy chief designer from the design bureau. This post was occupied by Boris Ye. Chertok from 1966 to 1968 and Yakov I. Tregub from 1968 to 1973.[63] This management hierarchy, in which a military officer headed flight control while his principal

60. Pokrovskiy, "Zarya"—pozyvnoye zemni, p. 227; I. Meshcheryakov, "The Center for Long-Range Space Communications" (English title), Aviatsiya i kosmonautika no. 6 (June 1988): 42–43. Object MV was augmented in 1979 by the seventy-meter-diameter RT-70 radio telescope, which allowed spacecraft tracking to extend to 1.5 billion kilometers. The RT-70 was designed by NPO Radiopribor (formerly NII-885). Three other large dishes for the deep space communications network were designed and built by the OKB of the Moscow Power Institute (OKB-MEI). These included two dishes (twenty-five and thirty meters) at Crimea and one dish (sixty-four meters) at Medvezhiy Lake near Moscow. The twenty-five-meter dish was evidently located at NIP-10 in Simferepol. See Chertok, Rakety i lyudi: Fili Podlipki Tyuratam, p. 301; A. V. Ponomarev, "2 June—75 Years From the Birth of Academician A. F. Bogomolov (1913)" (English title), Iz istorii aviatsii i kosmonautiki 59 (1989): 47–50.

61. Pokrovskiy, Kosmos nachinayetsya na zemlye, pp. 114–16; Semenov, ed., Raketno-Kosmicheskaya Korporatsiya, p. 355.

62. Pokrovskiy, Kosmos nachinayetsya na zemlye, pp. 279–80; Semenov, ed., Raketno-Kosmicheskaya Korporatsiya, p. 355. There was a third GOGU chief during the period 1966–73, Colonel M. S. Posternak.

63. Semenov, ed., Raketno-Kosmicheskaya Korporatsiya, p. 355. Whereas, at first, the GOGU was established unique to each mission, starting in 1968, Chief Designer Mishin established a specialized control subdivision in Tregub's testing department at TsKBEM to focus exclusively on mission control.

assistant was a civilian from the design bureau, was symptomatic of all flight control teams. It underscored not only the deeply enmeshed military nature of all Soviet space programs, but also the decades-long aftereffects of the actions of artillery officers who had pragmatically taken operational control over missile projects during the late 1940s. Ironically, Tregub had started his career as an artillery officer overseeing the early A-4 and R-1 launches from Kapustin Yar. He later moved on to direct launches of air defense and anti-ballistic missiles for the Soviet military during the 1950s and 1960s. In 1964, Korolev had invited him to join OKB-1 as the deputy chief designer responsible for flight testing.

The Rise and Fall of the UR-700

Through the mid-1960s, in the post-Korolev era, General Designer Vladimir N. Chelomey continued to push his own conception of a piloted lunar landing project. This proposal, involving the giant UR-700 booster, had gained ground in 1964 when Khrushchev had suggested that scientists carry out a detailed appraisal of the costs and advantages of the UR-700 over the N1 plan. Despite Khrushchev's ouster, Chelomey lined up a formidable array of supporters, including Chief Designers Glushko, Kuznetsov, and Barmin. By October 1965, the Ministry of General Machine Building had approved the development of a predraft plan at TsKBM. Perhaps realizing the absurdity of the situation, Korolev had evidently authored a letter to Minister Afanasyev, requesting that the government not waste money on duplicating the N1-L3 project. The letter never reached Afanasyev; days after preparing it, Korolev was dead.

Chelomey's engineers at his Branch No. 1 at Fili approached the UR-700 effort with some amount of humor. There was evidently a joke making the circles at the design bureau that because Korolev had died, his subordinates could not be expected to make anything out of the "hopeless" characteristics of the N1. Therefore, Chelomey's engineers were acting only out of kindness by offering "humanitarian" aid in the form of the UR-700.[64] Because they were working in a less-than-favorable post-Khrushchev climate, Chelomey's deputies developed a technical plan that significantly reduced cooperation with outside subcontractors and relied heavily on internal expertise. In addition, the actual design of both the UR-700 booster and its lunar payload, designated the Lunar Ship No. 700 (LK-700), was derived from already existing designs to minimize long lead times for development.[65] TsKBM completed the predraft plan (the mechanics of the proposal) for the UR-700 and its LK-700 lander in August–September 1966.[66] The achievement of this milestone served as a catalyst for action from the government. Minister Afanasyev finally fulfilled the deposed Khrushchev's original command by issuing an order on September 17, 1966, for the formation of a commission to conduct a comparative study between the UR-700 and the N1-L3 on "the reasonability of proceeding with further works on those projects."[67]

The "expert commission" to compare the UR-700/LK-700 and the N1-L3 proposals was headed by the ubiquitous Academician Keldysh. According to one observer, most of the thirty-four members of the commission were sympathetic to the late Korolev. Chelomey's relation-

64. Rudenko, "Designer Chelomey's Rocket Planes."

65. Sergey Khrushchev, Nikita Khrushchev: krizisy i rakety: vzglyad iznutri: tom 2 (Moscow: Novosti, 1994), p. 518.

66. Rudenko, "Space Bulletin: Lunar Attraction: Historical Chronicles" (English title), Vozdushniy transport 24 (1993): 11. One source states that in August 1966, "the predraft plan for the piloted ship for circling the Moon of the type 'LK-3' and the piloted ship for landing on the Moon 'LK-700' was finished." The source also implies that in September 1966, "the work on the mechanics of the proposal for the UR-700 rocket-carrier planned for landing the LK-700 piloted apparatus on the Moon" was finished. See Mikhail Rudenko, "Space Bulletin: Lunar Attraction: Historical Chronicles: First Publication" (English title), Vozdushniy transport 27 (1993): 8–9.

67. Vetrov interview, November 15, 1996; Georgiy Stepanovich Vetrov, "Development of Heavy Launch Vehicles in the USSR," presented at the 10th International Symposium on the History of Astronautics and Aeronautics, Moscow State University, Moscow, Russia, June 20–27, 1995.

ship with Keldysh had also evidently soured despite the latter's occasional support.[68] In late October 1966, Minister Afanasyev, accompanied by the commission, visited both TsKBM and TsKBEM to assess the pros and cons of both projects as explained by their respective creators.[69] Chelomey had set up a stunning display of posters in his huge sixth floor office room at Reutov, and the commission spent the day asking detailed questions. The visit to Mishin's design bureau differed only in the use of models instead of posters. Afanasyev was evidently uncertain of which project to favor. By this point, Chelomey felt that he was fighting a losing battle because Mishin had the backing of Keldysh and Ustinov. He told one of his assistants, "I don't want to fight with [the commission]."[70] He wanted instead to concentrate his time and resources on the UR-100 ICBM project, one of his few bright prospects for the future. Finally, on November 16, 1966, Chelomey presented the basic technical details of his competitive lunar landing proposal at a plenary session of "the advisory council reviewing the course of work being done in the N1-L3 program."[71]

The origins of the UR-700 booster can be traced back to 1961, when Chelomey tasked his Branch No. 1 to explore possible designs for a booster capable of lifting approximately seventy tons to low-Earth orbit. Serious work on the concept did not, however, begin until the collapse of the LK-1 circumlunar plan in 1965. Chelomey was also inspired to pursue the idea from Yangel's defunct heavy-lift R-56 rocket project offered briefly as a competitor to the N1-L3 program. Perhaps he did not want to be left out of this mother lode of space projects. Chelomey made sure that his UR-700 proposal would have radical differences with Korolev's N1-L3; if the two projects were only marginally dissimilar, any evaluation commission would have little reason to pick the UR-700 over the N1. Like a good politician, he made sure that the UR-700 proposal was not just incrementally, but significantly more superior to the N1 project in every relevant parameter.

When Chelomey formally presented his UR-700 lunar landing project in November 1966, he emphasized five major requirements for the overall plan, which he believed would give it the advantage over his principal competitor:

- His design bureau and only his design bureau, TsKBM, would be the primary contractor for the project. Mishin's TsKBEM would be completely excluded from any participation in the work.
- All subcontractors working on the N1-L3 should redirect all their work to the UR-700 project. (In addition, all ground equipment developed for the N1-L3 would be used for the UR-700 with minimum updating.)
- The UR-700 project could be accomplished in the shortest time possible with the most minimum of expenditures. Curiously, Chelomey made no mention whatsoever of a competition with Apollo; apparently, Chelomey believed that even in the most favorable of circumstances, the first landing mission would mostly likely be after an Apollo landing.
- All stages of the UR-700 and its LK-700 would use storable propellants (nitrogen tetroxide and unsymmetrical dimethyl hydrazine).
- All of the manufacturing of the UR-700 and the LK-700 would be carried out at TsKBM and its affiliate M. V. Khrunichev Machine Building Plant in Moscow.

68. Khrushchev, *Nikita Khrushchev: tom 2*, p. 519.
69. Rudenko, "Designer Chelomey's Rocket Planes."
70. Khrushchev, *Nikita Khrushchev: tom 2*, p. 521.
71. I. B. Afanasyev, "Unknown Spacecraft (From the History of the Soviet Space Program)" (English title), *Novoye v zhizni. Nauke, tekhnike: Seriya kosmonavtika, astronomiya* no. 12 (December 1991): 1–64.

For the UR-700 launch vehicle in particular, there were four design specifications:

- The booster would launch a payload about one and a half times the mass of the L3 payload of the N1 rocket.
- The booster would be built on the "block" principle—that is, its separate blocks could be transported by rail and assembled at the launch site. These blocks would be based in design on the individual rocket units of the smaller Proton booster.
- The booster would have a minimum number of stages and engines to increase reliability. The engines of the lower two stages would have very high thrusts per combustion chamber.
- Booster staging would be designed with a composite layout in mind—that is, the first stage would be connected in parallel like strap-ons, and the second and third stages would be linked in tandem.

The LK-700 lunar landing payload had two major requirements:

- Because of the selection of a direct ascent, the LK-700 would have a launch mass of one and a half times as much as the L3 payload.
- The design of the LK-700 would be such that maximal use would be made of already created space vehicles. This would significantly reduce development time. Engineers would make good use of already-built robotic spacecraft such as the "IS" and the "US," the abandoned piloted Raketoplan and LK-1 projects, and the UR-100 ICBM.[72]

In exploring various concepts of the LK-700 lunar landing spacecraft, Chelomey proposed using a "direct ascent" mission profile; it dispensed with both lunar-orbit rendezvous and Earth-orbit rendezvous. In the United States, NASA had foregone direct ascent in favor of lunar-orbit rendezvous in 1962, while Korolev's camp in the Soviet Union had done the same in 1964. Chelomey, however, did not want to deal with complex docking operations, which might introduce weak links in the system as a whole. His engineers also believed that a direct ascent profile would allow a wide range of landing sites on the Moon, up to as much as 88 percent of the lunar surface, as opposed to lunar-orbit rendezvous, in which landing sites would be limited only to the equatorial regions. A direct ascent profile necessitated the use of a very heavy launch vehicle—one with a lifting power about one and a half times more than that of the N1. Payload capability to Earth orbit of the UR-700 was in the range of 145 tons, sufficient for a translunar-injection stage, a lunar braking stage, and a large lunar lander. The mass of the latter two components—that is, the mass injected on an escape trajectory—was approximately fifty tons.[73] The increased mass of the lander would allow a crew of two persons to land on the Moon, unlike the L3's one cosmonaut. Two cosmonauts on the ground afforded significantly increased levels of safety and more scientific research. With high-energy stages, this number could be increased to three during later missions.

Unlike the N1-L3 plan, Chelomey outlined an extensive program of scientific research for his new project, to be carried out both en route to the Moon and on its surface. This program would include:

72. E-mail correspondence, Igor Afanasyev with the author, December 16, 1997.

73. The mass of 145 tons is from N. Kamanin, "A Goal Worth Working for" (English title), *Vozdushniy transport* 45 (1993): 8–9. Other figures have also been quoted, including 130 tons and 150–151 tons. For the former, see Afanasyev, "Unknown Spacecraft." For the latter, see V. Karrask, O. Sokolov, and V. Shishov, "Known and Unknown Pages of the Russian Khrunichev Center's Space Activity," presented at the 47th Congress of the International Astronautical Federation, Beijing, China, October 7–11, 1996.

- Research on radiation conditions in space
- Studies on micrometeoroids in space
- Research on solar plasma
- The study of the lunar surface for identifying optimal landing sites and refining seleno-graphic coordinates for purposes of navigation
- The collection of samples from the Moon at various depths
- Passive seismographic studies on the Moon
- Measurements of lunar surface temperatures
- Studies of lunar soil properties by spectroscopy
- Research on cosmic rays
- Research on electrical potentials in lunar soil caused by natural magnetic fields
- A precise determination of the Moon's movement relative to Earth with the use of lasers delivered to the Moon
- The study of variations in the lunar gravitational field
- Research on variations of the lunar magnetic field
- Extensive surface photography[74]

In making his report, Chelomey also offered up the somewhat ambitious prospect of gearing all UR-700 landing missions such that they would eventually lead to the establishment of permanent bases on the Moon. Initial landing sites would be chosen for their possible use as future "colonies." Work on these future prospects would be aided by Ye-8 robotic rovers on loan from the Lavochkin organization.

From a hardware perspective, the UR-700 booster was a behemoth. On the pad, the complete booster-payload stack would be approximately seventy-six meters in length (including the standard launch escape tower) and have a base diameter of about seventeen and a half meters (excluding four large aerodynamic stabilizers for use during the active portion of the ascent trajectory). For engines on the rocket, Chelomey had initially contracted his favorite subcontractor, Chief Designer Kosberg. In 1962, Kosberg's design bureau, OKB-154, had begun work on a 250-ton engine, the RD-0215. A number of other research organizations, including the Central Institute of Aviation Engine Building, the Scientific-Research Institute of Thermal Processes, and the All-Union Institute of Aviation Materials, were involved in the early work on the engine, which was the most powerful engine Kosberg had ever designed. Using technologies derived from engines of the UR-200 ICBM, in two years, Kosberg's engineers prepared a large volume of ground equipment for testing the unit at its own manufacturing plant. Two initial engines were built, one for cold testing and one for ground firings.[75] In 1965, Glushko stepped in. For several years, he had been working on a giant 680-ton (vacuum) thrust engine for possible use on a future Soviet booster. When Korolev rejected all his overtures to use this engine on the N1, Glushko turned to Chelomey and convinced the latter that his RD-270 would be a better choice for the UR-700 than Kosberg's RD-0215. All work on the Kosberg engine was terminated immediately.

The cooperation with Glushko led to two variants of the UR-700: one with a multitude of RD-253 engines, identical to the ones used on the first stage of the more famous UR-500K (or Proton) booster, and the second one with the massive RD-270s.[76] This second version of the rocket was a three-stage monster that dwarfed the N1 in size. Compared to the N1-L3's total mass of 2,750 tons, the UR-700/LK-700 would weigh a whopping 4,820 tons at launch. Its mass was more comparable to the giant Nova studies pursued by NASA in the early 1960s before the decision in favor of the Saturn C-5. The new system's specifications were:

74. Afanasyev correspondence, December 16, 1997.

75. KB KhimAvtomatiki: Stranitsy istorii: tom I (Voronezh: KB KhimAvtomatiki, 1995), pp. 57–58.

76. Vetrov, "Development of Heavy Launch Vehicles"; Telephone interview, Sergey Nikitich Khrushchev with the author, October 10, 1996. Lt. General N. P. Kamanin wrote in his journal entry for December 28, 1966, that "the first and second stages of the UR-700 are basically the same as those of the UR-500." It is possible that he was referring to the first variant of the UR-700 using the RD-253 engines. See Kamanin, "A Goal Worth Working for."

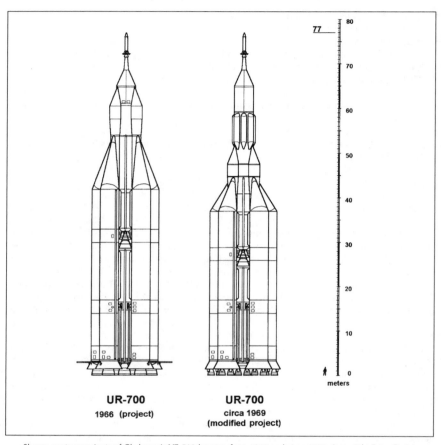

Shown are two variants of Chelomey's UR-700 booster, from 1966 and circa 1969. (copyright Peter Gorin)

Stage	Engines	Thrusts	Total Thrusts
Stage I	Nine RD-270s	640 tons each (sl)	5,760 tons (sl)
Stage II	Three RD-270s	680 tons each (v)	2,040 tons (v)
Stage III	Three RD-254s	170 tons each (v)	510 tons (v)[77]

The third stage's RD-254 engines were merely altitude versions of the Proton's RD-253 units.

In terms of design, the UR-700 held a superficial resemblance to the Proton and Vostok boosters, in that it looked like a core booster surrounded by strap-ons. The arrangement and use of the core and strap-ons were, however, vastly different. In the UR-700's case, Chelomey's engineers used both a tandem and a parallel strap-on scheme on the same booster. The core

---

77. E-mail correspondence, Igor Afanasyev with the author, December 17, 1997.

of the launch vehicle—the second stage—consisted of a two-stage booster. The lower portion was a cluster composed of three long cylindrical modules, each with a diameter of just over four meters, which was a limit from a rail transport perspective. These modules were derived from the same tanks used on the Proton booster. Each module was equipped at the base with a single RD-270 engine. The upper portion of the core consisted of three smaller diameter cylindrical tanks clustered together, each with a single RD-254 engine.

The core was surrounded through its entire length by three clusters, each with two identical cylindrical modules. This set of six cylinders was known collectively as the first stage of the booster. Like the core, the first stage also used single RD-270 engines on each module. All nine modules of the first and second stages were to fire at liftoff, giving a total sea-level thrust of 5,760 tons, far above both the N1 (4,620 tons) and the Saturn V (3,404 tons). The effectiveness of the excessively high thrust was tempered to a great degree by the use of low-performance propellants—unsymmetrical dimethyl hydrazine and nitrogen tetroxide—which significantly lowered the efficiency of the engines as compared to both its competitors. At a certain point in the trajectory, the strap-ons would be discarded, leaving the lower portion of the core to fire at a vacuum thrust of 2,040 tons. This section would eventually fall away, and the three RD-254 engines would fire at a total of 510 tons thrust to insert the 151-ton payload into Earth's orbit. Initial parameters would be 260 by 186 kilometers at a fifty-one-and-a-half-degree inclination.[78]

The entire LK-700 complex was a four-stage vehicle. The first stage was for translunar injection (TLI), the second for braking prior to landing on the Moon, the third for soft-landing on the Moon, and the fourth for liftoff and direct return to Earth. Their performance characteristics were:

Stage	Purpose	Engines	Number X Thrust	Design Bureau
Stage IV	TLI	11D23	Three X 23.5 tons	Kosberg
Stage V	Lunar braking	11D23	One X 23.5 tons	Kosberg
Stage VI	Lunar landing	11D416	Three X 0.75–1.9 tons	Isayev
Stage VII	Lunar takeoff	15D13	One X 13.4 tons	Izotov

After being put on a trajectory toward the Moon, the crew would discard the heavy TLI stage weighing about 100 tons and settle into their lunar lander, which would have a mass of fifty and a half tons, en route to the Moon. During this part of the mission, mid-course corrections would be effected by small 1.67-ton-thrust verniers on the side of the spacecraft. After a three-and-one-third-day coast to the Moon, the single lunar braking engine, similar to the ones used for TLI, would fire to reduce velocity to levels safe for the initiation of lunar landing maneuvers. After the use of this engine, this stage would be jettisoned, releasing the 18.3-ton lander proper. At this point, the two-man crew would use a set of three throttleable 1.9-ton-thrust engines for hovering over the lunar surface and selecting a site. At landing, the LK-700 lander would have a mass of just over seventeen tons. For initial landing sites, Chelomey's engineers picked two possible areas stemming from two different trajectories to the Moon: the Sea of Fertility after a six-and-a-half-day flight to the Moon or the Ocean of Storms after a three-and-a-half-day flight.

The cosmonauts would spend the majority of their trip in a cone-shaped return apparatus shaped similarly to the abandoned LK-1 circumlunar ship, but scaled up in size to hold two

---

78.   Karrask, Sokolov, and Shishov, "Known and Unknown Pages." The mass of the LK-700 complex on the ground was 154 tons. The three missing tons was the launch escape system, which would be discarded prior to insertion into Earth orbit.

On the right of this Russian drawing is one of the few publicly available representations of Chelomey's LK-700 lunar landing complex. The resemblance of the LK-700 to NASA's Gemini is clearly evident. Below the lander is the final stage of the UR-700 rocket. For comparison, Korolev's L3 lunar complex is shown on the left. (copyright Igor Afanasyev)

cosmonauts. The link in design between the LK-1 and the LK-700 would establish a genealogy of spaceship design across several generations of space vehicles designed at Chelomey's design bureau.[79] The return apparatus would set down on the Moon with its apex pointing upwards—looking much like an upright Apollo Command and Service Module. The crew would spend about twelve to fourteen hours on the lunar surface during early missions, sufficient for one excursion outside. At liftoff from the Moon, the cosmonauts would sever attachments to the descent stage of the LK-700 and launch from the surface using a single 13.4-ton-thrust engine firing at full thrust. Two different options were available to the crew in their 14.8-ton ascent stage: either flying directly toward Earth or entering lunar orbit and leaving for Earth at the most appropriate moment. After further mid-course corrections on the way back to Earth using three small 200-kilogram-thrust engines, the return apparatus would separate from the rest of the LK-700 spacecraft and reenter Earth's atmosphere. Looking remarkably similar to the Apollo Command Module, the 3.1-ton capsule would land by parachute on Soviet territory after a guided descent through the atmosphere. The total mission would last eight and a half days from start to finish.[80]

79.   A drawing of one variant of the LK-700 has been published in Afanasyev, "Unknown Spacecraft."
80.   Afanasyev, "Unknown Spacecraft"; Afanasyev correspondence, December 17, 1997.

The Kosberg and Isayev design bureaus were contracted to build most, but not all, the engines for the payload. One exception was the designer for the critical ascent stage engine of the LK-700 lander, which was contracted out to OKB-117 (later the Leningrad experimental design bureau named after V. Ya. Klimov). Like many other aviation design organizations, OKB-117, headed in the mid-1960s by Chief Designer Sergey P. Izotov, was trying to diversify into the missile and space business to preclude economic collapse. Izotov had primarily been famous for designing engines for Soviet military helicopters from the Mil and Kamov design bureaus.[81] Izotov's first foray into the missile business had been the creation of the 8D423, the second-stage engine for Chelomey's UR-100 ICBM. This single-chamber engine with a thrust of 13.7 tons also had four one-and-a-half-ton-thrust verniers.[82] Chelomey took this engine, modified it, and used it as the liftoff engine for his LK-700 lander. This sort of appropriation and cross-pollination was symptomatic of many of the elements of the UR-700/LK-700 project, a point that Chelomey repeatedly emphasized as one of its principal advantages.

When Chelomey presented his conception of the UR-700 project in November 1966, he did not mince words or hold back. He took every opportunity to firmly criticize various aspects of the N1-L3 project, bringing the arguments down to levels that were clear to industry leaders who had little or no engineering backgrounds. He also had some key supporters in tow, including Chief Designers Glushko, Barmin, and Kuznetsov, as well as Air Force Lt. General Kamanin. According to one respected Russian space historian:

> Chelomey tried to convince the leadership of the sector that with financial support and the research base that had been created in previous operations, his OKB would be able to execute the program quickly and make the USSR the first to land on the Moon. . . . The advisory council, however, considered such a declaration too bold and allowed only the performance of preliminary design work on the UR-700/LK-700 complex.[83]

Kamanin, with his own biases against the N1, wrote in his journal in late December 1966:

> Based on the UR-500 and [the UR-100] Chelomey has designed the UR-700 rocket, which has been approved by a panel of experts from the Ministry of General Machine Building, but so far the go-ahead has not been given for its implementation. Our leaders hesitate about simultaneously building Chelomey's UR-700 rocket and Korolev's N1 (hundreds of millions of rubles have already been spent to build the latter). But they are oblivious to the fact that the cost of building a UR-700 will be ten times less than the amount spent to build the N1. Because the first and the second "stages" of the UR-700 are basically the same as those of the UR-500 and, besides, it can use the same assembly and test building and launch equipment as the N1. . . . One would have thought that one should go ahead with UR-700 immediately, but L. V. Smirnov and D. F. Ustinov will hardly dare to take such a step because it was they who gave the green light to the N1. . . .[84]

Despite the compelling nature of Chelomey's arguments, several members of the evaluation commission were not thrilled by some of the weak links of the project, in particular the

---

81. A. N. Ponomarev, *Sovetskiye aviatsionnyye kosntruktory* (Moscow: Voyennoye izdatelstvo, 1990), pp. 312–13.

82. Ye. B. Volkov, ed., *Mezhkontinentalnyye ballisticheskiye rakety SSSR (RF) i SShA* (Moscow: RVSN, 1996), p. 148.

83. Afanasyev, "Unknown Spacecraft," pp. 39–40

84. Kamanin, "A Goal Worth Working for," p. 9.

development of the high-thrust RD-270 engines. Glushko had begun work on these in 1962, but by 1966, there had still been no ground firings of the engine. Commission members were also less than happy with the environmental dangers posed by such huge amounts of toxic propellants in the UR-700 rocket. The acoustic problems at liftoff were also unresolved. Finally, the actual return apparatus of the LK-700 had a very small volume. For cosmonauts who would have to wear EVA spacesuits the entire duration of the mission, comfort would have to be sacrificed. Despite Chelomey's protestations to the contrary, the commission members believed that the limited size and performance characteristics of the LK-700 would preclude long-duration landings on the Moon; such missions would have to use high-energy stages. The N1-L3 also had many of the same weaknesses as the UR-700, but at least work on the former had already been ongoing for several years. In the end, the Keldysh Commission declined to recommend serious work on the UR-700 project in November 1966, although it seems that a formal termination decision did not take place until August 1967, invoking the "unreasonability of continuation of further works on the UR-700."[85] Unfortunately for the Soviet lunar program, this was only a temporary respite. Like a phoenix, the specter of the UR-700 would rise again.

Deadline for the Moon

If, for the time being, the threat from Chelomey and his UR-700 had receded to the background, Mishin's N1-L3 effort had much more imposing problems; these involved funding, delays, and technical obstacles. His engineers had completed the final draft plan for the L3 complex in mid-1966, and it was only after that "with a six year delay the government issued the decision on subcontractors for the program."[86] Earlier, in April 1966, Mishin met with Soviet leader Brezhnev to inform him of the sequence of missions in the overall Soviet piloted lunar program. It would be a three-stage process involving the use of:

- The 7K-OK Soyuz to master rendezvous and docking in Earth orbit
- The UR-500K-L1 complex to perform a circumlunar mission with two cosmonauts
- The N1-L3 complex to land on the Moon

The N1-L3 complex would consist of three stages:

- Test the N1 booster and accomplish an automated lunar-orbital flight
- Test the L3 complex and accomplish piloted lunar-orbital flight with a robotic landing on the surface of the Moon
- Perform a piloted landing on the surface of the Moon

Within the framework of N1 missions for robotic lunar-orbital flights, in March 1966, Mishin's engineers emerged with a plan to launch the stripped-down Soyuz spacecraft known as the 7K-L1, which was intended for use in the circumlunar project on the N1 booster. In this variant, the spacecraft was known as the 7K-L1S, with the "S" standing for the Russian word for satellite ("sputnik"), indicating that its primary mission was to circle the Moon. Engineers believed that three N1-L1 launches early in the N1 launch test series would provide valuable experience in not only proving out problems in the N1, but also mastering operations in lunar orbit—an essential requisite for the ultimate piloted lunar landing.

---

85. For November 1966, see Khrushchev, *Nikita Khrushchev: tom 2*, p. 522. For August 1967, see Vetrov interview, November 15, 1996. See also Vetrov, "Development of Heavy Launch Vehicles."

86. V. P. Mishin, "The Development of Booster-Launchers in the USSR," presented at the 43rd Congress of the International Astronautical Federation, IAA-92-0197, Washington, DC, August 28–September 5, 1992.

By October 1966, the plan was to start with two to three launches of the automated N1-L1 complex. These would lead to three to four launches of the piloted LOK orbiter in lunar orbit, during which an automated LK lander would set down on the Moon, return to the lunar orbit, and link up with the LOK. Finally, it would be on the eighth, ninth, or tenth launch that cosmonauts would accomplish the actual piloted lunar landing.[87] With strong lobbying from senior engineers within the design bureau, such as Feoktistov, TsKBEM formulated its N1 flight plan in such a manner that there was a contingency plan to use a dual-launch Earth-orbit rendezvous mission profile to deliver the landing crew to the L3 complex in Earth orbit. The engineers would resort to this profile only in case there was little confidence in the ability of the N1 to safely launch cosmonauts into Earth orbit. All these slight modifications to the basic mission profile put forward by Korolev in late 1964 added layer after layer of complexity to the original vision of a Soviet lunar landing. Instead of simplifying matters, each modification threatened to topple the tenuous balance that barely kept the effort together.

The additions and modifications to the design of the L3 complex through 1967 meant that models designed for flight differed in many ways to the original technical plan on paper which was prepared by engineers in 1965.[88] For example, the use of three different vehicles on the lunar surface—the LK, the LK_R and the Ye-8 rover—necessitated having constant communications and telemetry from more than one spaceship. Additional communications systems for voice and telemetry, named *Foton* and *Mezon*, respectively, were added to the design of the ground stations by late 1967. Mishin also proposed a special ground communications station in Cuba specifically for lunar operations. Remarkably, the Soviets announced the existence of such a station by October 1968.[89] Power and mass limitations also affected the conceptions of the LK lander; in late 1967, Mishin was proposing the replacement of the lander's chemical batteries with solar panels on the fifth and sixth production models. There were other changes in the Ye-8 rover designed for lunar surface operations. In January 1967, Mishin and Babakin agreed to a tactical-technical requirement for the rover, stipulating that life support would be ensured on the lunar car for a full forty-eight hours. By early April, however, mass constraints deadlocked Babakin's engineers, and a variety of problems arose in the operation of the life support package on the rover. The problem evidently delayed the preparation of a final draft plan for the Ye-8 well beyond the expected time period.

The sequence of launches planned in October 1966 meant that, at conservative levels, the hundreds of contractors and subcontractors would have to sustain a launch rate of about one N1 every three months through 1967 and 1968. Any realistic assessment of the situation within the lunar program in late 1966, however, would have given pause even to the most superficial of observers that this pace would be impossible to maintain. Perhaps the most serious source of delays was the main engines for the N1 booster. Space program leaders such as Smirnov, Afanasyev, Dementyev, and Pashkov met in March 1966 to discuss problems with the development of the engines. One major source of anxiety was the NK-15V engine for the second stage. While the NK-15s for the first stage had been tested 153 times in static stands, there was still no test stand existing that allowed the NK-15Vs to be tested in altitude conditions. Chief Designer Kuznetsov's OKB-276, the lead developer of the engines, and several plants located at Kuybyshev were lagging in their work on the engines—a problem compounded by

87. According to production figures from October 1968, three to four N1s were to be manufactured in 1967, six in 1968, and six in 1969—a total of fifteen or sixteen. These numbers evidently included three for ground testing only: the articles 1M1, 1M2, and 1M3. The flight versions began with the designations 2L, 3L, 4L, and so on. By December 1966, the preparation of 1M1 was delayed from December 1966 to February 1967.

88. The "final" draft plan for the L3 complex was finished in mid-1966 according to Mishin. See Mishin, "The Development of Booster-Launchers."

89. *Soviet Space Programs, 1966–70*, p. 150.

a shortage of labor.[90] Pashkov reminded the participants that the engines were to have been delivered for use on flightworthy N1s in January 1965. It was clear that the primary bottleneck in the program was engine development, and it was this fact that determined the huge delays in the N1 program at the time.

The estranged Glushko also may have contributed to raised tensions among Kuznetsov's engineers. Astonishingly, as late as 1967, Glushko was still talking openly of revising the N1 rocket so as to use his old RD-253 engines, which by then were in use in Chelomey's UR-500K Proton booster. One engineer later recalled that "[It] was a difficult period of time for Kuznetsov: there was one accident after another on the test stands. Glushko followed all this jealously."[91] The final testing of the NK-15 engine occasionally displayed partial burnout of the firewall of the combustion chamber or the nozzle. Engineers at OKB-276 later introduced deliberate burn-throughs in the engines to test engine tolerance, and they were fortunate to discover that the units performed in a stable manner despite the burn-throughs. Before the NK-15 engines were released for series production, on one occasion, one of the experimental units "smoked out" during a test, bolstering Glushko's arguments against Kuznetsov's engines. At a meeting of a joint commission to investigate the accident, Glushko said, "You can see for yourselves that the engine is bad. It's not fit for work, and certainly not for installation on such a crucial piece of hardware like the N1."[92] Fortunately for Kuznetsov, the commission later ascertained that the fault had been caused by a production defect and not a design flaw; the engines were recommended for series manufacture.

The program to develop high-performance liquid hydrogen engines, so doggedly pursued by Korolev in the last years of his life, was also vigorously supported by his successor Mishin. It took a long time, and seven years after Korolev's first letters to the government requesting funds for liquid hydrogen engines, the Soviets tested such an engine. On April 8, 1967, engineers directed the first ground test of the first Soviet liquid oxygen–liquid hydrogen engine, the 11D56, designed and built by the Chemical Machine Building Design Bureau (formerly OKB-2) headed by Chief Designer Isayev.[93] By this time, the Soviets were a full six to seven years behind the United States in this critical area of rocket engine technology. While it was clear that liquid hydrogen would not be an integral part of the first N1 version, by September 1967, Mishin had sent proposals to the Ministry of General Machine Building on the use of Isayev's engine on an upper stage designated Blok R for a subsequent version of the N1.

There were delays in the development of the L3 complex. The late start of the Soviets in 1964 was finally beginning to have a significant long-range effect on competing with Apollo. By the end of 1966, neither the Blok I engine (for the LOK orbiter) nor the Blok Ye engine (for the LK lander) had been tested on the ground. The most optimistic forecast was that they would be tested in July and August 1967, respectively. The workload on TsKBEM was so severe in 1966 that Mishin and his deputies even considered handing over all development of the LK to Chief Designer Babakin's organization.[94] Naturally, such uncertainties did little to instill con-

---

90.   Within OKB-276, V. N. Orlov and V. S. Anisimov, two of Kuznetsov's deputies, were appointed to lead the N1 engine team. Several subgroups focused on specific areas, including high-thrust engines (headed by Deputy Chief Designer N. D. Pechenkin) and N1 third- and fourth-stage engines (Deputy Chief Designer N. A. Dondukov). Engineers Astakhov and Yelizarov were assigned to lead the development of gas generators and turbopumps, respectively. See Rudenko, "Space Bulletin: Lunar Attraction."

91.   M. Rebrov, "But Things Were Like That—Top Secret: The Painful Fortune of the N-1 Project" (English title), Krasnaya Zvezda, January 13, 1990, p. 4.

92.   Igor Afanasyev, "N-1: Absolutely Secret" (English title), Krylya rodiny no. 11 (November 1993): 4–5. The chair of this commission was Chief Designer A. D. Konopatov of the Chemical Automation Design Bureau (KB KhimAvtomatiki), formerly known as OKB-154.

93.   "Calendar of Memorable Dates" (English title), Novosti kosmonavtiki 8 (April 7–20, 1997): 59–60; Semenov, ed., Raketno-Kosmicheskaya Korporatsiya, p. 262.

94.   V. M. Filin, Vospominaniya o lunnom korablye (Moscow: Kultura, 1992), pp. 9–10.

fidence in the engineers who had worked on the vehicle for several years. TsKBEM's finances were also stretched to the limit in 1965 and 1966, which led officials to cut corners on various ground and in-flight systems. The design bureau was beset by a 51 million ruble shortage in 1965 that increased in 1966.

Construction of the launch complexes for the N1 was well under way by the time that Mishin took up his duties as chief designer. The original plan was to build two launch complexes, each with two pads. Financial constraints, however, forced engineers to plan for only a single launch complex, designed by GSKB SpetsMash led by Chief Designer Vladimir P. Barmin. It would be the culmination of Barmin's career in the space and missile business. A contemporary of Korolev's, Barmin graduated in 1930 as a mechanical engineer and had the ghoulish honor of creating a special refrigerating device for Lenin's mausoleum. In 1937, Barmin was dragged off to the Lubyanka prison to be questioned about a trip he and some other engineers had made to the United States in 1935 as part of a business delegation. When the group had come under suspicion, the head of the group committed suicide; most of the other members were arrested. Barmin was let go, but he lost his job. He made the leap from refrigerators to missiles in June 1941, when he was put in charge of production at the famous Kompressor Plant, where thousands of Katyusha missile launchers were manufactured during the war. For a brief period, Barmin had the dubious distinction of working for Andrey G. Kostikov, the engineer who had been instrumental in sending Korolev and Glushko to the GULag.

After the war, Barmin and Korolev struck up their acquaintance once again, and the former led the development of launch complexes for almost every single Soviet long-range ballistic missile, including the famous R-7 ICBM. Barmin also had his run-ins with the Soviet leadership. In 1959, when Khrushchev abruptly decided to terminate further work on the Mirnyy missile launch site near Plesetsk, Barmin asked permission to speak at a meeting and told Khrushchev to his face that such a decision would be in error. His persuasive arguments won the day. The Mirnyy site was completed, eventually becoming the most prolific space launch site in the world.[95] Although Barmin's GSKB SpetsMash organization did not retain its monopoly in the design and creation of launch complexes, it inherited a leading role in the field by the strength of its participation in the UR-500K and N1 programs. In January 1967, GSKB SpetsMash was renamed the Design Bureau of General Machine Building (KB OM).

Barmin's team began construction of the first launch pad (site 110 right) in September 1964 and completed it in August 1967. The second pad (site 110 left) was built between February 1966 and late 1968. The scale of construction associated with the launch complex, about thirteen kilometers to the northwest of the famous site 1, was huge. A large technical zone and living area was built seven kilometers from the launch pads at site 113 for personnel from the Progress Machine Building Plant who were on assignment from Kuybyshev to oversee the assembly and testing of flight-rated boosters. Technical and materiel supplies were brought to Tyura-Tam on a daily basis via two huge trains, each with several dozen wagons. The railcars were evidently so large that delegations from other socialist countries often came to the launch site to view the trains.[96]

When it was finished in 1968, site 110 consisted of two launch pads located 500 meters from each other, each with 145-meter-tall service towers for propellant loading, power supplies,

95.   Col. M. Rebrov, "To Do Tomorrow: Pages From the Life of the Chief Designer of Space Launch Complexes" (English title), *Krasnaya zvezda*, October 22, 1988, p. 3; Boris Khlebnikov, "Vladimir Barmin: One of the Top Six Designers," *Aerospace Journal* no. 2 (March–April 1997): 81–83.

96.   J. Villain, "A Brief History of Baikonur," presented at the 45th Congress of the International Astronautical Federation, IAA-94-IAA.2.1.614, Jerusalem, Israel, October 9–14, 1994; Leonard Nikishin, "Rough Going on Interplanetary Trajectories. How We Exerted Ourselves to the Utmost in the Lunar Race" (English title), *Obshchaya gazeta*, July 15, 1994, p. 9.

crew boarding, and thermal control. After the completion of prelaunch procedures, the tower would be moved away, leaving the N1 at the pad, "held down" by forty-eight pneumo-mechanical locks. In addition, four 180-meter-tall lightning rods were built around each launch pad. A total of ninety unique structures were eventually constructed at site 110 for N1 operations, dwarfing any other launch complex at Tyura-Tam.[97] In the early 1960s, engineers had originally proposed assembling the 105-meter-tall launchers vertically in a special assembly building. Because this would have necessitated the construction of a gigantic building 160 meters tall, the engineers decided to lessen the funding strain by opting to assemble the boosters horizontally in a "smaller" building. The latter was the gigantic assembly-testing building for N1 assembly at site 112, with the dimensions of forty-seven (height) by 240 by 250 meters. It was reputed to be the largest building on the Eurasian landmass. A second assembly-testing building at nearby site 2B was dedicated for assembling the L3 complex, while the fueling station was located at site 112A. During launch operations, the L3 would undergo preflight checking in its building, covered by cowling, and be transported by rail to the fueling station for propellant loading. From

Chief Designer Vladimir Barmin was one of the original members of the Council of Chief Designers. His organization was responsible for designing and creating launch complexes for a wide spectrum of Soviet missiles and space launch vehicles. In later years, Barmin expanded into other areas, such as designing lunar bases, lunar sample return scoopers, and space-based furnaces. (files of Peter Gorin)

there, the L3 stack would be transported to the larger assembly building, where it would be connected to the assembled N1 in a horizontal position. After further tests, the N1-L3 booster stack would be transported by two diesel locomotives moving on parallel tracks to the launch pad.[98]

With such an impressive level of construction at Tyura-Tam in the 1960s, it is not surprising that U.S. photo-reconnaissance satellites were able to pick up convincing signs that the Soviet Union was indeed running the race to the Moon. The first public indication that the USSR was engaged in building a massive rocket came in the fall of 1966 when a reporter from *The New York Times*, Evert Clark, quoted "official sources" that the "Soviet Union is believed to have finally begun developing a rocket of 7.5-to-10-million pounds of thrust enough to send men to the Moon. . . ."[99] A top-secret CIA report from early 1967, declassified twenty-five years later, indicates that U.S. intelligence services were well apprised of concurrent Soviet efforts. Designating site 110 at Tyura-Tam as "Complex J," the authors of the report wrote:

> The construction of Complex J at Tyuratam [sic] makes it clear that the Soviets have under development another and much larger booster [than the Proton]. Complex J is a very large launch facility which appears to be of the same magnitude as the U.S. Apollo launch facility at Merritt Island. It has been under construction for the past [three-and-

97. I. A. Marinin and S. Kh. Shamsutdinov, "Soviet Programs for Piloted Flight to the Moon" (English title), *Zemlya i vselennaya* no. 5 (September–October 1993): 77–85.

98. Ibid.; Villain, *Baïkonour*, pp. 65–66.

99. Evert Clark, "Soviet Is Reported Developing a Big, New Rocket," *New York Times*, September 13, 1966, p. 28.

a-half] years and we estimate it will be ready for initial launch operations in the first half of 1968 at the earliest.[100]

As for the actual piloted lunar landing, the CIA was evidently under the impression that the Soviets were not in it to beat Apollo:

Two years ago, we estimated that the Soviet manned lunar landing program was probably not intended to be competitive with the Apollo program as then projected, [that is,] aimed at the 1968–1969 time period. We believe this is still the case . . . we estimate that the earliest the Soviets could attempt a manned lunar landing would be mid-to-late 1969. We believe that the most likely date is sometime in the 1970–1971 time period.[101]

NASA Administrator James E. Webb joined the chorus of believers who were convinced that the Soviets were building a huge rocket—a belief no doubt bolstered by his access to classified reports from the CIA. During testimony to a House Appropriations subcommittee in August 1967, he stated that "the U.S.S.R. is building a larger booster and will shortly, I believe, in calendar year 1968, be flying a booster larger than the Saturn 5."[102] Webb's claims were dismissed by many, because he was unable to provide any supporting evidence. The complete lack of physical evidence would come in handy in later years when the Soviets engaged in one of the most successful deceptions in the history of space exploration.

The Soviets themselves were not being particularly coy at the time. Although they were shy about specifics, the general tone of Soviet public figures did not leave any doubt as to the ultimate goal of the Soviet space program. As one would expect, the cosmonauts were the most vocal in their pronouncements; although the Communist Party maintained strict control over each and every word uttered by these young men, they were more amenable to fits of spontaneity than their elder bosses. On April 12, 1965, during celebrations in honor of Gagarin's flight, cosmonaut Belyayev, fresh from his recent trip on Voskhod-2, spoke in hyperbolic terms about the lunar program: "Preparations are in full swing. The Americans speak broadly about their preparations to land a man on the Moon, but naturally, we in our country, are not idle either. We shall see who will be there first."[103] Less than a year later, Bykovskiy, praising NASA's lunar-orbit rendezvous mission profile, added that work was in full swing to develop maneuvering ships and suits needed for work on the lunar surface.[104] A few months later, in April 1966, Leonov spoke candidly in Hungary:

I think that I do not disclose any secret by saying [that] Soviet cosmonauts are preparing for such a journey [to the Moon]. I should very much like it if a Soviet man went to the Moon first because we were the first who made the most important steps in space. I believe we shall soon witness man's landing on the Moon. I cannot say when, but it will be during this five-year plan period.[105]

In the complete vagueness that surrounded Soviet pronouncements on the space program at the time, cosmonaut Komarov made one of the most specific statements during a visit to Japan in July 1966:

100. U.S. Central Intelligence Agency, "National Intelligence Estimate 11-1-67: The Soviet Space Program," Washington, DC, March 2, 1967, p. 11, as declassified December 11, 1992, by the CIA Historical Review Program.
101. *Ibid.*, p. 2.
102. Evert Clark, "New Soviet Shot is Expected Soon," *New York Times*, August 19, 1967.
103. *Soviet Space Programs, 1966–70*, p. 359.
104. *Ibid.*, p. 361.
105. *Ibid.*, p. 362.

There is no need to make haste about a Moon trip by human beings—and the important thing is how to carry out everything in safety. But I can positively state that the Soviet Union will not be beaten by the United States in a race for a human being to go to the Moon.[106]

Upon his return to Moscow, cosmonaut overseer Kamanin confronted Komarov about his unauthorized statement. Having deviated too much from the doctrinal line, there were calls from the Central Committee and the Council of Ministers regarding the "incident."[107] Remarkably, it was roughly at the same time that one of the most authoritative aerospace trade journals in the United States, *Aviation Week & Space Technology*, reported that the Soviets were *not* heading for the Moon. In a long article in November 1966, the author reported that the:

Soviet Union is showing increasing signs of having conceded the manned lunar landing race to the U.S. as part of a vastly revamped space program. The new space philosophy, which the Soviets consider better balanced though less dramatic than their previous one, could produce a much less complex manned circumlunar mission without landing within the next year.[108]

It was one of the best examples of how much Western analysts misread the intentions of the Soviet space program at the time, which as it happens was going through a transition, but one that was not clear to observers of that era.

In contrast to the early 1960s, the Soviet space program as a whole was not afforded relatively uncontrolled access to funding. Brezhnev was considerably less sympathetic toward the space program than his predecessor, and salaries in the space industry were said to have gravitated to more average levels during the early years of the post-Khrushchev era. As one senior official at the Central Scientific-Research Institute of Machine Building (TsNIIMash) recalled, Brezhnev "supported space only if brought political dividends."[109] While detailed figures on appropriations for space still remain classified, it is known that the Soviet Union spent 7.9 billion rubles on its space program during the period 1966–1970.[110] At the prevailing unofficial conversion rate, this amounted to approximately $24 billion, or 1.25 percent, of the Soviet Union's yearly gross national product during the same period.[111] The N1-L3 project was about 20 percent of the total space budget each year, amounting to roughly $4.8 billion of expenditure from 1966 to 1970 (in 1966 U.S. dollars).[112] Thus, although the Soviet Union's expenditures on space were close to twice the portion of its gross national product as in the United

106. Evert Clark, "Soviet Spaceship Hunting Quarks," *New York Times*, July 17, 1966, p. 55; *Ibid.*, p. 363.

107. Kamanin, "A Goal Worth Working for."

108. Donald C. Winston, "Soviets Revamp Lunar Space Plan," *Aviation Week & Space Technology*, November 28, 1966, pp. 22–23.

109. Stéphane Chenard, "Twilight of the Machine Builders," *Space Markets* 7(5) (1991): 11–19.

110. Yu. Koptev, "Space Fantasies: *Glasnost* vs. Rumors" (English title), *Ekonomika i zhizn* 38 (September 1990): 19.

111. The conversion rate used was $3 = 1 ruble, extrapolated from figures in *Soviet Space Programs, 1966–70*, pp. 108–09. Table 1 gives the Soviet gross national product for 1967 as $365 billion (in 1966 U.S. dollars). Table 2 gives the Soviet state budget for 1967 as 115.24 billion rubles. The figure of 1.25 percent has been extrapolated from totaling the Soviet state budget for 1966–70 and then determining the ratio of the space budget (7.9 billion rubles) as a percentage of the cumulative state budget (631.13 billion rubles). Note that the actual figures were remarkably close to those predicted in 1971 without the benefit of any "real" figures. Analysts hypothesized at the time that the Soviet space budget was 1.5 to 2.0 percent of the Soviet gross national product.

112. The ratio of the N1-L3 to the total space budget has been extrapolated from Stéphane Chenard, "Budget Time in Moscow," *Space Markets* 7(5) (1991): 10.

States, actual dollar expenditure on space and the lunar program in particular was far less than that of its primary competitor.[113]

The end of 1966 was a particularly critical decision-making point for the leaders of the Soviet space program. NASA had just completed ten highly successful Gemini missions, displaying a remarkable level of expertise in mastering complex operations in Earth orbit, while the Soviets had not launched a single cosmonaut into space. American successes were bolstered in 1966 by two launches of the Block I Apollo Command and Service Module, as well as a test launch of the S-IVB high-energy cryogenic upper stage.[114] By the end of the year, three astronauts were preparing for the first piloted launch in a Block I Command and Service Module aboard the Saturn IB to conduct a thorough testing of the entire spacecraft in Earth orbit. The giant Saturn V, meanwhile, was scheduled to take an automated Apollo spacecraft into Earth orbit by the summer of 1967. In early January 1967, Boris A. Stroganov, one of Serbin's deputies in the Central Committee's Defense Industries Department, told Mishin that the upper echelons of the Communist Party were extremely concerned about the Soviet lag behind the United States. All this warranted a response, especially given that many of the deadlines from the original August 1964 decree on the Soviet lunar landing had remained unfulfilled as a result of poor management and insufficient funding. There had already been a number of decrees through 1966 on the lunar program at the level of the Ministry of General Machine Building.[115] Speaking of a decree in late 1966, Lt. General Kamanin wrote in his personal journal on November 10:

> I read the [Military-Industrial Commission] decree which says that the 1964 decisions of the [Communist Party] and the Council of Ministers on orbiting the Moon and landing humans on the Moon are not being fulfilled properly. The resolution reiterates orders to the industry to give top priority to all work connected with spacecraft and rockets and to treat them as special state assignments. There are sure to be many more such resolutions, rebukes, and reprimands as the temperature over the Moon rises. But papers and reprimands don't get anywhere; too much time has been wasted. The bosses, however, won't hear about our problems and will demand new "spectacular" flights to mark the 50th anniversary of the October Revolution.[116]

In October 1966, the so-called "Council for the Problems of Mastering the Moon," which included the leading ministers, deputy ministers, academicians, chief designers, and military officers from the Soviet space establishment, was set up specifically to examine both the macro- and micro-level details of the Soviet program to land a human on the Moon. Headed by Minister of General Machine Building Afanasyev, the council heretofore was the primary advisory body to the Soviet Party and government on all affairs involving the N1-L3 project. Rumor had it that Ustinov and Smirnov had set up the council so as to insulate themselves from the possibility of blame if the Soviet lunar program failed. Another possible motive may have been

---

113. Central Scientific-Research Institute of Machine Building (TsNIIMash) Director Yu. A. Mozzhorin recalled, "The Americans had spent $15 billion on the creation of an experimental base: we had spent only about $1 billion." See Rebrov, "But Things Were Like That—Top Secret."

114. The two Command and Service Modules were launched on February 26 and August 25, 1966. The S-IVB test was on July 5, 1966. See Linda Neuman Ezell, *NASA Historical Data Book, Volume II: Programs and Projects 1958–1968* (Washington, DC: NASA SP-4012, 1988), p. 187.

115. There was a Ministry of General Machine Building order (no. 207ss) on May 16, 1966, on the N1. A Military-Industrial Commission decree (no. 428) was issued on September 14, 1966, on the N1-L3.

116. N. Kamanin, "A Goal Worth Working for," p. 8. Kamanin refers to two VPK decrees from this period: one on the "conclusions of the expert commission on [scientific-research work] to settle the organization of search and evacuation of lunar ships" and the other on the UR-500K-L1 and N1-L3 complexes. See also Kamanin, *Skrytiy kosmos: 1964–1966*, p. 388.

to circumvent the power of the Council of Chief Designers with regard to the lunar landing program.[117] The council in its deliberations returned to the original 1964 decree to discuss the issuing of a second decree to stipulate specific schedules for the achievement of a circumlunar and lunar landing mission. TsNIIMash Director Yuriy A. Mozzhorin, an individual who probably had much to do with determining the pace of the space program, recalled:

> It was clear to me that the objective was becoming unrealistic and that the volume of the work ahead exceeded the capacities of the sector by a factor of 2–2.5. At a conference of Chief Designers and curators, I expressed doubts. They were met with criticism.[118]

Mozzhorin evidently refused to approve the conditions of the new decree, but it seems that he eventually capitulated under pressure from Afanasyev.[119] At the same time, Mishin's principal deputy for the N1, Deputy Chief Designer Okhapkin, pleaded to Ustinov, "We want to solve this problem, we can solve it, and we will solve it on schedule if we receive assistance."[120]

These intensive discussions in late 1966 eventually led to the adoption of another important decree associated with the piloted lunar landing program—one that established goals competitive with the late President Kennedy's set for Apollo. On February 4, 1967, the Central Committee and the Council of Ministers issued a document (no. 115-46) titled "On the Progress of the Work on the Development of the UR-500K-L1."[121] The document, signed just eight days following the Apollo 1 fire, in which three U.S. astronauts were killed during a ground test, called for the consolidation of all national resources in support of the accomplishment of a piloted lunar landing on the Moon prior to the United States. The document was prepared by the four most powerful individuals in the Soviet space program: Ustinov, Serbin, Smirnov, and Afanasyev.[122]

The authors of the resolution, which still remains classified, described as "unsatisfactory" the work of the government in fulfilling the terms of the original 1964 decree on piloted lunar programs and stated that "a flight around the Moon by a manned spacecraft and the landing of a manned mission on the Moon shall be considered to be objectives of *national importance.*"[123] Implicitly at least, the resolution freed the purse strings of the Ministry of Defense for the program, but in reality, it seems that the attitude of the primary financiers of the project

---

117. Kamanin, "A Goal Worth Working for." The composition of the council is still unknown but presumably included all the major chief designers, such as G. N. Babakin (GSMZ Lavochkin), V. P. Barmin (KB OM), V. N. Chelomey (TsKBM), V. P. Glushko (KB EnergoMash), A. I. Iosifyan (VNII ElektroMekhaniki), A. M. Isayev (KB KhimMash), A. D. Konopatov (KB KhimAvtomatiki), N. D. Kuznetsov (KB Trud), V. I. Kuznetsov (NII Prikladnoy mekhaniki), A. M. Lyulka (KB Saturn), V. P. Mishin (TsKBEM), A. S. Mnatsakanyan (NII Tochnykh priborov), N. A. Pilyugin (NII Avtomatiki i priborostroyeniya), M. S. Ryazanskiy (NII Priborostroyeniya), G. I. Severin (KB Zvezda), S. K. Tumanskiy (MMZ Soyuz), G. I. Voronin (KB Nauka), and M. K. Yangel (KB Yuzhnoye). Initially, there were only two military representatives on the council: A. G. Karas (TsUKOS Commander-in-Chief) and A. I. Sokolov (NII-4 Director). Another source states that by December 1967, the council included "Marshal N. I. Krylov, Marshal Rudenko, Ministers of Aviation, Defense and Radio Industries, all the primary Chief Designers, the President of the Academy of Sciences Keldysh, and the Deputy Chairman of the VPK Pashkov." See Chertok, *Rakety i lyudi: goryachiye dni kholodnoy voyny*, pp. 476–77. There was evidently another "Moon Council," this one for automated exploration, whose chair and deputy chair were M. V. Keldysh (AN SSSR) and G. A. Tyulin (MOM), respectively. See Mozzhorin, *Dorogi v kosmos: I*, p. 162.
118. Rebrov, "But Things Were Like That—Top Secret," p. 4. For Mozzhorin's own account on his doubts on the decree, see Rudenko, "Space Bulletin: Lunar Attraction."
119. Rudenko, "Space Bulletin: Lunar Attraction."
120. Rebrov, "But Things Were Like That—Top Secret," p. 4.
121. Mishin, "Why Didn't We Fly to the Moon?"
122. Leonard Nikishin, "Inside The Moon Race," *Moscow News* 7 (April 11, 1990): 15.
123. N. Kamanin, "A Goal Worth Working for" (English title), *Vozdushniy transport* 46 (1993): 8–9. Author's emphasis.

remained unchanged. Less than two weeks after the document was issued by the leadership, new USSR Minister of Defense Marshal Andrey A. Grechko refused to provide money for a search-and-rescue service for returning cosmonauts from the Moon. When he was told by Air Force leaders that about 25 to 30 million rubles and 9,000 personnel would be required, he lashed back, "I won't give you personnel, I won't give you money. Do what you like but I won't raise this with the government. . . . And in general I am against Moon missions."[124] This lack of commitment was devastating to the project.

The February 1967 document detailed astonishingly ambitious timetables for both the L1 and the L3 programs:

Mission	Date
First piloted circumlunar flight of the UR-500K-L1	June–July 1967
First flight tests of the N1-L3	September 1967
First piloted lunar landing of the N1-L3	September 1968

In an extreme case, the landing could have been achieved between October and December 1968.[125] It remains unclear what prompted Ustinov and the others to aim for such an unrealistic schedule. By February 1967, the N1 had yet to fly while the L3 complex existed only on paper, and yet the Soviets were proposing that this highly complex mission be accomplished in less than two years. The only visible manifestation of any progress was the completion of the first full-scale N1 test vehicle, the 1M1, which was finally assembled at Tyura-Tam that same February, although it remained in the giant assembly-testing building. Actual flight models, although being manufactured, were well behind in the queue. Clearly, the senior staff of TsKBEM, including Mishin, were as much responsible for stipulating these outlandish deadlines as was the political leadership. These TsKBEM employees, after all, were the ones who made assessments of the state of the program in late 1966, on whose basis Ustinov and the others made their decisions. To have agreed to the late 1968 deadline seems in retrospect to have been professional suicide, but for reasons that are still not clear, the designer faction accepted them. Kamanin wrote in his journal entry for March 15, 1967: "There is no doubt in my mind that these deadlines are anything but realistic."[126] It was probably clear to most engineers that if past experience was any indicator, the government would be unwilling to back this near ridiculous deadline with any sort of financial commitment.

First Secretary of the Central Committee of the Communist Party Leonid I. Brezhnev and Chairman of the USSR Council of Ministers Aleksey N. Kosygin signed the February 1967 document and officially made it binding to all the hundreds of primary and secondary contractors working on the lunar program. Nearly six years after Kennedy's speech, the Soviet piloted lunar landing program was an objective of national importance. It was the Soviet leadership's belief that if the Soviet military-industrial complex performed as stipulated, a Soviet citizen would be standing on the surface of the Moon by the end of 1968. The fact that the United States' with all its industrial might, had been trying for the same objective for six years could not have escaped the notice of all involved. Speaking of the document that had appeared far too late and of the government that had ignored the pleas of designers for so many years, a Soviet journalist wrote years later:

---

124. *Ibid.*
125. See Tarasov, "Missions in Dreams and Reality," in which Mishin says that the timetable for the flight tests was to be in the second quarter of 1967 and the landing in the third quarter of 1968.
126. Kamanin, "A Goal Worth Working for," no. 46.

This shows the level of competence of the top Soviet leaders Brezhnev and Kosygin who signed the document [and] the honesty of the Party and government officials who prepared this document: Ustinov, Smirnov, Serbin, Afanasyev.[127]

Defining the Circumlunar Program

Through 1966, the L1 program to send Soviet cosmonauts around the Moon assumed primacy in importance over the L3 landing effort—a strategic shift motivated very much by the impending fiftieth anniversary of the Great October Revolution in late 1967. The basic elements of the project had been frozen by a document issued on December 31, 1965, titled "Initial Data on the L1 Payload Block (Product 11S824)," signed just two weeks before Korolev's death. The main points of this document described the changes necessary to the spacecraft and launch vehicle to accomplish the piloted circumlunar mission.[128] For the 7K-L1 vehicle in particular, there were three goals:

- Create a modification of the 7K-OK Soyuz spacecraft, designated the 7K-L1, capable of circumlunar flight with a crew launched in the vehicle
- Establish a phased realization of the goals:
    - Create the technological-model complex 1M1 with 7K-L1 no. 1P
    - Create automated variants for circumlunar flight on 7K-L1 nos. 4–9
- Prepare 7K-L1 nos. 11–14 for piloted circumlunar flight[129]

The 7K-L1 spacecraft (also called simply the "L1") was essentially a stripped down 7K-OK Soyuz, reduced to "fit" the 5.1- to 5.2-ton mass constraints for a circumlunar mission using Chelomey's UR-500K rocket and Mishin's Blok D upper stage combination. Depending on the particular variant, total mass varied from 5.2 to 5.7 tons (in Earth orbit) and 5.0 to 5.5 tons (after TLI). The primary difference between the Soyuz and the L1 was the omission of the spheroid living compartment in the latter, making the L1 a compact two-module spacecraft built for a singular objective with little room for upgrades. The two modules were the descent apparatus and the instrument-aggregate compartment.

The descent apparatus was a segmented-conical body with an improved heat shield sufficiently strengthened to withstand lunar return velocity reentries. This shielding would be cast off prior to the actual landing on Earth. The two-person crew would spend their entire eight- to ten-day mission within the confines of this capsule with an internal volume of only two and a half cubic meters, compared to the Soyuz, which afforded six and a half cubic meters. Apart from the crew couches, the descent apparatus also contained the ship's control panel, an on-board computer, scientific instrumentation, a camera, life support systems, portions of the ther-

---

127. Nikishin, "Inside the Moon Race," p. 15.
128. Semenov, ed., *Raketno-Kosmicheskaya Korporatsiya*, p. 235. The nine points related to the overall complex were: (1) use Blok D from the N1 booster as a TLI stage; (2) use the 7K-OK Soyuz as the spacecraft, but without the spheroid living compartment at the forward end, the descent apparatus would be modified for lunar speed atmospheric reentry, and a special supporting cone at the apex of the now-shortened vehicle would allow connections with the launch escape tower; (3) eliminate mooring and orientation engines from the 7K-OK Soyuz and transfer these functions to the SOZ system on Blok D; (4) develop a new payload shroud; (5) ensure that Blok D can refire in vacuum conditions; (6) ensure that Blok D can fire to allow TLI; (7) agree on a cycle of events for the UR-500K Proton booster during launch, allowing launch escape and rocket safety; (8) develop the details of a circumlunar trajectory with return at lunar velocities; and (9) create simplified 7K-L1 spacecraft numbers 2P and 3P for tests in Earth orbit, which would include two firings of Blok D.
129. *Ibid.*, p. 236.

mal regulation and communications systems, biological samples, an optical orientation device, and a storage battery. One of the improvements on the capsule compared to the Soyuz was doubling the number of thrusters for yaw during guided reentry. This augmentation in reentry capability was offset to a great degree by the omission of the reserve parachute from the descent apparatus because of both space and mass constraints. The single remaining parachute had a dome area of 1,000 square meters. The deletion of the living compartment prompted engineers to attach a special support cone to the apex of the spacecraft to allow a firm connection

The 7K-LI spacecraft was the final iteration of Korolev's repeated attempts to design a flight-capable piloted circumlunar ship. The vehicle, later publicly named Zond, was similar in terms of most systems to the Earth-orbital Soyuz. The major design difference between the two was the omission of the forward living compartment on the Zond spacecraft. Two cosmonauts would have to spend a cramped week within the confines of the tiny descent apparatus. (copyright VideoCosmos Co., via Dennis Newkirk)

with the nose fairing and the launch escape tower of the booster stack. The cone, weighing 150 kilograms, would be cast off from the vehicle prior to TLI. As with the Soyuz and the N1-L3, the launch escape system was equipped with a set of powerful solid-propellant engines to remove the descent apparatus far from an exploding rocket.

As in the 7K-OK Soyuz, the 7K-LI instrument-aggregate compartment was divided into three sections: the transfer compartment, the instrument compartment, and the aggregate compartment. The pressurized instrument compartment contained the primary and backup buffer storage batteries and additional ship instrumentation for on-board systems. The unpressurized aggregate compartment at the extreme aft of the ship contained the single high-thrust engine on the spacecraft, the S5.53, developed by the Design Bureau of Chemical Machine Building, led by Chief Designer Isayev. The engine used unsymmetrical dimethyl hydrazine and a mixture of nitric acid and nitrogen tetroxide (AK-27) and had a thrust of 425 kilograms—that is, it was identical to its counterpart on the Soyuz spacecraft. The 400 kilograms of propellant for the engine was contained in four spherical tanks at the aft of the aggregate compartment, which also included eight attitude control thrusters operating on hydrogen peroxide (of one and one and a half kilograms thrust). Thermal radiators covered the whole compartment on its outer surface. As with the Soyuz, primary power on the vehicle was provided by two large solar arrays, spread like bird wings from the aggregate compartment. Unlike the Soyuz's four segments on each panel, the 7K-LI had three per panel, with a wingspan of nine meters and a total surface area of eleven square meters.

Apart from the deletions, TsKBEM engineers supplemented or changed a number of systems from the basic 7K-OK Soyuz craft. These included the attitude orientation system, which had improved solar (the 99K) and stellar sensors (the 100K), gyroscopes and command instruments, memory devices, and so on. For transmitting telemetric information, the engineers introduced a pencil-beam parabolic antenna operating in the decimeter range, which was attached at the front of the descent apparatus. The antenna had its own self-contained optical sensor for aiming at Earth (the 101K). The antenna as a whole would be discarded once its work was finished. Other antennas included short-range ones at the end of the solar panels (for radio communications) and additional ones for ultra-shortwave telemetry and radiotelemetry.

The guidance system for the 7K-L1 spacecraft was developed cooperatively by the organizations of Mishin and Pilyugin based on earlier models used for deep space probes as well as control engines for earlier ships and rocket stages. For the first time in a Soviet piloted spacecraft, the guidance systems operated on the basis of a three-axis stabilized platform and a special computer named the *Argon-11*, developed by Scientific-Research Institute of Digital Electronic Computing Technology. It would serve as the prototype for all further models in the Soyuz spacecraft.

The 7K-L1 spacecraft had a total length of five meters with the support cone and four and a half meters without. Maximum diameter was 2.72 meters at the base and 2.2 meters around the main body. The total length on the pad of the UR-500K, Blok D, 7K-L1, and launch escape tower combination was just over sixty-one meters, far exceeding the length of Soyuz spacecraft stack.[130]

A nominal mission profile of the circumlunar mission would begin with the launch of the UR-500K Proton booster with its 7K-L1 and Blok D payloads. During the launch, the ship would be beneath a fairing, which would be cast off after passing through the dense layers of the atmosphere. The partially filled Blok D would fire for the first time to achieve sufficient velocity to lift itself and the 7K-L1 into an Earth orbit with the parameters of 220 by 190 kilometers inclined at fifty-one and a half degrees. The cosmonauts aboard would check the state of all systems for a period of one orbit or one day, depending on the circumstances, orient the stack for boost toward the Moon, and then separate the support cone from the apex of the spacecraft. Blok D would fire for a sufficient period of time to accelerate the stack to Earth escape velocity toward the Moon. The stage would then separate while the ship's solar orientation system would put the spacecraft in a one-degree-per-second turning mode while ensuring maximal solar panel exposure to the Sun. The 7K-L1 ship would circle around the Moon at a range of 1,000 to 12,000 kilometers while the cosmonauts would carry out photography and TV sessions. The scientific investigations planned for the automated precursor missions would include studying radiation through the flight path, studying cosmic rays, and performing experiments on small biological payloads. During the course of the seven days in flight, the S5.53 main engine of the ship would carry out three or four mid-course corrections: the first on the outbound trajectory at 250,000 kilometers from Earth and the second and third ones on the return trajectory at 320,000 and 150,000 kilometers, respectively, from Earth.

Before reentry back into Earth's atmosphere, the parabolic antenna and instrument-aggregate compartment would separate from the descent apparatus with its two-person crew. The precision-guided reentry had two endo-atmospheric phases and an intermediate exo-atmospheric portion to radically decrease the gravitational loads subjected to the crew. The first "dip" into the atmosphere would decelerate the vehicle to about just over seven and a half kilometers per second, after which the capsule would "bounce" out of the atmosphere along a ballistic trajectory and reenter the atmosphere again at a reduced velocity of 200 meters per second. A special guidance system would control the motion of the descent apparatus throughout this entire portion by changing the lift force via roll control of the capsule. The length of return trajectory would vary between 6,000 and 10,500 kilometers, depending on the angle between the horizontal plane and the ship at the moment of entry; this was also an important determinant of radio visibility with ground communications stations. After the double-dip reentry, the capsule would come down by parachute, discard its thermal shielding, and finally land in Kazakhstan by using soft-landing engines much like the Soyuz spacecraft. If for some reason the guided reentry

130. *Ibid.*, pp. 235–36; Afanasyev, "Unknown Spacecraft"; I. A. Marinin and S. Kh. Shamsutdinov, "Soviet Programs for Lunar Flights" (English title), *Zemlya i vselennaya* no 4 (July–August 1993): 62–69; Lardier, *L'Astronautique Soviétique*, p. 159; Glushko, *Kosmonavtika entsiklopediya*, pp. 129–30. The two batteries for the 7K-L1 were a silver-zinc battery and calcium-nickel battery (blok 800).

procedure failed, the descent apparatus would be able to accomplish a simple ballistic reentry into the atmosphere with a subsequent landing in the Indian Ocean.[131]

There was one additional cautionary element of the L1 circumlunar project, introduced to compensate for any potential troubles with the UR-500K Proton launch vehicle. From early discussions in the fall of 1965, Korolev's engineers had expressed reservations of launching cosmonauts on the still-untested Proton booster—concerns motivated primarily by the use of toxic storable propellants in the rocket. As insurance against the possibility of designers not being able to declare the Proton safe enough to launch humans, Mishin came up with a plan to launch the 7K-L1 on the Proton in an automated mode. The crew would be launched separately on a special variant of the Soyuz, which would dock with the 7K-L1 ship. The two cosmonauts wearing their *Yastreb* ("Hawk") EVA suits would exit the Soyuz and transfer into the 7K-L1 via "a curved tunnel in the . . . support cone."[132] The Soyuz would then automatically undock, while the cosmonauts in the L1 would carry out their circumlunar mission after a corresponding boost from the Blok D stage. For this plan to work, TsKBEM had to accommodate the manufacture of two special modifications of the 7K-OK and 7K-L1 vehicles. The 7K-OK's modification, designated 7K-OK-T, was equipped with a forward unit equipped for docking with a 7K-L1. The 7K-L1's modification not only had the "curved tunnel" but also a custom-built passive docking unit installed at the forward end of the spacecraft at the support cone. This heavy unit would be discarded once the transfer took place and before TLI.[133]

The Military-Industrial Commission, on April 27, 1966, adopted a decree (no. 101), titled "On Approving the Work Plan to Build the 7K-L1 Piloted Spacecraft," which addressed the entire spectrum of issues associated with the L1 circumlunar program. The commission approved the manufacture of fourteen such spacecraft: five in 1966 and nine in 1967. Ground testing was to finish and flight testing begin by the last quarter of 1966 or the first quarter of 1967. Among other things, the decree specified schedules for the development, manufacture, and delivery of L1 simulators and the establishment of a search-and-rescue service for a spaceship returning from the Moon.[134] According to the commission's decree, a specific schedule of operations was established for the program:

- September 1966—ground testing of one ship (7K-L1 no. 1P) at Tyura-Tam
- October 1966—two automated Earth-orbital tests (using 7K-L1 nos. 2P and 3P)
- November–December 1966—two automated circumlunar flights (using 7K-L1 nos. 4 and 5)
- December 1966–May 1967—five piloted circumlunar flight with crew transferred to the 7K-L1 in Earth orbit after being launched on the 7K-OK-T Soyuz (using 7K-L1 nos. 6 through 10)
- June–September 1967—four piloted circumlunar flight with crews launched in the 7K-L1 (using 7K-L1 nos. 11 through 14)

Such a schedule would ensure the fulfillment of the primary objective of a piloted circumlunar mission prior to the fiftieth anniversary of the Great October Revolution in November 1967.

---

131. Semenov, ed., *Raketno-Kosmicheskaya Korporatsiya*, pp. 238–39; Marinin and Shamsutdinov, "Soviet Programs for Lunar Flights"; Petrovich, ed., *The Soviet Encyclopaedia of Space Flight*, pp. 513–14.

132. Marinin and Shamsutdinov, "Soviet Programs for Lunar Flights."

133. The details of the "curved tunnel" and the special docking unit on the 7K-L1 in this variant remain unknown.

134. N. P. Kamanin, "A Goal Worth Working for," *Vozdushniy transport* no. 44 (1993): 8–9; Semenov, ed., *Raketno-Kosmicheskaya Korporatsiya*, p. 237. Note that one of the stipulations of the decree was the official termination of Chelomey's UR-500K-LK-1 program in the "full-scale modeling stage. The remaining stages, which envisioned the complete ground-based optimization of all the systems of the carrier and the vehicle, as well as the performance of 12 unmanned and 10 manned launches of the UR-500K-LK complex, were canceled by the same decree." See Igor Afanasyev, "Without the Stamp 'Secret': Circling the Moon: Chelomey's Project" (English title), *Krasnaya zvezda*, October 28, 1995.

As with most other timetables of the Soviet space effort of the period, there were delays. Many within TsKBEM believed the entire program to be a useless diversion from the main L3 landing project. Although the L1 project had moved into first priority over the L3, there were continuous postponements in issuing the technical documentation on the spacecraft, as well as testing delays in the construction of and upgrades to the two Proton launch complexes at Tyura-Tam. Being a matter of state importance, the status of the project was constantly examined at the ministry level throughout 1966. The concurrent work on the mainstream Soyuz effort was clearly a major drain on facilities and resources. If TsKBEM believed before that carrying out three full-scale piloted projects (Soyuz, L1, and N1-L3) was a manageable prospect, the employees were finding out that they were stretched to the limit. By December 1966, a single 7K-L1 spaceship had yet to get off the ground.

On December 9, 1966, at a meeting of the Council of Chief Designers, Mishin presented a new schedule of flights for the L1 program. Automated test flights of the first phase would include only four missions. Of these, the first two (2P and 3P) would be in Earth orbit to test out Blok D firings, while the remaining two (4L and 5L) would fly full-scale circumlunar missions and return to Earth. After these flights finished in March–May 1967, the first crew would fly to the Moon on June 25, 1967 aboard 6L. Kamanin noted about the meeting: "All the designers expressed doubts that the work could be accomplished within such a short timeframe. Mishin explained to them that he did not invent the schedule, but that it had been dictated to him by Ustinov and Smirnov."[135] An *ad hoc* twenty-member State Commission to guide the entire test program was established in mid-December with First Deputy Minister of General Machine Building Tyulin as its chair. Among its members were Mishin, Chelomey, and Keldysh.[136]

The State Commission for L1 met for the first time on December 24, when Mishin, Chelomey, and Barmin presented reports on the readiness of the spacecraft, the launch vehicle, and the launch pads, respectively. It was evidently the first time that all the heads of the various branches involved in the project discussed the project together. In accordance with the recommendation of the Council of Chief Designers, the new target date for the first piloted circumlunar mission was set for July 26, 1967. This would be preceded by the four automated flights. During the meeting, Mishin also presented his conception of the "fall-back" docking-in-Earth-orbit scenario to launch the crew not on the Proton booster, but rather in a 7K-OK-T Soyuz spacecraft. After the first few outbound piloted missions, once engineers had gained a modicum of faith in the Proton booster, the cosmonauts would fly directly into orbit on the Proton.[137]

During a second meeting of the commission on December 30, Mishin, Chelomey, and Barmin reported that all systems were on track for the first L1 launch at the end of January. All the members of the commission were due to arrive at Tyura-Tam on January 10–12, 1967. There was some discussion on the establishment of search-and-rescue services for vehicles returning from the Moon. Because, for the first time, the landing of a Soviet piloted spacecraft could be in the oceans, Marshal Matvey V. Zakharov, the Ministry of Defense General Staff Chief, had issued an order on December 21 that assigned the Air Force the responsibility for all land recov-

135. Kamanin, "A Goal Worth Working for," no. 45.

136. The members of the State Commission for L1 included: G. A. Tyulin (MOM), V. P. Mishin (TsKBEM), V. N. Chelomey (TsKBM), M. V. Keldysh (AN SSSR), V. A. Anfilatov (affiliation unknown), N. N. Gurovskiy (IMBP), N. P. Kamanin (VVS), A. G. Karas (TsUKOS), V. A. Kasatanov (affiliation unknown), V. A. Khazanov (affiliation unknown), A. A. Kurushin (NIIP-5), G. P. Melnikov (NII-4), N. K. Mordasov (affiliation unknown), Yu. A. Mozzhorin (TsNIIMash), A. G. Mrykin (MOM), D. A. Polukhin (TsKBM Branch No. 1), Ye. V. Shabarov (TsKBEM), I. I. Spitsa (TsKIK), Ya. I. Tregub (TsKBEM), and Yu. N. Trufanov (TsKBM Branch No. 1). See Semenov, ed., *Raketno-Kosmicheskaya Korporatsiya*, p. 238.

137. Kamanin, "A Goal Worth Working for," no. 45.

ery operations and the Navy the responsibility for all sea recovery operations. In addition, tracking stations at Feodosiya and Ussuriysk were being modified to communicate with returning spacecraft from the Moon.

The final point of discussion at the meeting was the selection of crews for the project. Cosmonauts had been unofficially grouped together to train for the LI mission by early September 1966. By early December, the main players had agreed on a list of fourteen men from the larger team at the Cosmonaut Training Center to train specifically for this project.[138] Because of the increasing requirements for cosmonauts in the mainstream 7K-OK Soyuz program, whose launches had already begun by this time, Kamanin and Mishin agreed to train cosmonauts by late December for the LI independently of Soyuz. Cosmonauts who would fly Soyuz missions would be added sequentially to the circumlunar program. The LI group was to undergo a five-month-long training program beginning on January 1, 1967. Each crew would include a commander who had experience from a previous space mission. By January 1967, eleven lucky men had been selected to train for the project, including Leonov, the spacewalker from Voskhod 2, and Popovich, the ebullient pilot from Vostok 4, both favorites for the first outbound flight.[139] The training for these men was impeded to a great degree by the absence of LI simulators, which, despite much discussion, the M. M. Flight Gromov Flight-Research Institute had not delivered by the end of 1966 to the Cosmonaut Training Center. The cosmonauts instead trained in 7K-OK Soyuz simulators equipped with new control instruments.

The LI Takes Flight

The first 7K-LI spacecraft was a model built specifically for ground testing at Tyura-Tam. These tests were completed successfully in conjunction with a UR-500K-Blok D combination in January.[140] The success did little to instill confidence that the planners would be able to maintain the compressed schedule handed down by Ustinov and Smirnov. The State Commission met twice on January 17, 1967, and heard reports from a number of chief designers involved in the program. There were "new difficulties" in the preparation of the first Earth-orbital mission, bringing Mishin and TsKBEM under fire from members of the commission. Some designers received reprimands; the commission decided to report the most glaring lags in work to the Central Committee. Chief Designers Grigoriy I. Voronin (of KB Nauka) and Gay I. Severin (of KB Zvezda), responsible for life support, emerged with an unlikely proposal to limit the number of cosmonauts in the 7K-LI crew to one, because of difficulties with the life support system. A final decision on the issue was delayed.[141]

At a meeting of the State Commission on February 14, the first test flight of the 7K-LI, originally scheduled for January 1967, was put back to late February or early March. The first two flights would primarily test the Blok D TLI stage with two firings: one to achieve Earth orbit and the second to boost the payload to escape velocity. No recovery was planned on either flight. Incredibly, the commission still hoped to carry out four automated missions prior to a piloted one set for June 26, 1967, despite the fact that within the same period, Mishin and the

---

138. *Ibid.* The cosmonauts for the 7K-LI program were G. T. Beregovoy, V. F. Bykovskiy, Yu. A. Gagarin, Ye. V. Khrunov, A. A. Leonov, V. M. Komarov, A. G. Nikolayev, V. A. Shatalov, and B. V. Volynov as crew commanders and G. M. Grechko, V. N. Kubasov, O. G. Makarov, V. N. Volkov, and A. S. Yeliseyev as flight engineers.

139. Kamanin, "A Goal Worth Working for," no. 46. The new 7K-LI training group had been agreed on as early as December 24, 1966. They were P. I. Klimuk, A. A. Leonov, P. R. Popovich, V. A. Voloshin, and B. V. Volynov (all commanders) and Yu. P. Artyukhin, G. M. Grechko, O. G. Makarov, N. N. Rukavishnikov, V. I. Sevastyanov, and A. F. Voronov (all flight engineers).

140. Afanasyev, "Unknown Spacecraft."

141. Kamanin, "A Goal Worth Working for," no. 46.

other chief designers were to carry out the first highly complex Soyuz mission of docking two such ships in Earth orbit with the subsequent transfer of cosmonauts. Through it all, Mishin, Chertok, and others tried to compensate for the poor management conditions by personally appealing to subcontractors to deliver parts on time. Unbelievably at this late stage, some contractors, such as Chief Designer Ryazanskiy, were not only behind schedule, but did not even know that they had been assigned to make a parts delivery in the first place.[142] Without a singular overseeing entity such as NASA, there was no coordinated plan for maintaining deadlines for dozens of subcontractors.

Some of the pressure on the Soviets to accelerate their lunar program was alleviated by a tragic accident half a world away. By early 1967, NASA was preparing for the first flight of the Apollo Command and Service Module, the spacecraft intended to take the first astronauts to the Moon. The first mission, Apollo 1, was planned to thoroughly test all the essential systems aboard the Block I class of modules. The fourteen-day mission, set tentatively for launch on February 21, 1967, was to be crewed by astronauts Lt. Colonel Virgil I. Grissom, Lt. Colonel Edward H. White II, and Lt. Commander Roger B. Chaffee. Both Grissom and White had flown previous space missions. In preparation for the launch, the crewmembers were simulating a countdown on January 27, when arcs from electrical wiring in an equipment bay in the Command Module began a fire. In the 100-percent oxygen atmosphere of the capsule, the crew succumbed to burns and asphyxia within minutes of the beginning of the fire.[143]

NASA immediately canceled all further missions in the Apollo program and established several teams to determine the causes of the accident. Outside analysts predicted that this would set back the Apollo program by at least a year, if not more. The accident inadvertently gave the Soviet Union an added probability to catch up with the United States following inactivity lasting almost two years. Despite the tragic nature of circumstances, the disaster no doubt instilled a glimmer of hope among the Soviets that perhaps the "race to the Moon" was still a race that had no clear winner. It would not have been surprising if Mishin, Chelomey, Keldysh, and others believed for this brief window that it was a foregone conclusion that the first human to fly around the Moon would be a Soviet citizen.

The first 7K-L1 spacecraft, vehicle no. 2P, was launched from Tyura-Tam at 1430 hours, 33 seconds Moscow Time on March 10, 1967, into a 190- by 310-kilometer Earth orbit inclined at fifty-one and a half degrees to the equator. It was the very first launch of the graceful silver four-stage Proton booster. The spacecraft was named *Kosmos-146* upon entering orbit, no doubt to hide the true mission of the vehicle. The Blok D stage, also in its first mission, performed flawlessly, firing both times—the second time boosting the 5,017-kilogram 7K-L1 vehicle to escape velocity. All except two on-board systems on the spacecraft operated without fault. The RDM-3 radio beacon did not turn off at the computed time because of a circuit error, and the unit worked continuously for forty-two hours instead of the nominal forty minutes. The second minor problem was a fault in the thermo-regulation system that led to an unexpected fall in pressure in one of the main lines.[144] Kosmos-146 remained in orbit for about nine days, while ground controllers maintained contact for at least five days.[145] The spacecraft probably reached lunar distance apogee before returning back to the vicinity of Earth and burning up on reentry.

The success of Kosmos-146 was no doubt a tremendous boost for engineers who had labored over the program for more than a year. The second spaceworthy 7K-L1 vehicle, space-

142. *Ibid.*

143. Ezell, *NASA Historical Data Book, Volume II*, p. 176.

144. Semenov, ed., *Raketno-Kosmicheskaya Korporatsiya*, pp. 240–41. The actual type of the 7K-L1 spacecraft was 7K-L1P, with the "p" indicating a simplified version not equipped for recovery.

145. Westerners tracked transmissions from the payload during March 11–15 at 20.008 megahertz. See Sven Grahn and Dieter Oslender, "Cosmos 146 and 154," *Spaceflight* 22 (March 1980): 121–23.

craft no. 3P, was quickly prepared for launch within less than thirty days. The vehicle would repeat the exact same profile as its predecessor, except Blok D's second firing would follow one day after entering Earth orbit instead of after one orbit. On April 6, Chelomey, Glushko, Barmin, and other chief designers arrived at Tyura-Tam to view the launch, along with ten cosmonauts who were training for the circumlunar flights. The latter group, including Leonov and Popovich, would study the equipment at the launch pad and get acquainted with all prelaunch operations involving the Proton booster. It was the first time that they physically saw the launch vehicle.

On April 8, the designers and guests watched the launch from site 92, the location of the assembly-testing building for the Proton, a distance of just over one and a half kilometers from the pad at site 81. Lt. General Kamanin described the scene:

> Unlike the [R-7,] the UR-500K rocket has a simple and well-designed service frame: the base of the frame is to one side of the rocket, but it "hugs it" with five service landings and has two elevators. After the frame is opened the rocket stands there like a beautiful white church. . . .[146]

At exactly 1200 hours, 33 seconds Moscow Time, the booster gracefully lifted off from its pad. Despite gusts as high as eighteen meters per second, the performance of all four stages, including the first firing of Blok D, was nominal. The 5,020-kilogram 7K-L1 ship entered a 186- by 232-kilometer orbit with a fifty-one-and-a-half-degree inclination to the equator. TASS announced the mission under the designation of *Kosmos-154*. About forty minutes following launch, all the members of the State Commission gathered at the office of Colonel Kirillov, the newly appointed Deputy Commander of Cosmodrome, to congratulate Chelomey on the success. Throughout the day, ground controllers monitored all systems aboard the Blok D-L1 stack in Earth orbit, conscious of the fact that this would be the first time in the Soviet space program when an upper stage would fire after a stay of twenty-four hours in weightlessness and vacuum.

The news turned sour on April 9, when telemetry proved that Blok D had failed to fire for the second time. After an analysis of incoming data, TsKBEM engineers believed that an instrument switch had been left in the wrong position because of negligence on their part. The instrument was used for triggering a system of engines that stabilize the propellant after the first firing of the Blok D main engine. The engines of this system were apparently prematurely jettisoned, disabling the main engine completely because it was unable to effectively use the propellants.[147] The blame for the error fell on Mishin's shoulders, and State Commission Chairman Tyulin gave him a dressing down. Kamanin recalled:

> Tyulin was furious and swore at him. In the evening, still fuming after the unpleasant experience of reporting to Ustinov and Smirnov, he gave a devastating but perfectly accurate assessment of Mishin: "He has five times more arrogance than Korolev and ten times less competence."[148]

The Kosmos-154 stack remained in its low-Earth orbit for about two days following launch before decaying naturally. The failure undoubtedly slowed the pace of the circumlunar program,

---

146. Kamanin, "A Goal Worth Working for," no. 46.
147. *Ibid.*; Afanasyev, "Unknown Spacecraft"; Semenov, ed., *Raketno-Kosmicheskaya Korporatsiya*, p. 241.
148. Kamanin, "A Goal Worth Working for," no. 46, p. 9.

and the prospect of carrying out the first piloted mission in June or July must have seemed shaky by any stretch of the imagination, especially given the intensive work concurrent in the Earth-orbital Soyuz program. At the same time, even if the June–July deadline seemed out of reach, there was still much hope that two Soviet men would circle the Moon by the November 1967 deadline. But this still vibrant hope was dealt a fatal blow just sixteen days after the launch of Kosmos-154. It would be one of the most devastating incidents in the history of the Soviet piloted space program—an event that crippled its run in the race for the Moon.

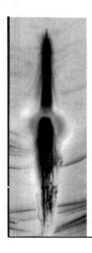

CHAPTER THIRTEEN
TRAGEDY

The Soyuz spacecraft was the centerpiece of the post-Korolev space program. Since Korolev's death in January 1966, the design, development, and testing of the 7K-OK Earth-orbital Soyuz were expected to lead to the most spectacular mission in the Soviet canon to date: the docking of two Soyuz spacecraft in Earth orbit, followed by the transfer of two crewmembers from one vehicle to the other via a spacewalk. Soviet space program leaders strongly believed that this one mission would overshadow the cumulative achievements of all ten of NASA's Gemini flights during 1965–66. Thousands of engineers worked toward this singular goal to reestablish Soviet preeminence in piloted space exploration. From a political, technical, and human perspective, the failure to do so was not an option. But as haste crept into the preparations, an atmosphere of unease began to pervade the program.

Civilians in Space

For many years before his death, Sergey P. Korolev had spoken of sending not only military officers into space, but also the young civilian engineers who actually designed and developed Soviet spacecraft, such as Vostok, Voskhod, and Soyuz. Intermittently throughout the early 1960s, several engineers at OKB-1 had passed through preliminary medical screening, but their candidacy as cosmonauts was never taken seriously by the Soviet Air Force, the service responsible for all cosmonaut training.[1] The impetus to include engineers on spacecraft increased significantly with the development of the 7K-OK Soyuz spacecraft, which afforded two to three extra seats for missions. In September 1965, eight military cosmonauts began training for the docking and EVA Soyuz mission, prompting Korolev to entrust one of his engineers to look into the matter of forming a parallel civilian training group.[2] At least eleven civilians from the design bureau passed the initial medical screening at the Ministry of Health's Institute of Biomedical Problems, but Korolev's death put the matter temporarily on the backburner.[3]

---

1.   In September 1961, the Air Force allowed Korolev to send civilian engineers through medical screening. See Rex Hall, "Soviet Civilian Cosmonauts," in Michael Cassutt, ed., Who's Who in Space: The International Space Year Edition (New York: Macmillan, 1992), p. 278

2.   The eight military cosmonauts who began training in early September 1965 for the first Soyuz mission were V. F. Bykovskiy, Yu. A. Gagarin, V. V. Gorbatko, Ye. V. Khrunov, P. I. Kolodin, V. M. Komarov, A. G. Nikolayev, and A. F. Voronov. See N. P. Kamanin, Skrytiy kosmos: kniga utoraya, 1964–1966gg (Moscow: Infortekst IF, 1997), pp. 347, 349; N. Kamanin, "A Goal Worth Working for" (English title), Vozdushniy transport 44 (1993): 8–9. These eight men were still in training for the first Soyuz mission by late August 1966.

3.   These eleven men were S. N. Anokhin, V. Ye. Bugrov, G. A. Dolgopolov, G. M. Grechko, V. N. Kubasov, O. G. Makarov, N. N. Rukavishnikov, V. A. Timchenko, V. A. Yazdovskiy, and A. S. Yeliseyev. See I. Marinin, "The First Civilian Cosmonauts" (in Russian), Novosti kosmonavtiki 12-13 (June 3-30, 1996): 81-87.

With unexpected vengeance new Chief Designer Mishin took up the gauntlet of training civilians, in part motivated by his hostility toward the Air Force, which coveted its total monopoly over cosmonaut training. A governmental decree six years previously had codified that all Soviet cosmonauts, regardless of their affiliations, should be trained exclusively at the Air Force's Cosmonaut Training Center.[4] But without the agreement of either the Ministry of General Machine Building or the Ministry of Defense, Mishin signed an official order (no. 43) on May 23, 1966, establishing the 731st Flight-Methods Department, which consisted of the first civilian "cosmonauts group" in the Soviet Union. The group members were:

- Sergey N. Anokhin (fifty-six years old)
- Vladimir N. Bugrov (thirty-three)
- Gennadiy A. Dolgopolov (thirty)
- Georgiy M. Grechko (thirty-four)
- Valeriy N. Kubasov (thirty-one)
- Oleg G. Makarov (thirty-three)
- Vladislav N. Volkov (thirty)
- Aleksey S. Yeliseyev (thirty-one)[5]

Anokhin was an odd selection for the group because he was more than twenty years older than the rest. A famous World War II pilot, he had gone on to be one of the most accomplished test pilots in the Soviet Union, flying out of the famous M. M. Gromov Flight-Research Institute outside of Moscow. Acquainted with Korolev since the wartime days, Anokhin had been invited to head up a flight testing department at OKB-1 in April 1964, ostensibly to oversee the training of future cosmonauts from the design bureau.[6] Given his age (he was six years older than Mishin), his inclusion in the group seems to have been more of a personal favor to Korolev's memory than to any serious plan to launch him into space.

Without official recognition from the Air Force, the eight candidates had little hope of actually flying in space and were known only as "cosmonaut-testers." Mishin, however, tried everything in his power to bypass official Air Force rules. On June 15, 1966, he forced through a formal Military-Industrial Commission decree (no. 144) that stipulated that his eight civilian cosmonaut-testers be considered as candidates for the forthcoming Soyuz flight.[7] By this time, the friction between the Air Force, represented by the ubiquitous Lt. General Nikolay P. Kamanin, and TsKBEM began to affect the course of the Soyuz program. Without any agreement on the crew, the engineers faced great difficulties in establishing timetables for the highly complex joint mission. In late June, Mishin even went so far as to propose completely civilian crews for the mission, although the eight military officers were finishing up several months of training.[8]

Throughout the month of July, the arguments went back and forth, with both Mishin and Kamanin refusing to budge on their positions. Although First Deputy Minister of General Machine Building Tyulin served as a mediator, Mishin convinced him and other officials,

---

4. This decree was issued on August 3, 1960. See *ibid.*

5. *Ibid.*; Yu. P. Semenov, ed., *Raketno-Kosmicheskaya Korporatsiya "Energiya" imeni S. P. Koroleva* (Korolev: RKK Energiya, named after S. P. Korolev, 1996), p. 426.

6. Semenov, ed., *Raketno-Kosmicheskaya Korporatsiya,* p. 425; Hall, "Soviet Civilian Cosmonauts," p. 287.

7. Kamanin, "A Goal Worth Working for," no. 44; Marinin, "The First Civilian Cosmonauts"; Semenov, ed., *Raketno-Kosmicheskaya Korporatsiya,* p. 426.

8. The civilian crews proposed by Mishin were Dolgopolov/Yeliseyev/Volkov (primary) and Anokhin/Makarov/Grechko (backup). See Marinin, "The First Civilian Cosmonauts." In early July, there was a new civilian crew proposal: Dolgopolov/Makarov (Soyuz 1) and Yeliseyev/Kubasov (Soyuz 2). See Kamanin, *Skrytiy kosmos: 1964–1966,* p. 348.

including Academy of Sciences President Keldysh and Deputy Minister of Health Avetik I. Burnazyan, to approve a program on July 30 to train a group of civilian cosmonauts for the LI circumlunar program. The implication was clear: Mishin would no longer use the Air Force's Cosmonaut Training Center. Kamanin, predictably, called the document "a piece of nonsense."[9] The acrimony came to a head in early August, when First Deputy Commander-in-Chief of the Air Force Marshal Sergey I. Rudenko, Kamanin's immediate boss, agreed on a compromise: to allow civilians to fly, but only if they passed through military medical screening and then trained at the Cosmonaut Training Center. Although Kamanin still objected, Mishin apparently found the plan agreeable, and on August 16, he wrote a letter to Kamanin explaining that civilian engineers should fly on the Soyuz spacecraft because "design solutions can only be checked by highly qualified specialists directly involved in designing and ground testing of the spacecraft. . . ."[10]

On August 31, the eight TsKBEM engineers led by Anokhin arrived at the Air Force's Central Scientific-Research Aviation Hospital for medical screening. Having passed through the tests, Grechko, Kubasov, and Volkov arrived at the Cosmonaut Training Center on September 5, the first group of civilians engineers in the Soviet space program to do so. The three, joined later by Yeliseyev, began training on October 1.[11] Makarov arrived in November. All five were accomplished engineers in their own right, participating in many of the historic events during the early space program. Grechko had helped fuel the early R-7s before launches in 1957. Makarov had been on the teams that designed the Vostok, Voskhod, and Soyuz spacecraft. For the Soyuz, LI, and L3 programs, each of these engineers were to occupy the flight engineer's seat— "the member of the crew . . . with responsibility for the correct operation of on-board systems and carrying out the flight program."[12] The remaining three from the group—Anokhin, Bugrov, and Dolgopolov—failed to pass the Air Force's medical screening and were never considered for these Soyuz missions.[13]

The addition of civilian engineers to train for the Soyuz flights, while it did not end the battle between TsKBEM and the Air Force on the issue of cosmonaut selection, did allow Soyuz training to proceed without further disruptions. The training regime was, however, incredibly compressed. Although all the civilians had the advantage of being intimately familiar with the 7K-OK vehicle, they still had a scant three months before the docking mission, then set for early January 1967. By mid-November, Kamanin was looking at a mixed crew composed of military officers, in training for more than a year, and the new civilians.[14] Ultimately, Mishin's insistence on training civilian engineers had a long-lasting legacy on the composition of crews for the next thirty years of the Soviet and later Russian space programs. During the late 1980s and throughout the 1990s, each and every crew to the *Mir* space station included a flight engineer who was a spacecraft designer from the design bureau, now known as the Energiya Rocket-Space Corporation (RKK Energiya).

Despite the arrival of the new civilian engineers at the Cosmonaut Training Center, Kamanin stubbornly remained resistant to allowing the engineers to *fly* on the immediate Soyuz missions. On his orders, the eight military officers continued to train for the flight, two of whom—Gorbatko and Khrunov—prepared for the EVA from one ship to another. Mishin,

9.   Kamanin, "A Goal Worth Working for," no. 44.
10.   *Ibid.*
11.   Marinin, "The First Civilian Cosmonauts."
12.   Semenov, ed., *Raketno-Kosmicheskaya Korporatsiya*, p. 425, first footnote.
13.   Grechko, Kubasov, Makarov, Yeliseyev, and Volkov were joined on January 8, 1967, by two more civilian engineers from TsKBEM: V. I. Sevastyanov and N. N. Rukavishnikov. Grechko dropped out of training temporarily when he broke a leg during parachute training on or about October 10, 1966. See Marinin, "The First Civilian Cosmonauts."
14.   N. Kamanin, "A Goal Worth Working for" (English title), *Vozdushniy transport* 45 (1993): 8–9.

however, insisted that Kubasov and Yeliseyev, two of his own men, be put on the flight for the spacewalk. On November 16, 1966, the Communist Party's Defense Industries Department Chief Serbin finally arbitrated a compromise: of the two EVA cosmonauts, one would be from the Air Force (Khrunov) and one from TsKBEM (Yeliseyev). There was some controversy over Yeliseyev's past. The Soviet security apparatus had discovered that Yeliseyev's original last name was Kureytis, a Lithuanian name. His father, Stanislav A. Kureytis, had been arrested in 1935 and spent five years in jail for "anti-Soviet agitation." Later, Yeliseyev had taken his wife's last name to put the past behind him. Evidently, the KGB let the issue go, although in past years such "tarnished" biographies had given pause to select cosmonauts for flight crews.[15]

The remaining cosmonauts on the docking flight would all be military officers. Since September 1965, four Air Force cosmonauts had been training for the commander's spot on the two Soyuz spacecraft: veterans Bykovskiy, Gagarin, Komarov, and Nikolayev. Of them, it seems that Vladimir Komarov had been the leading contender for the commander aboard the active Soyuz, and he distinguished himself with excellent grades during mission training. Of all the flown and unflown cosmonauts, there was little doubt that he was the most technically accomplished as well as the most intellectually sophisticated member. He had originally served as a fighter pilot in the Caucasus military district during the early 1950s before joining the prestigious N. Ye. Zhukovskiy Air-Engineering Academy in August 1954. He graduated five years later, in time to join the State Red Banner Scientific-Research Institute of the Air Force with the rank of "captain engineer" of the Air Force. When he joined the cosmonaut team—that is, military unit no. 26266—in 1960, he was only one of two individuals who had graduated from Air Force academies; the rest had only finished the equivalent of American junior colleges. Komarov nearly dropped out of training early on, because of the diagnosis of an irregular heartbeat, but he had persevered and flew into space as the commander of the historic three-person Voskhod crew in October 1964. Within less than two years, he had become the sole contender for the primary crew commander's spot for the first Soyuz flight.[16] At a State Commission meeting at Tyura-Tam on November 21, 1966, it was Komarov who announced the candidates for the two spacecraft: Soyuz 1 would fly with Komarov, and Soyuz 2 would fly with Bykovskiy, Yeliseyev, and Khrunov.[17]

Yeliseyev was the sole civilian engineer from Mishin's design bureau, Bykovskiy was the veteran from Vostok 5, and Khrunov was one of the remaining unflown cosmonauts from the famous "Gagarin group" of 1960. Gagarin was, for the first time in five years, back on a back-up crew. Since his first mission in 1961, he had served as more of a public relations linchpin for the Soviet space program than anything else. Some of his international duties were mitigated by his appointment in late 1963 as a deputy director of the Cosmonaut Training Center—a desk job that posited him as a leading member of the State Commissions for the Voskhod flights. In the intervening period, Gagarin had gained weight, and his flying skills seemed to have deteriorated. This was not simply Gagarin's fault; cosmonaut overseer Kamanin had continually opposed Gagarin's reassignment back to cosmonaut training. As early as April 1963, Kamanin emoted that "Gagarin hopes that someday he will fly new space missions. It is unlikely, however, that this will happen. Gagarin is too dear to mankind to risk his life for the sake of an ordinary space flight."[18] Gagarin, however, pursued a second flight with unfettered vigor

15.   Kamanin, *Skrytiy kosmos: 1964–1966*, pp. 385–86, 389, 390, 391, 394, 395, 399.

16.   I. Marinin, "Anniversaries—Vladimir Komarov—70 Years" (English title), *Novosti kosmonavtiki* 6 (March 10–23, 1997): 51–53. For details on the training program for the first Soyuz mission through 1966, see Viktor Mitroshenkov, *Zemlya pod nebom* (Moscow: Sovetskaya rossiya, 1987), pp. 385–98.

17.   Marinin, "The First Civilian Cosmonauts." The two backup crews were: Gagarin (Soyuz 1) and Nikolayev, Kubasov, and Gorbatko (Soyuz 2). Note that prior to the admission of civilian engineers (Yeliseyev and Kubasov) on the crews, two military engineers had trained for the EVA transfer: P. I. Kolodin and A. F. Voronov.

18.   N. Kamanin, "For Him, Living Meant Flying" (English title), *Vozdushniy transport* 9 (1994): 8.

and was even considered the primary contender for the Soyuz flight until April 1966, when a combination of political and personal factors forced officials to replace him with Komarov. Instead, Gagarin served as Komarov's backup.[19]

Stumbling Toward Piloted Flight

According to the Military-Industrial Commission decree from August 1965, the Soyuz program was to set off with the first automated missions in the first quarter of 1966. Upon Mishin's official appointment as Chief Designer in May, one of his first tasks was to evaluate the state of the project, and he was remarkably optimistic, scheduling the first piloted attempt in August 1966. The plan at the time was to launch two automated Soyuz spacecraft to check the operation of all systems in robotic mode.[20] Needless to say, this schedule was not maintained. Throughout 1966, engineers carried out ground testing of the spacecraft at a feverish pace. Apart from static testing on stands, the Soyuz was involved in intensive dynamic design testing, work on the nominal separation of the three component modules, testing of the payload shrouds, thermal testing, checking of the operation of the life support system in pressure chambers, docking of ground models by using suspended cables in a high-altitude chamber, testing of the engine units, flight testing of the landing system, and dynamic testing of the launch escape system.

The engineers began the ground testing of the first flight model of the Soyuz spacecraft on May 12, 1966. There were many problems. Instead of the anticipated thirty days, it took four months to debug the ship. There were as many as 2,123 defects in the vehicle, significantly affecting the pace of the project. The official history of the design bureau states that the testing of the Soyuz spacecraft:

> required the solution of a number of serious scientific-technical and management problems, which arose due to the considerable complexity, as compared to the "Vostok" and "Voskhod" in the composition and logic of the functioning of the on-board systems. . . .[21]

Among the factors that the engineers had to face were problems with the parachute system. Serious defects were identified when two out of seven drop tests from the An-12 aircraft at Feodosiya failed. After one test on August 9, when the reserve parachute failed to open, Kamanin prophetically wrote in his diaries:

> One has to admit that the 7K-OK parachute system is worse than the parachute system of the Vostoks. And the spacecraft isn't much to look at in general: the hatch is small, the communications equipment is outdated, the emergency rescue system is primitive and so on. If the automatic docking device turns out to be unreliable (which cannot be ruled out) our space program will be headed for an ignominious failure.[22]

---

19. On April 16, 1966, at a meeting at the Cosmonaut Training Center, officials proposed Komarov instead of Gagarin as the primary candidate for the first Soyuz flight. Gagarin was proposed as his backup. See Mitroshenkov, *Zemlya pod nebom*, p. 382. In January 1966, the primary crew for the first mission was Gagarin and Voronov.

20. There was also a Military-Industrial Commission decree in early 1966 that stipulated that the first two automated flights would be in August 1966, the joint piloted flight would be in September–October 1966, and the second joint piloted flight would be in November 1966. See Marinin, "The First Civilian Cosmonauts."

21. Semenov, ed., *Raketno-Kosmicheskaya Korporatsiya*, p. 177; B. Ye. Chertok, *Rakety i lyudi: goryachiye dni kholodnoy voyny* (Moscow: Mashinostroyeniye, 1997), p. 402.

22. Kamanin, "A Goal Worth Working for," no. 44. The parachute system was designed and built by NIEI PDS headed by Chief Designer F. D. Tkachev.

The political pressure to return to flight was immense, as official TsKBEM historians noted later:

> . . . it was impossible to allow a gap in the realization of piloted flights after the successful series of launches of the "Vostok" and "Voskhod" ships and it was necessary to maintain the priority in space research relative to the Americans . . . there was also pressure on the part of the government. Thus, Deputy Minister [of General Machine Building Valentin Ya.] Litvinov personally daily in the morning carried out operative meetings in the 44th assembly shop . . . and signed a list of bonuses for accelerating work.[23]

Maj. General Kerim Kerimov was the chair of the ad hoc State Commission for Soyuz from 1966 to 1991. A veteran Strategic Missile Forces officer, he officially served in several high positions in the Ministry of General Machine Building during the Soviet era. (files of Peter Gorin)

To oversee the test launch phase of the Soyuz spacecraft, the Soviet government established a new State Commission in October 1966, whose official title was the "State Commission for Flight-Testing of the Soyuz Spacecraft." Maj. General Kerim A. Kerimov, a forty-nine-year-old artilleryman, formerly of the Strategic Missile Forces, was appointed to head the commission apparently to honor the late Korolev, who had originally suggested Kerimov for the post.[24] He was an odd choice for the position. Unlike the State Commissions for the Vostok, Voskhod, L1, and N1-L3, it was the first occasion when a commission chair did not have a ministerial or even a deputy ministerial rank. In fact, the actual duties of the chairs of the N1-L3, L1, and Soyuz State Commissions show a progressive decline in state importance with Minister Afanasyev (for the N1-L3), First Deputy Minister Tyulin (for the L1), and Chief of the Ministry's Third Chief Directorate Kerimov (for Soyuz), respectively. The latter was yet another former artillery expert who had gone to Germany after the war to recover German A-4s. Throughout the 1950s the native Azerbayjani had worked at Kapustin Yar before heading the first space directorate at the Strategic Missile Forces. In 1965, he quit the Strategic Missile Forces under dubious circumstances before going on to the Ministry of General Machine Building.

Throughout the summer of 1966, senior space officials met on several occasions to agree on a manifest leading up to the ambitious docking mission. Because almost all the systems on board the 7K-OK Soyuz spacecraft were automated, some members recommended that instead of two automated solo flights, engineers carry out a full-scale rendezvous and docking mission

23. Semenov, ed., *Raketno-Kosmicheskaya Korporatsiya*, p. 178.

24. K. Isaakov, "Earth—Our Paternal Home" (English title), *Bakinskiy rabochiy*, August 19, 1987, p. 3; V. Ovcharov and L. Chernenko, "Recommended by Korolev" (English title), *Sovetskaya rossiya*, August 22, 1987, p. 2; S. Leskov, "Sputnik," *Komsomolskaya pravda*, October 4, 1987, p. 4; "Living History, Noteworthy Events: Rockets Go Into Space" (English title), *Sovety narodnykh* no. 4 (April 1988): 50–53; K. Isaakov, "Breakthrough into the Unknown: Today is Cosmonautics Day" (English title), *Bakinskiy rabochiy*, April 12, 1988, p. 3; Kamanin, *Skrytiy kosmos: 1964–1966*, pp. 380, 383.

between the two ships. Among those in favor of such a plan was Chief Designer Armen S. Mnatsakanyan of the Scientific-Research Institute for Precision Instruments, responsible for the design and development of the *Igla* rendezvous and docking radar system. After assessing the pros and cons, Kerimov and Mishin agreed to Mnatsakanyan's suggestion. The first two automated flight models of the 7K-OK Soyuz arrived at Tyura-Tam in August 1966 for their launches in September. Further problems, however, necessitated moving the launches to November 1966. This was to be followed in January or February of the following year with the piloted mission.

On the morning of November 18, the commission met at Tyura-Tam in preparation for the upcoming dual launches set for November 26–27. Spaceship no. 2, the active Soyuz, would be launched first, followed twenty-four hours later by Spaceship no. 1, the passive Soyuz. Upon orbital insertion, if the passive ship was within twenty kilometers of the active vehicle, then docking would take place between the two ships on the passive one's first or second orbit. If the distance was greater, then the docking would occur a day later. If all systems were operating ideally, then the two spacecraft would remain docked for three days; both would land on the fourth day of their respective missions.[25] Engineers believed that a piloted flight with the third and fourth Soyuz vehicles could be mounted as early as December 26–27. A lot of factors had to work perfectly to maintain the deadline—for example, both of the two pads (at sites 1 and 31) capable of launching the 11A511 booster would have to be available for launches. This meant that the commission would have to obtain permission from the military to delay the launch of a Zenit-4 photo-reconnaissance satellite scheduled for launch from one of those pads. The Soyuz launches would mark the first launches of the 11A511 booster, a marginal modification of the earlier 11A57 launch vehicle used for Voskhod.

A final State Commission meeting took place on November 25, by which time the two launches were set for November 28 and 29. On launch day, Kamanin wrote:

> We've been waiting for this to happen for more than four years (the industry delayed the manufacture of the spacecraft because they were overautomated: they have to be able to link up even if unmanned). Today and tomorrow will see launches on which the immediate future of our space program will hinge: all the Moon spacecraft are based on Soyuz.[26]

The first Soyuz spacecraft lifted off successfully at 1600 hours Moscow Time on November 28, 1966, from Tyura-Tam. It entered an initial orbit of 181 by 232 kilometers at a 51.9-degree inclination; the perigee was lower than expected because of the less-than-stellar performance from the new launch vehicle. The Soviet news agency TASS designated the spacecraft *Kosmos-133* and, as was customary, did not indicate that the flight had any connection with the piloted space program. Problems beset the mission almost immediately. As soon as the payload separated from the booster, the pressure in the tanks of the mooring and orientation engine system dropped from 340 atmospheres to thirty-eight atmospheres in 120 seconds. Within less than fifteen minutes, all or most of the propellant in the system had been used up, sending the spacecraft into a slow rotation of two revolutions per minute. Given that these engines were indispensable for attitude control during approach and docking, there was little hope of carrying out a docking with a second Soyuz. Kerimov and Mishin immediately decided to cancel the preparations for the second launch and instead focus efforts on bringing Kosmos-133 back to Earth.

The spaceship had more problems. The mooring and orientation system thrusters were required not only for rendezvous and docking but also to position the spacecraft into correct attitude to fire the main deorbit engine. On Deputy Chief Designer Chertok's suggestion,

---

25.   Kamanin, "A Goal Worth Working for," no. 45; Chertok, *Rakety i lyudi: goryachiye dni kholodnoy voyny*, p. 412; Kamanin, *Skrytiy kosmos: 1964–1966*, p. 396.

26.   Kamanin, "A Goal Worth Working for," no. 45, p. 8.

ground controllers at Yevpatoriya decided to use a backup set of attitude control engines linked to the backup main engine. A test of these small thrusters, however, showed that they turned the ship in an opposite direction to the one commanded—that is, they could not be used for reentry attitude orientation either. Kosmos-133 seemed to be stranded in orbit. Preliminary ballistics projections showed that the spacecraft would decay naturally after about thirty-nine orbits, in which case the automatic self-destruct system would blow up the vehicle during descent because of an incorrect orientation.

Given these almost insurmountable problems, the Chief Operations and Control Group found an ingenious way to work around them. The flight control team decided they could use a third set of tiny thrusters, the orientation engines, which were used only for minor attitude control, to position the vehicle correctly for short time periods. Thus, instead of firing the main S5.35 engine for a full 100 seconds for reentry, the controllers would fire it in short bursts of about ten to fifteen seconds while the orientation engine system maintained proper attitude. The cumulative effect of several of these short firings would be the same as one long burn— that is, sufficient to deorbit the spacecraft safely. There was, however, little hope of bringing the ship back to a preselected target area.[27]

In the early morning of November 29, after extensive consultations with Chertok at Yevpatoriya and with Deputy Chief Designer Bushuyev in Moscow, the State Commission opted to try for a reentry on the seventeenth orbit using a combination of the automatic solar orientation system, the orientation engine system thrusters, and the main engine. Controllers apparently doubted whether all the correct commands had been sent to the spacecraft at the time, and Mishin decided to call off the attempt and not take the risk. Attempts to bring the ship down on the eighteenth and nineteenth orbits using ionic attitude control sensors did not succeed either. Kamanin in his journal recorded that the controllers fired the engine two times, but each time the unit cut off after ten and thirteen seconds, respectively. A third burn to change Kosmos-133's orbit to shift its landing track over Soviet territory also prematurely cut off after twenty seconds, apparently because the ship was not properly stabilized during the firing. It remains unclear whether these aborted burns were deliberate firings to guide the ship in for deorbit or failed attempts at reentry. Ultimately, the State Commission decided to delay the landing for another day to wait for the following opportune landing opportunity on Soviet soil.[28]

On the morning of November 30, on the spacecraft's thirty-second orbit, the controllers carefully sent commands for new engine firings to be carried out on the succeeding orbit. But on the thirty-third orbit, the main engine apparently shut down prematurely again. A fifth engine firing on the thirty-fourth orbit using the ionic sensor system did the job; the spacecraft was sufficiently slowed down to begin orbital decay. Kosmos-133 separated into its three component modules and began reentry, but the descent apparatus abruptly disappeared from radar screens about seventy to 100 kilometers over Earth, 200 kilometers southeast of the city of Orsk. An extensive visual search by the Air Force's search service ended without result. Later, the State Commission ascertained that the descent trajectory had been too flat and the capsule had begun to overshoot Soviet territory and head toward China. The self-destruct system, consisting of twenty-three kilograms of TNT, exploded automatically and destroyed the capsule. Debris apparently rained down on the Pacific Ocean east of the Mariana Islands.[29] The mission had lasted about one day, twenty-one hours.

---

27.   Chertok, *Rakety i lyudi: goryachiye dni kholodnoy voyny*, pp. 416–19.

28.   Kamanin, *Skrytiy kosmos: 1964–1966*, pp. 414–16.

29.   *Ibid.*, pp. 416–17; Leonard Nikishin, "Soviet Space Disaster on the Revolution's Anniversary: How and Why Cosmonaut Komarov Died," *Moscow News* 9 (March 1–8, 1992): 16; Kamanin, "A Goal Worth Working for," no. 45; I. Marinin, "'Soyuz': 30 Years Since the First Flight" (English title), *Novosti kosmonavtiki* 24 (November 18–December 1, 1996): 64–65.

Although the flight could hardly have been considered successful, the mission did give engineers and controllers on the ground a chance to evaluate the operation of all the Soyuz systems in realistic conditions. The ionic orientation system was stable, the main engine could be fired repeatedly in vacuum, and the spacecraft could be reentered despite faulty stabilization. Based on an analysis of the problems, many State Commission members, including Chairman Kerimov, Mishin, and Ryazanskiy, believed that Kosmos-133 could have been safely recovered if there had been a cosmonaut on board instead of a mannequin. Four investigation commissions, which included Chief Designer Ryazanskiy, Deputy Chief Designer Tsybin, and Department Chief Raushenbakh, reported their findings on December 8. There had been three major failures on the ship: the complete spurious exhaust of the propellant in the mooring and orientation engine system; insufficient stabilization of the spacecraft when the deorbit engine was fired; and a failure of the *Tral* telemetry instrument on the fifteenth orbit. They found that the failures had nothing to do with design flaws but rather problems in assembling and testing that particular model on the ground. The service lines for the jet vane controls of the orientation engines were evidently tangled up, and a faulty system was installed on the vehicle. During reentry, the retro-engine had fired for less than a nominal period because of the lack of vehicle stabilization, which itself was a result of the faulty orientation system. The State Commission recommended that the second Soyuz, the passive 7K-OK, be launched no later than December 18 on a solo flight. Igla system Chief Designer Mnatsakanyan opposed a solitary launch and continued to insist on an automated docking flight, but he was overruled by Mishin, who apparently regretted following Mnatsakanyan's advice to mount a joint flight on Kosmos-133. If all went well, cosmonauts would fly into space aboard two different Soyuz vehicles in late January or early February.[30]

The pace at Tyura-Tam was intense. A little more than two weeks later, the remaining Soyuz spacecraft, vehicle no. 1, was ready for launch, this time from the pad at site 31. The launch was set for 1430 hours local time on December 14, 1966. At the count of zero, shards of flame shot out from the base of the 11A511 booster, but they were suspiciously smaller and less powerful than normal. The rocket remained fixed on the pad, and those present assumed that computers had aborted the launch at the last minute because of a then-unknown glitch. The flames at the base died down soon, and steam filled the area as thousands of gallons of water poured onto the launch mount. Approximately twenty-seven minutes after the abort, observers saw the launch escape system suddenly start firing. At this point, there were many pad workers who were engaged in "safing" the booster, as was customary following a launch abort. Although the rocket seemed to remain inert, within a few seconds, the flames from the escape system directly engulfed the lower portion of the Soyuz spacecraft and the booster's third stage below. As the fire spread, scores of workers near the pad took cover in their bunkers. Kamanin described the scene:

> I ran to the cosmonauts' house and ordered everyone who was there to quickly go from the rooms into the corridors. It proved to be a timely measure: within seconds a series of deafening explosions rocked the walls of the building which was located 700 meters from the pad. Stucco fell down and all the windows were smashed. The rooms were littered with broken glass and pieces of stucco. Fragments of glass hit the walls like bullets. Clearly, if we had remained in the rooms a few seconds longer we would all have been mowed down by broken glass. Looking out through the window openings I saw huge pillars of black smoke and the frame of the rocket devoured by fire. . . .[31]

---

30. Kamanin, "A Goal Worth Working for," no. 45; Semenov, ed., *Raketno-Kosmicheskaya Korporatsiya*, p. 179; Marinin, "'Soyuz': 30 Years Since the First Flight."
31. Kamanin, "A Goal Worth Working for," no. 45, p. 9.

State Commission members met about twenty minutes later at the Soyuz assembly-testing building, but among those missing were Mishin, Kerimov, and Maj. General Kirillov, the Chief of the First Directorate at the Baykonur Cosmodrome. As concern mounted for the missing individuals, Baykonur Commander Maj. General Aleksandr A. Kurushin quickly sent an emergency medical team to the launch pad area to search for survivors. Within a short time, Mishin, Kerimov, and Kirillov turned up safe at another command bunker. The Soyuz descent apparatus miraculously landed safely at a distance from the pad without incident.

On December 16, an investigation commission reported on the probable causes of the terrible accident. It seems that when the command to ignite had been sent to the booster, only the second stage of the 11A511 launcher (that is, the strap-ons) had fired, and computers had instantly aborted the launch. After the announcement for pouring water around the launch mount, Mishin and Kirillov had concluded that it was safe to egress from their bunkers because all engines on the booster were shut down. Ground control then sent a command to relocate the escape frames of the pad structure onto the vehicle to prevent the launcher from swaying in the gusty winds present at the time. By this point, many service personnel had already arrived at the pad to climb up the service tower to inspect the rocket. As the frames were lifted near the booster, one of these touched the booster prematurely and tilted it. This occurred because the launch vehicle had moved very slightly from its original position at the launch abort. As soon as the booster tilted, the emergency rescue system was automatically triggered by gyroscopes, which detected a vertical angle exceeding seven degrees. The ninety-ton solid-fuel engine of the system fired on command, and its long exhaust penetrated the Soyuz propellant tanks on top of the booster; at that point, all service personnel fled the area in panic. It took almost two minutes between the firing of the system and the final explosion of the first and second stages of the booster—a length of time that no doubt saved the lives of most of those who were close to the booster, including Mishin, Kerimov, and Kirillov. Most managed to run as much as 150 to 200 meters to safety, while Mishin and the others fled to a nearby bunker. A Major Korostylev unfortunately took refuge behind the concrete walls of the launch assembly and, as a result, became the sole fatality in the accident. Several others were severely injured. The entire pad complex and associated structure was completely destroyed.[32]

At the meeting on December 16, Mishin admitted that the design of the emergency rescue system had been fundamentally faulty because the gyroscopes could trigger the operation of the system even when all power was cut off to the booster. Remarkably, just three days prior to the explosion, engineers carried out a test of the rescue system at the Air Force's test site at Vladimirovka near Kapustin Yar. Because the goal of the test was not to check fire safety, the tanks of the spacecraft were left empty for the firing of the rescue system engines. A fueled spaceship could have easily precluded such a disaster. Engineers introduced a number of design changes on the rescue system based on the recommendations of the accident commission, including ensuring that the solid-propellant engine of the system could be turned off manually or remotely immediately after aborts.[33]

The explosion and destruction of an 11A511 booster, a Soyuz spacecraft, and the pad at site 31 significantly delayed any hope of mounting an early piloted Soyuz mission. Another automated Soyuz flight was inserted into the schedule, to be carried out on January 15, 1967, from the other remaining pad at site 1. Mishin had ordered re-equipping one of the piloted versions for the solitary robotic mission. The piloted mission was postponed to March—a delay accounted for by the time needed to transform the pad at site 1 to support dual Soyuz launches. In the meantime, on December 21, Kamanin sent the eight primary and backup cosmonauts for

32. Ibid.; Nikishin, "Soviet Space Disaster."
33. Semenov, ed., Raketno-Kosmicheskaya Korporatsiya, pp. 179–80.

the first mission, who had been intensively training through November and December, on a short vacation.[34] The year would end without a single Soviet piloted flight, the first such year since crewed spaceflight was inaugurated by Gagarin in 1961.

The next Soyuz spacecraft, a passive 7K-OK, vehicle no. 3, was prepared for its two-day mission in late January 1967. The State Commission met on January 19 in Moscow before flying to Tyura-Tam starting January 23.[35] Mishin was evidently ill for the two weeks preceding the launch, set for February 6, and was not present at many of the technical meetings. Due to minor technical reasons, the launch was delayed exactly twenty-four hours, and the vehicle lifted off successfully from the pad at site 1 at 0620 hours Moscow Time on February 7, 1967. Initial orbital parameters were 170 by 241 kilometers at a 51.7-degree inclination. TASS announced the flight as Kosmos-140, another in a long series of nondescript generic satellites with no particular mission. One of the unusual payloads aboard the ship was a cryogenic superconducting magnet on board for the analysis of charged particles. The Soviets later claimed that this was the first such instrument launched into space to study cosmic rays.[36] Communications were interrupted briefly during the powered ascent, but they were restored once in orbit, which was once again lower than intended because of the less-than-nominal booster performance.

Trouble began to appear on the fourth orbit. The vehicle failed to respond to a command to orient itself to turn the solar panels to face the Sun to recharge the on-board batteries. The astro-orientation sensor system used for this maneuver had evidently malfunctioned. Worse, propellant levels in the attitude control system had dropped to 50-percent levels during this test. After anxious consultations, the State Commission decided to raise the orbit and try one more time to test the sensor system, which used the 45K solar-stellar sensor. On the twenty-second orbit, the Soyuz main engine fired for fifty-eight seconds, but the spacecraft failed to respond to the "spin up to the Sun" command, and all the propellant in the main attitude control system was spent. By the end of the day, commission members were looking to terminate the flight early. Once again, most members believed that the failures on Kosmos-140 were only in systems that had duplicates for manual control, such as "spinning up" and the astro-orientation system. All of these malfunctions could have been compensated by cosmonauts.[37] The remaining systems such as life support, the main engine, thermal control, and so on, worked without problems.

The State Commission decided to use the ionic sensor system of orientation to posit the vehicle in the correct attitude prior to retrofire. The designer of the system, TsKBEM Department Chief Raushenbakh, had little confidence in the device, because he believed that the main engine might misfire as a result of exhaust, which could disorient the ionic sensors. Luckily for everyone, the system worked without a flaw, and the descent apparatus of Kosmos-140 began its reentry.

Following deorbit, the search-and-rescue service received faint signals from the descent apparatus, which were evidently originating from the Aral Sea, far west of the intended landing site. It was apparent by then that the capsule had automatically changed its landing profile from a guided reentry to a ballistic return. About four hours after landing, searchers discovered the descent apparatus eleven kilometers from Cape Shevchenko, lying on an iceberg in the Aral Sea.

34.   Mitroshenkov, Zemlya pod nebom, p. 397. Among the cosmonauts training for 7K-OK missions by late December were the eight men for the first mission (Bykovskiy, Gagarin, Gorbatko, Khrunov, Komarov, Kubasov, Nikolayev, and Yeliseyev), as well as four other cosmonauts training for future missions (Beregovoy, Makarov, Shatalov, and Volkov).

35.   Ibid., pp. 399–400.

36.   V. P. Glushko, ed., Kosmonavtika entsiklopediya (Moscow: Sovetskaya entsiklopediya, 1985), pp. 201–02; Yu. A. Mozzhorin, ed., Kosmonavtika (Moscow: Mashinostroyeniye, 1981), p. 446.

37.   Kamanin, "A Goal Worth Working for" (English title), Vozdushniy transport 46 (1993): 8–9.

It was the first sea landing for a Soviet piloted vehicle. Unfortunately, soon after the rescue teams discovered the capsule, it sank through the ice to a depth of about ten meters. It seems that the capsule had crashed through the iceberg and floated in the resulting hole until it became water-logged and simply sank. Engineers back in Moscow were naturally alarmed by the news, because the descent apparatus had been repeatedly tested for floatation in case of a water landing.

The recovery of the capsule proved to be extremely difficult, and the Air Force had to call in a team of divers. Helicopters were not able to lift the capsule because it was too heavy. With much difficulty, an Mi-6 helicopter managed to accrue sufficient horizontal velocity to drag the thing the three kilometers back to the shore. In their postflight analysis, engineers discovered a thirty- by ten-millimeter hole at the center portion of the bottom of the vehicle, which was sufficient for loss of pressure and the subsequent sinking. The investigation showed that the hole was the result of an infringement of the unity of the heat shield, which had been cast off. The heat shield itself had a maintenance hole with a plug attached with special glue for a thermal gauge pipe. The plug was incorrectly mated to the heat shield, resulting in a chain of events that led to the hole in the spacecraft. If a crew had been on board, they would have died, since Soyuz crewmembers would not be wearing spacesuits during reentry. To address the problem, engineers eliminated the plug completely from the heat shield, and they also made the heat shield a monolithic structure instead of being assembled piece by piece. In addition, all "suspect" areas of the heat shield were reinforced with extra material as a cautionary measure. At a meeting on February 16, Mishin and Bushuyev reassured the State Commission that the necessary measures would be carried out to preclude such an accident from happening again.[38]

Soyuz I

From an outsider's perspective, the natural course of action for the State Commission would have been to add another precursor Soyuz mission to the schedule. The two spacecraft flown in 1966 and 1967 had significant problems, primarily in their reentry phase, and certainly there would have been the need to verify the operation of all the components of reentry, such as the heat shield, parachute system, reentry orientation systems, and so forth. Despite the three attempts to launch the 7K-OK Soyuz, Mishin and his engineers recovered only a single descent apparatus after a space mission—one whose thermal protection system had a catastrophic failure. This is not to say that Mishin did not undertake a thorough analysis of the situation. The results of the three Soyuz attempts were the subject of intense discussion; the main decision for the engineers was whether to carry out another automated mission or whether to go directly to a piloted mission. Deputy Chief Designers Konstantin D. Bushuyev and Yakov I. Tregub of TsKBEM led this analysis in February and March 1967. Mishin invited a host of representatives from all organizations involved in the Soyuz program to hear their individual assessments of the status of their particular system and its potential readiness for a piloted flight. Remarkably, most of the other designers and engineers recommended crewed flight. Among the dissenters was TsKBEM Department Chief Ivan S. Prudnikov, who based his objections on the insufficient testing of the new, improved heat shield. The majority of engineers, however, expressed confidence in the work of the heat shield.[39]

38.   Ibid.; Semenov, ed., *Raketno-Kosmicheskaya Korporatsiya*, p. 180; Nikishin, "Soviet Space Disaster"; G. Salakhutdinov, "Once More About Space" (English title), *Ogonek* 34 (August 18–25, 1990): 4–5.

39.   Semenov, ed., *Raketno-Kosmicheskaya Korporatsiya*, pp. 180–81.

On March 25, 1967, Chairman Smirnov's Military-Industrial Commission met to discuss the preparations for the mission. Representing the State Commission, five men spoke on the flight, including Chairman Kerimov, Mishin, and Kamanin.[40] Smirnov asked several questions, including: "Do you think the equipment will work smoothly?" Kamanin replied:

Three launches of Soyuz spaceships and the completion of all ground tests have made us confident that the flight will be successful, although at one point some of the cosmonauts had certain doubts about the ship's bottom. We know that following the burn-out of the bottom of ship no. 3, the Central Design Bureau of Experimental Machine Building has worked hard to reinforce it. Chief Designer Mishin has said on more than one occasion that now there should be no doubts about the bottom. We believe Mishin.[41]

Kamanin introduced all the cosmonauts preparing for the flight: the eight primary and backup crew members—Bykovskiy, Gagarin, Gorbatko, Khrunov, Komarov, Kubasov, Nikolayev, and Yeliseyev—as well as four additional understudies who were expected to fly a subsequent Soyuz mission after finishing their training on June 1.[42] Although there was no formal decision on the primary crew, Komarov (for Soyuz 1) and Bykovskiy, Yeliseyev, and Khrunov (for Soyuz 2) were the leading candidates. Mishin personally met with Ustinov two days later to discuss the flight, setting in motion a series of events that would cripple the Soviet space program.[43]

The decision to move ahead with the docking mission has been obfuscated and mired in controversy and speculation for thirty years. One TsKBEM engineer, who later emigrated to the United States in the 1970s, added to the rumor mill by recalling that:

The management of the Design Bureau knew that the vehicle had not been completely debugged; more time was needed to make it operational. But the Communist Party ordered the launch despite the fact that four preliminary launches had revealed faults in coordination, thermal control, and parachute systems. It was rumored that Vasiliy Mishin, the deputy chief designer who headed the enterprise after Korolev's death in 1966, had objected to the launch.[44]

There was clearly much political pressure from Brezhnev and Ustinov to get the flight off the ground. It had been almost two years since a piloted Soviet spaceflight, while the Americans had flown ten Gemini missions. In addition, May Day, one of the most important holidays in Soviet culture, was imminent, and there is reason to believe that the Soyuz flight was timed to roughly coincide with the anniversary. A simple automated flight of the vehicle

40.   Kamanin, "For Him, Living Meant Flying." The other two speakers were Maj. General A. G. Karas (Commander of the Central Directorate of Space Assets) and Maj. General A. I. Kutasin (the Air Force head for rescue and recovery operations).

41.   *Ibid., p. 8.*

42.   The four understudies were probably Beregovoy, Makarov, Shatalov, and Volkov.

43.   There was apparently also a State Commission meeting the same day. See Mitroshenkov, *Zemlya pod nebom,* p. 404.

44.   Victor Yevsikov, *Re-Entry Technology and the Soviet Space Program (Some Personal Observations)* (Falls Church, VA: Delphic Associates, 1982), p. 4. See also Dmitriy Payson, "Eternal 'Soyuz'—Today Marks the 25th Anniversary of the First Docking in Orbit" (English title), *Nezavisimaya gazeta,* January 15, 1994, p. 6, in which the author states, "The Soyuz was hastily prepared for launching and it was launched (an unprecedented act!) despite the categorical refusal of Vasiliy Mishin. . . ."

would have hardly mattered for such an auspicious occasion. When asked in an interview in 1990 whether the he had been pressured to carry out the mission, Mishin replied:

> Truly, there never was a time when we worked in peace, without being hurried or pressured from above. The unskilled, totally bewildered, high-ranking bureaucrats believe that they are fulfilling their duties if they are shouting "Let's go, let's go!" at people who don't even have time to wipe the sweat off their brows.[45]

Asked about the possibility that his deputies may have committed errors during the preparations, Mishin emphasized:

> No, the deadlines and the pressure from above have nothing to do with that. Not a single supervisor for any of the Soyuz systems would have given the "go-ahead" to the flight if he were not certain of that system's satisfactory operation.[46]

Ultimately, it was a decision motivated by the apparently huge lag in piloted space exploration accrued through 1965 and 1966 as compared to the United States. Throughout 1966, both the political and technical managers of the Soviet space program banked on the inauguration of the Soyuz program to take some steam out of U.S. space achievements, which finally seemed to have gained momentum after years of humiliation. When Mishin, Bushuyev, Tregub, and others recommended a go-ahead with the flight, clearly they did not have full confidence in their ship. Korolev, of course, had also taken his own risks, particularly with the two Voskhod missions, which were highly risky endeavors. The EVA mission of Voskhod 2, for example, was not preceded by a successful precursor mission. But Soyuz was a far more complex spacecraft; it was a completely untested quantity in terms of crewed operations. The Soyuz mission was a gamble of extraordinary levels.

The intensive discussions on Soyuz in February and March 1967 were mirrored by the slowly increasing number of rumors emanating from "unofficial" sources from the Eastern bloc that a Soviet space spectacular was imminent. On March 7, a commentator on Prague Radio reported that "much more complicated manned operations in Earth orbit are about to begin which have taken over two years to prepare."[47] Just two days later, Lt. General Kamanin, in a long interview with Warsaw Radio, said that piloted flights would begin again that spring. He added that the Soviets were not locked onto any particular date and that the flight would come only when they were assured of success. He implied that the deaths recently of the three American astronauts were the result of unnecessary haste in the U.S. space program, a factor absent in the Soviet space program.[48]

After an unusually grueling training program involving countless hours in simulators on the ground, the eight primary and backup cosmonauts for the mission took their final exams for the flight on March 30, and all passed with excellent marks. On April 6, the men visited the depths of the Kremlin to meet with high Central Committee officials and receive wishes of good luck. The same day, Kamanin, accompanied by veteran and rookie cosmonauts, flew into Tyura-Tam.

---

45.   Salkahutdinov, "Once More About Space," p. 4.
46. · Ibid.
47.   Soviet Space Programs, 1966–70: Goals and Purposes, Organization, Resources, Facilities and Hardware, Manned and Unmanned Flight Programs, Bioastronautics, Civil and Military Applications, Projections of Future Plans, Attitudes Toward International Cooperation and Space Law, prepared for the Committee on Aeronautical and Space Sciences, U.S. Senate, 92d Cong., 1st sess. (Washington, DC: U.S. Government Printing Office, December 1971), p. 364.
48.   Ibid.; Peter Smolders, Soviets in Space (New York: Taplinger Publishing Co., 1973), p. 150.

The crews of Soyuz 1 and Soyuz 2 present themselves before the State Commission in front of the launch pad in April 1967. In the foreground from left to right are the primary crew of Vladimir Komarov, Valeriy Bykovskiy, Yevgeniy Khrunov, and Aleksey Yeliseyev (in civilian clothes) and the backup crew of Yuriy Gagarin, Andrian Nikolayev, Viktor Gorbatko, and Valeriy Kubasov (also in civilian clothes). Chief Designer Valentin Glushko is visible in the background between Yeliseyev and Gagarin. (copyright Christian Lardier)

Komarov followed on April 8 and Gagarin on April 14.[49] For many, it was the first time that they had spent the celebrated "Cosmonautics Day," the anniversary of Gagarin's pioneering flight, at the Baykonur Cosmodrome.

There was a meeting of the State Commission on April 14 at which the members decided to begin fueling the two launch vehicles and spacecraft. Assuming an eight-day period for complete preparation, the first launch was tentatively set for April 24–25. Mishin telephoned both Ustinov and Brezhnev later; Ustinov evidently expressed some anxiety over the impending flight. The mission would be inaugurated by the launch of the active 7K-OK Soyuz 1, on the first day, with Komarov. The following day, as the ship was flying over Tyura-Tam, the passive 7K-OK Soyuz 2 would be launched with Bykovskiy, Yeliseyev, and Khrunov. The two spacecraft would dock on the very first orbit of Soyuz 2; it would be the first docking of two piloted spaceships. After docking, Yeliseyev and Khrunov would exit from their depressurized living compartment and crawl over to the depressurized living section of Soyuz 1. Following the transfer, Soyuz 1, with a crew of three, would return the following day. Soyuz 2, with a crew of one, would also return that same day. Apart from the dramatic nature of the flight, the mission had significant value for future operations in the N1-L3 project as well as possible Earth-orbit rendezvous profiles for the circumlunar L1 program.

---

49.   Mitroshenkov, *Zemlya pod nebom*, pp. 405–06. Among the cosmonauts accompanying Kamanin were rookie G. T. Beregovoy and veterans A. A. Leonov, and P. R. Popovich. Note that another source states that the primary and backup crews arrived at the launch site on April 10, 1967. See Grigoriy Reznichenko, *Kosmonavt-5* (Moscow: Politicheskoy literatury, 1989), p. 97.

The EVA itself had been the subject of much discussion for months. In November 1966, two of Mishin's Deputy Chief Designers, Sergey O. Okhapkin and Pavel V. Tsybin, proposed having one cosmonaut move away from the docked vehicles to a distance of about ten meters to photograph the complete complex and the second cosmonaut. After opposition by some of the cosmonauts, TsKBEM opted for the use of a ten-meter boom to ensure that the vehicles would be photographed—a problem entrusted to Deputy Chief Designer Bushuyev.[50] By the time of the actual mission, Bushuyev had abandoned the idea, possibly projecting its use on a later Soyuz docking mission. The cosmonauts on this first flight would simply crawl from ship to ship. There were other changes to the spacewalk schedule. TsKBEM engineers had apparently designed the hatch on the Soyuz ship with too small a diameter (0.66 meters). This would be barely enough for a spacesuited cosmonaut to egress from the ship and make it all but impossible for the men to get back into the second ship. Mishin and his boss, Deputy Minister Litvinov, were categorically opposed to redesigning the hatch to a larger size for the first few Soyuz vehicles, believing that a redesign would delay the initial launches by months. Instead, at a meeting on August 4, 1966, attended by Chief Designers Mishin (spacecraft) and Severin (spacesuits), officials decided to move the *Yastreb* EVA suit backpacks from the cosmonaut's back to the waist. Mishin promised that future Soyuz ships, beginning from vehicle no. 8, would have larger hatches.[51] Such changes added an extra level of tension to an already hurried situation. Just a week prior to the launch, on April 15, Kamanin wrote in his journal:

> I am personally not fully confident that the whole program of flight will be completed successfully, although there are no sufficiently weighty grounds to object to the launch. In all the previous flights we believed in success. Today there is not such confidence in victory. The cosmonauts are prepared well, and the ships and the instruments have gone through hundreds of tests and verifications, and all seems to have been done for successful flights, but [still] there is no confidence. This can perhaps be explained by the fact that we are flying without Korolev's strength and assurances; we were spoilt by Korolev's optimism.[52]

The fueling of the Soyuz 1 launch stack began at 2300 hours Moscow Time on April 15. The morning of April 17, the cosmonauts attended a final five-hour class under Raushenbakh's supervision to study once again the modes of docking, orientation, and so on. In the afternoon, Mishin arrived to talk personally with the crews about various portions of the mission. Even at this late point, there seems to have been some disagreement over which mode of operation to use for the crucial docking maneuver. Mishin favored a completely automatic docking, believing in the infallibility of the ship, but he was opposed by Kamanin and some of the cosmonauts, including Komarov and Gagarin. For more than two years, Bykovskiy, Gagarin, Komarov, and Nikolayev, the four commanders, had been training for an automatic approach followed by a manual docking and were reluctant to let automation do the whole thing. At the meeting, Komarov argued that the Igla system could automatically bring the active vehicle within 200 to

50. Kamanin, *Skrytiy kosmos: 1964–1966*, pp. 396–97, 400.

51. *Ibid.*, pp. 355, 361. Present at the August 4 meeting were S. M. Alekseyev (Chief Designer of KB Zvezda), K. D. Bushuyev (Deputy Chief Designer of TsKBEM), N. P. Kamanin (Space Aide to the Commander-in-Chief of VVS), V. A. Kazakov (Deputy Minister of MAP), V. M. Komarov (Cosmonaut of TsPK), V. Ya. Litvinov (Deputy Minister of MOM), V. P. Mishin (Chief Designer of TsKBEM), G. I. Severin (Chief Designer of KB Zvezda), N. S. Stroyev (Director of the M. M. Gromov Fight-Research Institute), and P. V. Tsybin (Deputy Chief Designer of TsKBEM).

52. Lev Kamanin and Aleksandr Nemov, "Komarov's Star: The Tragic Details of the Testing of the 'Soyuz-1' Space Ship" (English title), *Poisk* 5 (June 1989): 4–5.

300 meters of the passive vehicle, following which he could manually dock the two spacecraft. Mishin listened to their arguments and delayed a final decision on the matter until the following day. By the end of the day, the fueling of the Soyuz 1 launcher had concluded while the fueling of the Soyuz 2 booster had begun. Thus, the launching was informally set for April 24–26.[53]

The Council of Chief Designers met on the morning of April 18 to discuss the docking issue. State Commission Chairman Kerimov supported an automatic approach via the Igla to fifty to seventy meters, followed by manual docking, although many engineers still defended the fully automatic variant. TsKBEM Department Deputy Chief and cosmonaut Feoktistov mediated the issue and argued in favor of the semi-automatic profile, and the council accepted his recommendations. Later in the day, Feoktistov discussed various contingency measures for emergency situations with the cosmonauts. The final State Commission meeting prior to launch took place on April 20 at site 2. The launch of Soyuz 1 was set for 0335 hours Moscow Time on April 23, while the launch of Soyuz 2 was set for 0310 hours Moscow Time the following day. All the Chief and Deputy Chief Designers confirmed that the launch vehicles, space ships, and support services would be completely ready to accomplish the launch on time. The commission also formally approved the crews for the two missions and gave the official go-ahead for the flight.[54]

On April 22, the 11A511 rocket was already at the launch pad at site 1. In the late morning, the primary and backup crews had their customary meeting with the launch command and industrial representatives. A number of chief designers met with the crews and informed them that after the Soyuz 1 launch, there would only be two reasons for a postponement or cancellation of the Soyuz 2 launch: if there was a failure in the Igla system or if there was a low charge in the solar batteries on Soyuz 1. Kamanin counseled Komarov that the most important factor on the mission would be safety; in the case of any malfunctions, there would be no need to proceed with the complicated docking procedure. Later in the day, Komarov attended a press conference for journalists with special access. Komarov dedicated his flight to the fiftieth anniversary of the Bolshevik Revolution.[55]

A final meeting of the State Commission, lasting forty-five minutes, began one-half hour before midnight and concluded with recommending a full go-ahead for the flight. Komarov woke up about two hours after midnight, and doctors attached sets of medical sensors to his body. He was dressed in a plain light woolen gray suit and a blue jacket. At 0300 hours, he arrived at the pad to give a short speech to State Commission Chairman Kerimov and then embraced senior officials goodbye. Mishin, Kamanin, and Gagarin accompanied him to the rocket; Gagarin went up with him all the way to the top of the rocket and remained there until the hatch closed.

There were no anomalies prior to launch. The spacecraft, 7K-OK no. 4, lifted off exactly on time at 0335 hours Moscow Time on April 23, 1967, with its sole passenger, forty-year-old Colonel Engineer Vladimir M. Komarov. He was the first Soviet cosmonaut to make a second spaceflight. It took 540 seconds for the ship to successfully enter orbit. The official Soviet news agency TASS released a brief statement, calling the flight *Soyuz 1*, and announced the orbital parameters and some vague objectives of the program. Characteristically, there was no mention of the impending Soyuz 2 mission. Rumors in the West, however, had reached crescendo proportions, some clearly indicating that a docking with a second ship was planned.[56] Cosmonaut Popovich

53.   *Ibid.* There was a minor delay on April 18, when a valve on one of the systems for loading nitric acid into the spacecraft failed. The problem was fixed without much delay.

54.   *Ibid.*

55.   *Ibid.*; *Soviet Space Programs, 1966–70,* p. 18; Mitroshenkov, *Zemlya pod nebom,* p. 407.

56.   For a summary of these rumors, see James Oberg, "Soyuz-1 Ten Years After: New Conclusions," *Spaceflight* 19 (May 1977): 183–89.

informed Komarov's wife, Valya, that her husband was in orbit about twenty-five minutes after launch. She told reporters that "my husband never tells me when he goes on a business trip."[57]

For the first time on a Soviet piloted mission, the Chief Operations and Control Group—that is, the flight control team—was located at the Scientific-Measurement Point No. 16 at Yevpatoriya in Crimea. A team of twenty controllers, including TsKBEM Deputy Chief Designers Chertok and Tregub and Department Chief Raushenbakh, assisted Chief Operations and Control Group Chief Colonel Pavel A. Agadzhanov, the "flight director." The flight control team would actively communicate with the spacecraft in orbit while maintaining continuous contact with the State Commission, all of whose members remained behind at site 2 at Tyura-Tam. Additional ballistics support was provided by NII-4's military control center in Moscow.

The initial incoming report from telemetry streams from two ground stations indicated that the Soyuz spacecraft's left solar panel had not opened upon entering orbit. As Agadzhanov's team examined the data, they found other anomalies. A backup antenna in the telemetry system was inoperable and the 45K solar-stellar attitude control sensor's optical surface had probably been contaminated by engine exhaust. While the antenna was a minor annoyance, the sensor malfunction was serious because without it, Soyuz 1 would be unable to orient the ship properly to change orbital parameters in preparation for the rendezvous and docking. Telemetry indicated that current orbital parameters were 196.2 by 225 kilometers at a 51° 43' inclination. It was on the second orbit that controllers first established stable communications with Komarov on ultra-shortwave frequencies; for reasons unknown, the shortwave system was inoperable. Komarov calmly reported:

> I feel well. The parameters of the cabin are normal. The left solar battery has not opened. There's been no spin toward the Sun. The "solar current" is 14 amperes. Shortwave communications are not working. Attempted to manually perform spinning. Spinning did not occur, but pressure in the [orientation engines] dropped to 180.[58]

Unconfirmed reports suggest that Komarov even tried to knock the side of the ship to jar open the recalcitrant panel. Already, the situation had deteriorated dramatically. Because one solar panel was not operative and the ship had failed to automatically orient the other toward the Sun, power on board the ship was far below normal. Power experts at Yevpatoriya had calculated that the buffer batteries could operate with the current levels of power up to the seventeenth orbit, after which Komarov could use reserve batteries for up to two more orbits. This meant that Soyuz 1 could safely operate for about a day, significantly less than the three days needed for a docking mission. In the meantime, Agadzhanov told Komarov to shut down nonessential systems and to try at all costs to orient the right panel toward the Sun. On the third orbit, Komarov told ground control that the left panel was still folded against the ship and that the vehicle had not oriented toward the Sun. Current had stabilized at a low fourteen amperes, far below that required for a nominal flight. The 45K attitude control sensor was still inoperative. Despite the troubles, the State Commission believed that the orientation problem would be solved, and it recommended that preparations for the launch of Soyuz 2 be continued. Kamanin meanwhile sent Gagarin directly to Yevpatoriya to assist the Chief Operations and Control Group in its operations.[59]

57. Smolders, Soviets in Space, p. 156.
58. Kamanin and Nemov, "Komarov's Star"; Chertok, Rakety i lyudi: goryachiye dni kholodnoy voyny, pp. 444–45.
59. Kamanin and Nemov, "Komarov's Star"; M. F. Rebrov, Kosmicheskiye katastrofy (Moscow: IzdAT, 1993), p. 27; Chertok, Rakety i lyudi: goryachiye dni kholodnoy voyny, pp. 445–46.

On the fifth orbit, Komarov attempted to manually orient the ship by using Earth's horizon to position the vehicle at correct attitude, but he found it difficult to do so, partly because it was difficult to keep a target hold on the moving Earth. In addition, his attempt seems to have been overruled by the on-board control system. Apart from the astro-orientation system, which used the 45K solar-stellar sensor, and the manual orientation system, the vehicle was also equipped with ionic sensors. The use of these, however, also met with little success on the fifth orbit. From the seventh to the thirteenth orbits, Komarov was outside radio visibility using ultra-shortwave communications because the spacecraft would pass over the Atlantic and the American continent. As planned earlier, Komarov was ordered to sleep during this period, while consultations among Moscow, Tyura-Tam, and Yevpatoriya continued throughout the day at a feverish pitch.

Most of the senior members of the State Commission, including Chairman Kerimov, Keldysh, and Kamanin, recommended the immediate postponement of the Soyuz 2 launch and the return of Komarov at the earliest possible opportunity—that is, the seventeenth orbit. Incredibly, Mishin still had hope and believed that the commission should make a final decision on the thirteenth orbit, once Yevpatoriya reestablished contact with Komarov. There was even a plan to have the two EVA cosmonauts, Yeliseyev and Khrunov, manually unfurl the jammed solar panel during their spacewalk from one ship to the other. But on the thirteenth orbit, Komarov reported that his second attempt to use the ionic orientation system had failed.[60] He added that the left solar panel was still jammed; current on the ship had remained static at twelve to fourteen amperes. Mishin later recalled that "because of the emergency, the shortage of power on board caused a chain of problems [including] a change in the temperature conditions."[61] Immediately, the State Commission unanimously canceled the Soyuz 2 launch. Evidently, the Soyuz 2 cosmonauts were bitterly disappointed, blaming the commission for "excessive caution and indecisiveness."[62]

The problem at that point was how to return the spacecraft from orbit, nominally on the seventeenth orbit, but with the eighteenth and nineteenth orbits as reserve. Agadzhanov's team at Yevpatoriya considered the matter carefully. There were three main failures on board Soyuz 1: the unopening of the left solar panel, the failure of the ionic orientation system, and the mal-function of the 45K solar-stellar attitude control sensor. The recalcitrant solar panel not only deprived the spacecraft of much-needed power, but also caused an asymmetry in the ship, which prevented the open solar panel from facing the Sun. Because of this mechanical imbalance, engineers were all but sure that all of Komarov's efforts to spin the ship in the direction of the Sun would fail and, in fact, would simply waste the precious propellant in the orientation engine system. If there was too little fuel in this system, then during retrofire, Komarov might not be able to compensate for moments arising from the mass displacement because of the single opened panel.

The Soyuz had three orientation systems. If all three orientation systems were inoperative, it would be practically impossible for Komarov to return his ship. With an incorrect attitude, Soyuz 1 would either burn up in the atmosphere or fly into a higher orbit. The ionic orientation system had already failed to perform twice. Engineers also believed that the system would be

60.   Kamanin and Nemov, "Komarov's Star"; *Russian Space History, Sale 6516* (New York: Sotheby's, 1993), description for Lot 46; Reznichenko, *Kosmonavt-5*, p. 97. One source suggests that Komarov may have actually tried to fire his main engine to change his orbit. In Nikishin, "Soviet Space Disaster," the author notes: "The first orbital correction was widely off mark because the maneuver thrusters' exhaust affected the operation of the attitude control system's ion sensors."
61.   Salakhutdinov, "Once More About Space."
62.   Rebrov, *Kosmicheskiye katastrofy*, pp. 27–28.

unreliable during the morning hours when the return was planned because of ion pockets, which could disrupt the work of the sensors. As for the 45K solar-stellar sensor, it was not functioning at all. This left manual orientation, which was working, but as Komarov reported, it was extremely difficult to manipulate in Earth's shadow because it would be difficult to locate Earth's horizon. Normally, using manual orientation, the cosmonaut would cross Earth's terminator into lighted areas. In Komarov's case, with a reentry at the earliest opportunity, he would still be in the dark.[63]

Time was already running short for Komarov. If he was to perform a successful reentry on the seventeenth orbit, then Agadzhanov's team needed to transmit a precise set of commands to Komarov on the sixteenth orbit. It was already the fifteenth orbit, and officials at both Yevpatoriya and Tyura-Tam were still arguing over a proper choice of orientation for reentry. It had been almost twenty-four hours since the launch, and not one member of either the State Commission or the Chief Operations and Control Group had slept. In their state of alarm, members continuously violated established rules to communicate only via secret channels between the two centers. On the fifteenth orbit, Komarov reported that he believed that the ionic system and its associated attitude control engines were in working order. Based on his recommendations and assessment from data on the ground, the State Commission recommended that the ship be landed on the seventeenth orbit using the automatic ionic orientation with the backup set of orientation engines. Agadzhanov, Raushenbakh, and Chertok carefully checked over the set of instructions that Gagarin personally transmitted to Komarov. In the final seconds before loss of contact, Mishin and Kamanin both wished Komarov good luck.[64]

At the appointed time, Soyuz 1 initiated the reentry sequence. The main engine was supposed to fire for deorbit at 0256 hours, 12 seconds Moscow Time on April 24, but nothing happened. Ballistics reports pouring into Yevpatoriya indicated that Soyuz 1's orbital parameters had remained the same. Once communication with Komarov was reestablished, the cosmonaut reported that the ion system seemed to have worked fine, but evidently, as the ship had crossed the equator, it had flown into an "ion pocket" in Earth's shadow where the concentration of the ions was less than what the sensors could detect. The ship's control system correctly issued a command to prohibit the firing of the retro-engine.[65] State Commission members decided to immediately begin preparations for another landing attempt on the eighteenth orbit. As the seventeenth orbit was ending, however, the flight control team did not have any new instructions ready to transmit to Komarov. Finally, the State Commission decided to land Komarov on the nineteenth orbit.

With the use of both the ionic and solar-stellar orientation systems out of the equation, the only remaining option was for Komarov to manually orient the ship prior to retrofire, but using a very complex series of operations in orbit. Komarov would have to orient the ship manually to Earth's horizon in the light portion of the orbit. Just before entering Earth's shadow, he would transfer attitude control to the spaceship's KI-38 gyroscope system. Once he was out of the shadow, he would check to see whether Soyuz 1 was still correctly oriented for retrofire. If not, he would once again take over manual control and issue all the commands to complete the retrofire sequence for a landing on the nineteenth orbit. It was an incredibly difficult task—one for which none of the cosmonauts had ever trained on the ground. One of the power specialists warned at the time that Komarov had one to two orbits at the most—that is, he might not have very many more chances to attempt reentry. Gagarin once again transmitted the new set of instructions to the Soyuz 1 cosmonaut. Komarov seemed calm and agreed to carry

63. Chertok, *Rakety i lyudi: goryachiye dni kholodnoy voyny*, pp. 446–47; Kamanin and Nemov, "Komarov's Star."
64. Chertok, *Rakety i lyudi: goryachiye dni kholodnoy voyny*, p. 447.
65. *Ibid.*; Nikishin, "Soviet Space Disaster"; Kamanin and Nemov, "Komarov's Star."

out all the operations on time, which would lead to a 150-second retrofire with engine ignition at 0557 hours, 15 seconds on April 24.

Komarov performed skillfully and carried out his assigned program almost to the letter. He replied through the increasing static, "The engine worked for 146 seconds. Switch-off occurred at 0559 hours 38.5 seconds. At 0614 hours 9 seconds, there was the command 'Accident-2'."[66] The "Accident-2" signal threatened to give controllers a collective heart attack, but Raushenbakh gathered his resolve and explained to the team not to worry. The attitude control system had been unable to handle the strong moments because of the asymmetry of the vehicle, and the gyroscopes had issued the "Accident-2" command after the spacecraft deviated from its set angle by eight degrees. That only meant that instead of a guided reentry, Komarov would perform a direct ballistics return. All other parameters, such as the length of the burn, were well within range for a successful reentry.

At Tyura-Tam, the members of the State Commission were huddled together on the second floor of the administrative portion of the huge assembly-testing building at site 2. Journalists at the launch site were excluded from the meeting but were able to overhear voices. Cosmonaut Leonov served as an intermediary to brief reporters on the ongoing situation. Mishin, Kerimov, Keldysh, Minister Afanasyev, and Air Force First Deputy Commander-in-Chief Marshal Rudenko all exchanged brief comments as they heard Komarov's report. About fifteen minutes after retrofire, there was the expected break in communications as Komarov's capsule entered an ionization layer. A few minutes later, Komarov's voice cut through the radio silence; he evidently sounded "calm, unhurried, without any nervousness."[67] By this time, Kamanin and a group of Air Force officers had already taken off from Tyura-Tam in an Il-18 aircraft to head for the projected landing range—the reserve landing area for the mission, about sixty-five kilometers east of Orsk, far west of the planned site for a guided reentry. According to ballistics data, Soyuz 1 had landed at 0624 hours Moscow Time.

Once search services determined the landing site, the reserve search-and-rescue service at the town of Orenburg was called into operation to locate the descent apparatus. It was a beautiful and sunny morning at the landing site, and visibility was evidently very good. Members of the rescue service recalled that:

> The commander of one of the An-12 search aircraft reported to the helicopter commander that he could see Soyuz-1 in the air. All the group members were immediately at the windows. But we couldn't see the reentry vehicle descending in the air. The helicopter commander began a rapid descent. Then the helicopter turned sharply to the right, and many of the group members saw the reentry vehicle down in a green field. It was lying on its side, and the parachute could be seen right next to it. And then the soft-landing engines kicked in. That alarmed the specialists on the helicopter, because the engines were supposed to switch on just before the landing of the reentry vehicle, right above the ground.[68]

The first helicopter landed seventy to 100 meters from the capsule, which was surrounded by a cloud of black smoke. The fire inside the vehicle was still very intense, while the bottom of the ship, where the soft-landing engines were, had completely burned through. Witnesses claimed that streams of molten metal were falling on the ground. Along with foam fire extinguishers, they used dirt around the ship to temper the fire: "The vehicle was completely

---

66. Chertok, *Rakety i lyudi: goryachiye dni kholodnoy voyny*, p. 448.
67. Rebrov, *Kosmicheskiye katastrofy*, p. 28.
68. Iosif Davydov, "How Could That Have Been?: Slandered Space" (English title), *Rossiyskaya gazeta*, June 11, 1992, p. 5. Author's emphasis.

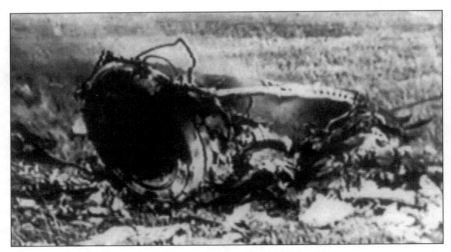

This shows the wreckage of the Soyuz I descent apparatus immediately after the crash. Cosmonaut Vladimir Komarov's body was still buried within the wreckage at the time of this photograph. (Rudy, Inc., via Quest magazine)

destroyed while the fire was being extinguished, and the spot looked like a small earthen mound, beneath the peak of which was the cover for the hatch-crawlway."[69]

The rescue service originally communicated on an open channel with ground controllers at Moscow, Tyura-Tam, and Yevpatoriya, although they spoke in code. Once the rescuers had seen the ship on the ground and on fire, one of the pilots had cryptically reported, "I see the object, the cosmonaut needs urgent medical attention out in the field."[70] At that point, perhaps to preclude rumors, the search service terminated all communications with the three control centers. For the next few hours, there was no news from the site as Mishin, Kerimov, and others anxiously waited for any scrap of news.

Kamanin, meanwhile, landed at Orsk airport about two hours after the Soyuz I impact, fully expecting to meet Komarov there. Once out of his plane, he was told that the ship had landed sixty-five kilometers away, that it was burning, and that the cosmonaut had not been found. Another unconfirmed report came in that Komarov was wounded but alive in a hospital at a town three kilometers from the landing site. The Air Force general decided to go directly to the landing site first, although he had been explicitly ordered to wait for a call from Moscow to report on Komarov's status. Back at the three control centers, there was complete confusion. Ustinov in Moscow was frantic for information. He began calling up Party secretaries in Orenburg and Orsk on special lines, but could not reach anyone. Although the vehicle had landed at 0624 hours, Ustinov received no information on the state of the cosmonaut for the next three and a half hours.

When Kamanin arrived at the landing site, the Soyuz I descent apparatus was still on fire. He was not the first high-ranking space official on the scene. Academician Georgiy I. Petrov, the Director of the Space Research Institute of the Academy of Sciences, had arrived there first and was directing efforts to assess the situation. There was still no sign of the cosmonaut. Local

---

69.   *Ibid.*
70.   Nikishin, "Soviet Space Disaster." Note that in Semenov, ed., *Raketno-Kosmicheskaya Korporatsiya,* p. 181, the first detection is said to be have been from an Il-14 aircraft, while Nikishin, "Soviet Space Disaster," suggests that it was from a helicopter.

residents reported that the ship had fallen toward Earth at a great speed and that the parachute was turning and not filled up with air. They confirmed the observations of the search-and-rescue service that at the moment of landing, there were some explosions followed by the fire. Kamanin recalls:

> *A cursory examination of the ship convinced me that Komarov was dead and was still in the remains of what used to be his ship. I ordered to clear out the debris on the ground and search for Komarov's body. Simultaneously I sent one of the workers by helicopter, and others by automobile to the local hospital in order to verify the story of the injured cosmonaut. After an hour of excavations [that is, at around 0930 hours] we discovered the body of cosmonaut Komarov among the remains of the ship. . . .[71]*

Finding the body had been a difficult job. One of the rescuers recalled:

> *The group's physicians set to work—they shoveled away the top layer of dirt from the top of the mound from the hatch cover. After the dirt and certain parts of instruments and equipment were removed, the cosmonaut's body was found lying in the center chair. The physicians cleaned the dirt and the remnants of the burned helmet phone from the head. They pronounced the death to be from multiple injuries to the cranium, spinal cord, and bones.[72]*

Meanwhile, Kamanin flew back to Orsk and personally telephoned Central Committee Secretary Ustinov with the following short message:

> *I was at the location, cosmonaut Komarov has died, the ship burnt up. The primary parachute of the ship did not open, and the reserve parachute did not fill with air. The ship hit the ground at a speed of 35–40 meters per second; after impact there was an explosion of the braking engines and a fire started. I was not able to report on the fate of the cosmonaut earlier since nobody could see anything, and during that time we extinguished the fire in the ship by covering it with dirt. Only after carrying out excavations were we able to find Komarov's body.[73]*

At noon on April 24, Ustinov called Soviet General Secretary Brezhnev, who was at an international conference of communist parties in Czechoslovakia, with information on the accident. Ustinov also edited a TASS report, which was issued after a full twelve hours of silence from the Soviet press. The official line was that although the flight had been eventless until reentry, "when the main parachute was deployed at a height of 7 kilometers, the spaceship, according to preliminary reports, crashed at great speed as a result of the parachute cords getting entangled, [and] killed Komarov."[74]

In the early afternoon, State Commission members Kerimov, Keldysh, and Chief Designers Mishin, Tkachev, and Severin arrived at the impact point escorted by KGB agents. Soon, senior engineers from TsKBEM, including Deputy Chief Designer Tsybin and specialists involved in Soyuz development, arrived to catalog and inspect the entire landing area. Komarov's remains were taken in a coffin back to Moscow, arriving an hour after midnight on April 25. Aboard the aircraft were Keldysh, Kamanin, and the other cosmonauts who had trained for the mission:

71.   Kamanin and Nemov, "Komarov's Star," p. 5.
72.   Davydov, "How Could That Have Been?," p. 5.
73.   Kamanin and Nemov, "Komarov's Star," p. 5.
74.   Smolders, *Soviets in Space*, p. 159.

Bykovskiy, Gagarin, Gorbatko, Khrunov, Kubasov, Nikolayev, and Yeliseyev. They were met in Moscow at the airport by Komarov's widow Valentina Yakovlevna Komarova. His remains were then cremated and the urn placed in the Red Banner Hall of the Central House of the Soviet Army for mourners to pay homage. The next day, the Soviet Party and government gave him a state funeral with full honors, and his ashes, like Korolev's, were interred in the Kremlin Wall. In a grisly aside to his death, not all of Komarov's remains were found during the initial search, and a group of Young Pioneers, the equivalent of Boy Scouts in the Soviet Union, discovered additional remains that were later buried at the crash site itself. Reportedly, Party officials took great pains to hide this fact from the general public.[75]

The death of Vladimir Mikhaylovich Komarov was a catastrophic blow to the Soviet space program. Apart from the pure psychological cost of losing a cosmonaut on a space mission, the disaster immediately stopped all three major Soviet piloted space projects—the Soyuz, the L1, and the L3. Any hope of accomplishing a circumlunar flight by late 1967 was in great doubt, while landing a Soviet cosmonaut on the Moon by late 1968 was sheer fantasy at this point. The blow to morale was incalculable, not only to the design bureaus, institutes, and military units involved in the project, but also to the nation as a whole. It was bitter news to swallow that the first Soviet piloted spaceflight after two years had ended in tragedy, in the process losing perhaps the Soviet Union's most accomplished spacefarer. At the spot where Komarov landed, Party officials later collected the remaining tiny fragments of his last ship and erected a small hill. Sergey N. Anokhin, the famous Soviet test pilot, who at the time was the head of the testing department at TsKBEM, placed Komarov's officer's cap in the hill, after which a gun salute sounded out, paying tribute to what many considered a fallen hero of the Soviet Union.

All further piloted flights were indefinitely canceled at the time. On April 27, Ustinov met with the leading space industry representatives and established a special governmental commission headed by himself to determine the causes of the accident. This commission included seven subcommissions. One of them, to investigate the landing itself, was headed by the recently appointed Director of the M. M. Gromov Flight-Research Institute, Viktor V. Utkin, a respected aeronautical engineer. The commission included two representatives from TsKBEM, Chief Designer Mishin and Deputy Chief Designer Bushuyev. Soyuz 1 and 2 backup cosmonauts Gagarin and Bykovskiy also served as members.[76]

Utkin's subcommission finished its work, which included some experimental analyses, by June 20 and emerged with the cause of the accident: a release failure of the container block of the primary parachute. The parachute was packed in a container whose hatch was jettisoned, releasing a "braking" or drag parachute, slowing down the vehicle to a manageable forty meters per second, sufficiently slow to allow the primary parachute to fill up with air instead of shredding. The drag parachute itself was supposed to pull out the main parachute, but it did not do so because the latter had gotten jammed in the container. Under nominal circumstances, automated instruments on board the capsule would have detected an increase in velocity, discarded the primary drag and main parachutes, and activated the backup system. On Soyuz-1, once instruments detected the velocity increase, the capsule was unable to discard the primary chute because it was still stuck in the container. This meant that the primary drag chute was still deployed above the spacecraft. Once the single backup parachute was released, it was to have come out in the shape of a long, thin cylinder and then unfurl to its dome shape. In Komarov's case, the backup chute began to extend under the still attached drag parachute from

75.   Semenov, ed., *Raketno-Kosmicheskaya Korporatsiya*, p. 181; Kamanin and Nemov, "Komarov's Star"; Nikishin, "Soviet Space Disaster."

76.   Semenov, ed., *Raketno-Kosmicheskaya Korporatsiya*, p. 182; Chertok, *Rakety i lyudi: goryachiye dni kholodnoy voyny*, p. 453. For Gagarin, see Mitroshenkov, *Zemlya pod nebom*, p. 411. For Bykovskiy, see Reznichenko, *Kosmonaut-5*, p. 97.

the primary system, and it never filled with air. Without any means of braking, the ship plummeted and hit the ground at a velocity of 144 kilometers per hour (forty meters per second). An autopsy of Komarov confirmed that he died on impact with the ground and that the effects of the fire were secondary. Despite rumors to the contrary, Komarov did not cry or scream before the impact, although during the last seconds, he was surely aware that he had little chance to live.[77] Because of the rapid velocity of descent, the frontal heat shield was never discarded at an altitude of three kilometers, and the soft-landing engines never fired prior to touchdown. Those engines, in fact, detonated after landing, burning with the thirty kilograms of concentrated hydrogen peroxide from the capsule's attitude control engines. From launch to impact, Komarov's ill-fated flight had lasted one day, two hours, forty-seven minutes, and fifty-two seconds.

The commission discovered that the reason that the primary parachute never issued was because of friction within the container between the parachute and the inside walls of the container. The increased pressure within the parachute container relative to the low pressure outside the vehicle caused the parachute to simply block up against the insides of the container. This effect was never detected on four drop tests of the parachute system prior to the flight. As late as 1990, however, Chief Designer Mishin continued to believe that the parachute had been incorrectly packed during preparations. The solar panel failure was later traced to the panel getting snagged on the external vacuum-shield cover of the spacecraft. The 45K attitude control sensor had failed because of a "steam-up" of its optical surface. The commission recommended redesigning the parachute container by making it conical instead of cylindrical, increasing its internal volume, and polishing the inside walls. Additional measures would include installing an autonomous node for separating the primary drag chute and photographing the assembly of the parachute packages.[78]

There was also an unofficial and more likely version of the cause of the accident—one that attributed the accident to gross negligence on the part of technicians at TsKBEM's manufacturing plant. During preflight preparations, the two Soyuz ships had been coated with thermal protection materials and then delivered into a high-temperature test chamber to polymerize the synthetic resin. In the case of the two Soyuz ships for the April 1967 mission, technicians tested the vehicles in the chamber with their parachute containers, but apparently *without* the covers for the containers. In Deputy Chief Designer Chertok's investigation of the matter in the early 1990s, he could not find anyone still alive who could remember why the covers had been left off. Because of the omission of the covers, the interiors of the parachute containers were coated with a polymerized coating, which formed a very rough surface, thus eventually preventing the parachute from deploying on Soyuz 1.[79] Clearly, the most chilling implication of this manufacturing oversight was that *both* Soyuz spacecraft were doomed to failure—that is, if Komarov had not faced any troubles in orbit and the Soyuz 2 launch had gone on as scheduled, all four cosmonauts would have certainly died on return.

The unofficial cause of the accident was never included in the official report on Soyuz 1, partly because those at the manufacturing plant who knew of the violation of testing procedure chose to remain silent on the issue so as not to incriminate themselves. The one major casualty of the post-Soyuz 1 investigation was Chief Designer Tkachev of the Scientific-Research and

77. Rebrov, *Kosmicheskiye katastrofy*, p. 29; Semenov, ed., *Raketno-Kosmicheskaya Korporatsiya*, p. 182; Nikishin, "Soviet Space Disaster." Note that Davydov, "How Could That Have Been?," gives the impact velocity as twenty-six to thirty meters per second (ninety-four to 108 kilometers per hour). Most Western sources quote the incorrect 450 kilometers per hour.

78. Semenov, ed., *Raketno-Kosmicheskaya Korporatsiya*, p. 182; Chertok, *Rakety i lyudi: goryachiye dni kholodnoy voyny*, p. 457; Salakhutdinov, "Once More About Space."

79. Semenov, ed., *Raketno-Kosmicheskaya Korporatsiya*, p. 182; Chertok, *Rakety i lyudi: goryachiye dni kholodnoy voyny*, p. 457.

Experimental Institute of the Parachute-Landing Service who had designed the Soyuz parachute system. Although the unofficial version clearly exonerated his organization of any blame, Tkachev was fired from his job in 1968, ending his role in designing the parachute systems for Vostok, Voskhod, Zenit, Soyuz, and many other Soviet spacecraft of the era. Two parachute testing failures following Soyuz I apparently sealed his fate.[80] He was replaced by Chief Designer Nikolay A. Lobanov.

In retrospect, the Soyuz I flight should not have been carried out at that time. The spacecraft was insufficiently tested in space conditions, and it was certainly not ready for the ambitious first mission it was scheduled to accomplish. Although participants continue to deny that there was explicit pressure from Brezhnev, Ustinov, and Serbin to accomplish the flight as soon as possible, the implicit pressure had a much more imposing effect. It was not just a matter of Soviet prestige in space exploration, it was also the fact that perhaps many of the leading designers' jobs were on the line. When Brezhnev or Ustinov complained about the lack of Soviet successes in space, it translated into political pressure on Mishin, Kerimov, Keldysh, and others. Thus, both sides made decisions that were counterproductive and eventually had fatal consequences for the Soviet space program. All told, the responsibility and guilt for the accident lay not on the conscience of any one person, but rather on a technological culture that considered high risks acceptable in the cause of satisfying political imperatives.

A Diamond . . .

The Soyuz I disaster crippled the three major Soviet piloted space programs in the mid-1960s: the Soyuz, the L1, and the L3. While these were the central components of Soviet efforts to compete with the United States in space, these were not the only ones. There was, in fact, a huge parallel effort aimed at piloted military operations in space—one that was completely hidden from view, and whose existence, as with most other Soviet space projects, was unknown until the late 1980s. The Soviet military, left out of the Soyuz, L1, and L3 programs, had promoted its own participation in space research by financing projects dedicated to establishing a Soviet military human presence in space. These efforts were motivated to a great extent by perceptions about the U.S. Department of Defense's well-publicized conceptions of a military space program. After several years of intensive research, President Lyndon B. Johnson canceled the X-20A Dyna-Soar spaceplane program in December 1963. Opinions at the time were moving in favor of a military space station in Earth orbit capable of supporting multicrewed long-duration missions. Preliminary work on such a vehicle, later named the Manned Orbiting Laboratory (MOL), began in late 1963, concurrent with the termination of the X-20A program, although official approval did not come until President Johnson's announcement on August 25, 1965.[81]

The underlying concept behind the U.S. Air Force's MOL was the use of a modified Gemini spacecraft named the Gemini-X (later referred to as the Gemini-B), which would be launched together with the Mission Test Module (later the Laboratory Module) as a single unit by a Titan IIIC launch vehicle. Once in orbit, astronauts would open a hatch in the heat shield of the Gemini-B vehicle and crawl into the Laboratory Module for a month-long mission. By the time that Johnson made his announcement, MOL's primary goal was overhead reconnaissance, primarily over the Soviet Union. Other tasks emerged later, including satellite inspection, accuracy testing of orbital bombardment systems, command and control over military operations during wartime, assessing the effects of month-long missions on humans, and electronic intelligence reconnaissance.[82]

80.   Chertok, *Rakety i lyudi: goryachiye dni kholodnoy voyny*, p. 458.
81.   Paul B. Stares, *The Militarization of Space: U.S. Policy, 1945–1984* (Ithaca, NY: Cornell University Press, 1985), p. 98.
82.   William E. Burrows, *Deep Black: Space Espionage and National Security* (New York: Berkley Books, 1988), p. 227; Donald Pealer, "MOL Part I: Manned Orbiting Laboratory," *Quest* 4(3) (Fall 1995): 4–16.

Plans for MOL caused of much anxiety in the USSR Ministry of Defense. On August 24, 1965, the day before Johnson's announcement, the Central Committee and the Council of Ministers issued a joint decree calling for the expansion of military research in space.[83] By this time, the Soviet Union had already begun the development of a specialized, piloted vehicle exclusively for military purposes, the Soyuz-R, which was a small "space station" consisting of a modified Soyuz docked to another modified Soyuz. Work on the Soyuz-R had proceeded from about 1963 to 1965 at Korolev's Branch No. 3 at Kuybyshev under the direct command of branch chief Dmitriy I. Kozlov, one of Korolev's protégés. The appearance of the MOL seems to have quashed Kozlov's hopes as the Ministry of Defense's General Staff began looking for a more substantial military presence in space. They found a willing provider in General Designer Chelomey, whose proposals seem to have originated from a combination of the Soviet's own desire for crewed reconnaissance and their fear of MOL. It was rumored that Khrushchev had a "fixation" on U.S. aircraft carriers and desired a Soviet response, perhaps some way to keep track of them. Apprised of the MOL effort, Chelomey emerged with a mirror concept: a space station containing sophisticated reconnaissance equipment, including powerful radars to track U.S. naval forces.[84]

On October 12, 1964, just two days before Khrushchev's overthrow, Chelomey gathered all his deputies and proposed the creation of a new Earth-orbital complex named Almaz ("Diamond"). The twenty-ton station would be crewed by two to three military officers on a rotating basis and launched by a three-stage UR-500K booster, better known as the Proton. The station was intended for operation for about one to two years, during which time cosmonauts would conduct experiments and scientific activities formulated by the Ministry of Defense, primarily consisting of photographic and visual reconnaissance.[85] With the MOL project clearly accelerating, Kozlov's modest Soyuz-R proposal was no match for Chelomey's Almaz. In early 1966, the Scientific-Technical Council of the Ministry of Defense's General Staff reviewed both projects on a competitive basis and decided to recommend Almaz for formal approval. All the technical documentation on Soyuz-R was turned over to Chelomey for planning and designing the Almaz complex.[86]

As projected in 1966–67, the Almaz complex consisted of two elements: a space station proper called the Orbital Piloted Station (OPS), or 11F71, and a transport ship to bring crews back and forth between Earth and the station. Originally, Chelomey had proposed a large cargo ship based on the design of the Almaz and about as large, but this proposal was not adopted by the Scientific-Technical Council. As an alternative, Chelomey used Kozlov's transport ship for the Soyuz-R complex, a modified 7K-OK Soyuz spaceship named the 7K-TK. On March 30, 1966, Minister of General Machine Building Afanasyev formally assigned TsKBEM's Branch No. 3 under Kozlov to design and build this modified Soyuz to serve as a ferry vehicle for the Almaz complex. It was the second occasion on which the Mishin and Chelomey design bureaus would undertake significant cooperation with each other despite a competitive rivalry extending back to 1960. Kozlov, using the 7K-OK vehicle as a basis, quickly completed the draft plan for the 7K-TK the same year and began working on preparing the technical documentation for the manufacture of the ship.[87]

---

83.   K. Lantratov, "Dmitriy Kozlov's 'Zvezda': Part II" (English title), *Novosti kosmonavtiki* 4 (February 10–23, 1997): 82–84.

84.   Roald Z. Sagdeev, *The Making of a Soviet Scientist: My Adventures in Nuclear Fusion and Space From Stalin to Star Wars* (New York: John Wiley & Sons, 1993), pp. 206–07.

85.   I. B. Afanasyev, "Unknown Spacecraft (From the History of the Soviet Space Program)" (English title), *Novoye v zhizni. Nauke, tekhnike: Seriya kosmonavtika, astronomiya* no. 12 (December 1991): 1–64.

86.   K. Lantratov, "Dmitriy Kozlov's 'Zvezda': Part I" (English title), *Novosti kosmonavtiki* 3 (January 27–February 9, 1997): 50–55. The chair of the Scientific-Technical Council of the Ministry of Defense's General Staff at the time was Lt. General N. N. Alekseyev.

87.   Ibid.

One of the major bottlenecks in the Almaz program was incorporating a wide variety of systems as specified by various factions within the Ministry of Defense. Technical requirements were revised over and over again, causing significant delays. For example, on December 28, 1966, the Military-Industrial Commission adopted a decree (no. 304) to change the timelines for the 7K-TK transport ship's development. By 1967, Chelomey completely dropped Kozlov's transport ship from the Almaz plan—a decision perhaps partly motivated by a reluctance to cooperate with the old Korolev design bureau. The Almaz space station, the OPS, would include its own large return capsule for the crew. At the same time, Chelomey continued to promote his old idea of a separate transport craft to deliver crews to the station at a later date. During this period, the Soviet government established an "interdepartmental" commission of seventy renowned scientists, heads of design bureaus, and research institutes from the aviation industry and the Ministry of Defense to evaluate the design of the Almaz complex. Their recommendation and high appraisal of the technical characteristics of the plan were critical to the further progress of the project. The final details of the Almaz design were frozen by June 21, 1967, when Chelomey signed the draft plan for the spacecraft, which consisted of more than 100 volumes of technical documentation from twenty-five major design bureaus. Two months later, on August 14, 1967, the Central Committee and the Council of Ministers issued a joint resolution fully committing to the project.[88]

The central component of the Almaz complex was the OPS (11F71), a space station just under twenty tons that was composed of three sections:

- The return apparatus (11F74)
- The station proper (11F75)
- A small recoverable capsule (11F76)

The station proper was shaped like a long cylinder with sections of two different diameters: a large-diameter (4.15 meters) portion and a small-diameter portion. It had a mass of fifteen tons and a length of 11.61 meters. The small-diameter section was in the forward portion of the station and would be enclosed during launch by a conical nose fairing. The large-diameter area was at the aft of the station and ended in a spherical airlock with a passive docking port, called Konus, along the main axis of the station for visiting spacecraft. There was a hatch between the airlock and the large-diameter area, allowing for depressurization for spacewalks. EVAs would be carried out via a large hatch at the upper portion of the spherical airlock. There was a second smaller hatch at the lower end of the airlock that connected to a chamber containing a small drum-shaped recoverable capsule, the 11F76, which was capable of being ejected from the station and returning to Earth with the exposed film and other scientific materials. Once the capsule was packed with its payload, the crew would spin-stabilize the pod and then eject it from the OPS. The one-meter-long capsule had its own solid-propellant propulsion system for reentry, a parachute system, a jettisonable heat shield, and the actual descent compartment equipped with a radio beacon for recovery forces on the ground.

There were antennas as well as two main engines positioned around the airlock on the end of the large-diameter portion for orbital corrections. Each RD-0225 engine with a thrust of 400 kilograms was developed by the Chemical Automation Design Bureau (formerly OKB-154) under Chief Designer Aleksandr D. Konopatov. Power for the station was provided by two large

88. *Ibid.*; Vladimir Polyachenko, "The 'Pep' of Almaz" (English title), *Krylya rodiny* no. 1 (January 1992): 18–19; Afanasyev, "Unknown Spacecraft"; Olaf Przybilski, *Almaz: Das supergeheimer militärische Orbitalstationsprogramm der UdSSR* (Dresden, Ger.: Institut für Luftfahrt, 1994), pp. 16–17. The Ministry of General Machine Building had issued an earlier decree on Almaz on February 9, 1967.

This is a model of the military Almaz space station on display. This model is of the original variant of the space station, with the large Gemini-shaped reentry capsule attached on the forward end (to the right). This capsule was later deleted from the station. The folded-up solar panels are on the left, surrounding the single docking port of the station. (copyright Dietrich Haeseler)

solar panels spread like wings to a span of 22.8 meters, whose bases were attached to the spherical compartment. The panels would provide 3.12 kilowatts of power. The entire aft end of the station was surrounded by a cone-shaped shield made of vacuumed thermal insulation.

Cosmonauts would dock at the aft docking port, open the hatch into the spherical airlock, and crawl through a short tunnel into the large-diameter area. The tunnel itself was enclosed all around by a stubby instrument compartment containing spherical propellant tanks for the OPS main engines, the engines themselves, pressurized gases, and small attitude control thrusters.

Going back toward the station, there was the large-diameter area that had a control console, a work station, an optical sight allowing the cosmonauts to "freeze" the movement of Earth below and observe specific details, and periscopes allowing for the inspection of the space around the station. Instruments were designed and installed as detachable modules to facilitate easy repair. The compartment also included athletic instrumentation and the toilet. The centerpiece of the large-diameter area was the *Agat-1*, optical telescope, a large device that occupied a considerable portion of the compartment. The telescoping camera had a focus length of 6.375 meters and was certainly one of the largest mirrors ever put into orbit. In the open media, Russians have claimed that the resolution was less than three meters, but given the size of the mirror, it is more likely that the telescope was capable of distinguishing targets smaller than one meter. The cosmonauts would use Agat-1, in conjunction with the ASA-34R wide-film camera, to photograph targets on Earth, develop the film on the station, conduct an analysis, and send back the more militarily important ones directly to Earth via a TV link, all within about thirty minutes. The remaining photographs would arrive on Earth in the 11F76 recoverable capsule. Other optical instruments on the station included the OD-5 optical viewfinder, the *POU-11* panoramic instrument for wide coverage of Earth's surface, topographic and stellar apparatus, and the *Volga* infrared instrument with a resolution of 100 meters.

Heading further to the aft of the station, the cosmonauts would enter the smaller diameter section, which was the crew living compartment containing sleeping areas with deployable bunks, a dining table and chair, a food storage area, and viewports for photography. For the first time on a Soviet piloted spacecraft, the life support system included a device, designated *Priboy* ("Surf"), with the capability to recycle water from air humidity.

One of the most interesting components on the station was motivated by concerns among Soviet military leaders that the United States might attack such an explicitly military space station in orbit. Given the paranoia about U.S. military space plans, Chelomey agreed to the military's proposal to install a means to defend the station in case of such an attack. Under a contract, the Design Bureau of Precision Machine Building (formerly OKB-16) under Chief Designer Aleksandr E. Nudelman designed a twenty-three-millimeter-caliber rapid-fire cannon for the station. Cosmonauts would be able to use a gunsight to turn the station and aim the cannon at a selected target. Nudelman's previous claim to fame had been as the designer of several major anti-tank guns and missiles for the Soviet armed forces. The Soviets evidently considered the weapon more of a defensive system rather than an offensive one, given the limited maneuvering capabilities of the Almaz OPS.

Because its primary mission was overhead reconnaissance, the OPS would have a low operational orbit (220 by 270 kilometers) and be oriented toward Earth's surface for long periods. The search and observation of targets on the ground thus posed complex demands on the guidance system. As per the original requirements, Chelomey's engineers designed a guidance system that would control the station continuously from the moment it separated from the launch vehicle to orbital decay many months later. What they emerged with was a "decentralized" system, with subsystems for orientation, stabilization, movement control of the center of mass of the vehicle, navigation, and programmed control of the on-board apparatus. The primary flight control system was based on an analog system because a digital device that was continuously operable for a year was not in existence in the Soviet Union at the time. Instead, the All-Union Scientific-Research Institute for Electromechanics (formerly NII-627) headed by Chief Designer Andronik G. Iosifyan developed a new low-power electromechanical stabilization system using a spherical ring flywheel with a large kinetic movement. Unlike conventional orientation systems, there was almost no propellant consumption for this device. Cosmonauts would be able to carry out rapid roll control at one degree per second to expand their field of view. Precision would be achieved by a system that corrected the gyroscopic orientation system with a Doppler signal from a radar instrument, which itself was part of the radar observation gear for the station. This gyroscopic orientation system was developed by the Scientific-Research Institute for Applied Mechanics (formerly NII-944) under Chief Designer Viktor I. Kuznetsov, one of the original members of Korolev's old Council of Chief Designers from the 1940s.

The control system had various modes of operation, including precise orientation and stabilization, restoration of orientation from a disoriented position, and the spinning of the station into "storage" position. Cosmonauts themselves could also manually orient the station when observing targets by putting the target in the cross-hairs of their optical sight with a turn of the control stick. As a result, the guidance system would allow all the optical instruments on board to inspect the selected target. Although analog computers were used on the overall station's guidance system, Chelomey's engineers designed a digital system based on the *Argon-12A* computer for the observation instrumentation, a first for a Soviet piloted space vehicle. The computer was developed by the All-Union Scientific-Research Institute for Digital Computer Technology.[89]

The first version of the Almaz OPS was equipped with a large return apparatus (11F74), which was similar to the LK-1 and LK-700 lunar spacecraft. Apart from its shape, the OPS return

89. Afanasyev, "Unknown Spacecraft"; V. Polyachenko and A. Tumanov, "From the History of Space Science: The Controllable 'Almaz'" (English title), *Aviatsiya i kosmonavtika* no. 8 (August 1993): 41–43; I. Marinin, "AO Krasnogorsk Plant Named After S. A. Zverev" (English title), *Novosti kosmonavtiki* 19 (September 9–22, 1996): 44–49; *Officially Never Disclosed*, Moscow Ostankino Television, First Channel Network, Moscow, November 26, 1994, 1105 GMT; Igor Tsarev, "A 'Diamond-Studded' Sky: Should the Military, Who Maintain They Have Stopped Preparing for 'Star Wars,' Be Trusted" (English title), *Trud*, September 28, 1993, p. 4; S. A. Zhiltsov, ed., *Gosudarstvennyy kosmicheskiy nauchno-proizvodstvennyy tsentr imeni M. V. Khrunicheva* (Moscow: RUSSLIT, 1997), pp. 78–79; Christian Lardier, *L'Astronautique Soviétique* (Paris: Armand Colin, 1992), pp. 204–05; Dietrich Haeseler, "Original Almaz Space Station," *Spaceflight* 36 (October 1994): 342–44; B. A. Pokrovskiy, *Kosmos nachinayetsya na zemlye* (Moscow: Patriot, 1996). p. 405.

apparatus had two striking similarities to MOL's Gemini-B: the Soviet vehicle was designed to have a hatch in the center of the heat shield for transfer to and from the station proper; and the spacecraft was designed for reuse on subsequent stations. Originally, it seems that Chelomey intended to launch the return apparatus and the OPS separately and assemble the two in orbit, but this plan was abandoned later in favor of launching the crew in Almaz on a Proton rocket. The return apparatus consisted of three sections: a conical crew capsule with a flat top shaped remarkably like the Apollo Command Module; a second longer cone with a sharper angle attached at the apex of the crew capsule; and a short, thin cylinder at the very forward end containing a powerful deorbit engine. The length of the return apparatus was 3.64 meters, and the base diameter was 2.79 meters.

On the OPS, the truncated spherical base of the return apparatus was fixed at the forward end of the station on the opposite end from the docking unit. The 4.9-ton module had three seats in its internal volume as well as control panels for operations during mission end. The longer cone section of the return apparatus was equipped with a set of attitude control thrusters for use prior to reentry, as well as the primary and backup parachutes. At launch, the entire OPS-return apparatus complex was topped off by a long thin escape tower equipped with two sets of solid-propellant rocket engines for emergency situations during passage through the lower atmosphere. Once in orbit, the crew would vacate their seats and remove the center seat to open a hatch at the base of the return apparatus and crawl into the small-diameter area in the Almaz OPS. There were evidently many engineers who believed that having a hatch in the middle of a heat shield—that is, the most stressed part of a spacecraft—was akin to suicide, but Chelomey was confident that this was a workable design. For return to Earth, the cosmonauts would secure themselves in the return apparatus, close the heat shield hatch, and undock from the station. After they fired the deorbit engine, the conical capsule would separate from the cylinder and brake into Earth's atmosphere. Independent flight was limited to about thirty hours. The return apparatus was capable of returning at least 360 kilograms of equipment, film, and other materials to Earth after a long-duration flight. It was designed to have a lifetime of five flights.[90] Some of these missions would be as part of a future projected delivery vehicle to the Almaz station, named the Transport-Supply Ship, which was at a very early stage of planning in 1967. By this time, the first Almaz space station launch was set for sometime in 1968–69. The first cosmonaut training group for the Almaz station was established as early as September 1966, although crew training proved to be of a very preliminary nature through 1967.[91]

The early Almaz station's design and capabilities were quite similar to the American MOL. This was partly attributable to the ancestry of both complexes. The Almaz OPS descent apparatus emerged from the LK-700 and LK-1 capsules, which were based to a great degree on Gemini. Similarly, MOL Gemini-B was simply an uprated Gemini. Chelomey clearly had access to information on MOL. During the 1960s, the Soviet government used to publish a classified weekly journal entitled *Raketno-kosmicheskaya tekhnika* (*Rocket and Space Technology*) containing abstracts of articles published in the open media in the West. In 1964 and 1965, the journal evidently published numerous articles on the MOL.[92] While there is no clear evidence

90. Afanasyev, "Unknown Spacecraft"; Polyachenko and Tumanov, "From the History of Space Science: The Controllable 'Almaz'"; Lardier, *L'Astronautique Soviétique*, p. 203; Nina Chugunova, "Chelomey's Cosmonauts: Why There Are No Crews From NPO Mashinostroyeniya in Outer Space" (English title), *Ogonek* 4–5 (January 1993): 24–29; Haeseler, "Original Almaz Space Station."

91. There were a total of seven cosmonauts selected for the Almaz group on September 2, 1966: P. I. Belyayev, L. S. Demin, V. G. Lazarev, A. N. Matinchenko, G. S. Shonin, L. V. Vorobyev, and D. A. Zaykin. Belyayev, the only spaceflight veteran, was the commander of the group. See Vadim Y. Molchanov, "Soviet Manned Lunar Programs," *Quest* 2(4) (Winter 1993): 43; Kamanin, *Skrytiy kosmos: 1964–66*, p. 374.

92. K. Lantratov, "Dmitriy Kozlov's 'Zvezda': Part III" (English title), *Novosti kosmonavtiki* 5 (February 24–March 9, 1997): 81–86.

to suggest that Chelomey took MOL plan wholesale, macro-level design characteristics of Almaz were probably influenced significantly by the American project.

. . . a Star . . .

As befits the story of any Soviet space project from the 1960s, the Soviet Union did not respond to a singular U.S. space project, such as MOL, with a singular response. Almaz, in fact, had a complementary military piloted project that, while a little more modest, was also a response to the MOL. When the Military-Industrial Commission approved the initial plans for the Almaz station in 1965, the first flight was expected in 1968. Motivated by concerns of having a Soviet crewed military presence in the intervening three years, the commission looked into other options. In early August 1965, Commission Chairman Smirnov signed an order to develop a military version of the 7K-OK Soyuz for missions involving visual and photographic reconnaissance, satellite inspection, the testing of early warning technologies, and the verification of the operation of weapons in orbit. The Central Committee and the Council of Ministers, in its decree of August 24, 1965, approved a timetable for the development of such a vehicle, officially named the *Zvezda* ("Star"). Coincidentally or not, by this time, OKB-1's Branch No. 3 in Kuybyshev under Deputy Chief Designer Kozlov had, on his own authority, completed the draft plan that fulfilled the government's requirements. After further discussions, on July 7, 1966, the Ministry of General Machine Building signed an order (no. 296ss) selecting Kozlov's branch as the lead developer of the Zvezda ship. Kozlov proposed a modification of the original 7K-OK Soyuz named the 7K-VI.[93]

In its original conception, the design of the 7K-VI was very similar to Korolev's 7K-OK. It had three major components arranged from the front to the aft: the living compartment, the descent apparatus, and the instrument-aggregate compartment. The first section would have carried a full complement of military instrumentation. By late 1966, Kozlov began to rethink this design, motivated by the two failures in the Soyuz precursor program, including the catastrophic launch failure in December 1966 when a military officer had been killed. To preclude such problems from occurring on his ship, Kozlov prepared a new design for the 7K-VI, which departed significantly from the 7K-OK. In the new design, the descent apparatus and the living compartment switched places. This meant that just as in Almaz and MOL, there would be a hatch in the middle of the crew compartment's heat shield to allow cosmonauts to move into the main experiment module of the ship. The new ship had a heavier mass of just over six and a half tons and could fly thirty-day-long missions in Earth orbit with two crewmembers. The heavier ship required an uprated version of the basic 11A511 Soyuz launcher, called the 11A511M. The Ministry of Defense found the new design worth pursuing, and in a governmental resolution on July 21, 1967, set a formal timetable for the first launch, targeted for 1968. The system would reach operational status a year later.[94]

As with the early Almaz station, the 7K-VI was equipped with a weapon designed by Chief Designer Nudelman's Design Bureau of Precision Machine Building. The complement consisted of a single rapid-firing gun modified for use in vacuum, mounted on the descent apparatus. Cosmonauts would be able to aim the gun by maneuvering the entire spacecraft using a special visor. Skeptics believed that pilots would not be able to aim the gun; they also believed that the recoil from gunfire would send the entire ship into a spin. To eliminate such problems, Kozlov's engineers built a dynamic test stand at Branch No. 3 in mid-1967, consisting of the descent

93.   Lantratov, "Dmitriy Kozlov's 'Zvezda': Part II." Its production index was 11F73.
94.   *Ibid.* Initially, Kozlov wanted to have one crewmember aboard the 7K-VI to compensate for the heavier mass, but the Ministry of Defense believed that one cosmonaut would not be able to accomplish all the planned work in orbit.

apparatus, an optical visor, control systems, and seats set on a platform resting on an air cushion. Subsequent tests dispelled any doubts about the capability of both the pilot and the ship during a shooting match. As in the Almaz OPS, Zvezda's gun was insurance against the possibility that American satellites on anti-satellite or inspection flights would engage the Soviet spaceship.

The descent apparatus, although shaped like the basic Soyuz version, had two seats in it, but facing in slightly different directions, like a "v"-shaped pattern. The hatch was positioned underneath the seats. Tests at the time verified the hatch-in-the-heat-shield design, which was the subject of much concern, both in the Zvezda and Almaz programs. The living compartment of the 7K-VI contained the primary reconnaissance instrument, the OSK-4 optical visor and camera, installed on a side porthole. The cosmonaut would sit in a saddle, looking somewhat like a cyclist, and use a visor to observe and photograph Earth's surface. Cosmonauts could also mount other instruments on the porthole, including the *Svinets* device, a throw-over from the abandoned Voskhod 3 flight, for observing ballistic missile launches. They would also use a long mast extending from the outside of the living compartment for electronic intelligence and the detection of any approaching enemy spacecraft.

Dmitriy Kozlov was the First Deputy Chief Designer at TsKBEM under Vasiliy Mishin. He headed the Korolev design bureau's Branch No. 3 in Kuybyshev starting in 1959. As chief of the branch, which later became independent in 1974, Kozlov oversaw the development of robotic and piloted reconnaissance spacecraft for the USSR Ministry of Defense. Kozlov remains the head of his organization to this day, remaining one of the last chief designers from the Korolev era who are still active in the Russian space program. His design bureau continues to build almost all of Russia's photo-reconnaissance spacecraft. (files of Peter Gorin)

One unusual attribute that set the 7K-VI apart from any previous piloted vehicle was its power source. Kozlov dispensed with solar arrays, believing them to be a potential source of problems (confirmed on Soyuz 1). He proposed the use of two radio-isotope generators, which would convert heat produced by the radioactive decay of plutonium into the large amount of electricity required for the extensive instrument complement aboard the vehicle. To preclude accidents upon reentry, the generators were encased in landing capsules capable of surviving reentry. Once they were recovered, engineers would reuse them for subsequent missions.

A final design objective of the 7K-VI spaceship was to serve as a transport ship for future crews to the Almaz space station, much like the terminated 7K-TK from Kozlov's early plans for a military space vehicle. Branch No. 3 engineers looked into the possibility of installing a docking unit at the forward end of Zvezda to allow it to dock with the Almaz station, thus establishing quite a formidable military space complex in Earth orbit, designated imaginatively the 11F711.[95]

Given the several years of work on the abandoned Soyuz-R variant, progress on the 7K-VI Zvezda program was swift. By mid-1967, Kozlov had defended a revised draft plan for the ship and its launch vehicles, based on a tactical-technical requirement for the spaceship issued by the Ministry of Defense in March 1967. His engineers had also transferred all technical

95.   Lantratov, "Dmitriy Kozlov's 'Zvezda': Part III."

documentation to the Progress Plant for the manufacture of the first models. The Air Force Commander-in-Chief's Aide for Space Matters Lt. General Kamanin established the first 7K-VI cosmonaut training group in September 1966, comprising six cosmonauts headed by the veteran Pavel R. Popovich.[96] Through 1967, Popovich spent much time in Kuybyshev training on the ship and testing out its rapid-fire gun in simulators. In addition to career cosmonauts, the Ministry of Defense was also intent on including scientists from its various research institutes. Three researchers from NII-2 of the Air Defense Forces joined the team at the Cosmonaut Training Center on April 12, 1967. NII-2 was the leading institute developing strategy for anti-satellite operations on automated Soviet satellites, such as the IS system.[97]

Zvezda (7K-VI, 11F73)

1967 (project)

This is a drawing of Dmitriy Kozlov's 7K-VI piloted reconnaissance spacecraft developed in the 1960s. Although the spacecraft design was based on the original Soyuz spacecraft, there were major differences in the layout of the main modules of the vehicles. (copyright Peter Gorin)

Schedules for the program were also set at that time. On August 31, 1967, Military-Industrial Commission Deputy Chairman Georgiy N. Pashkov chaired a meeting to discuss the course of the Zvezda project, calling it a program of "extraordinary importance."[98] Kozlov optimistically predicted that the first automated flight would take place in the second half of 1968, although Progress Plant Director A. Ya. Linkov believed 1969 was more realistic.

That military piloted operations were of great concern not only to the Ministry of Defense but also to the Soviet leadership was underlined by a meeting of the Council of Defense on July 15, 1967. The council, a shadowy body attached to the Politburo, was the supreme arbiter for all defense issues in the Soviet Union. At the meeting, Brezhnev and Kosygin expressed dissatisfaction with delays in the Soviet piloted space program and ordered an expansion of military operations in space. The breadth of Soviet plans for the late 1960s and early 1970s was astonishing. In a diary entry for September 16, 1967, Lt. General Kamanin summarized his notes on the next eight-year plan for Soviet space operations, covering 1968 to 1975. According to his calculations, the military would need twenty Almaz space stations and fifty Zvezda ships, in addition to 400 "transport ships," presumably the Soyuz. The total annual launch rate of crewed ships would reach forty-eight.[99]

Soviet plans for the military piloted dominance of space were not limited to conventional systems such as Almaz and Zvezda. As more evidence of an almost unprecedented military buildup in space, the USSR had a third, much more ambitious, piloted space project approved in the mid-1960s. Since the beginning of the space era, a host of Soviet aviation designers, such

---

96. N. Kamanin, "A Goal Worth Working for," no. 44. The six cosmonauts were Yu. P. Artyukhin, B. N. Belousev, A. A. Gubarev, V. I. Gulyayev, G. M. Kolesnikov, and P. R. Popovich. They were later joined by A. F. Voronov and D. A. Zaykin.

97. These three cosmonauts were V. B. Alekseyev, M. N. Burdayev, and N. S. Porvatkin. See V. Semenov, I. Marinin, and S. Shamsutdinov, *Iz istorii kosmonavtiki: vypusk I: nabory v otryady kosmonavtov i astronavtov* (Moscow: AO Videokosmos, 1995), pp. 10, 12.

98. Lantratov, "Dmitriy Kozlov's 'Zvezda': Part III."

99. *Ibid.*

as Tsybin, Myasishchev, and Chelomey, had doggedly pursued a dream of building a reusable spaceplane, one that could eventually fly from airport into space and land back on a runway. Thwarted mostly by the winds of political change, none of their three projects ever got off the ground. By 1965, the Soviet Air Force gave it yet another try, in a project that would eventually span thirteen long years.

. . . and a Spiral

General Designer Vladimir N. Chelomey's Raketoplan project, consisting of the R-1 and R-2 spaceplanes, had died an ignominious death around 1965—a result of the technological limitations and the political exigencies of the period. At the same time, the primary *raison d'être* for the project, the U.S. Air Force's X-20A Dyna-Soar, had long been consigned to history. For the immediate future, there were no serious plans by the U.S. armed forces to pursue the creation of such vehicles. Only some test vehicles were flown. Under a joint NASA-Air Force program, lifting bodies such as the M2-F2 and HL-10 were tested at NASA's Flight Research Center (later the Dryden Flight Research Center) at the Rogers Dry Lake in California.[100] The lack of U.S. support for spaceplanes did not deter the Soviets. Unlike almost any other Soviet piloted space project of the Cold War era, something prompted the Soviets to push the development of a piloted spaceplane well after the Americans had abandoned such hopes. Historical precedent suggests two reasons: either the Soviets believed that secretly the United States was developing such a vehicle, or it was insurance against the possibility of the United States developing such a vehicle in the future. Both rationales, of course, hinge critically on the assumption that in their Cold War–era space projects, the Soviet Union and the United States were doing things in a parallel and responsive manner instead out of a unilateral need to do such things. Whether this is a hypothesis that will hold up to historical scrutiny remains to be seen. The record from the former USSR still remains vastly incomplete.

In the early 1960s, the Air Force contracted two aviation industry design bureaus, OKB-156 headed by Andrey A. Tupolev and OKB-155 headed by Artem I. Mikoyan, to propose elements of an integrated reusable aerospace transportation system.[101] Little is known about the Tupolev proposal. Scientific research on lifting bodies had apparently begun during the late 1950s at the famous N. Ye. Zhukovskiy Central Aerohydrodynamics Institute (TsAGI). Based on this research, OKB-156 had initiated work in the late 1950s and the early 1960s on a suborbital lifting body using "hot" construction—that is, frames using heat-resistant alloys without special thermal shielding. In the 1960s, General Designer Tupolev apparently designed and built a full-scale hypersonic vehicle capable of Mach 2 to 5 to verify ground research on developing a winged space glider. Research conducted in cooperation with the famous M. M. Gromov Flight-Research Institute helped engineers experimentally verify data already obtained from wind tunnels on such parameters as aerodynamic quality, characteristics of longitudinal and lateral static stability, and balance at different angles of attack during reentry. The engineers discovered that for a winged hypersonic vehicle with a relatively large stern area, air resistance could reduce aerodynamic quality by 30 to 40 percent. The overall research helped identify changes in further research on the basic layout of a reusable spaceplane.[102]

100. Richard P. Hallion, *On The Frontier: Flight Research at Dryden, 1946–1981* (Washington, DC: NASA Special Publication (SP)-4303, 1984), pp. 147–72.

101. Mikhail Rebrov, "The Moor Did Its Business . . . The Fate of 'Buran,' as Dramatic as Our Lives" (English title), *Krasnaya zvezda*, November 29, 1997, p. 5; Kamanin, *Skrytiy kosmos: 1964–1966*, p. 262.

102. K. K. Vasilchenko, *et al.*, eds., *Letnyye issledovaniya i ispytaniya: fragmenty istorii i sovremennoye sostoyaniye: nauchno-tekhnicheskiy sbornik* (Moscow: Mashinostroyeniye, 1993), p. 55; *TsAGI-Osnovnyye etapy nauchnoy deyatelnosti, 1968–1993* (Moscow: Nauka, 1996), p. 155.

This early work was to lead to the development of a complete two-stage reusable space transportation system. The first stage would be a hypersonic carrier aircraft, and the second stage a small plane for short jaunts into space. Between 1961 and 1966, Tupolev's engineers apparently built a small automated prototype of the winged space launcher designated "product 130." Although details still remain classified, the aircraft was developed on the basis of the Tu-95 bomber as part of a large-scale study of hypersonic flying vehicles in the 1950s and 1960s. Work on the 130 was to have led to the creation of a rocket-propelled spaceplane named Zvezda, which would have been launched into orbit by some modification of the UR-200 ICBM. The launch system for the 130 would have been similar to the one used on the American B-52A aircraft for "drop-launching" the X-15 rocket-plane.[103] Unlike his competitor Mikoyan, Tupolev apparently had a "cool" attitude toward the spaceplane program in general. By 1966, whatever work had been accomplished at OKB-156 was terminated. Instead of a unilateral spaceplane program, it seems that Tupolev joined up with Mikoyan for a cooperative project, which proved to be the most famous Soviet spaceplane of the early Soviet space era.

General Designer Mikoyan, the head of the MiG design bureau, had had a long interest in such topics. He had publicly expressed an interest in space as early as 1962, when in an article in the Soviet military newspaper Krasnaya zvezda (Red Star), he described a spaceplane design:

The spaceplane is an intermediate link between aviation and rocket technologies, a combination of a ballistic rocket and airplane; viewed as a whole, the spaceplane will have the general outlines of a modern airplane with elements of a spaceship. The spaceplane will be launched as is a ballistic missile and will fly at altitudes of 100 to 200 km. After acceleration to a speed of 7.9 km/sec, the spaceplane will follow a ballistic trajectory with deceleration.[104]

It seems that Mikoyan had begun exploratory studies on such topics in the early 1960s, possibly derived from Chief Designer Tsybin's research on the abandoned PKA spaceplane from the late 1950s. It would be 1965, however, before Mikoyan initiated any productive work on the spaceplane project.[105] At the time, Mikoyan inherited a secondary source of information to accelerate his efforts. When the new Brezhnev administration terminated Chelomey's R-1/R-2 spaceplane project in 1965, much of the database was transferred to Mikoyan's Moscow-based OKB-155, along with a number of engineers who had worked on Chelomey's project. This information proved invaluable for Mikoyan's designers to quickly advance from a research to an experimental stage in the development of a new aerospace system.[106] Chelomey, of course, had inherited his spaceplane research from Myasishchev's work on the promising but ultimately abandoned M-48 design. Mikoyan also was favorably placed to take advantage of the massive research work at the prestigious TsAGI during the early 1960s on various Chelomey and Tupolev research projects. In the topsy-turvy world of space politics, Mikoyan had thus inherited the

103. Bill Gunston, The Osprey Encyclopedia of Russian Aircraft: 1875–1995 (London: Osprey Aerospace, 1996), p. 435; E-mail correspondence, Mark Hillyer to the author, April 30, 1998.

104. A. Mikoyan, "Future Aerospace Technology" (English title), Krasnaya zvezda, January 9, 1962, pp. 2–3, quoted in Soviet Space Programs: Organization, Plans, Goals, and International Implications, prepared for the Committee on Aeronautical and Space Sciences, U.S. Senate, 87th Cong., 2d sess. (Washington, DC: U.S. Government Printing Office, May 1962), p. 333.

105. Gleb E. Lozino-Lozinskiy and Vladimir P. Plokhikh "Reusable Space Systems and International Cooperation," Aerospace America (June 1990): 37–40; Afanasyev, "Unknown Spacecraft."

106. Interview, Gerbert Aleksandrovich Yefremov with the author, March 3, 1997; Anatoliy Kirpil and Olga Okara, "Designer of Space Planes. Vladimir Chelomey Dreamed of Creating a Space Fleet of Rocket Planes" (English title), Nezavisimaya gazeta, July 5, 1994, p. 6.

complete database for most prior spaceplane research in the Soviet Union. It put him in an extremely favorable position to move quickly on the project.

Less than two years after the cancellation of the X-20 Dyna-Soar, on July 30, 1965, the Ministry of Aviation Industry approved work on a new spaceplane project named Spiral.[107] The head of the Spiral project at OKB-155 was one of Mikoyan's principal deputies, Chief Designer Gleb Ye. Lozino-Lozinskiy, a fifty-five-year-old engineer who had played a tremendously significant role in the development of numerous MiG fighters. During Khrushchev's downsizing of aviation in favor of rockets, Lozino-Lozinskiy had stood up to the Soviet leader's tirades against airplanes, suggesting that "in spite of all the enthusiasm with regard to rockets, one should not forget the little wings. They are still of use to us."[108] As chief designer of the Spiral project, Lozino-Lozinskiy signed off on the preliminary design of the system on June 29, 1966, just a year after work had begun.[109] To expand the work profiles at his design bureau, Mikoyan subsequently established a branch of OKB-155 (by this time renamed MMZ Zenit) dedicated specifically to space themes at the premises of the Dubna Machine Building Plant near Moscow. Coincidentally, it was at this same plant that former Chief Designer Tsybin had directed his work on spaceplanes in the late 1950s. Mikoyan's new Dubna branch, created in 1966, had its own design bureau, headed by Yuriy D. Blokhin, who supervised all of Lozino-Lozinskiy's work on Spiral. A third man, Petr A. Shuster, served as the chief of the branch.[110]

The primary goal of Spiral was piloted spaceborne reconnaissance, satellite inspection, and anti-satellite operations. To do this, engineers needed to create a system capable of operating within very short lead times, one that was reusable, and one that could be launched from a variety of locations. Thus, Mikoyan dispensed with the idea of launching the spaceplane on conventional rocket boosters and, in fact, adopted a design that was in some ways very similar to the Chelomey and Tupolev concepts—that is, launching the spaceplane into orbit from a mother aircraft. Rummaging through the extensive database on spaceplane research available to them, Mikoyan's engineers firmly believed that this would be the most efficient option. Early analyses showed that with an air-launched system, effective payload increased by about 9 percent over standard ballistic models, while the associated costs were projected to be three to three and a half times lower for launching one kilogram of payload into orbit over conventional single-use launch designs. There were also operational advantages. Soviet engineers believed that an air-launched system would afford them all-weather and twenty-four-hour launch capability. Space visionaries, of course, continue to debate to this day the advantages and disadvantages of such systems for delivering payloads to orbit, but in the heyday of the

107. A Central Committee and Council of Ministries decree on Spiral was issued in late 1965.

108. Col. M. Rebrov, "The Revolutions of 'Spiral': A Biography and Portrait of the Chief Designer of the Buran Space Plane" (English title), *Krasnaya zvezda*, July 31, 1991, p. 4.

109. Vyecheslav Kazmin, "The 'Quiet' Tragedy of EPOS" (English title), *Krylya rodiny* no. 11 (November 1990): 25–26.

110. *Ibid*. The Dubna Machine Building Plant (MZ Dubna) was formerly known as Plant No. 256. P. V. Tsybin's OKB-256 moved here on April 25, 1956. After OKB-256 was dissolved on October 1, 1959, the plant was subordinated to OKB-2-155 headed by Chief Designer A. Ya. Bereznyak. OKB-2-155 was a branch of the Mikoyan design bureau at the time and had produced a number of short-range cruise missiles after its establishment on October 12, 1951. When OKB-2-155 was separated from its parent entity in 1966, part or all of the facilities of the Dubna Machine Building Plant remained subordinate to the Mikoyan design bureau. It was here that OKB-155's Space Branch was established. From 1966 on, OKB-2-155 was known as the Raduga Machine Building Design Bureau (MKB Raduga). There is evidence to suggest that MKB Raduga cooperated with OKB-155's Space Branch on the Spiral project. See Vladimir Nikolayevich Trusov, "45! MKB 'Raduga'" (English title), *Vestnik vozdushnogo flot* 1 (1997): 16–18; Stepan Mikoyan, "'Molniya': From 'Spiral' to MAKS" (English title), *Vestnik vozdushnogo flot* 1 (1997): 60; Lardier, *L'Astronautique Soviétique*, p. 100; Piotr Butowski, "Steps Towards 'Blackjack'," *Air Enthusiast* 73 (January–February 1998): 36–49; E-mail correspondence, Mark Hillyer to the author, March 29, 1998.

mid-1960s, to a generation of old-school aeronautical engineers such as Mikoyan and Lozino-Lozinskiy, there was no question that air-launched spacecraft were the wave of the future.

Lozino-Lozinskiy's 114.8-ton Spiral system was a two-stage system consisting of the reusable Hypersonic Booster-Aircraft ("product 50-50") and a two-stage payload. The payload consisted of an expendable two-stage booster rocket and the Orbital Aircraft ("product 50"). The engineers proposed two near-identical Spiral systems—a primary and a secondary model, each differentiated only by the choice of propellants:

Component	Primary Model Propellants	Secondary Model Propellants
Hypersonic Booster-Aircraft	Liquid hydrogen	Kerosene
Booster rocket	LOX–liquid hydrogen	LOX-kerosene
Orbital Aircraft	Nitrogen tetroxide–unsymmetrical dimethyl hydrazine	Nitrogen tetroxide–unsymmetrical dimethyl hydrazine

The Hypersonic Booster-Aircraft (the 205) was a large tailless aircraft built somewhat like a "flying wing," with sweptback wings and vertical stabilizer surfaces on the wing tips. It was equipped with four multimode air-breathing turbojet engines operating on kerosene (on the secondary variant) or on liquid hydrogen (on the primary variant). The aircraft's turbojets were under the main long fuselage and had a common, regulated supersonic air intake. The actual orbital payload was fixed on top of the aircraft to a pylon, with its forward and rear ends covered by fairings. The Hypersonic Booster-Aircraft had a total length of thirty-nine meters, a wingspan of sixteen and a half meters, and a mass of fifty-two tons (primary version) or seventy-two tons (second variant).[111] One of the more imposing technical challenges was the development of a hydrogen-fueled carrier aircraft. Much of this research was carried out at TsAGI near Moscow in cooperation with the Institute of Theoretical and Applied Mechanics of the Academy of Sciences, based in Novosibirsk, Siberia. Beginning in 1967, Institute Director Academician Vladimir V. Struminskiy was instrumental in laying the foundation for this work, which was not only in support of the Spiral carrier aircraft but also for future transport and bomber aircraft.[112]

The payloads—the two-stage rocket and the Orbital Aircraft—were attached on top of the Hypersonic Booster-Aircraft's fuselage from the rear, to two-thirds of the way toward the front of the carrier aircraft. The booster rocket was a classical cylindrical rocket with a mass of 52.3–52.5 tons consisting of two stages, both fueled on either liquid oxygen (LOX)–kerosene or LOX–liquid hydrogen. Unconfirmed reports suggest that this rocket, designed to accelerate the Orbital Aircraft into orbit, may have been a contribution from Korolev's OKB-1. Other contradictory evidence suggests that Lozino-Lozinskiy may have considered using one of Chelomey's ICBMs, the UR-100, for the role. If indeed the UR-100 was actually under consideration for the Spiral system, Mikoyan and Lozino-Lozinskiy must have factored in a significant amount of redesign to accommodate the new propellant combinations because the UR-100 used storable hypergolic combinations. In the Spiral conception, the

111. Afanasyev, "Unknown Spacecraft"; Lardier, L'Astronautique Soviétique, p. 248; E-mail correspondence, Igor Afanasyev to the author, December 6, 1997.

112. Another participant in this program was OKB-165 headed by General Designer A. M. Lyulka. See also Lardier, L'Astronautique Soviétique, p. 175; G. P. Svishchev, ed., Aviatsiya entsiklopediya (Moscow: Bolshaya Rossiyskaya entsiklopediya, 1994), p. 546.

This is a model of the complete Spiral system on display. The high-speed 50-50 carrier aircraft would have returned to an airport after accelerating its combined payload to a velocity of about Mach 5–6. The actual Spiral spaceplane is mounted on top of the carrier's fuselage backed by a two-stage cylindrical rocket at its base. (files of Asif Siddiqi)

booster rocket would have a first-stage thrust of 100 tons, a little more than the eighty tons on the UR-100 ICBM. Second-stage thrust would be twenty-five tons.[113]

The main component of the Spiral system was the Orbital Aircraft (the 105). The relatively small vehicle was built on a triangular base and had wings swept back at fifty-five degrees. The vehicle had a length of eight meters and a wingspan of just under seven and three-quarters meters. Four meters of the wingspan covered the width of the fuselage. The mass of the ship was only 10.3–10.5 tons. The useful payload of the ship would be two tons. The shapes of the lifting body, the wing, and the rear fin were designed for optimum performance in any given flight regime and potential shell temperatures as a result of frictional heating. The rear fin was swept back at sixty degrees and was attached at the rear of the spacecraft on top of the vehicle's turbojet engine. Additional airbrakes were hinged on the upper surface of the fuselage. The wings themselves could be rotated to a vertical position during orbital injection and the initial portion of reentry to reduce thermal stresses. In the subsequent gliding phase through the atmosphere, these panels would be folded out to provide maximum surface area and better lift-drag ratios.

The single pilot's cockpit consisted of a pressurized metallic capsule lined with insulating material. In case of an emergency in orbit that might prevent the entire vehicle from deorbiting, the pilot could detach the headlight-shaped capsule from the main fuselage and use its own engine to reenter and land by parachute. The rear part of the cockpit thus had its own self-contained heat shield. To facilitate ejection, the capsule was mounted on two rails anchored to the fuselage structure with a pyrotechnic ejection device. Internal pressure and temperature would be maintained at 760 mm Hg and ten to fifty degrees Centigrade, respectively. While the pilot

---

113. Kazmin, "The 'Quiet' Tragedy of EPOS"; Afanasyev, "Unknown Spacecraft"; Lardier, *L'Astronautique Soviétique*, p. 175; Afanasyev correspondence, December 6, 1997.

could control most operations manually, including the elevons and rudders as well as the main turbojet engine, there was an on-board computer for navigation and automatic flight control.

For landing, instead of wheels, Lozino-Lozinskiy chose to use four skids retracting via compressed air stored at the front of the struts. With a high angle of attack, the ship would land on the rear skids first, before tipping forward onto the forward ones. Each skid strut was equipped with special shock absorbers.

For propulsion, the Orbital Aircraft had three different sets of engines. The primary engine for maneuvering in orbit and deorbiting was a one-and-a-half-ton-thrust rocket engine positioned at the rear of the fuselage. In addition to the main thrust chamber, the engine also had two auxiliary combustion chambers at forty kilograms each for use in case of primary engine failure. The propellants for the engine, unsymmetrical dimethyl hydrazine and nitrogen tetroxide, were carried in tanks positioned at the fuselage's center, near the ship's center of gravity. A second set of engines with an independent feed system would be used for attitude control in both space and the atmosphere. It consisted of six engines at sixteen kilograms thrust and ten engines at one kilogram thrust. The higher powered ones were the primary means of controlling pitch, yaw, and roll, while the lower powered units were for precise orbital stabilization. The final propulsion unit on the Orbital Aircraft was the powerful RD-36-35K turbojet engine created by the Rybinsk Design Bureau of Engine Building (formerly OKB-36) under Chief Designer Petr A. Kolesov, the famous aviation engine designer who had up to that point developed jet engines for Tupolev, Sukhoy, and Yakovlev. Rated at two-ton thrust, the kerosene engine could be used both at takeoff for test flights to reach Mach 0.8 and at landing.

As in the previous Soviet spaceplane programs, much of the research and development effort surrounding Spiral was focused on the development of reusable thermal protection for the spacecraft. For high-speed aircraft of the period, the Soviets were moving slowly from aluminum and aluminum alloys to titanium alloys and eventually to beryllium and niobium alloys. In creating the Orbital Aircraft, the engineers designed the vehicle in such a manner as to compensate for thermal stresses not by a resilient heat shield, but rather by its aerodynamic design. Tests showed that with a special heat shielding screen, the maximum temperature at stressful points, such as the front of the fuselage, the edges of the wings, and the tail, did not exceed 1,500 degrees Centigrade. Consequently, Lozino-Lozinskiy's engineers used titanium alloys and in some places aluminum alloys without any expensive coatings, such as tiles. The heat "screen" itself was not solid, but composed of a set of sheets, much like a fish's scales, suspended on ceramic bearings. Given deviations in temperature, these scales automatically changed shape while preserving the stability of the shield's relative position to the main body of the craft.[114]

Each flight of the Spiral system would begin with the use of a "launch truck" to boost the stack into the sky. In the case of the carrier aircraft using kerosene, the Hypersonic Booster-Aircraft was to take the complex to Mach 5.5–6 hypersonic velocities until the Orbital Aircraft with its two-stage booster separated at an altitude of twenty-eight to thirty kilometers. In using the hydrogen carrier variant, the separation was to occur at twenty-two to twenty-four kilometers altitude and at Mach 4. The two-stage booster would then come into operation and accelerate the vehicle to near-orbital velocity. Burn times would vary between 387.2 (liquid hydrogen) to 281.5 seconds (kerosene), depending on the propellant combination used. Then the Orbital Aircraft's own engine would kick in to inject the spaceplane into a low-Earth orbit at approximately 130 by 150 kilometers altitude. Orbital inclination would vary between forty-five and 135 degrees. The carrier aircraft would then flew back to its originating airport, ready for another flight.

114. Kazmin, "The 'Quiet' Tragedy of EPOS"; Afanasyev, "Unknown Spacecraft"; R. A. Belyakov and J. Marmain, MiG: Fifty Years of Secret Aircraft Design (Annapolis, MD: Naval Institute Press, 1992), pp. 417–21.

The flight of the Orbital Aircraft was short in duration, geared to its specific missions of interception, inspection, or reconnaissance. During the course of its two or three orbits in flight, the pilot could effectively change altitude and inclination of the orbit. After accomplishing its primary goal, the aircraft could dive into the atmosphere at a very high angle of attack (up to fifty-three degrees) with its wings folded at the standard forty-five degrees to the vertical and drop to hypersonic speed. When folded during reentry, the wings would remain in an aerodynamic "shadow," significantly reducing thermal stresses on critical areas while also improving stability. The spaceship was designed to have a 1,500- to 1,800-kilometer cross-range maneuver capability, allowing it much flexibility in choosing landing sites. After further reductions in speed, the pilot would unfold the spaceplane's wings to a near-horizontal position (ninety-five degrees to the vertical), glide down, and land at the chosen airport on its skids. In case the pilot was unable to land on the first pass over the runway, he would fire up the turbojet engine to steer the vehicle back for another try, at a landing speed of about 250 kilometers per hour.[115]

The Spiral project, as proposed in 1965–66, was to be performed in four distinct phases. During the first stage, MMZ Zenit was to build a suborbital analog of the Orbital Aircraft with a rocket engine for launch from a variant of the Tu-95 bomber named the Tu-95KM, apparently derived from the earlier Tupolev studies for the "product 130." The purpose of such tests was to evaluate the basic aerodynamic and power performance characteristics of the actual Orbital Aircraft in conditions close to spaceflight (altitudes of up to 120 kilometers and speeds up to Mach 6–8), as well as reentry into the atmosphere. Lozino-Lozinskiy planned to build three analogs, with subsonic flights beginning in 1967 and supersonic and hypersonic flights starting a year later.

In the second stage, engineers were to design and build the Experimental Piloted Orbital Aircraft (EPOS) for further improvement of design and flight characteristics of the Orbital Aircraft. The two vehicles were to be externally identical, differing only in some internal systems. The launch of the EPOS was planned on a standard Soyuz-type 11A511 booster. When Korolev and Lozino-Lozinskiy first discussed the use of an R-7–derived booster for use in the Spiral program, Korolev apparently pushed the idea hard. One of Lozino-Lozinskiy's deputies remembered later that Korolev's motivations for offering the Soyuz rocket for the Spiral program was "so he could get a big order for R-7's to make them cheaper."[116] After launch by the Soyuz booster, the spaceplane was to enter a 150- by 160-kilometer orbit with a fifty-one-degree incli-nation, make two to three orbits, and then perform a reentry and landing nearly identical to that planned for the Orbital Aircraft. According to the initial plan, MMZ Zenit was to build four mod-els of the EPOS for automated orbital missions beginning in 1969 and piloted missions the year after.

The third stage was to focus on the creation of the Hypersonic Booster-Aircraft, probably contracted out to Tupolev's OKB-156. The work on the Hypersonic Booster-Airplane was to begin with the creation of four models of the kerosene variant by 1970. After further experi-mental testing at hypersonic speeds, Tupolev's engineers were to proceed to the construction of the more complex hydrogen variant, with flight tests beginning in 1972. Four models were slated for production in the initial plan.

The final stage of the Spiral program included integrated testing of the entire system, with the Hypersonic Booster-Aircraft, the two-stage booster rocket, and the Orbital Aircraft. Automated

---

115. Afanasyev, "Unknown Spacecraft"; Kazmin, "The 'Quiet' Tragedy of EPOS"; Lozino-Lozinskiy and Plokhikh, "Reusable Space Systems and International Cooperation"; Andrey Batashev, "Steep Turns of the Spiral. A Quarter-Century Did Not Suffice for Implementing the Project Created by the 'Father' of the Soviet Shuttle" (English title), *Trud*, June 30, 1994, p. 4; Afanasyev correspondence, December 6, 1997.

116. James Harford, *Korolev: How One Man Masterminded the Soviet Drive to Beat America to the Moon* (New York: John Wiley & Sons, 1997), p. 275.

flights in the kerosene variant were to begin in 1972, leading to full-scale testing of a piloted variant using liquid hydrogen in 1973. It was, in all senses, a long-range program and one not tied
to meeting unrealistic deadlines arising from a necessity to respond to a similar U.S. project.

The Spiral project was huge, much larger than any of the previous spaceplane programs in
the Soviet Union, certainly rivaling and perhaps exceeding the amount of effort the U.S. Air Force
had invested in the Dyna-Soar program. The rich historical legacy of spaceplane research in
the USSR, leading all the way back to the Sänger-Bredt studies in the late 1940s, served as a
springboard for the new project. Apart from MMZ Zenit, another important player in the program
was the famous TsAGI, whose director ironically at the time of Spiral's birth was former General
Designer Myasishchev. Earlier, during 1961–64, Myasishchev had initiated a program under
the codename *Tayga* to study complex phenomena associated with hypersonic flight, inspired
apparently by concurrent American projects such as PRIME. Throughout 1965–69, TsAGI
scientists conducted extensive tests in wind tunnels to refine the design of the Spiral Orbital
Aircraft. Here, scientists used the MK-105 stand for determining the architecture of the complex
guidance system for the spaceplane. The institute also conducted tests in support of Spiral in
specially re-equipped L-18 flying laboratories. In 1967, a team of TsAGI scientists also began
research on determining the layout for a single-stage-to-orbit aerospace system using hydrogen
engines. Engineers studied the possibility of extrapolating the results of the Spiral program from
a one-person spaceplane to a multicrewed orbital transport vehicle. Remarkably, the Orbital
Aircraft's excellent lift-drag ratio and thermal characteristics were retained in the large model.

Based on the research at TsAGI, especially on the Tayga program, three institutions—the
M. M. Gromov Flight-Research Institute at Zhukovskiy, Plant No. 166 at Omsk, and MMZ
Zenit—cooperated in the design of a series of test beds to prove the basic technologies of the
new Spiral spaceplane program. Under the name Unpiloted Orbital Rocket-Glider (BOR),
the engineers set out to study the various critical points in a spaceplane's trajectory during both
suborbital and orbital flights. The early BOR vehicles came in three different variants, scale models of the EPOS at one-half and one-third size for launch on suborbital ballistic trajectories.
BOR-1, BOR-2, and BOR-3 were to be used primarily to study stability and controllability
characteristics at supersonic and subsonic speeds and also to evaluate the performance of
thermal shielding to be used on the EPOS.[117]

Some cosmonauts also got into the act. As early as December 1965, three pilots,
including veteran cosmonaut Titov, began preliminary studies in connection with the Spiral project. They performed more intensive flight training than was usual for other cosmonauts at the
time, first flying MiG-17s and then moving on to MiG-21s in 1966. By the following year, they
were flying fighter-interceptor aircraft of all types currently in operation with the Soviet Air Force.[118]
Perhaps not coincidentally, fifteen Air Force officers were at the time completing their graduate
degree work at the prestigious N. Ye. Zhukovskiy Military-Air Engineering Academy in Moscow.
At Korolev's behest, the entire group, which included most of the 1960 and 1962 cosmonaut
enrollments, were studying the development of a single-seat reusable spaceplane.[119] Among their
study duties was to analyze the performance characteristics of the defunct Dyna-Soar spacecraft.
The cosmonauts later named their own project "Buran-68," which as it turned out differed

117. E-mail correspondence, Igor Afanasyev to the author, December 11, 1997; Kazmin, "The 'Quiet' Tragedy
of EPOS"; *TsAGI-Osnovnyye etapy nauchnoy deyatelnosti, 1968–1993*, pp. 156, 244.
118. G. Titov, ". . . This is Needed for All of Us" (English title), *Aviatsiya i kosmonavtika* no. 4 (April 1993):
2–3. The other two cosmonauts training with Titov were A. V. Filipchenko and A. P. Kuklin, both rookies. See also
Kamanin, *Skrytiy kosmos: 1964–1966*, pp. 295, 306, 347.
119. S. M. Belotserkovskiy, *Gibel Gagarina: fakty i domysly* (Moscow: Mashinostroyeniya, 1992), p. 19. The
fifteen cosmonauts were V. F. Bykovskiy, Yu. A. Gagarin, V. V. Gorbatko, Ye. V. Khrunov, A. A. Leonov, A. G.
Nikolayev, T. D. Pitskhelauri, P. R. Popovich, Zh. D. Sergeychik, G. S. Shonin, I. B. Solovyeva, V. V. Tereshkova,
G. S. Titov, B. V. Volynov, and D. A. Zaykin.

significantly from Dyna-Soar, but was very similar to the Spiral EPOS spaceplane. Through complex mathematical modeling and theoretical research, each cosmonaut developed a particular part of the spaceplane. Gagarin was responsible for the general layout, the aerodynamic design of elements ensuring landing, and control systems. Titov developed the emergency rescue system. Nikolayev created the aerodynamic form for hypersonic and supersonic speeds as well as the thermal protection.[120] The Air Force's decision to have all of these cosmonauts focus on the space-plane theme underscored the fact that they were indeed very serious about the program.[121]

The Almaz, Zvezda, and Spiral projects were critical to Soviet plans to militarize space operations. Adding to the concurrent Soyuz, L1, and L3 programs, there were six major Soviet crewed projects by 1967, an impressive contrast to the two U.S. piloted space programs of the time, Apollo and MOL. From a political and pub-lic relations perspective, the military projects were, perhaps, less important than the three

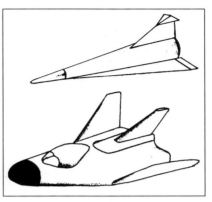

At top is a winged rocket-glider developed by the Tupolev design bureau in the early 1960s to carry out research at Mach 2–5 on the aerodynamic characteristics of a hypersonic winged vehicle. At bottom is the BOR-2 lifting body developed by the M. M. Gromov Flight-Research Institute in the late 1960s within the framework of the Spiral program. (copyright Asif Siddiqi)

major efforts in support of crewed lunar operations. The military and civilian programs ran parallel with each other with some modicum of interdependence, but all were affected by cosmonaut Komarov's tragic death in April 1967. For those involved in Soyuz, L1, and L3, in particular, the disaster paralyzed their efforts with uncertainty and doubt. Numerous deadlines fell through the cracks as engineers from TsKBEM began their long, hard road back to recovery.

---

120. *Ibid.*, pp. 16–17, 20. The topics of focus for some of the other cosmonauts were: Zaykin (work on com-ponents and computation of mass characteristics), Popovich (power sources), Khrunov (orientation systems), Bykovskiy (propellant system for the liquid-propellant rocket engine), and Sergeychik (safety systems on the flight).

121. There may have been a third competitor in the Soviet spaceplane programs apart from Mikoyan's OKB-155 and Tupolev's OKB-156: General Designer P. O. Sukhoy's OKB-51, whose proposal was evidently based on an existing high-speed bomber design named the T-4. In the early 1960s, Sukhoy had proposed the creation of a new-generation strategic supersonic bomber, which was part of a competition with the Tupolev and A. S. Yakovlev (OKB-115) design bureaus. On May 21, 1963, Sukhoy presented his conception of the T-4, also known as the "prod-uct 100" because it weighed 100–120 tons. The forty-four-and-a-half-meter-long aircraft had a maximum design speed of 3,200 kilometers per hour (Mach 3.01) and a supersonic range of about 6,000 kilometers. The T-4 bomber made only ten test flights between August 1972 and January 1974, one of which achieved supersonic speed. The Soviet Air Force, however, soured on this technological marvel by the early 1970s, believing that its goals could be performed by more conventional and reliable aircraft, such as the famous Tu-145, also known as the Tu-22M Backfire bomber. Three prototypes of the T-4 were scrapped, while a fourth one was consigned to an air museum after work was stopped in 1975. According to an interview with test pilot Maj. General V. S. Ilyushin on December 23, 1990, the T-4 was planned as a booster for a spaceplane. E-mail correspondence, Sergey Voevodin to the author, September 2, 1997; letter, Peter Pesavento to the author, August 15, 1997. See also Piotr Butowski, "Steps Towards 'Blackjack'," *Air Enthusiast* 73 (January–February 1998): 36–49; L. L. Selyakov, *Maloizvestnyye stranitsy tuorcheskoy deyatelnos-ti aviatsionnogo konstruktora Vladimira Mikhaylovicha Myasishcheva* (Moscow: AO ANTK im. Tupoleva, 1997), p. 112; Gunston, *The Osprey Encyclopedia of Russian Aircraft*, pp. 352–53; Mikhail Rebrov, "The Unknown 'One Hundred'" (English title), *Krasnaya zvezda*, September 13, 1995, p. 4; Svishchev, *Aviatsiya entsiklopediya*, pp. 550–51.

Chapter Fourteen
Getting Back on Track

The road out of the quagmire of the Soyuz 1 disaster was a difficult one. Because all three major piloted space projects—the Soyuz, the L1, and the L3—depended greatly on the vagaries of the basic Soyuz spacecraft, the accident had a widespread effect on the Soviet space program. Throughout 1966–67, the most important goal for the Soviets had been the celebration of the fiftieth anniversary of the Great October Revolution in November 1967 with a circumlunar flight of two cosmonauts in the L1 spacecraft. Because the L1 shared the same design as the Soyuz spacecraft that had killed Komarov, the disaster had grave implications for an early circumlunar flight. Technical issues were the primary determinant to any plans for lunar flyby in November 1967, but remarkably, the leading Soviet space officials still held out hope for meeting that increasingly elusive deadline.

The Tough Road Ahead

In late May 1967, two veteran NASA astronauts, Lt. Colonel Michael Collins and Lt. Colonel David R. Scott, arrived at the Paris Air Show to make a joint appearance with two Soviet cosmonauts, Colonel Pavel I. Belyayev and Konstantin P. Feoktistov. It was only a month after Komarov's death, but the unexpected meeting provided a brief but illuminating view of the Soviet space program. Over numerous toasts of vodka, what the astronauts found out was not so surprising: the cosmonauts indicated "that there would be several Earth orbital flights and then . . . a circumlunar flight."[1] As Collins later recalled, "Belyayev himself expected to make a circumlunar flight in the not-too-distant future."[2] The revelation was noteworthy precisely because of the almost complete information blackout on future plans in the Soviet space program. What was particularly astonishing was that despite the Soyuz 1 disaster, the Soviets were being remarkably optimistic in public of their circumlunar plans.

In October 1967, Academician Obraztsov stated with unusual explicitness that "the very next milestone in the conquest of space will be the manned circumnavigation of the Moon, and then a lunar landing."[3] But as if to cover their bets, in their typically confusing way, Soviet

1.   "Soviet Plans Manned Trip Round Moon," *Washington Post*, June 4, 1967, p. A9.
2.   Michael Collins, *Carrying the Fire: An Astronaut's Journeys* (New York: Farrar, Straus and Giroux, 1974), p. 280.
3.   *Soviet Space Programs, 1966–70: Goals and Purposes, Organization, Resources, Facilities and Hardware, Manned and Unmanned Flight Programs, Bioastronautics, Civil and Military Applications, Projections of Future Plans, Attitudes Toward International Cooperation and Space Law*, prepared for the Committee on Aeronautical and Space Sciences, U.S. Senate, 92d Cong., 1st sess. (Washington, DC: U.S. Government Printing Office, December 1971), p. 366.

spokespersons of the period ensured against the possibility of failure. Academician Leonid I. Sedov, the chairman of the "Commission for the Promotion of Interplanetary Flights" under the Academy of Sciences, was particularly notorious for brilliant obfuscations of the Soviet reach for the Moon. Because Western observers found it difficult to identify any single individual with real power within the Soviet space program, by default, many of Sedov's statements were magnified out of proportions, despite the fact that he had almost no connection whatsoever with the space program's operation. In September 1967, Sedov confidently told journalists that "manned flight to the Moon is not in the forefront of Soviet astronautics, as the problems of return from the Moon have still to be solved."[4] It was a typically disingenuous statement that was symptomatic of the Soviet public relations effort of the time.

One of the more prominent pronouncements of the period was a cryptic news item in August 1967 that ten Soviet cosmonauts were practicing sea landing tests for future space missions.[5] Unlike standard Earth-orbital flights, cosmonauts flying back from the Moon would potentially land in water areas because of the nature of their return trajectories. Among the group were four Air Force officers preparing for the commander's seat on the first lunar missions: veterans Leonov and Popovich and rookies Klimuk and Voloshin.[6] Remarkably, because of poor planning and bureaucratic gridlock, the trainees did not have the luxury of a 7K-L1 spacecraft simulator throughout 1967. One interesting component of their training regime in 1966–67 was to rehearse for the possibility that it would not be sufficiently safe to launch cosmonauts on the Proton booster, and, therefore, they would have to transfer to the 7K-L1 in Earth orbit from a Soyuz ship launched on a more reliable 11A511 rocket. The cosmonauts flew on parabolic trajectories in a Tu-104 aircraft and used a special curved tunnel to carry out the transfer. The results of the training were not too encouraging, and it proved to be a very difficult exercise.[7]

Immediately after the Soyuz 1 accident, despite pervasive uncertainty, TsKBEM engineers had assumed that the problem with Soyuz 1 would be quickly identified and eliminated. Just six days after Komarov's death, Chief Designer Mishin set a new tentative plan for the circumlunar project, with four automated 7K-L1 spacecraft flying around the Moon between June and August 1967. They would be followed by three piloted flights on spacecraft 8L, 9L, and 10L in sufficient time to make the November 1967 deadline. By June, however, a one-month delay had already accumulated, possibly because of the extensive and time-consuming work of the Soyuz 1 accident investigation commission. The Komarov disaster had other repercussions on the L1 program. It was clear to most senior space program leaders that the Soyuz docking and EVA mission would be delayed possibly to early 1968. This meant that the cosmonauts would not have an opportunity to rehearse an extravehicular transfer prior to a dual-launch circumlunar flight. During a meeting of the L1 State Commission in early June 1967, Chairman Tyulin officially decided to abandon the docking-in-Earth-orbit option for the circumlunar project and opt for launching cosmonauts on the new UR-500K Proton booster. As

4.   *Ibid.*, p. 365.

5.   "Soviet Describes Splashdown Tests," *New York Times*, August 25, 1967; "Cosmonauts Train for Water Recoveries," *Aviation Week & Space Technology*, September 11, 1967, p. 31; Viktor Mitroshenkov, *Zemlya pod nebom* (Moscow: Sovetskaya rossiya, 1987), p. 424.

6.   There were twelve cosmonauts training for the L1 program in May 1967. They were pilots V. F. Bykovskiy, P. I. Klimuk, A. A. Leonov, A. G. Nikolayev, P. R. Popovich, and V. A. Voloshin, as well as engineers Yu. P. Artyukhin, G. M. Grechko, O. G. Makarov, N. N. Rukavishnikov, V. I. Sevastyanov, and A. F. Voronov. See Vadim Y. Molchanov, "Soviet Manned Lunar Programs," *Quest* 2(4) (Winter 1993): 43. Other sources give a slightly different composition. See, for example, I. A. Marinin and S. Kh. Shamsutdinov, "Soviet Programs for Lunar Flights"· (English title), *Zemlya i uselennaya* no. 4 (July–August 1993): 62–69. Scientist V. G. Yershov is said to have joined the L1 training group in May 1967.

7.   Marinin and Shamsutdinov, "Soviet Programs for Lunar Flights."

a compensatory measure, he introduced two additional automated circumlunar missions into the flight sequence, making a total of six robotic flights before a piloted one. Of the six precursor missions, two had already flown in March and April 1967 with mixed success. The results of the remaining four would make or break the ability of the space program to make the sacred November 1967 deadline. The immense pressure to celebrate the anniversary with a piloted circumlunar mission was such that the first of the four remaining L1 ships would fly in July with the old parachute system because there was simply no time to install a modified version, corrected following Komarov's death.[8]

If there was any hope left for a circumlunar flight before the end of 1967, by mid-July, it was clear to most in the State Commission that the engineers would simply be unable to make the deadline. The first fully equipped 7K-L1 vehicle, spacecraft no. 4, had only just finished its experimental testing in July after a long four months.[9] TsKBEM Deputy Chief Designer Yevgeniy V. Shabarov, overseeing the preparation of the vehicle, spent many long days ensconced at the Kaliningrad plant eliminating problems from the vehicle. Preflight testing, usually lasting several weeks, had yet to even begin. Top Communist Party and government leaders, such as Ustinov, Serbin, and Smirnov, were simply in a state of panic, knowing that the first launch of the Saturn V was slated for late 1967, while the N1 was still many months away from flight. At a meeting of top officials in August 1967, Secretary of the Central Committee for Defense and Space Ustinov was infuriated. He told Mishin: "We have a celebration in two months, and the Americans are going to launch again, but what about us? What have we done? Imagine October 1967. Please understand this! We must suppress all personal interests and partiality!"[10]

On September 7, the L1 State Commission met to set a date for the launch of the first automated circumlunar flight of a 7K-L1 spacecraft. Several chief designers, including Mishin, Ryazanskiy (radio-control systems), and Barmin (launch complexes), reported on the readiness of the booster and the spacecraft.[11] Although many of the participants believed that their systems were 99 to 99.9 percent reliable, Mishin himself believed that the complete booster-payload system had a reliability rate of 60 percent, illustrating a remarkable lack of faith in the equipment. According to the plan, after flying around the Moon and heading toward Earth, the spacecraft would have the option of two different reentry profiles: a direct ballistic reentry into a 100- by 2,000-kilometer area in the Indian Ocean or the more preferable guided reentry in Kazakhstan. As a precautionary measure, the Soviet government signed an agreement with the Indian government in early September that would allow Soviet spacecraft to be brought to Indian soil following recovery.[12]

There were several malfunctions during the days leading up to the planned launch, but nothing critical enough to delay an automated flight. The 7K-L1 vehicle, spacecraft no. 4L, lifted off precisely on time in the dark night at Tyura-Tam at 0111 hours, 54 seconds Moscow Time on September 28, 1967. Air Force representative Lt. General Kamanin recalled the scene:

8.   N. Kamanin, "A Goal Worth Working for" (English title), *Vozdushniy transport* 46 (1993): 8–9.

9.   Yu. P. Semenov, ed., *Raketno-Kosmicheskaya Korporatsiya "Energiya" imeni S. P. Koroleva* (Korolev: RKK Energiya, named after S. P. Korolev, 1996), p. 241.

10.   A. Tarasov, "Missions in Dreams and Reality" (English title), *Pravda*, October 20, 1989, p. 4.

11.   Others reporting included Yu. N. Trufanov (TsKBM Branch No. 1 responsible for the UR-500K Proton) and P. A. Agadzhanov (Chief of the Chief Operations and Control Group and Deputy Chief of TsKIK).

12.   NASA Science and Technology Division, *Astronautics and Aeronautics, 1967: Chronology of Science, Technology, and Policy* (Washington, DC: NASA Special Publication (SP)-4008, 1968), p. 321. Another source says that the agreement between the two countries was signed on November 18, 1967. See Christian Lardier, *L'Astronautique Soviétique* (Paris: Armand Colin, 1992), p. 161.

This still from a movie shows the transport of a 7K-L1 circumlunar spacecraft on its Proton booster on the way from the assembly building to the launch pad at Tyura-Tam. Note the cluster of solid-propellant rocket engines at the top of the launch escape tower. The hatch on the external fairing for cosmonaut entry into the actual spacecraft can be seen in the foreground as a dark oblong shape. (files of Asif Siddiqi)

It immediately seemed to me, as well as other observers, that the rocket was going up slower than usual. But none of us counted seconds, and we all hoped that it was the rocket's unusual night launch that inhibited our ability to assess the takeoff adequately. When the first stage's side units decoupled, we were prepared to cast off doubts, but suddenly the automatic rescue system came into action, and the burning mass abruptly changed its path and began moving down to Earth. . . .[13]

It later transpired that one of the six main engines of the Proton first stage had failed to fire at launch. Remarkably, the ascent was steady for sixty-one seconds before diverting from a nominal path, which provoked the emergency rescue system into action. The booster itself crashed about sixty-five kilometers from the pad amid the thunder of loud explosions. The L1 descent apparatus separated from the wandering launch vehicle on time. Although the capsule was destabilized at the moment of separation because of an unexpected pressure shock, the vehicle landed safely in one piece not far from the exploded booster. When rescuers arrived, they were greeted by a strange scene: from one end of the horizon to the other, there was an eerie yellowish-brown cloud of nitrogen tetroxide and unsymmetrical dimethyl hydrazine all over the steppes. The descent apparatus lay majestically on top of a hill amid the toxic vapors.[14] The difficulty in rescuing the capsule was a nagging reminder of the dangers of using storable

13.   N. Kamanin, "A Goal Worth Working for" (English title), *Vozdushniy transport* 47 (1993): 8–9.
14.   Semenov, ed., *Raketno-Kosmicheskaya Korporatsiya*, p. 241.

propellants on a booster intended for launching humans into space. If there had been a crew aboard the descent apparatus, they might possibly have been exposed to the dangerous propellants.

With the foregone conclusion that there would not be any piloted circumlunar missions in 1967, the engineers trudged on with their work on the next 7K-L1 spacecraft. Late on the day of the launch failure, some members of the State Commission met to discuss the preliminary results of the accident investigation. Chief Designer Mishin, perhaps to lift the rapidly falling spirits of his engineers, told those present that they should not be discouraged and should work even more energetically for the next flight of the L1 spacecraft, tentatively set for the next lunar launch window in two months. It would be a busy time for TsKBEM engineers because Mishin had also scheduled the first post-Komarov flights of the Soyuz spacecraft in October. These would be followed by the L1 launch on November 21–22.[15]

On October 7, there was a major meeting at the Kremlin presided by Ustinov to discuss various aspects of the troubled L1 program. Chief Designer Glushko reported on the reasons for the unfortunate Proton failure on September 28. The single engine failure on the first stage had occurred because of the blocking of the propellant supply system by a rubber plug. The plug had evidently fallen into the engine during its assembly at Plant No. 19 at Perm, where the units were manufactured on order from Glushko's Design Bureau of Power Machine Building (formerly OKB-456). Ustinov castigated Minister of Aviation Industries Petr V. Dementyev for his negligence in the matter, telling his audience that the Proton failure had cost the Soviet government 100 million rubles and a two- to three-month delay in the circumlunar program. All the reports, from Minister of General Machine Building Afanasyev, Mishin, Tyulin, Chelomey, and others, were filled with recriminations against subcontractors who were inefficient in their deliveries.[16]

The fiftieth anniversary of the Great October Revolution passed with much fanfare in the first week of November 1967 all over the Soviet Union. But for those involved in the space program, it was a time marked by the acknowledgment that their handiwork had failed the task given them by the Soviet government. Since 1964–65, numerous decrees and decisions from the Central Committee, the Council of Ministers, the Military-Industrial Commission, and the Ministry of General Machine Building had all aimed for this date as the holy grail of Soviet cosmonautics—the month when two Soviet citizens would fly around the Moon and bring their hammer-and-sickle flags back to parades and celebrations in honor of the Bolsheviks. It, of course, never happened that way. Engineers, cosmonauts, chief designers, ministers, and military officers all dug back into preparations for the next circumlunar launch attempt. A success would bring some consolation to a beleaguered effort.

In mid-November, L1 State Commission Chairman Tyulin arrived at Tyura-Tam to oversee the prelaunch testing of the flight vehicle, the 7K-L1, spacecraft no. 5L. Several of the lunar cosmonauts, including Leonov, Popovich, and Dobrovolskiy, were escorted to the launch site by Kamanin on the morning of November 18.[17] After the launch, they were evidently to fly to Yevpatoriya to participate in the control of the vehicle during its weeklong circumlunar mission. The only prominent chief designer present at the launch range to oversee preparations was Glushko; Mishin and Chelomey did not arrive until 36 and 11 hours, respectively, before launch, probably because of numerous prior commitments in several other concurrent projects. It was a particularly chilly launch night at Tyura-Tam, with the Moon beautifully suspended over the Proton

15. Kamanin, "A Goal Worth Working for," no. 47.
16. *Ibid.* Cosmonaut A. A. Leonov has also described the reason for the Proton failure: "It turned out that a rubber plug had fallen into the manifold ahead of the turbopump assembly. Having gotten stuck in the line, it cut off the fuel feed." See Major I. Kuznetsov, "The Flight That Did Not Occur" (English), *Aviatsiya i kosmonavtika* no. 8 (August 1990): 44–45.
17. Mitroshenkov, *Zemlya pod nebom,* p. 426.

launch pad. The 7K-L1 spacecraft lifted off just after midnight local time, 2208 hours Moscow Time, on November 22, 1967. Everything seemed to be working perfectly until second-stage operation, when one of the four engines of the second stage failed to ignite. The remaining three engines continued to fire for four additional seconds until an automatic signal from the ground detecting trajectory deviation shut them off. Once again, the emergency rescue system fired on time and shot the L1 descent apparatus away from the launch vehicle. The descent apparatus crashed about 300 kilometers from the pad, while the automated crew capsule flew eighty kilometers southwest of the town of Dzhezkazgan. Because of a spurious command from the vehicle's altimeter, the soft-landing engines fired at an altitude of four and a half kilometers instead of just prior to touchdown, causing the capsule to perform a "hard" landing. Engineers later added a filter to the gamma-ray altimeter to preclude such malfunctions, in both the L1 and Soyuz spacecraft.[18]

At the end of 1967, the pressure was off Mishin a little bit. No longer chasing after an impossible target, his immediate goal was to beat the Americans in a circumlunar flight. Given that piloted Apollo operations were not expected to resume prior to the fall of 1968, the Soviets could be forgiven for being optimistic about doing just that. The accumulated delays allowed engineers to continue fine-tuning the 7K-L1 spacecraft design. One of their ultimate goals was to replace the original *Argon-11* computer by the more improved *Salyut-1* model sometime in 1967–68. The engineers also continued to shave off weight from the vehicle in an attempt to optimize its capabilities. The major changes introduced into the Soyuz spacecraft parachute system were also incorporated into the L1. The results of the testing were, however, not very encouraging. On January 26, during a test of the L1 landing system at the Air Force range at Vladimirovka near Kapustin Yar, the parachute shot out and filled with air but abruptly collapsed, and the capsule crashed on to the ground and exploded.[19]

In January, the L1 cosmonauts finally began training in a specially built simulator delivered by the Special Experimental Design Bureau of the M. M. Gromov Flight-Research Institute at Zhukovskiy near Moscow. The simulator, known as *Volchok* ("Top"), was installed at the Air Force's Institute of Aviation and Space Medicine to allow cosmonauts to train for the return to Earth at lunar velocities. The simulator was part of a complex that included an M-220 computer, a centrifuge, the L1 cockpit, and an instructor's control panel. The L1 group conducted at least seventy runs on the simulator using precise methodologies for the circumlunar training program consisting of the two different reentry profiles: one ballistic and the other guided. The favorite to command the first circumlunar mission, cosmonaut Leonov, later recalled:

> We had to learn to choose the angle of entry after the last [mid-course] correction using the star-tracker and sextant. [The angle] depended on the magnitude and direction of the deceleration burn. It was possible to "bury" oneself in the atmosphere with a large angle and to "slip through" it with a small angle. The optimum version was an entry with a "pop-up": enter, exit the atmosphere after extinguishing great speed, and reenter, already knowing the angle of incidence at which the craft had to be held to get to the calculated landing point. The "manual firing input" instrument highlighted the number of burns after passage of the first sector. From that we figured the distance to the calculated landing point, then converted distance into angle of incidence. . . . As a result we learned to make a "landing" with an accuracy to one kilometer.[20]

18.   Semenov, ed., *Raketno-Kosmicheskaya Korporatsiya*, p. 241; Kamanin, "A Goal Worth Working for," no. 47.
19.   N. Kamanin, "A Goal Worth Working for" (English title), *Vozdushniy transport* 48 (1993): 8–9.
20.   Kuznetsov, "The Flight That Did Not Occur," p. 44.

General L1 training consisted of studying the 7K-L1 ship's on-board systems, the dynamics of its motion, mathematical support, programming, ballistics, and astro-navigation. Included in the cosmonauts' training program was a ten-day trip to Mogadishu, Somalia, in the summer of 1968 to familiarize themselves with the constellations in the southern sky; returning L1 vehicles would fly over Antarctica, then Africa, before heading toward Soviet territory. On an actual flight, the vehicle would use its star-tracker and sextant for autonomous navigation, and the cosmonaut would take over in case of sensor malfunction.[21]

By early February 1968, Mishin and Kamanin had agreed on the selection of four crew commanders to train for the first few missions: cosmonauts Bykovskiy, Leonov, Popovich, and Voloshin.[22] They, along with eight others, were engaged in an intensive program throughout 1967–68, but it seems that they did not have much confidence in the spacecraft. Kamanin recalled in early March that:

> [The cosmonauts] are working diligently and know the craft well. Perhaps, it is precisely because the cosmonauts excellently know all the strong and weak points of the craft and the carrier rocket that they no longer have their initial faith in the space hardware.[23]

In their training in the L1 simulators, the cosmonauts remarked that although it was quite easy to work with the new instrumentation, it would be a very trying job to spend about seven days cramped in the tiny descent apparatus of the 7K-L1 vehicle.[24] The two recent launch failures of the Proton booster did not do much to raise their spirits.

The next L1 launch was set for March 1–2, 1968. The unusually long gap between the fourth and fifth L1 flight attempts was partly a result of the poor results of the emergency rescue system's ground testing of the UR-500K-L1 booster stack, carried out under Deputy Chief Designer Tsybin's direction. There were evidently repeated parachute failures in the escape system in January and February, but the necessity to maintain deadlines prompted him to recommend launches despite complete confidence in the systems. On February 20, the L1 State Commission met, presided over by an ill Tyulin. General Designer Chelomey and Chief Designer Aleksandr D. Konopatov of the Design Bureau of Chemical Automation, responsible for the Proton's second-stage engines, briefed the attendees on the possible reasons for the two consecutive failures in late 1967. While the specific cause of the November 1967 malfunction was still unknown, the two designers believed that the premature ignition of propellant because of local heating to more than 200 degrees Centigrade led the suspect engine to fail. Chelomey, Konopatov, and Mishin proposed a number of changes to the engine design—suggestions that were approved by the remaining members of the State Commission. At this point, the State Commission still planned to carry out four more fully automated L1 flights before proceeding with a crewed flight.

A number of the cosmonauts training for the L1 program, including Bykovskiy, Leonov, Popovich, and Sevastyanov, flew to Tyura-Tam in a Tu-124 aircraft on February 28, 1968,

---

21. *Ibid.*; Marinin and Shamsutdinov, "Soviet Programs for Lunar Flights."
22. Kamanin, "A Goal Worth Working for," no. 48. By this point, five tentative crews had been formed for the L1 program: A. A. Leonov/O. G. Makarov, V. F. Bykovskiy/N. N. Rukavishnikov, P. R. Popovich/V. I. Sevastyanov, V. A. Voloshin/Yu. P. Artyukhin, and P. I. Klimuk/A. F. Voronov. In addition, one civilian scientist, V. G. Yershov, and another civilian engineer, G. M. Grechko, also trained with the core group of ten. For crew complements, see S. Shamsutdinov and I. Marinin, "Flights Which Never Happened: The Lunar Program" (English title), *Aviatsiya i kosmonavtika* no. 2 (February 1993): 30–31.
23. Kamanin, "A Goal Worth Working for," no. 48, p. 9.
24. I. B. Afanasyev, "Unknown Spacecraft (From the History of the Soviet Space Program)" (English title), *Novoye v zhizni. Nauke, tekhnike: Seriya kosmonavtika, astronomiya* no. 12 (December 1991): 1–64.

accompanied by cosmonaut overseer Col. General Kamanin and first cosmonaut Gagarin, who, although he was not preparing for a mission, was closely involved in the L1 cosmonauts' training program.[25] It was very windy and cold at the launch site, and the snow cover gave the area a beautiful sheen. Later that day, the State Commission held a meeting to discuss the specific plans for the next launch, set for March 2. Besides Mishin and Chelomey, their deputies for the 7K-L1 and the Proton booster—Yevgeniy V. Shabarov and Yuriy N. Trufanov, respectively—spoke on the readiness of all the preparations. Because there was no lunar launch window at the time, Mishin and Chelomey had agreed to launch the spacecraft out to a distance of about 330,000 kilometers into deep space—that is, out to lunar distance—and then bring the vehicle back to Earth, thus simulating an actual circumlunar flight. The nonlunar objective also gave launch controllers the luxury of having launch windows lasting more than just a few seconds. The next 7K-L1 launch, slated at the time for April 23, would fly to the Moon.[26]

There was a remarkable lack of confidence during the preflight preparations. Even State Commission Chairman Tyulin had misgivings about the launch. Kamanin wrote in his journal on March 1: "All of us need a successful launch like a breath of fresh air. Another failure would bring innumerable troubles and may kill the people's confidence in themselves and the reliability of our space equipment."[27] The 7K-L1 ship, spacecraft no. 6L, lifted off at 2129 hours, 23 seconds Moscow Time on March 2, 1968, into a circular Earth orbit at around 200 kilometers altitude at a fifty-one-and-a-half-degree inclination. Exactly one hour, eleven minutes, and fifty-six seconds after launch, the Blok D stage fired for 459 seconds to boost the spaceship into a highly elliptical orbit with an apogee of 354,000 kilometers. The Soviet news agency TASS did not announce anything of note about the launch, except to name the spaceship Zond 4 ("zond" being the Russian word for "probe"). The Zond designation had previously been used for three completely unrelated deep space probes in the early 1960s, and it was a curious excavation of an obsolete moniker. Retroactively, the Soviets would call the entire circumlunar effort the Zond program.

The day after launch, a group of cosmonauts led by Gagarin flew to the flight control center at Yevpatoriya to support the activities of the Chief Operations and Control Group. The L1 crew of Popovich and Sevastyanov, one of the leading contenders for an early mission, spent long periods in a special "bunker" at Yevpatoriya, playing the role of an actual flight crew. Communications between the two were routed through Zond 4 back to Yevpatoriya to simulate as closely as possible realistic conditions during an actual mission.[28]

The first minor sign of trouble on the flight appeared on the morning of March 4. At 0753 hours Moscow Time, the controllers attempted to carry out the first mid-course correction, but they failed to do so because of a failure in the attitude control system: the 100K stellar sensor (using minimum shading) correctly tracked the Sun, but failed to find Sirius. The first mid-course correction was, however, not a necessary factor for a successful mission, and engineers were confident that everything would work fine. All systems on Zond 4, including the communications systems, were working without serious disruptions, although the main omnidirectional antenna had evidently not unfurled properly. A second attempt to use the stellar orientation system on March 5 was also a failure; the sensor tracked Sirius for only a few seconds (with maximum shading) before losing it, suggesting some sort of malfunction in the astro-orientation sensor built by

---

25. Mitroshenkov, *Zemlya pod nebom*, p. 436.
26. Kamanin, "A Goal Worth Working for," no. 48. The March mission was timed to be launched a half lunar month outside the nominal lunar launch window and was, in fact, aimed in the exact opposite direction of the Moon.
27. *Ibid.*, p. 9.
28. Marinin and Shamsutdinov, "Soviet Programs for Lunar Flights"; Shamsutdinov and Marinin, "Flights Which Never Happened: The Lunar Program."

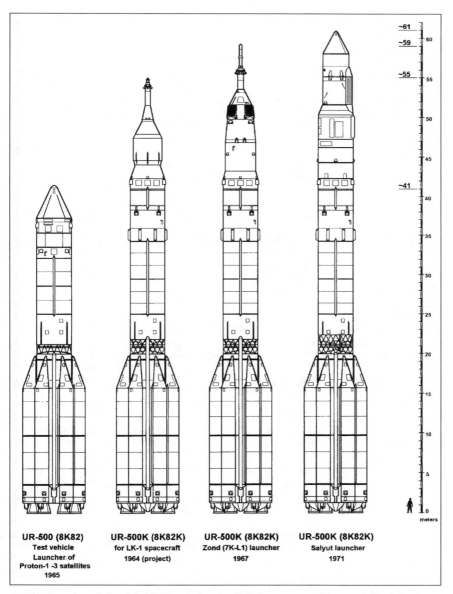

UR-500 (8K82)	UR-500K (8K82K)	UR-500K (8K82K)	UR-500K (8K82K)
Test vehicle	for LK-1 spacecraft	Zond (7K-L1) launcher	Salyut launcher
Launcher of	1964 (project)	1967	1971
Proton-1 -3 satellites			
1965			

This shows the evolution of the UR-500 space launch vehicle. For many years, Westerners believed that the clustered modules of the first stage were strap-on boosters. These were in fact only propellant tanks and not self-contained stages. (copyright Peter Gorin)

the Geofizika Central Design Bureau. The engineers finally declared success the next day when a medium-density filter on the sensor proved to be the right solution to the stellar tracking problem. The vehicle was oriented properly and fired its main engine to sharpen its trajectory.[29] Ballistics calculations showed that Zond 4's trajectory was perfect and that there would be no need for further mid-course corrections. The vehicle was expected to enter Earth's atmosphere down to an altitude of only 45.8 kilometers, then bounce out to 145 kilometers and then reenter again.

In the complex schema of Soviet ground control over spacecraft, the Zond flights were controlled from Yevpatoriya, but supported by ballistics centers at NII-4 in Bolshevo and a new Coordination-Computation Center at the premises of the Central Scientific-Research Institute for Machine Building (TsNIIMash) located right next to TsKBEM in Kaliningrad. The Coordination-Computation Center had provided only ballistics support for space missions since January 1963, but it had steadily expanded its activities in the mid-1960s to support the piloted lunar program. It would eventually form the basis for the famous Flight Control Center (TsUP) that controlled all missions to the Mir space station.[30] Some of the Air Force officers involved with the Zond 4 flight were in attendance at the Coordination-Computation Center during the return portion of the spacecraft's trajectory as they saw the projected "pop-up" trajectory mapped out on giant screens in front of them. But the projections were unfortunately markedly different from the true path of Zond 4 on March 11. After the vehicle separated into its two component parts, the descent apparatus was evidently in the wrong attitude because of the "unpreparedness of the orientation system." Thus the spacecraft entered the atmosphere into the correct corridor, but then never left it. Instead, it entered into an uncontrollable ballistic trajectory. It evidently passed through the atmosphere safely and was about to deploy its parachutes, when at an altitude of ten to fifteen kilometers over the Gulf of Guinea near the west African coast, the emergency destruct system of the descent apparatus was commanded to explode the capsule. The destructive charge had been included on the spacecraft for precisely such a contingency: "for fear that the Americans may get hold of it."[31] The Soviet press refrained from commenting on Zond 4's fate, although in later years, official Soviet publications would say that the spacecraft was in heliocentric orbit.[32] The order to destruct had strong support: Tyulin and Mishin evidently cleared the decision through Central Committee Secretary Ustinov and Military-Industrial Commission Chairman Smirnov.

A crew in the spacecraft would have endured up to twenty g's during the descent, but would probably have survived the splashdown. The main problem on the Zond 4 spacecraft was traced to the 100K stellar sensor, whose surface had evidently been contaminated. For future vehicles, engineers introduced a special cover for the sensor, which would be cast off before use. The State Commission for the L1 program met on the afternoon of March 26, 1968, to discuss the status of the project. TsKBEM Deputy Chief Designer Chertok summarized all the failures of the stellar attitude control sensor on Zond 4 as well as the results from the flight.

29. Kamanin, "A Goal Worth Working for," no. 48.
30. The genealogy of this center can be traced back to May 13, 1959, when the Council of Ministers issued a decree for the formation of a Computation Center (VTs) at the premises of NII-88 in Kaliningrad. In January 1963, it assumed the role of one of the many ballistics centers for space missions. In October 1964, this ballistics center served as the chief ballistics center for the Voskhod mission. A second decree of the Central Committee and the Council of Ministers on October 25, 1965, led to the creation of the Coordination-Computation Center (KVTs) on the basis of the ballistics center. See V. I. Lobachev, V. N. Pokuchayev, and N. P. Shcherbakova, "3 October—30 Years From the Beginning of Functioning of the Computation Center of the NII-88 (TsNIIMash), Assumed as the Start of Creation of the Soviet Flight Control Center (1960)" (English title), Iz iztorii aviatsii i kosmonavtiki 64 (1993): 98–106.
31. Kamanin, "A Goal Worth Working for," no. 48; Semenov, ed., Raketno-Kosmicheskaya Korporatsiya, p. 241. The mission duration was about ten days, nineteen hours.
32. See, for example, Yu. A. Mozzhorin, ed., Kosmonavtika (Moscow: Mashinostroyeniye, 1981), p. 446.

Mishin reported that the next 7K-L1 vehicle and its Proton booster would be ready for the next launch by April 20–22, in time for the next lunar launch window just after midnight local time on April 23.[33]

The L1 spacecraft arrived at the Baykonur Cosmodrome on April 12, the anniversary of Gagarin's Vostok flight in 1961. State Commission members Tyulin and others flew into the launch range four days later in preparation for the launch. Hopes were high that this would be the first fully successful automated circumlunar mission in the Soviet space program. The preparations for the launch proceeded without significant problems. The unusually cold April temperatures, down to minus five degrees Centigrade at night, did not deter work, which was concurrent with an unrelated Soyuz precursor flight in Earth orbit. The cosmonauts and officials were housed for the first time in the new Kosmonavt Hotel, a fully furnished abode for crews to spend their days before launch. On the morning of April 20, the State Commission met to go over all the changes in the 7K-L1 vehicle since the flight of Zond 4, including the modifications to the critical stellar sensor, responsible for the demoralizing failure at the end of the mission.[34]

At a last meeting on April 22, one of the topics of discussion was whether to blow up future 7K-L1 spacecraft if they returned to Earth in uncontrolled trajectories. Chief Designer Mishin, along with Deputy Chief Designer Shabarov, vigorously supported such a contingency but were opposed by Chief Designer Barmin, Kamanin, and all the cosmonauts. Many, including Chelomey, remained neutral, perhaps unwilling to take a stand on an issue that had implications for national security. In the end, a final decision seems to have been postponed; Mishin evidently believed that a ballistic landing would be unlikely on this particular flight.

It was another cold night launch for the program. The UR-500K rocket lifted off precisely on schedule at 2301 hours, 7 seconds Moscow Time on April 22 with the 7K-L1, spacecraft no. 7L. The rocket flew gracefully into the dark skies as observers watched the exhaust become smaller and smaller. About seven minutes after launch, at T+260 seconds, the flame abruptly disappeared, although the third stage had yet to fire. It was clear that there had been some malfunction and that the emergency rescue system had been activated. The controllers at Tyura-Tam received a report from the rescue service about four hours after launch that the L1 descent apparatus had landed 520 kilometers from the launch site, about 110 kilometers east of the town of Dzhezkazgan in Kazakhstan. The initial reports were distressing: a helicopter commander relayed that he had located the capsule but that it was on fire, an impression confirmed by search service commander Air Force Maj. General Aleksandr I. Kutasin. In the morning, it turned out that both had been mistaken; the 7K-L1 capsule landed without problems, and all elements of its rescue system had worked flawlessly. By the afternoon, the capsule was back at Tyura-Tam, a stop on its trip back to Moscow the following day.[35]

A cursory investigation into the accident indicated that the failure was not because of a booster problem. A sensor on the spacecraft had erroneously detected a breakdown and ordered the booster's second-stage engines to shut down and abort the flight. By the late morning of April 23, engineers were leaning toward some sort of failure in the 7K-L1's power supply system. The failure laid to rest any hope that there would be a crewed circumlunar flight before the fall of 1968 at the earliest. Of the four L1 attempts in 1967–68 to fly to lunar distances, only one, Zond 4, had been a partial success. The remaining three had failed to reach even Earth orbit, underlying serious problems in the launch vehicle. The entire program was already more than a year behind schedule, with many tests still to be carried out. With little

33. N. Kamanin, "For Him, Living Meant Flying" (English title), *Vozdushniy transport* 9 (1994): 8.
34. N. Kamanin, "For Him, Living Meant Flying" (English title), *Vozdushniy transport* 12 (1994): 12.
35. N. Kamanin, "A Goal Worth Working for" (English title), *Vozdushniy transport* 49 (1993): 8.

hope of an impending piloted mission, the L1 cosmonauts were sent on leave on June 1, 1968. On May 20, Mishin held a meeting at his design bureau and targeted July 17 as the next launch opportunity for a circumlunar flight, putting a three-month gap between missions. The accident investigation of the last launch failure was evidently a big factor in the long interval.

Not surprisingly, the political leadership at this time was extremely disconcerted by the continuing series of failures in the program. Mishin met with Military-Industrial Chairman Smirnov in May 1968 to discuss the status of the project. The latter asked Mishin to accelerate the pace of work on the L1 as much as possible to launch a crew around the Moon by October 1968. Smirnov's boss, Ustinov, had also set the same deadline, which took into account three more automated launches in July, August, and September, leading to a flight by two cosmonauts in October.[36] Despite the spate of setbacks, publicly the Soviets continued to maintain their interest in a piloted circumlunar flight. On a tour of Hungary in February 1968, cosmonauts Belyayev and Bykovskiy were remarkably explicit in their pronouncements. The latter, one of the leading candidates for commanding the first circumlunar flight, told journalists:

> The Soviet Union will send men to the Moon only when there is no longer any risk, and there is every guarantee that a safe return can be made. One of our next steps is not a Moon landing, but the orbiting of the Moon by a manned space vehicle. Naturally [the death of Komarov] had a certain retarding effect. It took many weeks to investigate and learn the causes of the accident. However, it caused no essential revisions in the space research and spaceship development program which had been worked out.[37]

In a hint of the troubles facing the circumlunar project, Academician Vasiliy V. Parin, one of the leading space biomedicine specialists in the Soviet space program, did admit that precursor "pathfinder" flights could delay the first Soviet piloted lunar mission.[38]

U.S. observers were also getting in on the act. Through the spring of 1968, U.S. government officials and the American press were unusually vocal about imminent Soviet space plans. Noted journalist John Noble Wilford wrote in February that among the immediate goals of the Soviet space program was "[a]n unmanned flight of the Soyuz around the moon and back to earth, without attempting a landing on the lunar surface . . . this summer."[39] That U.S. intelligence was clearly cognizant of the troubles plaguing the Soviet space program at the time was confirmed by articles in the U.S. media, clearly noting the two recent L1 launch failures in November 1967 and April 1968, which were covered up by the Soviets.[40] The knowledge of these failures does not seem, however, to have given pause to exclamatory pronouncements in the American media. In a prominent page-one article in *The New York Times* on May 5, a reporter claimed: "A mass of public and private evidence about the Soviet Union's recent space exploits has led analysts to believe that the American public is in for a series of space surprises."[41] No one could guess at the paramount level of managerial, technological, and funding chaos plaguing the Soviet piloted space effort.

36. N. Kamanin, "For Him, Living Meant Flying" (English title), *Vozdushniy transport* 16 (1994): 11.

37. *Soviet Space Programs, 1966–70*, p. 368; "Moon Fly-Around by Soviet Likely," *Baltimore Sun*, February 26, 1968, p. A3.

38. NASA Science and Technology Division, *Astronautics and Aeronautics, 1968: Chronology of Science, Technology, and Policy* (Washington, DC: NASA SP-4010, 1969), p. 105.

39. John Noble Wilford, "Renewed Soviet Space Drive Likely," *New York Times*, February 18, 1968, p. 18.

40. See, for example, Evert Clark, "Soviet Resumes Tests of Orbital Bombing System," *New York Times*, April 26, 1968, p. 35.

41. Evert Clark, "Soviet Advances in Space Awaited," *New York Times*, May 5, 1968, pp. 1, 50. For other articles claiming a big Soviet push in space, see Evert Clark, "Manned Flight Expected," *New York Times*, April 16, 1968, p. 37; Raymond H. Anderson, "Soviets in Space: A New Glamour Phase," *New York Times*, April 28, 1968, Section IV, p. 11.

A little more than a month after that article, on June 26, 1967, the LI State Commission met to discuss preparations for the next launch. Engineers from TsKBEM admitted that they, and not Chelomey's engineers responsible for the Proton booster, had been to blame for the most recent LI launch failure in April. A short circuit in the power supply system of the spacecraft's computer resulted in the "Accident in the Autonomous Guidance System" command being sent from the vehicle to the booster. Consequently, the engines in the second stage of the Proton automatically switched off. The problem was traced to a design error on the part of Department No. 212 at the TsKBEM, which had incorrectly mounted the three-axis stabilized platform in the descent apparatus of the LI.[42] Mishin and Tyulin agreed to attempt the next circumlunar launch on July 19. This flight would be followed by similar launches in August, September, and October. After three to four automated flights of the UR-500K-LI system, cosmonauts would fly to the Moon in November–December 1968, well over a year later than originally intended.

This schedule was again put into jeopardy as a result of a near-catastrophic accident at Tyura-Tam during the summer of 1968. On July 15, four days prior to the intended launch, the 7K-LI spacecraft, the Proton booster, and the Blok D upper stage were undergoing combined testing at the launch pad at the Baykonur Cosmodrome. The stack had already been fully loaded with propellant when the oxidizer tank of the Blok D stage exploded. The first reports suggested that the rocket, the spacecraft, and the pad were destroyed, killing three pad technicians. Later, it transpired that although the Blok D stage was destroyed, both the UR-500K launcher and its LI payload were relatively intact. One person, a Captain I. D. Khridin, had been killed and another seriously injured. The accident had occurred because of an erroneous electrical command from a malfunctioning ground cable network, which resulted in excess pressure in Blok D. The situation after the accident was extremely dangerous. The LI spacecraft and part of Blok D tipped over to one side, supported only by the emergency rescue system tower, which was stuck on a service girder on the pad structure. Blok D's fuel tank, with five tons of kerosene and two attitude control engines with their own oxidizer and fuel, had broken away from the girder and had pushed deep into the third stage of the Proton. Observers watched in terror as the seriousness of the situation became deathly clear. At the time of the accident, the payload contained five tons of fuel in Blok D, one and a half tons of solid propellant in the emergency rescue system tower, more than one and a half tons of toxic propellants for Blok D's attitude control system, thirty kilograms of highly concentrated hydrogen peroxide in the LI's guided reentry system, four and a half liters of triethylamine for the ignition of the Blok D propellants, benzine-based fuel for the thermo-regulation system connected to more than 150 pyrocar-tridges, and twenty-five kilograms of explosive for the payload's self-destruct system. It was a highly toxic explosion waiting to happen as more than 150 pad technicians stood in shock on trusses and girders all around the booster. Fortunately, not one of the pipes in any of the systems punctured.[43]

Because the situation was so serious, Minister of General Machine Building Sergey A. Afanasyev headed up an emergency commission to save the pad, the booster, and the space-craft. Afanasyev's First Deputy Tyulin supervised the general work of cutting the payload block to begin slowly pouring out propellants. Mishin personally directed all operations at the launch pad to separate, painfully and slowly, each component of the payload from the launch stack in the unbearably hot temperatures at the launch site. It took two weeks of concerted effort to finally dismantle the complex, based on thorough calculations on each component's center of

42. Semenov, ed., *Raketno-Kosmicheskaya Korporatsiya*, p. 242; Kamanin, "A Goal Worth Working for, no. 49.

43. Semenov, ed., *Raketno-Kosmicheskaya Korporatsiya*, pp. 242–43; Kamanin, "A Goal Worth Working for, no. 49.

gravity after the accident. Both the July and August lunar launch windows were abandoned as a result, reducing further the odds of a piloted circumlunar mission before the end of 1968. The best-case scenario was a December launch, although unofficially many engineers believed that January 1969 was a more realistic target. Maintaining this new deadline was complicated further by plans to concurrently run Soyuz missions in Earth orbit, which were indispensable to advancing the Soviet lunar *landing* program. Unlike the L1, however, the Soyuz had a less painful road back to recovery after the Komarov tragedy in 1967.

Docking in Orbit

In April 1967, when cosmonaut Komarov set off on his last mission, there were fairly distinct plans for at least two further Soyuz missions to follow. Both would have been solo Earth-orbital missions, the first (Soyuz 3) commanded by Gagarin and the second (Soyuz 4) commanded by rookie Beregovoy.[44] For Gagarin's career, the Soyuz 1 disaster was a severe setback. Having lost one of Soviet Union's best and brightest, cosmonaut overseer Lt. General Kamanin was not about to jeopardize Gagarin's life in grueling training programs. On April 29, 1967, five days after the accident, Kamanin met with a number of cosmonauts, including Gagarin. Beregovoy recalled that:

> . . . Kamanin, who looked aged by the tragedy, called us all together and laid out the future flight programme. He told Gagarin straight out that there was practically no chance he would be allowed to fly again. Kamanin himself would recommend that Gagarin not be permitted to participate in any other flights. Yuri listened to this terrible pronouncement in silence.[45]

The most immediate matter at hand for Kamanin was to reestablish a training plan for Soyuz, contingent upon a new schedule of flights set by Chief Designer Mishin. In revising the Soyuz manifest, all agreed that the first subsequent crewed mission should be a repeat attempt to carry out the aborted docking and EVA flight from Soyuz 1. By May 5, Kamanin had tapped test pilot Beregovoy to pilot the active vehicle. As plans stood at the time, the old Bykovskiy crew from Komarov's mission would remain as a team to fly the passive Soyuz spacecraft. They began training with the *Volga* rendezvous simulator by the fall of 1967.

Ironically, by the time that the Soyuz 1 disaster paralyzed the Soviet piloted space program, the cosmonaut corps was welling to its greatest number. Traditionally, most cosmonaut trainees were military pilots or engineers. Mishin's insistence on including engineers from TsKBEM had forced the Air Force to accept civilians who had participated in the design of the Soyuz spacecraft. Although such a group of eight engineers had begun training in late 1966, they did not receive official status as "cosmonaut-testers" until an order of the Ministry of

---

44.   E-mail correspondence, Sergey Voevodin to the author, January 30, 1997. At the time, the Soyuz 3 crew consisted of Yu. A. Gagarin/V. N. Volkov (primary) and A. G. Nikolayev/V. N. Kubasov (backup). The Soyuz 4 crew was G. T. Beregovoy/L. S. Demin/G. S. Shonin (primary) and D. A. Zaykin/A. N. Matinchenko/G. T. Dobrovolskiy (backup). See also V. Molchanov, "First Selection" (English title), *Apogey* 8 (March 1994): 2. In his diary entry dated December 7, 1966, N. P. Kamanin provides a slightly different crew composition. The Soyuz 3/4 crews would have been G. T. Beregovoy (Soyuz 3) and V. A. Shatalov/V. N. Volkov/O. G. Makarov (Soyuz-4). These would probably have been the backup crews for Soyuz 3/4. Kamanin also writes that Soyuz 5 would be commanded by one of Beregovoy, Bykovskiy, Gagarin, Komarov, Nikolayev, and Shatalov. The remaining two crewmembers would be one of four candidates: V. G. Fartushniy, P. I. Kolodin, Yu. N. Lapkin, and an unnamed engineer from TsKBEM. See N. P. Kamanin, *Skrytiy kosmos: kniga vtoraya, 1964–1966gg* (Moscow: Infortekst IF, 1997), p. 420.

45.   Georgi Beregovoi, "Not to Be Forgotten," in Viktor Mitroshenkov, ed., *Pioneers of Space* (Moscow: Progress Publishers, 1989), pp. 298–99; Mitroshenkov, *Zemlya pod nebom*, pp. 413–14.

Machine Building on May 27, 1968. Of the eleven men inducted at this time, ten were from TsKBEM and one, Vladimir G. Fartushniy, was a senior scientist at the Ye. O. Paton Institute for Electro-Welding based at Kiev.[46] His selection was primarily motivated by plans to carry out welding experiments in space, an idea that had originated as early as November 1964 when Korolev had instructed his deputies to draw up plans for the work. Paton Institute Director Academician Boris Ye. Paton was also very supportive of the project and had initiated the development of an instrument named *Vulkan* to allow Soyuz cosmonauts to carry out such experiments in space.[47]

In addition to engineers, the Soviets, like NASA, also looked into the matter of training career scientists for future space missions. In January 1965, Academy of Sciences President Keldysh set in motion the process of selecting scientist-cosmonauts, despite the almost customary resistance from the Air Force on the issue. What little science had emerged in the early 1960s was only after much lobbying by numerous highly placed academicians. While science was a junior partner in the U.S. space program, in the Soviet Union, it was considered an irritation at best. After the formation of the academy's Institute of Space Research, many scientists expected an expansion of scientific activities in space, but judging by the number of scientific satellites launched as part of the Kosmos cover name, it seems that the situation had not changed much. The only major components of scientific research were the continuing projects to send automated probes to Mars and Venus, but these efforts were to a great degree motivated by competition with the United States. Roald Z. Sagdeyev, later the Director of the Institute of Space Research, summarized the situation as one in which "the guiding philosophy behind Soviet space launches reflected the interests of the space industry to the complete neglect of science per se."[48]

In this climate, Keldysh sent the files of twenty-four scientists to the Air Force. Of them, the military allowed nineteen to undergo medical screening in September 1966. By November, only four passed the rigorous testing at the Air Force's Central Scientific-Research Aviation Hospital. Finally, on May 22, 1967, a month after Komarov's death, they arrived at the Cosmonaut Training Center to begin training. They were:

- Mars N. Fatkullin (twenty-eight years old)
- Rudolf A. Gulyayev (thirty-two)
- Ordinard P. Kolomiytsev (thirty-two)
- Valentin G. Yershov (thirty-nine)[49]

These four men were joined by Georgiy P. Katys, the accomplished scientist who had been passed over for several Voskhod missions because of his "questionable" background. Of the four new scientists, Fatkullin, Gulyayev, and Kolimiytsev were all researchers from the academy's

---

46. I. Marinin, "The First Civilian Cosmonauts" (English title), *Novosti kosmonavtiki* 12–13 (June 3–30, 1996): 81–87. The other ten TsKBEM selectees were K. P. Feoktistov, G. M. Grechko, V. N. Kubasov, O. G. Makarov, V. I. Patsayev, N. N. Rukavishnikov, V. I. Sevastyanov, V. N. Volkov, V. A. Yazdovskiy, and A. S. Yeliseyev.

47. Korolev's letter to his deputies, dated November 29, 1964, has been published as S. P. Korolev, "On a Program of Work on Welding in Space Conditions" (English title), in M. V. Keldysh, ed., *Tvorcheskoye naslediye Akademika Sergeya Pavlovicha Koroleva: izbrannyye trudy i dokumenty* (Moscow: Nauka, 1980), p. 520. For an account of a conversation between Korolev and Paton in the autumn of 1965 on the welding issue, see Aleksandr Romanov, *Korolev* (Moscow: Molodaya gvardiya, 1996), pp. 503–09. Korolev and Paton signed a formal agreement on the project on December 1, 1965.

48. Roald Z. Sagdeev, *The Making of a Soviet Scientist: My Adventures in Nuclear Fusion and Space From Stalin to Star Wars* (New York: John Wiley & Sons, 1993), p. 154.

49. I. Marinin, "Russian Cosmonaut-Scholars" (English title), *Novosti kosmonavtiki* 3 (January 28–February 11, 1996): 49–54; Kamanin, *Skrytiy kosmos: kniga vtoraya, 1964–1966gg*, pp. 132, 204, 378, 382.

Institute of Terrestrial Magnetism, Ionosphere, and Radio Wave Propagation, while Yershov was from the famous Institute of Applied Mathematics, which Keldysh headed at the time. Yershov was chosen specifically to provide navigational support on LI circumlunar missions. He, in fact, participated in the development of the LI autonomous navigation system. By coincidence, NASA selected its second group of scientist astronauts a little more than two months after the Soviet selection. These eleven new astronauts would be unofficially known as the "Excess 11" to indicate their less than hopeful chances of ever making it into space.[50] Under the command of Katys, the Soviet scientists finished their initial training program in July 1968 to await formal assignment to a flight.

Scientists were not the only civilians considered for spots on a Soviet spaceship. Decades before NASA considered sending a journalist into space, the late Korolev had given the idea some thought. One of those in the running was Yaroslav K. Golovanov, a writer for the newspaper *Komsomolskaya pravda*, who would thirty years later publish a biography of Korolev. Golovanov, one of the few Soviet journalists allowed into the inner sanctum of the Soviet space program, had spoken to Korolev in January 1965 on the possibility of beginning cosmonaut training. On February 12, 1965, the chief designer signed papers permitting him to begin initial medical screening tests. He was joined by a second reporter, Yuriy V. Letunov of the TV program *Vremya* (*Time*). In July–August 1965, both passed their initial medical tests, but the journalist-in-space idea receded into the background after Korolev's death. Golovanov tried to pursue the matter with a letter to the Central Committee in the spring of 1968, but the space leadership politely rejected the idea, no doubt because the Soyuz at the time was still a raw, untested machine, better to be flown by experienced pilots.[51]

Declaring the Soyuz safe took a considerable amount of time. Based on the recommendations of the Utkin subcommission, engineers at TsKBEM, the Scientific-Research Institute for Automated Devices (responsible for designing parachutes), and the M. M. Gromov Flight-Research Institute carried out an intensive series of corrective tests on the Soyuz capsule throughout 1967. The tests resulted in some supplementary modifications to the Soyuz parachute system, including changes in the operations schedule of the reserve parachute during launch aborts up to six kilometers altitude. Engineers built several boilerplate models of the descent apparatus to test these modifications; the Utkin subcommission evidently had the authority to recommend changes in design.

The process to declaring the 7K-OK Soyuz vehicle safe for automated flight was fraught with difficulties and accidents. Two new Soyuz spacecraft were the subject of vigorous testing for an automated docking flight in the fall of 1967. During a ground test of the solar panels on one of them, electric equipment burnt out, forcing engineers to dismantle the ship and replace the damaged instruments. Of the twenty tests at the Air Force site at Feodosiya by late September 1967, nearly half had malfunctions; three were complete failures.[52] Despite the setbacks, by the autumn of 1967, the Utkin subcommission declared the 7K-OK Soyuz vehicle safe for automated missions.[53] Parachute testing would continue until commission members were satisfied that the complete system was safe for humans.

50.   For a discussion of the events and controversy behind the selection of scientist astronauts in the U.S. space program, see William David Compton, *Where No Man Has Gone Before: A History of Lunar Exploration Missions* (Washington, DC: NASA SP-4214, 1989), pp. 57–58, 65–72. See also Donald K. "Deke" Slayton with Michael Cassutt, *Deke! U.S. Manned Space: From Mercury to the Shuttle* (New York: Forge, 1994), pp. 143–44, 152–53, for a more personal account.

51.   A. A. Tarasov, *Neizvestniy kosmodrom* (Moscow: Orbita, 1990), pp. 8–10.

52.   Kamanin, "A Goal Worth Working for," no. 47.

53.   Semenov, ed., *Raketno-Kosmicheskaya Korporatsiya*, p. 183.

The two Soyuz spacecraft finished their testing at the Baykonur Cosmodrome by mid-October 1967 and were prepared for launch soon after. On October 16, at a meeting of the State Commission, Mishin announced that the flight profile for the new launches would be slightly different than the one planned for the aborted Soyuz 1/2 mission. The primary goal of this test would be to check the reliability of all major spaceship systems of both spacecraft. The active Soyuz would spend almost three days flying solo in orbit, while controllers at Yevpatoriya would pore over incoming data. If the "health" of the ship was still acceptable, then the Strategic Missile Forces would launch the passive Soyuz at the end of the third day. The two spacecraft would merely approach each other in space using their *Igla* radar systems. Docking was not completely excluded from the plan, but it was not considered a primary goal. The first ground training simulation for the plan was held on October 19, with cosmonaut Gagarin participating as a member of the Chief Operations and Control Group. Later, he flew into Leninsk near the test site the day before the scheduled launch. Coincidentally, his Air Force boss Kamanin was promoted from lieutenant general to colonel general the same day.[54] For Kamanin, his rank was not the only good news of the week.

The active spacecraft, vehicle no 6, simulating the role of the lunar orbiter in the lunar landing mission, was launched successfully from site 31 at Tyura-Tam at 1230 hours Moscow Time on October 27, 1967. The initial orbital parameters of the spaceship, named *Kosmos-186* in the Soviet press, were announced as 209 by 235 kilometers at a 51.7-degree inclination. Naturally, TASS neglected to mention that the flight had any relation to the piloted space effort. For the first time in the Soyuz program, all systems were working without fault in orbit. The solar panels deployed, and the Igla system was operational.[55] There was some sign of trouble on the second day of the mission when controllers discovered that the spacecraft was unable to change its orbit on the seventeenth orbit, apparently because of a malfunction in the 45K stellar-solar attitude control sensor. There were also disruptions in the work of the ion sensor system the following day. Engineers dug into their work and managed to overcome the most serious problems by the third day of the flight, prompting the State Commission to give a go-ahead for the second Soyuz launch.

Before the launch of the passive Soyuz, Mishin, perhaps motivated by the relatively good state of Kosmos-186 in orbit, decided to attempt not just rendezvous, but full docking between the two vehicles.[56] Thus, with a new mission, the passive Soyuz, spacecraft no. 5, was launched at 1212 hours Moscow Time on October 30 and entered a 200- by 276-kilometer orbit, also at a 51.7-degree inclination. The vehicle was named *Kosmos-188* in the Soviet press. The performance of both vehicles fulfilled all expectations. The launch of the second spacecraft was performed in such a way as to insert the vehicle within twenty-four kilometers of the active ship. The latter then fired its engine twenty-eight times (over three minutes of burn time) on completely automatic commands from the Igla system. Within just sixty-two minutes of the launch of Kosmos-188, both vehicles were successfully docked to each other on the target's first orbit. At the time of docking, the two ships were out of communications range with Soviet surface tracking stations, but once they were over Soviet territory, ground controllers began receiving clear video pictures from the ships showing their docked configuration. These images

54. Mitroshenkov, *Zemlya pod nebom*, p. 425; B. Ye. Chertok, *Rakety i lyudi: goryachiye dni kholodnoy voyny* (Moscow: Mashinostroyeniye, 1997), pp. 467–69.

55. Kamanin, "A Goal Worth Working for," no. 47.

56. *Ibid.* Interestingly, Chief Designer A. S. Mnatsakanyan of the Scientific-Research Institute of Precision Instruments (NII TP), responsible for developing the Igla system, recalls that an unnamed deputy chairman of the Soyuz State Commission expressed reservations about going for a full docking only an hour before launch. Mnatsakanyan, however, gave his full support to Igla. See Yu. A. Mozzhorin, *et al.*, eds., *Dorogi v kosmos: II* (Moscow: MAI, 1992), p. 32.

were later shown on Soviet TV, giving the public their first brief look at the Soyuz spacecraft. It was an impressive display of automation, bolstering somewhat the argument that cosmonauts were mere passengers in the Soyuz spacecraft. It was also the first docking of two robot spaceships in history.

After the two ships were linked, the controllers discovered that there had not been full "hard" docking because, for reasons unknown, there was still an eighty-five-millimeter gap between the two ships. This was considered a minor problem, and after three and a half hours of connected operations over two and a half orbits, Kosmos-186 and Kosmos-188 separated. Both ships were to finish off their missions with guided reentries, but both ran into problems. In Kosmos-186's case, on October 31, the failure of the 45K sensor changed the reentry profile into a direct ballistic return. The descent apparatus, however, was recovered safely. The following day, Kosmos-188 was unable to perform a guided return because of incorrect attitude; the ship had flown into an ion pocket, confusing the ion attitude control sensor. The ship entered on a steep trajectory, and its self-contained explosive automatically destroyed the descent apparatus to prevent a landing on foreign territory. It was proved later that if the explosive had not been carried on board, the capsule would have landed 400 kilometers east of Ulan-Ude north of Mongolia, but in Soviet territory.[57]

The Kosmos-186/188 flight was timed to occur a week before the fiftieth anniversary of the Great October Revolution. It was a poor substitute for a piloted circumlunar mission, but it was a minor advance for a space program beleaguered by failures and catastrophes. The confidence imparted by the docking mission was, however, tempered by the two unrelated L1 launch failures before and after Kosmos-186/188. Immediately after the docking success, the Soyuz State Commission met on November 15 to discuss the future manifest for the project.[58] With no authorization from the Utkin subcommission to carry out piloted flights, it seems that Mishin had planned a repeat performance of the automated docking mission in early 1968, which would allow further testing of the problematic attitude control sensors on the Soyuz spacecraft. In the meantime, crews training for upcoming Soyuz flights continued their training program at a less intensive pace.

For "Cosmonaut No. 1," Yuriy A. Gagarin, the post–Soyuz 1 period was a particularly transitional time. Having been denied flight status, in November 1967, he was subjected to the additional humiliation of being grounded from flying aircraft solo. Apart from his important role in various State Commissions, he continued to serve as an international ambassador for the Soviet space program. His various obligations took their toll. Kamanin wrote in his journals in 1968:

There were many situations when Gagarin miraculously escaped big troubles. These situations often occurred when he attended parties, drove in cars or boats, or when hunting with the big bosses. I was particularly concerned about his driving cars at high speeds. I did a lot of talking with Yura on this issue. The active life style, endless meetings and drinking sessions were noticeably changing Yura's image and slowly, but steadily erasing his charming smile from his face.[59]

Training for the Soyuz 1 flight and an assignment to the subsequent Soyuz 3 mission apparently curbed his extracurricular activities. The cosmonaut lost weight, trained regularly, and eventually mastered the Soyuz spacecraft. In addition, by late 1967, he was finally wrapping up work on his graduate degree at the N. Ye. Zhukovskiy Military-Air Engineering

57. Chertok, Rakety i lyudi: goryachiye dni kholodnoy voyny, pp. 473–74.
58. Mitroshenkov, Zemlya pod nebom, p. 425.
59. N. Kamanin, "For Him, Living Meant Flying" (English title), Vozdushniy transport 11 (1994): 13.

Academy in Moscow dealing with a reusable single-seat military spaceplane. At Gagarin's own request, Kamanin temporarily relieved the young cosmonaut of his duties as training center deputy director to allow him to focus exclusively on his dissertation. At the same time, Kamanin and Center Director Maj. General Nikolay F. Kuznetsov promised Gagarin that he would be allowed to resume flight training once his academic work was finished.[60]

On January 8, 1968, several of the fifteen cosmonauts pursuing higher degrees graduated with their "Candidate of Technical Sciences." Gagarin and Titov defended their dissertations on February 17 at the academy, and both passed with excellent grades.[61] Immediately afterwards, Gagarin threw himself back into flying in training aircraft to gain enough experience to resume flying solo. After passing his medical tests on March 12, he was cleared to fly, and he did so jointly with another pilot the following day for a one-hour, fifty-two-minute jaunt. He flew several times the following days, always with other more experienced pilots who kept their hands on the controls. On March 23, Kamanin expressed some reservations about Gagarin's frenzied training pace, but could not dampen the cosmonaut's enthusiasm.[62]

On his flight on March 27, Gagarin was escorted by Colonel Vladimir S. Seregin, a forty-five-year-old test pilot with impeccable credentials, who had been assisting flight training for cosmonauts since 1963. The two took off from the Chkalovskaya airfield near Moscow a little after 10 a.m. in the morning for a flight over the town of Kirzhach. A few minutes after takeoff, Gagarin requested permission to alter course: "This is 625. Mission accomplished. Altitude 5,200. Request permission to approach."[63] It was the last communication from the UTI-MiG-15 trainer aircraft. Communications abruptly ended at 1030 hours, 10 seconds Moscow Time. As alarm began to rise back at the Cosmonaut Training Center, Air Force officials put together a search team to determine the fate of the two men. About four hours and twenty minutes after loss of contact, a helicopter commander finally reported back that he had found the wreckage of the airplane about sixty-four kilometers from the airfield. Debris was scattered in a very woody area, with snow as much as one meter deep. The engine and the cockpit were evidently buried six to seven meters in the ground, indicating that the plane had hit the ground at a velocity of 700 to 800 kilometers per hour. It was not long before searchers found a fragment of an upper jaw, which doctors identified as belonging to Seregin. Air Force officials immediately informed Soviet leaders Brezhnev and Kosygin of Seregin's fate, although they had no incontrovertible proof of Gagarin's death.[64]

Throughout the night, an emergency commission held meetings to establish what had happened. It was a long torturous night for many, as it was becoming increasingly clear that there was almost no chance that Gagarin had survived. One cosmonaut recalled, "We saw Kamanin with his lips pressed tightly together, Kuznetsov struggling to control his trembling chin, Leonov with his face to the wall and Popovich repeatedly leafing through flight documents."[65] As soon as dawn broke on March 28, a search party led by Kamanin was back

---

60.   The letter requesting that Gagarin be relieved of his duties as deputy commander of military unit no. 26266 (the Cosmonaut Training Center) and the letter permitting him solo flying privileges in the spring and summer of 1968 have been published in Kamanin, "For Him, Living Meant Flying," no. 9.

61.   Among the cosmonauts defending in January were V. F. Bykovskiy, V. V. Gorbatko, A. G. Nikolayev, P. R. Popovich, G. S. Shonin, and B. V. Volynov. See Mitroshenkov, *Zemlya pod nebom*, p. 429. Others who graduated in 1968 were A. A. Leonov, Ye. V. Khrunov, I. B. Solovyeva, and D. A. Zaykin. The three remaining cosmonauts of the group of fifteen—all women—graduated in 1969.

62.   Mitroshenkov, *Zemlya pod nebom*, pp. 437–41. Gagarin's wife recalls of this period: "He talked about another spaceflight and began to train for it." See Yevgeniya Malakohovskaya, "Tell Me About Him," in Mitroshenkov, ed., *Pioneers of Space*, p. 147.

63.   Beregovoi, "Not to Be Forgotten," p. 320.

64.   Kamanin, "For Him, Living Meant Flying," no. 9.

65.   Beregovoi, "Not to Be Forgotten," p. 318.

at the crash site. At around 8 a.m., Kamanin saw a piece of cloth hanging from a birch tree about ten to twelve meters in the air; the cloth was identified conclusively as a piece of Gagarin's flight jacket. By then there was no doubt: Gagarin was dead. Both pilots' bodies were found soon after. Gagarin's wallet contained his ID, a driver's license, 74 rubles, and small photo of Sergey P. Korolev. Both bodies were cremated by 2115 hours the same night. In contrast to the deaths of Korolev and Komarov, the outpouring of grief from the average Soviet citizen was unprecedented. The urns with the two pilots' ashes were laid at the Central House of the Soviet Army the following day for 40,000 people to pay their respects. On March 30, the urns of Gagarin and Seregin were escorted by Soviet leaders Brezhnev, Kosygin, Podgorniy, and others to the Kremlin Wall to be interred in their final place. Hundreds of thousands of Muscovites were on hand to view the dour funeral march for a man they considered a fallen national hero.[66]

The investigation commission into the disaster discerned a cause of the accident by late July 1968, although it was a process fraught with diverging opinions because of the absence of "a smoking gun" despite the thousands of hours spent poring over the evidence. The official report, issued in December 1968 by the Central Committee, hinted at pilot error:

> The most probable cause of the death of Gagarin and Seregin was a sudden turn of the aircraft to avoid a collision with a sounding balloon; a less probable cause was turning of the aircraft from the upper edge of the clouds. As a result of the sudden turn, the airplane entered critical flying angles; the adverse meteorological situation complicated aircraft control; and the crew died.[67]

Both the senior cosmonauts and Kamanin seem to have objected vigorously to attributing the accident to pilot error; they even sent a letter to Central Committee Secretary Ustinov on the issue. On the other side, many of the Air Force members investigating the accident were evidently reluctant to admit that there were defects in the UTI-MiG-15 aircraft.

Almost twenty years later, the files for the crash were reopened, and a number of researchers carried out a detailed investigation using computer modeling to determine the causes of the crash. The new study found that the accident did not occur because of pilot error or from a mid-air collision. There were a number of cumulative causes. Ground equipment was evidently faulty at the time of the accident and thus was unable to track the UTI-MiG-15 in flight. In addition, Gagarin and Seregin did not have accurate information regarding the altitude of the ceiling in that area. Other violations of safety regulations included the flight of two MiG-21s and a MiG-15 in the same area at the same time. As for Gagarin and Seregin, after receiving their last instruction to fly home, they began a turn and descent to 700–1200 meters. At that time, they were flying between two layers of clouds and could not see the horizon. The other MiG-15 then passed Gagarin's plane at a distance of only 500 meters, although the pilot of the other craft did not notice Gagarin's aircraft. Soon after, Gagarin's plane entered a trailing vortex created by the second MiG and flew into a spin. Gagarin and Seregin managed to pull out of the spin after five full revolutions but only in thick cloud cover, which disoriented the pilots. They overestimated their altitude by 200–300 meters and exited the cloud cover assuming their altitude was much higher than the actual 400–600 meters above the ground. Their angle of attack at the time was seventy degrees. The pilots were unable to activate the emergency ejection system in the less

66.   Kamanin, "For Him, Living Meant Flying," no. 9.
67.   N. Kamanin, "For Him, Living Meant Flying" (English title), *Vozdushniy transport* 18–19 (1994): 12.

than five seconds remaining and crashed into the ground. An extra two seconds or 250–300 meters altitude would have easily saved them.[68]

Clearing the Soyuz

Gagarin's death was an unprecedented psychological blow to the Soviets, especially because it came at a time when the Soviet piloted space program was reaching a nadir of sorts—a situation that no one could have anticipated a few years before. From the days of consecutive victories in the early 1960s, the Soviets witnessed an almost unending series of setbacks, tragedies, and failures. Perhaps the only bright spot in the quagmire was the recent successful docking-in-Earth-orbit Soyuz flight in October 1967. Since then, tests had continued slowly on the parachute and landing systems of the 7K-OK vehicle in preparation for a repeat attempt of the original Soyuz 1 mission. There were, however, a number of landing failures that progressively delayed plans—malfunctions that in retrospect were critical in moving piloted Soyuz flights downrange at a time when NASA was beginning to finally recover from the Apollo 1 disaster. The State Commission for Soyuz, under Lt. General Kerim A. Kerimov, met on March 26, 1968, the day before Gagarin's death, to discuss immediate plans. Mishin and Chief Designer Fedor D. Tkachev of the Scientific-Research Institute of Automated Devices, which was responsible for parachute design, reported that the 7K-OK ship's primary parachute system was already cleared for flight while the reserve system would be ready by launch time, then set for April 9–14.[69]

On April 10, exactly two weeks after Gagarin's death, several cosmonauts, including rookie Beregovoy, slated to command the Soyuz 1 repeat docking flight, flew to the Baykonur Cosmodrome accompanied by Air Force First Deputy Commander-in-Chief Marshal Sergey I. Rudenko. Many officials remained in Moscow, because of the investigation into the causes of Gagarin's death and also to celebrate April 12 or "Cosmonautics Day," the seventh anniversary of Gagarin's pioneering first flight. After arrival at Tyura-Tam, the State Commission set the two Soyuz launches for 14 and 15 April. Unlike the Kosmos-186/188 mission, this particular joint flight was to simulate an actual piloted flight as closely as possible. Consequently, the primary and backup crews training for the docking and EVA mission were sent to the Flight Control Center at Yevpatoriya to follow the flight on the ground and train in such a manner as to simulate their actions on a real mission. Both ships were also equipped with new infrared attitude control sensors to augment the chronically faulty ionic sensor system on the early Soyuz spacecraft.[70]

The active 7K-OK vehicle, spacecraft no. 8, was launched from Tyura-Tam at 1300 hours Moscow Time on April 14, 1968. Initial orbital parameters were 210 by 239 kilometers at a 51.7-degree inclination. The Soviet press announced the mission as *Kosmos-212*. A day later, on April 15, engineers successfully launched the passive Soyuz spacecraft, vehicle no. 7, at 1234 hours Moscow Time, with only a two-second delay. The target vehicle, named *Kosmos-213*, entered an initial orbit of 205 by 291 kilometers at a 51.4-degree inclination. At the point of orbital insertion, the active spacecraft was only four kilometers away from the passive one, a remarkable achievement in precision. With great economy of propellant, Kosmos-212 approached Kosmos-213 and automatically docked at 1331 hours, just fifty-seven

---

68. S. Belotserkovskiy and A. Leonov, "Two Seconds Was All They Needed—Yu. Gagarin and V. Seregin's Final Flight" (English title), *Pravda*, March 23, 1988, p. 4. As remarkable as it may seem, an entire book has been dedicated to Gagarin's death and the new investigation in 1987–88. See S. M. Belotserkovskiy, *Gibel Gagarina: fakty i domysly* (Moscow: Mashinostroyeniya, 1992).
69. Kamanin, "For Him, Living Meant Flying," no. 9.
70. Kamanin, "For Him, Living Meant Flying," no. 12.

minutes after the target spacecraft's launch. Ground controllers at Yevpatoriya were able to view the docking on their consoles via a live TV feed from both spacecraft. The two spacecraft remained connected for three hours and fifty minutes before continuing autonomous flight; each vehicle clocked up about five days in space. The major remaining objective of the flight was to verify the complete reentry procedure. Kosmos-212 successfully carried out the first guided reentry in the Soyuz program (with an aerodynamic efficiency ratio of 0.3) and landed near Karaganda in Kazakhstan on April 19. Winds were very high at the landing site, up to twenty-two to twenty-three meters per second, and although the descent apparatus landed safely, winds dragged the capsule about five kilometers from its landing spot, damaging the outside coating.[71]

Kosmos-213 remained in orbit for another day and conducted some unusual scientific experiments. On board the spacecraft was an extensive scientific payload, including a new type of luminescent micrometeoroid detector, an ultraviolet photometer, and a radiation-sensing package. The photometer measured ultraviolet and visual spectrographic night sky brightness, while the *Luch-1* instrument measured cosmic ray positrons and electrons. In addition, a cryogenic superconducting magnet, first tested on the Kosmos-140 Soyuz precursor, was used to detect cosmic rays in conjunction with scintillation, gas discharge, and Cherenkov detectors. The spacecraft's descent apparatus landed on April 20 near Tselinograd after another guided reentry. All systems worked without fault, but once again the descent apparatus was dragged after touchdown by twenty-five-meter-per-second wind speeds. Rescuers had to wait for the dust storm to subside before they could recover the capsule.[72]

The successful conclusion of two consecutive automated docking missions raised the question of moving on with piloted flights. One of the biggest factors were the results of ongoing ground testing of the redesigned parachute system. Throughout 1967–68, engineers carried out a series of approximately forty drop tests of mock-ups of the descent apparatus from Tu-16 aircraft to verify the parachutes and elements of its design. In addition, they also conducted six test drops from An-12 aircraft and carried out special "controlled" experiments using Mi-6 helicopters by introducing a maximum of eighteen-meter-per-second *horizontal* velocity during the drops. There were a number of major failures, especially in the operation of the reserve parachute.[73] The cosmonauts training for the docking and EVA mission completed their full training program by the end of May 1968, after many delays related to updating the Soyuz simulators concurrently with the actual Soyuz spacecraft. By February, Kamanin had tentatively tapped Beregovoy to command the active vehicle, and Volynov, Khrunov, and Yeliseyev to fly the passive vehicle, although as with many other earlier crews, the process of crew selection was caught up in an almost pointless conflict between Kamanin and Mishin.[74]

71.   *Ibid.*

72.· *Ibid.;* Joel Powell, "Research From Soviet Satellites," *Spaceflight* 25 (January 1983): 33–34.

73.   Semenov, ed., *Raketno-Kosmicheskaya Korporatsiya*, p. 183. The testing was a joint effort among TsKBEM, the M. M. Gromov Flight-Research Institute, the Scientific-Research Institute for Automated Devices (NII AU), the Zvezda Design Bureau (KB Zvezda), and the Iskra Design Bureau (KB Iskra), and was carried out at the Air Force's testing station at Feodosiya.

74.   The center of this disagreement was over Mishin's insistence that TsKBEM Department Deputy Chief K. P. Feoktistov, a civilian, be included as the crew commander of the first post–Soyuz 1 flight. Feoktistov himself eagerly supported this position and took great pains in 1967–68 to promote his candidacy, despite his relatively poor health and reluctance to commit to parachute training. The issue culminated in March 1968 during several State Commission meetings, when the Air Force, led by Marshal S. I. Rudenko and Col. General N. P. Kamanin resisted Mishin, and their highly placed supporters in the government, which included M. V. Keldysh (AN SSSR), G. N. Pashkov (VPK), G. A. Tyulin (MOM), K. A. Kerimov (MOM), and B. A. Stroganov (TsK Defense Industries Department). It was not until mid-June 1968 when Mishin finally withdrew Feoktistov's candidacy. See also Beregovoi, "Not to Be Forgotten," p. 299.

It was not until May 6, 1968, that the Council of Ministers formally approved the above crews. An additional four cosmonauts—Nikolayev, Shonin, Kubasov, and Gorbatko—would fly an exact repeat of the docking and EVA mission at a later date.

The debate over the next step after the Kosmos-212/213 missions was colored to a great degree by Central Committee Secretary Ustinov's pronouncement before the docking flight in early April 1968 that "irrespective of the results of the upcoming flights of two Soyuz spaceships, two more spacecraft should be prepared for an experimental flight."[75] After the success of the Kosmos-212/213 mission, Ustinov's decision was called into question by other space program officials, including Mishin and Kamanin, who were more confident of the Soyuz spaceship's safety. On April 21, the day after Kosmos-213's landing, the State Commission met in Moscow; Commission Chairman Kerimov and Chief Designer Mishin graciously allowed the cosmonauts' views to be aired on the issue. All four primary crew cosmonauts favored a piloted flight as the next step. Kerimov, Mishin, Chertok, and others thanked the cosmonauts for their work and seem to have been very pleased that they supported a piloted mission. At least tentatively, Kerimov and Mishin scheduled the flight for late June or early July 1968.

Those advocating another automated mission were a powerful lobby—that is, the leaders of the Soviet military-industrial complex—Ustinov, Afanasyev, Smirnov, and Dementyev—all of whom were clearly playing it safe after the Komarov tragedy. Their viewpoint had some basis because by early May, although all the major problems with the 7K-OK spacecraft had been eliminated, it still had two weak spots: the backup parachute and the emergency rescue system. Throughout the twenty-three drop tests after Soyuz 1, the backup parachute had evidently performed below par, while the rescue system malfunctioned more frequently. Kamanin wrote in his diary about the dilemma facing the managers of the Soviet space program:

> . . . under the circumstances Korolev would have assumed responsibility and given a go-ahead for the flight. Cosmonauts and Air Force specialists would have gone along with such a decision. But unfortunately, Mishin is not Korolev and he is hedging: "I am not going to propose a manned flight myself, but if the Central Committee tells me to, I will agree."[76]

The climate had clearly changed after the Soyuz 1 disaster. Kerimov and Mishin were definitely more conservative with their decisions. No one, from Ustinov down to Mishin, was gutsy enough to recommend a decision for flight and risk losing their jobs over a hasty decision. The decision would have important implications and, in retrospect, was a critical juncture in the Soviet space program. By mid-1968, NASA had meticulously modified its Apollo Command and Service Module and was close to declaring the spacecraft ready for piloted flight. Every month was desperately important as the two countries were closing in on their final goals. For the Soviets, including another automated mission would add yet another two months before they saw a return to piloted flight. For many, apart from the issue of safety, there were also exogenous considerations.

On May 7, 1968, Mishin held a meeting at TsKBEM in Kaliningrad. The engineers concluded at the end of the meeting that with the exception of the backup parachute system, the 7K-OK spacecraft was completely ready for piloted flight. Mishin believed that the parachute system would be cleared for flight by the first half of August. Troubles with the backup parachute system, however, forced Mishin and his deputies to rethink their strategies for an early August

---

75. Kamanin, "For Him, Living Meant Flying," no. 12.
76. *Ibid.*, p. 12.

flight. The major problem with the backup parachute was that with three crewmembers in the descent apparatus (an excess of 1,300 kilograms), it had a tendency to rip off upon deployment. Parachute Chief Designer Tkachev and Mishin proposed instead to reduce to the crew of the passive vehicle to two men, by 150–200 kilograms, to declare the system safe for operation. In addition, perhaps to avoid any unnecessary risk, Mishin proposed that during the August flight, the cosmonauts would dock the two Soyuz ships and only depressurize the living compartment of the passive Soyuz. In the interest of time, most of the cosmonauts as well as Kamanin agreed, at least tentatively, to the deletion of the spacewalk, leaving the more complex EVA transfer to a subsequent Soyuz mission.[77]

The uncertainty with the backup parachute system, combined with a general sense of conservatism, introduced a modicum of uncertainty throughout the month of May 1968 as different engineers proposed different variants of the flight. Some supported having one member transfer via EVA from one ship to the other, while others suggested merely having one cosmonaut from the passive ship carry out an EVA without transfer. Another controversial issue was the number of crewmembers on each ship; several different combinations were considered at the time, including one on the active ship and two on the passive one, one on the active ship and three on the passive one, and two on both ships.

The group supporting an early return to piloted flight expanded by mid-May 1968, with the addition of Chief Designers Voronin and Severin. Academy of Sciences President Keldysh dissented, however, clearly still influenced by Soyuz 1. He cautioned, "It seems to me that we are too hasty, and the question of technological launchings should still be discussed. I reserve my opinion on the selection of piloted flights without preliminary additional technological [that is, robotic] launchings."[78]

The issue seemed to reach some kind of resolution on May 29 at a meeting of the Council of Chief Designers. Pressured by Ustinov, Keldysh, and Smirnov, Mishin proposed a compromise variant for the initial Soyuz piloted flight: a docking of two 7K-OK vehicles in Earth orbit with a single cosmonaut in the active vehicle. At least a dozen other chief designers supported Mishin, and the Air Force agreed to the new proposal.[79] A second flight in September would have the full docking plus EVA mission with cosmonauts Khrunov and Yeliseyev performing the critical transfer spacewalk. With the support of Minister of General Machine Building Afanasyev, this plan seemed to be the most promising, but, within a few days, the imposing hand of the Communist Party's Central Committee intervened. In early June, Ustinov blocked the proposal, giving orders that regardless of what the chief designers believed, *another* automated docking flight of the Soyuz was required before a piloted flight. With that final blow, the Soviet space program lost two critical months.

On June 10, 1968, the Soyuz State Commission met to discuss a response to Ustinov's demands. Commission Chairman Kerimov approved a plan to launch a single automated Soyuz

---

77. N. Kamanin, "For Him, Living Meant Flying" (English title), *Vozdushniy transport* 15 (1994): 11. According to Mishin, the crews for the two Soyuz spacecraft would be Beregovoy/Kubasov (Soyuz 2) and Yeliseyev/Khrunov (Soyuz 3). Kamanin did not agree to Mishin's crew proposals and continued to resist efforts to posit a civilian (Yeliseyev) as a crew commander.

78. *Ibid.*

79. Among the chief designers and other officials present at this meeting were V. P. Barmin (Chief Designer, KB OM), V. P. Glushko (Chief Designer, KB EnergoMash), Maj. General A. G. Karas (Commander, TsUKOS), M. V. Keldysh (President, AN SSSR), K. A. Kerimov (Chief, Third Chief Directorate, MOM), V. I. Kuznetsov (Chief Designer, NII PM), V. P. Mishin (Chief Designer, TsKBEM), N. A. Pilyugin (Chief Designer, NII AP), M. S. Ryazanskiy (Chief Designer, NII Priborostroyeniya), G. I. Severin (Chief Designer, KB Zvezda), F. D. Tkachev (Chief Designer, NII AU), G. A. Tyulin (First Deputy Minister, MOM), I. I. Utkin (Chief Designer, NII IT), and G. I. Voronin (Chief Designer, KB Nauka).

vehicle in July, carry out a joint docking flight between two Soyuz spacecraft with a single cosmonaut in the active vehicle in September, and finally a full-scale docking and EVA mission in November–December 1968. The Military-Industrial Commission formally approved this plan in late July 1968. Ustinov had one more demand: that the third flight include a transfer of *two* cosmonauts from one vehicle to the other. This meant that Mishin and his engineers would have to come up with a solution to the reserve parachute problem before the end of the year. Because they could not reduce the mass of the reentry capsule below 2,750 kilograms (a low limit for three cosmonauts), the engineers had to search for other options to reinforce the reserve parachute system.[80]

The robot 7K-OK, spacecraft no. 9, was launched into orbit at 1300 hours Moscow Time on August 28, 1968, more than a month behind schedule because of a variety of problems related to the vehicle's parachute system. The spacecraft, named *Kosmos-238* by the Soviet press, entered an initial orbit of 199 by 219 kilometers at a 51.7-degree inclination. The vehicle was a passive variant of the Soyuz spacecraft. Little is known about the mission, although Western observers tracked at least one major orbital maneuver during its flight.[81] The descent apparatus returned to Earth without any significant anomalies on September 1, after a flight lasting one hour short of four days. Ustinov was satisfied, and the path was finally clear for piloted Soyuz missions after a break of close to one and a half years. This last flight, Kosmos-238, was critical not only because it finally instilled sufficient confidence for resuming crewed operations, but also because of the widespread importance of the 7K-OK spacecraft. The viability of almost all Soviet piloted space projects of the period, including the L1, the L3, the Soyuz, and the military 7K-VI, depended very much on the success and health of the 7K-OK vehicle. As evidenced by later declassified materials, the 7K-VI military reconnaissance offshoot of the Soyuz was suffering some major birth pains at the very same time that Mishin and his associates were trying to bring the Soyuz spacecraft back into crewed operations.

The Soyuz-VI

Looking back at the history of Soviet piloted space programs in the 1960s, what is most surprising is the unprecedented amount of work that was invested into projects that never saw the light of day. What the public saw at the time was only the tip of a supremely diverse space program; many projects were canceled prior to reaching flight status. In some cases, programs emerged and disappeared within the same year, inexplicably changing the direction of the Soviet space effort for a few months. One such program was the Zvezda military spaceship project, which had emerged in 1966–67 at TsKBEM's Branch No. 3 at Kuybyshev under the leadership of First Deputy Chief Designer Kozlov. Consisting of a completely redesigned Soyuz spacecraft named the 7K-VI, the vehicle was to provide military cosmonauts experience in activities such as reconnaissance and combat prior to the advent of the large Almaz space station in the late 1960s. By late 1967, Kozlov's immediate boss, Chief Designer Mishin, was evidently having second thoughts. For reasons that are not completely clear, Mishin countered with a new military station proposal at the time—one that would supersede Kozlov's Zvezda and in fact serve as a direct competitor to Chelomey's ambitious Almaz space station project, which had already received full support.

The situation was complicated by the relationship between the central headquarters of TsKBEM and its Branch No. 3. Although the latter reported nominally to Mishin, the branch

80. Kamanin, "For Him, Living Meant Flying," no. 16; N. Kamanin, "For Him, Living Meant Flying" (English title), *Vozdushniy transport* 17 (1994): 11.

81. Phillip Clark, *The Soviet Manned Space Program: An Illustrated History of the Men, the Missions, and the Spacecraft* (New York: Orion, 1988), pp. 48–49.

seems to have had some degree of autonomy with regard to its own programs. For example, in developing newer military photo-reconnaissance satellites such as Zenit-2M, Zenit-4M, and Yantar-2K, Kozlov's engineers for the most part worked without much interaction with Mishin's engineers. At the same time, Kozlov, as the organization's First Deputy Chief Designer, ultimately reported to Mishin on the progress of all his projects.

In October 1967, Mishin wrote a letter to Military-Industrial Commission Chairman Smirnov and Minister of General Machine Building Afanasyev to terminate Kozlov's 7K-VI program and use the freed-up resources to build an additional eight to ten Soyuz ships during the following year. Air Force Lt. General Kamanin, who clearly disliked Mishin both personally and professionally, wrote in his journal at the time:

> Work on developing the [7K-VI] ship is in full swing and it promises to be much better than the Soyuz. This is apparently exactly the thing that is tormenting Mishin. He didn't have anything against 7K-VI as long as he counted on the fact that it would be an exact replica of the Soyuz, but when he saw that Kozlov had refrained from blindly copying Soyuz and was developing a principally new and significantly better ship, he abruptly changed his opinion of Kozlov and his ship.[82]

Although recent accounts of the history of the 7K-VI portray Mishin as the "evil" figure in the attack against the vehicle, it is clear that he had the strong support of most of his leading deputies on the matter. Their criticism of Kozlov's spaceship centered on two factors—the use of radio-isotope generators and the use of a hatch in the heat shield—both of which they considered very weak design choices.

As an alternative to the 7K-VI, Mishin and his deputies instead proposed a new concept, the Orbital Research Station, better known simply as the *Soyuz-VI*, with the "VI" being the abbreviation in Russian for "military research." Within a few weeks of the new proposal, Kozlov capitulated to Mishin's new proposal, evidently because of intense pressure from Minister Afanasyev, and abandoned his coveted Zvezda project. In November 1967, Mishin and Kozlov signed a document titled "Basic Provisions for the Development of the Soyuz-VI Military-Research Space Complex," which officially testified to Kozlov's capitulation to Mishin on the matter.[83] Kozlov's abrupt change of direction put the military in the difficult position of having to support a program whose chief designer was no longer interested in it. In this climate, many military officers, including Commander of the Central Directorate of Space Assets Lt. General Andrey G. Karas, who had invested much time and resources in Zvezda, consolidated their forces to put up a resistance against Mishin's new Soyuz-VI. The standoff came to a head on December 8, 1967, at a meeting on the premises of TsKBEM. Mishin was on vacation at the time, and Kerim A. Kerimov, the Chief of the Third Chief Directorate of the Ministry of General Machine Building, presided over the deliberations. All of the leading deputy chief designers at TsKBEM, including Bushuyev, Chertok, and Okhapkin, came out in favor of terminating Zvezda. Predictably, most of the military officers were against it, raising a particularly relevant question: "Why do we need a small Almaz if we're already building a big one?"[84]

It seems that the Mishin faction had lined up its ducks in a row. By instructions from Minister of General Machine Building Afanasyev, on January 9, 1968, Kozlov signed an order

82.   K. Lantratov, "Dmitriy Kozlov's 'Zvezda'" (English title), *Novosti kosmonavtiki* 6 (March 10–23, 1997): 74–80. Translation by Bart Hendrickx.

83.   Semenov, ed., *Raketno-Kosmicheskaya Korporatsiya*, p. 210. Mishin met with Kozlov on October 12, 1967, to discuss the cancellation of the Zvezda project.

84.   Lantratov, "Dmitriy Kozlov's 'Zvezda'"; Kamanin, "A Goal Worth Working for," no. 48.

terminating all work on the Zvezda spacecraft to commence developmental work in support of Mishin's Soyuz-VI. The military did not give up. On January 27, Kamanin enlisted the support of six veteran cosmonauts and met with USSR First Deputy Minister of Defense Marshal Ivan I. Yakubovskiy, who promised to assist on the matter. The disagreement finally came to some kind of resolution on February 17, 1968, during a meeting of the Scientific-Technical Committee of the General Staff of the Ministry of Defense, the authoritative consultative body for all new military programs in the country. Chaired by Committee Chairman Col. General Nikolay N. Alekseyev, the meeting was called to discuss the joint proposal of Mishin and Kozlov to terminate Zvezda in favor of Soyuz-VI. Although all the attending high-ranking officers came out in favor of continuing with Zvezda, it was becoming increasingly difficult for them to offer support to the project when Kozlov himself had changed sides. In addition, the military's word on the issue may have been overruled by someone in the Communist Party's Central Committee. With little hope for victory, Alekseyev essentially dropped the matter, effectively closing the Zvezda program. Although Kozlov was shut out as a "prime contractor" in the piloted space program, he was able to use many of the basic systems from the Zvezda space complex to develop subsequent automated reconnaissance satellites in the Yantar ("Amber") series.[85]

The new Soyuz-VI program was clearly a competitor of sorts to Chelomey's Almaz, and therein may lie the answer to how Mishin was able to gain support for his project in the face of such imposing resistance from the military. Central Committee Secretary Ustinov, the *de facto* head of the Soviet space program, was known as being extremely hostile to Chelomey's ambitions. By supporting Soyuz-VI, he may have been trying to sabotage Chelomey's Almaz.

The Soyuz-VI complex consisted of a small space station, named the orbital block (OB-VI) and a crew delivery spacecraft (7K-S), which was to be developed on the basis of the original 7K-OK Soyuz spacecraft. Augmenting the entire Soyuz-VI complex would be three other spacecraft: two Soyuz-type ships for short- and long-duration independent missions (7K-S-I and 7K-S-II, respectively) and a robot cargo ship (7K-G), which was also a modification of the basic Soyuz spacecraft.[86]

Very little is known about the station proper of the Soyuz-VI complex; it was apparently very similar to the orbital block of the long-abandoned Soyuz-R project from the mid-1960s (that is, shaped like a cylinder about the size of a 7K-OK spacecraft). The OB-VI was to carry about 700 to 1,000 kilograms of scientific and military apparatus. Instead of radio-isotope generators to provide power as on the Zvezda, the OB-VI had solar panels. One of the requirements of the Soyuz-VI's design was that it allow cosmonauts to transfer from a ferry to the station via internal means. Thus, unlike the regular 7K-OK Soyuz vehicle, which had a system that prevented internal transfer, Mishin's engineers for the first time began work on a more flexible pin-cone system to allow through passage. Like much of the station, this system was also evidently based on the earlier Soyuz-R concept. The Soyuz-VI complex was to fly in an operational orbit of 250 by 270 kilometers at an inclination of 51.6 degrees. Piloted flights would last approximately thirty days.[87] For a brief period, Mishin evidently considered the idea of testing advanced particle accelerators on the Soyuz-VI complex. In June 1968, representatives of TsKBEM met with famous Soviet physicist Andrey I. Budker, one of the founders of the Institute of Nuclear Physics, to discuss the issue. The idea was probably dropped soon after because of the limited capabilities of the Soyuz-VI.

---

85.   Lantratov, "Dmitriy Kozlov's 'Zvezda'."

86.   *Ibid.* The production designations for these spacecraft were: the complete Soyuz-VI complex (11F730), the OB-VI (11F731), the 7K-S (11F732), the 7K-S-I (11F733), the 7K-S-II (11F734), and the 7K-G (11F735).

87.   Semenov, ed., *Raketno-Kosmicheskaya Korporatsiya,* p. 210.

The 7K-S crew supply ship was an improved version of the basic 7K-OK Soyuz vehicle. Under Mishin's direction, engineers addressed all the weak points of the original Soyuz ship and tried to replace systems and eliminate shortcomings. The official design bureau history adds that:

> . . . with the goal of improving the tactical-technical, technological, and operational characteristics in the ship's design and on-board systems, important changes were introduced, which affected the course of development and ultimately resulted in the creation of a new ship.[88]

When work began on the Soyuz-VI in the second half of 1967, it was overseen by Deputy Chief Designers Bushuyev and Tsybin; both men were principally responsible for piloted spaceships at the organization. The USSR Ministry of Defense issued a new tactical-technical requirement for the Soyuz-VI complex in May 1968, which supplemented a similar document issued in support of the canceled Zvezda. A month later, on June 21, amid the intense preparations for piloted lunar flights, TsKBEM and its Branch No. 3, jointly issued the first version of the draft plan for the Soyuz-VI. Mishin subsequently approved the "theoretical drawings" of the 7K-S Soyuz spaceship on October 14, 1968.[89] As part of the general change in direction from Zvezda to Soyuz-VI, many of the cosmonauts training for the former were reassigned to the latter. The group was originally commanded by veteran Popovich, but upon his transfer to the lunar program, he was replaced by Major Aleksey A. Gubarev.[90]

The project may have accelerated quickly, but it is clear that by 1968, Kozlov had lost much interest in the Soyuz-VI. His branch was intensively busy with the development of more important photo-reconnaissance satellites. Mishin, perhaps pragmatically, seems to have been more focused toward creating an improved version of the Soyuz, the 7K-S, than the actual OB-VI station itself. And without doubt, the target of all his energies was focused not on the Soyuz-VI station, but on the programs he had inherited from his late mentor Korolev—the Soyuz, the UR-500K-L1, and the N1-L3 projects. As the Moon seemed to loom close enough to reach, the year 1968 would have Mishin and his engineers set out on the penultimate lap of the race to the Moon by finishing up an extensive testing program for the N1-L3 rocket complex, certainly the most intensive such effort to date in the history of the Soviet space program.

Preparing for the Landing

Through 1968, U.S. television and the press were full of rumor and hearsay on the impending introduction of a super-heavyweight Soviet launch vehicle comparable to the Saturn V. While some of this reporting was pure speculation, much of it was trickled down and leaked information from U.S. intelligence services, which were continuing to monitor activities at Tyura-Tam for clues to Soviet plans. During testimony in support of NASA's fiscal year 1969 authorizations to the House Committee on Science and Astronautics in February 1968, NASA

---

88.   *Ibid.*, p. 211.
89.   *Ibid.*
90.   The other cosmonauts in the Soyuz-VI group, established in early 1968, were V. B. Alekseyev, M. N. Burdayev, Yu. N. Glazkov, L. D. Kizim, A. Ya. Kramarenko, M. I. Lisun, A. Ya. Petrushenko, N. S. Porvatkin, G. V. Sarafanov, E. N. Stepanov, and V. D. Zudov. See Lantratov, "Dmitriy Kozlov's 'Zvezda'." Other sources suggest that three other cosmonauts—V. A. Grishchenko, V. I Gulyayev, and D. A. Zaykin—were also training for Soyuz-VI. Note that Grishchenko and Gulyayev resigned from cosmonaut training on February 5, 1968, and March 6, 1968, respectively.

Administrator James E. Webb told his distinguished audience: ". . . there are no signs that the Soviets are cutting back as we are. New test and launch facilities are steadily added . . . and a number of spaceflight systems more advanced than any heretofore used are nearing completion."[91] Webb also forecast the introduction of a Soviet booster more powerful than the Saturn V. Five months later, George E. Mueller, the NASA Associate Administrator for Manned Space Flight, added fuel to the fire in a private memorandum distributed to Apollo contractor personnel in which he stated that the Soviets were developing a "large booster, larger by a factor of two, than our Saturn 5."[92] In May 1968, one American journalist encapsulated the tone of these sporadic reports on the giant Soviet super-booster:

> This booster, like the Loch Ness Monster or Soviet submarines seen off the East Coast when the American Navy's budget is under review, tends to be mentioned by witnesses who are considered unreliable or prejudiced. But students of Soviet space trends say there is direct evidence that the booster will appear when the Russians are ready to show it. This conviction is apparently based on evidence—reconnaissance photographs of rocket engine test stands or perhaps new launching pads.[93]

As was customary, Soviet officials never once mentioned the N1 rocket, although through the first part of 1968, they continued to make repeated allusions to the possibility of Soviet cosmonauts flying and even landing on the Moon in the near future.[94]

Behind the veil of secrecy, the N1 was indeed emerging in metal, but it was months behind the latest schedule. As stipulated by the February 1967 decree from the Central Committee and the Council of Ministers, the first test flight of the launch vehicle was set for the third quarter of the same year. Cosmonauts were to lift off in the N1-L3 complex in April 1968. Slowly, deadlines shifted month by month, until engineers lost another year engaged in a very broad ground testing program carried out at more than a dozen different locations.

Engineers built more than thirty-five full-scale experimental assemblies of the most intricate, heavily loaded elements of the rocket's frame, many of which were tested at the Central Scientific-Research Institute of Machine Building next door to TsKBEM in Kaliningrad. In addition, individual sections of the booster structure were verified for strength and stability at specially built test stands built in 1967 at the Experimental Machine Building Plant belonging to TsKBEM. The comprehensive ground testing included: work on precision and pressurization; testing in deep vacuum and in weightless conditions; work on the mechanical and pyrotechnical systems of separation and docking and on the pneumo-hydraulic systems of the rocket stages; work on the command instruments and measurement systems, power sources, armature, and life support systems of the L3 complex; testing in high temperatures and vacuum; static testing of the rocket stages (including work on the thermodynamic processes associated with fueling the stages, storage, and preparation for launch); and work on the booster at the launch position (including checking the thermodynamic processes of the propellant

---

91.   NASA Science and Technology Division, *Astronautics and Aeronautics*, 1968, p. 34.

92.   John Noble Wilford, "NASA Aides Fear Soviet Space Gain," *New York Times*, August 14, 1968, p. 11.

93.   Clark, "Soviet Advances in Space Awaited," p. 50.

94.   Perhaps the only hint by a Soviet official during 1968 on the existence of the N1 rocket was a statement by Academician L. I. Sedov on West German television on March 20, 1968: "Special rockets are now available, very large rockets which have been built exclusively for space research purposes. These rockets make it possible to consider practically many things of which formerly one could dream. Flights to the Moon and space flight to the planets are now quite feasible." See *Soviet Space Programs, 1966–70*, p. 369.

systems of the ground complex, the system of docking the rocket to the launch complex, and the technological processes of preparing the launch complex and the rocket for launch).[95]

Among the many problems engineers encountered at the time was how to protect the bottom part of the rocket from the thermal and mechanical effects of the exhaust coming from the array of liquid-propellant engines. Specially developed materials were subsequently tested in various simulated conditions, although they would not be ready until the fifth launcher manufactured for launch, vehicle no. 7L.[96] Testing the booster's propellant tanks proved to be more difficult than anticipated. During some tests in 1967, the tanks were completely destroyed when internal pressure reached three atmospheres despite the fact that they were rated to handle to over that limit during emergencies.[97] Another problematic issue involved dynamic precision with regard to pulsation pressure in the rocket's tanks, which seemed to have thwarted work in the late spring of 1967. As late as July 1968, TsKBEM Deputy Chief Designer Sergey O. Okhapkin, the man responsible for much of the work on the N1, reported that there was still much about the dynamic precision of the rocket's first three stages that was unresolved.

If earlier the development of the N1's engines threatened to be the major bottleneck in the program, by 1967–68, the Trud Design Bureau (formerly OKB-276) was finally able to report good progress. By September 1967, Trud, under the direction of its Chief Designer Kuznetsov, had completed the construction of two major engine static stands at Kuybyshev, the EU-28 and the EU-29, for ground tests of individual engines of the first and second stages of the N1 in both nominal and adverse conditions. The testing at Trud was followed by a second series at the mammoth testing facilities of the famous Scientific-Research Institute for Chemical Machine Building (formerly NII-229), the premier rocket engine test facility in the Soviet Union, located at Zagorsk. Stands originally built for R-7–based boosters were redesigned to fire all of the N1's stages except, of course, the important first stage, which remained an unknown quantity and would have to be flown "green." The testing at Zagorsk began with "cold" firings of the N1 stages, followed by:

- Firings on the EU-87 test stand of individual tests of the NK-15 first-stage engines
- Three firings on the EU-16 test stand of the Blok B (second) stage
- Four live firings on the EU-16 of the Blok V (third) stage
- Firings on the EU-15 test stand of the Blok G (fourth) stage
- Firings of the Blok D (fifth) stage[98]

"Interdepartmental testing" of all the engines as separate units was carried out between September and December 1967, opening the way to the firing of complete prototypes of the second and third stages, which were completed by June and August 1968, respectively.[99]

95.   Semenov, ed., *Raketno-Kosmicheskaya Korporatsiya*, p. 255.

96.   R. Dolgopyatov, B. Dorofeyev, and S. Kryukov, "At the Readers' Request: The N1 Project" (English title), *Aviatsiya i kosmonavtika* no. 9 (September 1992): 34–37.

97.   Kamanin, "A Goal Worth Working for," no. 46.

98.   Semenov, ed., *Raketno-Kosmicheskaya Korporatsiya*, p. 256–57; Boris Arkadyevich Dorofeyev, "History of the Development of the N1-L3 Moon Program," presented at the 10th International Symposium on the History of Astronautics and Aeronautics, Moscow State University, Moscow, Russia, June 20–27, 1995. The EU-16 firings were carried out on February 2, 1967, April 13, 1967, August 23, 1967, and November 25, 1970. The EU-15 firings were carried out on June 23, 1968, August 29, 1970, and December 15, 1973. EU-15 was the site of a major accident on June 15, 1968.

99.   Igor Afanasyev, "N1: Absolutely Secret" (English title), *Krylya rodiny* no. 11 (November 1993): 4–5; Dolgopyatov, Dorofeyev, and Kryukov, "At the Readers' Request: The N1 Project"; Vad. Pikul, "The History of Technology: How We Conceded the Moon: A Look by One of the Participants of the N1 Drama at the Reasons Behind It" (English title), *Izobretatel i ratsionalizator* no. 8 (August 1990): 20–21.

Progress on the L3 lunar complex was much slower than that of the N1, partly because of continuing modifications to the design through 1968 as a result of ground testing and monetary restrictions. Engineers carried out three major ground firings of Blok D in 1967 in support of L3 operations; these were in addition to the two Earth-orbital launches of the acceleration stage as part of the circumlunar L1 project. One of the major concerns regarding Blok D was its operation for powered descent initiation from lunar orbit. During discussions in January–February 1968, the engineers and Chief Designers expressed reservations that after finishing its part of the deorbit firing, the subsequent ejection of Blok D from the Lunar Ship (LK) lander could be dangerous because of a Blok D explosion upon impact on the lunar surface. Among the options explored were the possibility of increasing the propellant of the lander engine to raise the altitude of separation, or even re-igniting Blok D to move the stage further away from the lander. To be perfectly sure of Blok D operation during the entire landing phase, Mishin and his deputies tabled plans at the time to carry out a series of "rehearsal" tests in Earth orbit using the Proton booster. For this, the engineers proposed creating the L1E vehicle, which would consist of a simplified automated 7K-L1 circumlunar vehicle, an experimental Blok D upper stage, and a special payload fairing for the complex. During its mission, the L1E would specifically test two major operations: lunar-orbit insertion and powered descent from lunar orbit, both crucial maneuvers on the landing flight.[100]

During 1968, the engineers were still debating over the docking radar for the LK, choosing from two competitive variants, Igla and Kontakt. Despite the better performance characteristics of the former, for inexplicable reasons, the engineers chose the latter, designed by the Scientific-Research Institute of Precision Instruments under Chief Designer Mnatsakanyan, for the LK. It seems that the lander's *Planeta* radar was, however, based on Igla. Perhaps the most critical element of the LK, the Blok Ye main engine, was suffering severe delays in its development program at the time. Full-scale ground tests of the lander engine had been scheduled for 1966, then 1967, but the timelines were continually moved back. At a meeting in March 1968, Ivan I. Ivanov, the leading designer for the engine at the KB Yuzhnoye (formerly OKB-586), reported that the engine was displaying a specific impulse three seconds lower than needed during test runs— a serious problem that would affect the mass of the LK, which had already been reduced down to an absolute minimum.[101]

In the United States, NASA had plans to test the Apollo Command and Service Module and the Lunar Module in Earth orbit before declaring them safe for lunar operations. Not surprisingly, the Soviet Union had similar plans for their two analogous spacecraft, the Lunar Orbital Ship (LOK) and the LK. In 1967, Mishin had approved plans to design and build Earth-orbital versions of both vehicles, called the T1K and T2K, respectively. The two spacecraft would be equipped with fully functional life support systems to carry a single crewmember each. As was customary for the Soviets, the piloted flights would be preceded by joint automated flights of the T1K-T2K, also in Earth orbit. The T1K would be launched into orbit by the powerful UR-500K Proton booster, while the T2K would use a modified version of the Soyuz launch vehicle designated the 11A511L. Once in orbit, the T2K would simulate a descent to and an ascent from the lunar surface, followed by docking with the T1K. The two vehicles would then separate, with the descent apparatus of the T1K returning to Earth for recovery.[102] Despite the uncertainty regarding the Blok Ye engine, TsKBEM engineers were optimistic in their

100. Semenov, ed., *Raketno-Kosmicheskaya Korporatsiya*, p. 228.

101. In December 1966, the first test of Blok I (LOK) was set for July 1967, and the first test of Blok Ye (LK) was set for August 1967.

102. K. Lantratov, "The Fall From Orbit of the Last Soviet Lunar Ship" (English title), *Novosti kosmonavtiki* 25 (December 3–16, 1995): 32–36.

schedules for the TIK and T2K missions in Earth orbit. In March 1968, Mishin was planning for the first T2K launch in October 1968, with the second and third models a month later. In August, Mishin discussed with Chief Designer Gay I. Severin of the Zvezda Design Bureau, the man responsible for all spacesuits in the Soviet space program, the possibility of using *Yastreb* EVA suits on the TIK and T2K for a possible spacewalk. The idea seems to have been dropped soon after because of the added complexity of such a mission.

Much of the testing on the LK and LOK was carried out at the TsKBEM plant or at the imaginatively titled Scientific-Research Institute for Chemical and Construction Machines at Sergeyev Posad. These tests included those for the separation of the LOK and the LK in nominal and emergency situations, the docking systems, and the separation of Blok D. The same institute was also the location of landing tests of the lunar lander mock-ups to refine the design of the LK. At least 200 drop tests of the descent framework were conducted, half of them with full-sized prototypes. Engineers devised different simulated lunar landscapes for a variety of situations and introduced various landing profiles. For example, three different parameters, including the horizontal velocity (zero to one and a half meters per second), the height of the fall (several meters), and the angle of contact with the surface (thirty degrees to negative values), were considered. Designers also experimented with craters of various dimensions, repeating tests over and over to eliminate random results. Engineers carried out pyrotechnic separation tests to verify the operation of liftoff from the Moon, a problem made more difficult by temperature deformations in the ascent stage and none in the descent portion.[103]

Of the many potential hazards facing the LK during operations near the Moon, one of the most imposing was the influence of lunar gravitational anomalies. During the early robotic lunar probe missions in the mid-1960s, lunar satellites such as Luna 10, Luna 11, and Luna 12 deviated significantly from their expected trajectories around the Moon, raising the specter of such errors during piloted operations. To map out magnetic and mass anomalies on the lunar surface that could affect orbital vectors, engineers at the Lavochkin State Union Machine Building Plant under Chief Designer Babakin designed small lunar satellites designated the Ye-6LS to assist in mapping gravitational anomalies on the Moon. The first such spacecraft was launched on May 17, 1967, by a four-stage 8K78 booster (better known as the Molniya-M). Unfortunately, its Blok L translunar injection malfunctioned and was not able to impart sufficient velocity to the probe. As a failed deep space probe, the Soviet press referred to it by the nondescript name of *Kosmos-159*.[104] A second Ye-6LS probe failed to reach orbit on February 7, 1968, when the third-stage engine cut off prematurely at T+524.6 seconds because it ran out of propellant. Babakin was third time lucky, when vehicle no. 113 was launched successfully on April 7, 1968, and arrived at the Moon a few days later, officially named *Luna 14* in the Soviet press. Communications with the probe was carried out by the large TNA-400 dish at Simferopol in Crimea. Apart from successfully mapping gravitational anomalies, Luna 14 also carried motor

---

103. V. Filin, "At the Request of the Reader: The NI-L3 Project" (English title), *Aviatsiya i kosmonautika* no. 2 (February 1992): 40–41; Semenov, ed., *Raketno-Kosmicheskaya Korporatsiya*, p. 256. Note that the descent stage was not temperature controlled.

104. "Calendar of Memorable Dates" (English title), *Novosti kosmonavtiki* 10 (May 5–18, 1997): 51–53; Timothy Varfolomeyev, "Soviet Rocketry that Conquered Space: Part 6: The Improved Four-Stage Launch Vehicle, 1964–1972," *Spaceflight* 40 (May 1998): 181–84. Although Kosmos-159 did not reach lunar orbit, it did attain a highly elliptical orbit with an apogee of 60,637 kilometers, the highest for any satellite in the Kosmos series. Western analysis of the Kosmos-159 launch suggests that based on its launch time, it was launched directly away from the Moon, much like Zond 4. See Phillip S. Clark, "Obscure Unmanned Soviet Satellite Missions," *The Journal of the British Interplanetary Society* 46 (October 1993): 371–80. More recent evidence suggests that Kosmos-159 did indeed enter its originally planned orbit. See Timothy Varfolomeyev, "Soviet Rocketry that Conquered Space: Part 5: The First Planetary Probe Attempts, 1960–1964," *Spaceflight* 40 (March 1998): 85–88.

drives for testing different materials, lubrications, and coating for the wheels of the future Ye-8 lunar rover.[105]

The rover, for transporting cosmonauts from one lander to another on the Moon's surface, was Babakin's most important contribution to the Soviet lunar flotilla of the 1960s. In early 1967, Soviet space officials tabled a new proposal to build upon the Ye-8 rover: why not build a compact spacecraft capable of landing on the Moon, recovering a tiny portion of lunar soil, and then returning to Earth?[106] The idea was clearly motivated to a great extent as insurance against losing the race to the Moon. If all else failed—and Apollo was about to land on the Moon—then Babakin could dispatch one of these robots to recover soil before any American astronaut. It was a pragmatic public relations exercise, but one that obviously had important scientific payback. The proposal apparently originated from Babakin's design bureau, and it was the subject of "a brief but heated debate" before being approved for implementation.[107]

As with the L3 program, the primary limitation for the soil return spacecraft was mass. Instead of developing a completely new vehicle, Babakin chose to model his sample returner on the Ye-8 rover by using its descent platform, the so-called KT stage. But instead of the lunar rover as a payload, the KT would carry a vehicle capable of scooping some soil, lifting off from the Moon, heading for Earth, and reentering into Earth's atmosphere for subsequent recovery. Babakin designated the spacecraft Ye-8-5 to distinguish it from its antecedent, the Ye-8 rover.

When beginning to design the Ye-8-5 vehicle, the engineers assumed that it would be necessary to correct the return trajectory of the capsule on its trip back to Earth—that is, it would require complicated optical and gyroscopic devices, command radio links, and a rocket engine, all exceeding the mass requirements for the spacecraft. A solution to the problem came from Dmitriy Ye. Okhotsimskiy, one of the star scientists at the Institute of Applied Mathematics at the Academy of Sciences; he had helped optimize the design of the first Soviet ICBM in the early 1950s and later worked on many early Soviet space projects. Okhotsimskiy's mathematical analysis showed that among the possible trajectories on the return flight from the Moon, there were a small class of passive flight trajectories that do not require correction and exist only on the "Moon-to-Earth" trip because of the strong influence of Earth's gravity. He found that with these passive trajectories, the landing point on Earth depends on the starting point on the Moon. This meant that the landing point had to be very exact, to within plus or minus ten kilometers of a specified point on the lunar surface. The study of lunar gravitation anomalies on Luna 10, Luna 11, Luna 12, and Luna 14 proved to be extremely useful for mathematical analyses of landing profiles from lunar orbit.[108]

Babakin's engineers fought long and hard with the mass constraints. The launch vehicle for the Ye-8-5 was the same as that for the Ye-8, a four-stage Proton booster that could put a mass

---

105. N. G. Babakin, A. N. Banketov, and V. N. Smorkalov, *G. N. Babakin: zhizn i deyatelnost* (Moscow: Adamant, 1996), pp. 54–55; Lardier, *L'Astronautique Soviétique*, pp. 182, 267. The TNA-400 was designed by OKB MEI under Chief Designer A. F. Bogomolov. Luna 14 also carried scientific instruments, including one for measuring charged particles from the Sun. See Kenneth Gatland, *Robot Explorers* (London: MacMillan, 1972), p. 140.

106. Note that K. Lantratov, "The 'Late' Lunar Soil" (English title), *Novosti kosmonavtiki* 15 (July 16–29, 1994): 41–43, states that the sample return effort began in 1968, not 1967.

107. Yu. A. Mozzhorin, *et al.*, eds., *Dorogi v kosmos: I* (Moscow: MAI, 1992), p. 163.

108. B. V. Rauschenbach, "Soviet Program of the Moon Surface Research," presented at the 45th Congress of the International Astronautical Federation, IAA-94-IAA.2.2.626, Jerusalem, Israel, October 9–14, 1994. There were additional limitations on the Ye-8-5 lunar probe. With a passive trajectory, the predicted landing point on Earth was too wide for effective search. This required transmission of the actual post-takeoff trajectory from the Moon, which meant that the return ship needed to carry complex radio-technical equipment. Okhotsimskiy bypassed this problem by proposing the use of radio equipment working in the meter range instead of the standard decimeter range, thus reducing the mass of the communications instrumentation. These transmissions would also be augmented by ground observations to measure angular velocity of the returning ship from a distance of 150,000 kilometers from Earth.

of only 5,550 kilograms on a translunar trajectory from Earth orbit. This would include both the KT descent stage and the actual scooper with its returning spacecraft. Despite a widespread and intensive effort to reduce the mass of the Ye-8-5 sample returner, Babakin was able to produce a vehicle with a mass of only 5,880 kilograms. With the project in jeopardy, Babakin convinced both Chelomey and Mishin to optimize the capabilities of the Proton and the Blok D stage, respectively, to allow the rocket to carry the increased mass. Chelomey and Mishin evidently were able to fulfill Babakin's requirements by reworking several systems and reducing reserve propellant.[109]

The Earth-to-Moon trip for the Ye-8-5 sample returner was identical to that of the Ye-8 rover. A nominal flight for the Ye-8-5 would begin with its launch into a low-Earth orbit by the Proton. About seventy minutes after launch, Blok D would fire a second time to insert the payload on a trajectory toward the Moon. After two mid-course corrections, the Ye-8-5 would fly into a 120-kilometer-high lunar orbit four days and seven hours after launch. In lunar orbit, the ship would conduct two further corrections: the first to reduce perilune down to twenty kilometers over the landing point and the second to straighten out the plane of approach. After seven days and sixteen hours in space, the Ye-8-5 would fire its 11D417 engine to initiate powered descent from lunar orbit, landing on the lunar surface on its KT descent stage within six minutes.

The KT stage for the sample collector was identical to the one on the lunar rover except for the addition of a 0.9-meter-long remote arm with a drill appendage, stored in an upright position. After landing on the Moon, the arm would be rotated down to the target area. Electric motors, tested on Luna 14, would allow the arm to sweep over a 100-degree arc, while the drill itself could be swiveled in elevation. The latter consisted of a hollow rotary/percussion bit to drive into the surface. The Ye-8-5 ascent stage consisted of three spherical tanks for nitric acid and unsymmetrical dimethyl hydrazine for the ascent stage engines, which was composed of the S5.61 with a thrust of 1.92 tons placed in the center and four outbound verniers attached to the tanks. A pressurized cylinder above the central tank contained control, communications, and power equipment including gyroscopes and accelerometers. Four antennae were placed orthogonally on the horizontal plane on the outside of the cylinder. The central component of the ascent stage was a small thirty-nine-kilogram spherical capsule with a diameter of fifty centimeters placed at the top of the cylinder. Internally, in the upper portion, the capsule carried parachutes and descent antennas. The middle part had a receptacle for the sample, and the lower part had batteries and transmitting equipment that produced a displaced center of gravity toward the bottom where the ablative heat shield was the thickest. Once the remote arm had collected the soil, the arm would raise the drill and insert the soil into the small capsule at the top of the craft, pressurize it, and then seal it. The capsule as a whole was attached to the rest of the ascent stage via straps.[110]

After one day and two hours on the lunar surface, the ascent stage would lift off from the Moon and enter a direct trajectory toward Earth. There would be no mid-course corrections on the return trip, and its ultimate destination would depend on the precision of the trans-Earth injection burn. After a flight lasting eleven days and six hours, the small capsule would land on Soviet territory.

Preparations for both the Ye-8 and the Ye-8-5 accelerated through 1968. During the middle of the year, the lunar rover was subjected to ground simulations at a specially constructed lunar landscape near Simferepol in Crimea. At least five firing tests of the KT lander stage took place in late 1968 at Zagorsk, one of which was less than successful because of a premature engine cutoff.[111]

109. Babakin, Banketov, and Smorkalov, G. N. Babakin, p. 54.
110. Andrew Wilson, Solar System Log (London: Jane's Publishing Co., 1987), pp. 61–62.
111. O. A. Sokolov, "The Race to the Moon: A Look from Baikonur," presented at the 45th International Astronautical Congress, IAA-94-2.1.610.

Babakin's Ye-8-5 sample scooper may not have been an integral part of the N1-L3 lunar landing program, but it added to the burden of the Soviet lunar effort of the period. The repeated additions and modifications to the N1-L3 plan in 1965–67 also complicated mission design. Even after the ink was dry on a final draft plan for a particular element of the L3 complex, months later, engineers would propose modifications based on new anticipated needs. This not only made it impossible to manufacture flight models of the spacecraft, but also added layer after layer of complexity to the N1-L3 mission. By 1968, the following components were part of the entire program:

- Ye-6LS (two robot probes to map lunar gravitational anomalies)
- Ye-8LS (two robot lunar satellites to photograph the lunar surface)
- T1K-T2K (automated and piloted flights of the LOK and LK in Earth orbit)
- L1E (automated test of the Blok D stage in Earth orbit);
- N1-L1 (two lunar orbital L1 flights as test payloads for early N1 launches)
- Ye-8 (two lunar rovers to serve as transport for cosmonauts);
- N1-L3 (one N1 launch with the backup LK)
- N1-L3 (one N1 launch with two cosmonauts to land on the Moon)

This was in addition to the huge effort expended on the separate L1 circumlunar project. For a launch profile that was to originally include a single launch to the Moon, the Soviet program to land cosmonauts on the Moon now included a multitude of weak links that could seriously disrupt the schedule. Perhaps one of the few confidence boosters for Soviet space engineers at the time was the majestic sight at Tyura-Tam of the first N1 rocket as it was wheeled out to its launch pad.

The N1 Arrives . . . and Leaves

During late 1967, the Soviets could not have ignored the hoopla surrounding a significant milestone in the U.S. space program. On November 9, 1967, the first Saturn V booster lifted off from Launch Complex 39 at the John F. Kennedy Space Center at Cape Kennedy, Florida. *Apollo 4*, as it was called, was a magnificently successful mission, vindicating the so-called "all-up" philosophy, coming on the heels what one observer called "the most exhaustive ground-test program in aerospace history."[112] Coincidentally or not, the Soviet government issued a new decree five days after the Apollo 4 launch—one that amended the unrealistic targets laid down in the important February 1967 resolution on landing Soviet cosmonauts on the Moon. The new decision, adopted on November 14, called for the initiation of flight testing of the N1 booster in the third quarter of 1968, almost a year behind the Saturn V. A date for a landing was apparently not specified; the authors of the decree merely stated that it would take place "in a period ensuring the preeminence of the Soviet Union in the exploration of space"—that is, before the Americans.[113] Mishin recalled decades later that "by then, it was already clear that the dates set by these directives were unrealistic. They were not backed up by funds, or production capacities, or resources."[114] According to the chief designer, spending on

112. Roger E. Bilstein, *Stages to Saturn: A Technological History of the Apollo/Saturn Launch Vehicles* (Washington, DC: NASA SP-4206, 1996), pp. 347–48.
113. Mikhail Rudenko, "Space Bulletin: Lunar Attraction: Historical Chronicles: First Publication" (English title), *Vozdushniy transport* 28 (1993): 10; V. P. Mishin, "Why Didn't We Fly to the Moon?" (English title), *Znaniye: tekhnike: seriya kosmonavtika, astronomiya* no. 12 (December, 1990): 3–43.
114. Mishin, "Why Didn't We Fly to the Moon?"

the NI-L3 at *its* peak in 1967–68 amounted to about $1.5 billion, compared to Apollo's nearly $3 billion at its peak in 1966–67.[115]

When the Saturn V blasted off from Cape Kennedy, half a world away in the Kazakhstan desert at Tyura-Tam, Soviet engineers were putting the finishing touches on the first NI mock-up. The supervisory body over the entire NI-L3 program, the so-called Council for the Problems of Mastering the Moon, met on October 9, 1967, to discuss these preparations as well as the overall status of the Soviet lunar landing program. Mishin reported that the first NI flight model would only be able to lift seventy-six tons, while a slight modification of the second stage would allow the attainment of the nominal ninety-five tons required for a lunar landing for a single cosmonaut. More improvements in the first and second stages, including raising the thrust of the NK-15 engines from 154 to 170 tons, would provide a payload capability of 105 tons, sufficient to carry two instead of one cosmonaut down to the surface. Such a plan had been discussed among the senior staff in mid-1967, apparently prompted by continuing grave concerns over the safety of having a single cosmonaut on the surface of the Moon. Academy of Sciences President Keldysh was one of the strongest supporters of the two-cosmonaut plan, making the somewhat implausible proposal at the October 1967 meeting that the council should seriously consider landing two cosmonauts on the Moon on the very *first* launch of the NI. If that was impossible, then the mission should try and land a lone cosmonaut.[116] Keldysh's voice was not the only one touting this absurd idea. Communist Party General Secretary Brezhnev was rumored to have said: "We should prepare for a manned mission to the Moon straight after the first successful launch of the NI, without waiting for it to be finally developed."[117] Mishin understandably reasoned that it would be absolutely impossible to land two cosmonauts on the Moon on the first or second NI.

Brezhnev's ludicrous demands underline to a great degree the incredible gap between the people building the spacecraft and those who controlled the purse strings. If there were expectations that the creation of the Ministry of General Machine Building in 1965 would put an end to the institutional chaos in the space program, they were never fulfilled. The managerial chaos was underlined at an important meeting after the Apollo 4 mission. On January 23, 1968, Minister Afanasyev hosted a large conference with the senior staff at TsKBEM, including Mishin, Bushuyev, Chertok, Okhapkin, and Tregub, at which the primary subject of discussion was the NI-L3. Afanasyev pulled no punches and bluntly blamed Mishin for all the troubles in the Soviet space program. Going down the litany of delays and failures in the program, Afanasyev spared no words in criticizing the performance of TsKBEM and Mishin in particular. While the poor results of the NI program could not be attributable to the incompetence of one man, Afanasyev had good reason to single out Mishin. In the two years since he had assumed the post of chief designer of the design bureau, there had been nothing but failure. Mishin was also stubborn and ill-tempered, and he constantly alienated those around him, from his deputies to the other chief designers. Of the original five chief designers who were alive, only Pilyugin and Ryazanskiy had "normal" relationships with Mishin. The three others had some form of complaints against what they considered his rude behavior and poor leadership qualities.

115. The figure of $1.5 billion is extrapolated from "The Moon Programme That Faltered," *Spaceflight* 33 (January 1991): 2–3, in which Mishin gives a figure of "half a billion" rubles. The conversion rate used was $3 = 1 ruble, which was the *unofficial* rate at the time. The figure for Apollo is taken from Jane van Nimmen and Leonard C. Bruno with Robert L. Rosholt, *NASA Historical Data Book, Volume I: NASA Resources 1958–1968* (Washington, DC: NASA SP-4012, 1988), p. 148. The precise figures for 1966 and 1967 were $2.9713 billion and $2.8779 billion, respectively.

116. Kamanin, "A Goal Worth Working for," no. 47.

117. *What Stars Are We Flying to?* (English title), Moscow Teleradiokompaniya Ostankino Television, First Program Network, Moscow, April 9, 1992, 0825 GMT.

Despite the rising complaints against Mishin, he was not dismissed. Some believed that Ustinov kept him on as the "fall guy" to take the blame for a program that was all but doomed to fail. The chief designer may have also had powerful supporters in key positions, one of them being Politburo member Andrey P. Kirilenko.

At the meeting in January 1968, Mishin clearly articulated some of the inherent managerial problems at TsKBEM. In some ways, his two basic points were more substantive than Afanasyev's introductory tirade. The chief designer strongly believed that his design bureau was overburdened with extraneous tasks, which prevented it from concentrating on such space projects as the N1-L3. Primary among these was the solid-propellant RT-2 ICBM project, which swallowed a lion's share of the design bureau's resources in the late 1960s. Mishin also complained about having to work on subsystems, such as launch escape towers and spacecraft landing systems, simply because subcontractors were unable to do so. His second point was aimed at the organization of the Soviet space program, and in particular Afanasyev's Ministry of General Machine Building. He bluntly accused the ministry of not controlling the completion of items that were subcontracted out by TsKBEM—that is, not helping in having subcontractors meet deadlines, a job that was increasingly falling on already taxed engineers at the design bureau.

Mishin's deputies also spoke. Chertok and Bushuyev both admitted that it was TsKBEM's own fault that they were so overloaded with projects. They mentioned the 7K-L1 circumlunar program in particular, inherited from the Korolev days, as one that was a needless burden. The hasty and often personality-driven decisions of 1964–65 were finally having the negative consequences many had feared. In the end, as with many other meetings, nothing changed. Afanasyev refused to disrupt military programs, such as the RT-2 ICBM effort, in favor of "civilian" projects, such as the N1-L3. The missile project stayed at TsKBEM. Relations between the design bureau and its subcontractors remained just as chaotic. The engineers at TsKBEM shrugged their shoulders and went back to work.[118]

Through the tumultuous events of the lunar program in the late 1960s, there was one curious politically motivated episode that threatened to derail the N1-L3 program as late as 1967. On November 17, 1967, the Central Committee and the USSR Council of Ministers issued decree no. 1070-363, which assigned General Designer Vladimir N. Chelomey to design and develop the UR-700 heavy-lift booster and the LK-700 lunar spacecraft to land two Soviet cosmonauts on the surface of the Moon by 1972 or 1973.[119] To any observer with even cursory familiarity with the history of the Soviet piloted lunar program, this decision remains one of the most inexplicable—one that even the most intricate machinations of political intrigue fail to explain. How could the Soviet government commit to a second lunar landing program at a time when millions had been expended on the N1-L3? How did the UR-700 program reemerge after an official interdepartmental commission had already passed it over in favor of the N1-L3? According to Sergey N. Khrushchev, the former Soviet leader's son, the action was partly motivated by the astonishing delays in the N1-L3 program. He hints that the idea belonged to Minister of General Machine Building Afanasyev, who was increasingly at odds with his boss Ustinov over support to Chelomey's organization.[120] Cool in his promotion of the late Korolev's dreams, Afanasyev began to shift his allegiance to Chelomey's programs with the formidable backing of new USSR Minister of Defense Andrey A. Grechko. The UR-700 may have had other supporters,

---

118. There is a detailed account of this meeting in Chertok, *Rakety i lyudi: goryachiye dni kholodnoy voyny*, pp. 479–87.

119. A second decree (no. 472) was issued by the Ministry of General Machine Building on November 28, 1967.

120. Sergey Khrushchev, *Nikita Khrushchev: krizisy i rakety: vzglyad iznutri: tom 2* (Moscow: Novosti, 1994), p. 524.

specifically Chief Designer Glushko and Air Force Col. General Kamanin, both of whom were vocal and vociferous opponents of Mishin.[121]

The new order tasked Chelomey to produce a draft plan for the UR-700 and the LK-700 within a one-year period. According to Khrushchev, Chelomey was very reluctant to take on the order, and he did not believe that any program at this late stage could be competitive with Apollo. Perhaps expecting another accident to delay Apollo, Chelomey sank his teeth into reviving the UR-700 proposal, tasking the development of the booster to his Branch No. 1 at Fili under his First Deputy Viktor N. Bugayskiy. Having already worked on the project for several years, Chelomey and Bugayskiy were able to produce the draft plan for the LK-700 lunar landing ship as early as September 30, 1968. Engineers finished the draft plan for the gigantic UR-700 rocket on November 15, just two days before the stipulated deadline.[122] They may have worked on time to produce the results desired by Afanasyev, but the second coming of the UR-700 slowly sank into oblivion. The Americans were racing ahead with Apollo, there was already a huge commitment to the N1-L3, and Chelomey himself had little interest in forcing through this last-minute gasp. Perhaps understandably, Mishin's faction was less than pleased with the entire debacle. According to one of Bugayskiy's deputies:

> [W]e received the order for the 200 ton rocket and began working. And suddenly the specialists from Korolev's Design Bureau were writing a memo to the Minister of General Machine Building S. A. Afanasyev. Soon they "killed" our 200 ton rocket, and Korolev's people were left without any competitors.[123]

Chelomey's engineers never built their gargantuan booster: "All the work on the UR-700 was limited to the design and the mock-ups of certain sections of the rocket."[124] Like so many of Chelomey's dreams, the UR-700 never left Earth. By early 1969, Chelomey had abandoned work on his alternative lunar landing project.

As for the N1, components for the first batch of rockets were produced initially in February 1967 at the Progress Plant at Kuybyshev. After production, the parts were then transported to Tyura-Tam, where they were assembled at the giant assembly-testing building. The first group included two mock-ups for ground testing and fourteen models for flight testing. Later operational batches would be manufactured based on the results of the first set of launches. The first N1 mock-up, vehicle no. 1M1, was designed and built to allow engineers to refine the dynamic characteristics of all the ground assemblies and the rocket itself and was not meant for flight. They used the mock-up, a complete engineering model with a nose section, to carry out integrated final ground testing of the N1-L3 complex as well as to perform procedures for prelaunch preparations. The results of these tests would clear the way for releasing the first flight article, N1 vehicle no. 3L, for launch. Just two weeks after the Saturn V launch, on November 25, 1967, the 1M1 was moved on rail tracks from the assembly-testing building to the first completed launch pad at site 110P.[125] At the pad, giant cranes raised the booster to a vertical

---

121. In a diary entry on August 31, 1974, Kamanin recalls that he and Glushko, in 1967, proposed the cancellation of the N1-L3 program, presumably in favor of the UR-700 project. See N. Kamanin, "I Feel Sorry for Our Guys" (English title), *Vozdushniy transport* 15 (1993): 12.

122. Khrushchev, *Nikita Khrushchev: tom 2*, p. 524.

123. Dmitriy Khrapovitskiy, "Absolutely Unclassified: The Ground Waves of Space Politics" (English title), *Soyuz* 15 (April 1990): 15.

124. *Ibid.*

125. V. A. Lebedev, "The N1-L3 Programme," *Spaceflight* 34 (September 1992): 288–90; Afanasyev, "N1: Absolutely Secret." Note that there were originally three mock-ups of the N1: 1M1, 1M2, and 1M3. The 1M3 eventually became the first flight article, 3L.

position. It seems that the magnificent view of the graceful rocket lifted spirits considerably. U.S. spy satellites were also watching. In a classified report at the time, the CIA reported:

> [O]n several occasions since December 1967, [the N1] has been erected on the pad while on other occasions the pad has been empty, suggesting the Soviets are testing the erection and checkout facilities of the system. The vehicle has not been flown but there is no evidence that the program is experiencing major difficulties.[126]

On December 1, the Moon council met once again under Afanasyev's tutelage. Almost all the luminaries of the Soviet space program, including Minister of Aviation Industries Petr V. Dementyev, Commander-in-Chief of Strategic Missile Forces Marshal Nikolay I. Krylov, Tyulin, Kerimov, Mishin, Barmin, Kamanin, and many other chief designers, were present. The reports were fairly positive. Save for a few items on the service tower and some systems adjustments, the first launch pad was prepared for an actual launch. The 1M1 mock-up had been placed on the pad, while all its operational parameters were measured during three complete cycles, after which the booster was transported back to the assembly-testing building. The plan was to take the rocket out again to the pad to fuel it completely three times. Ground workers would then train for thirty days to master all operations in preparation for the first flight model of the N1. The flight article, rocket no. 3L, would then be moved to the pad and prepared for launch in the first half of March 1968, although all finishing work on the launch pad would not be completed until March 30. There apparently had been problems with the mock-up, for it was returned to the assembly-testing building on December 12, 1967, and moved back out once again in January. The official history of TsKBEM notes that the work highlighted the requirement for better technical documentation.[127]

As workers labored to prepare the first N1 flight model, focus shifted to the L3 complex. On January 15, 1968, the Moon council met to specifically discuss piloted lunar operations, both in lunar orbit and on the lunar surface. Apart from Mishin, Chief Designer Severin responsible for spacesuits and Deputy Minister of Health Avetik I. Burnazyan reported on the health safety measures for lunar surface operations. The news was not good. Severin, for example, told his audience that he would need two more years to clear his *Krechet-94* suit for operations on lunar landing missions. One of Mishin's demands for the suit was that it be sufficiently robust for up to five kilometers of movement on the lunar surface *and* allow EVA operations for up to seventy-two hours, perhaps to enable the cosmonaut to survive decompression in the lander. Like most other chief designers, Severin's primary problem seems to have been the severe mass limits on the suit. At the time, the suit had a mass of approximately ninety kilograms. A large conference on the Krechet-94 and Orlan suits for the lunar mission was held on March 19, 1968, at Severin's Zvezda plant at Tomilino. Severin apparently had confidence in meeting Mishin's requests, reporting that the Krechet-94 would ensure EVA life support for six hours of work on the lunar surface, while the Orlan would provide two and a half hours, sufficient for the spacewalks in Earth or lunar orbit from one ship to another. Because the replenishment of oxygen and water would be possible from the LK or from the Ye-8 lunar rover on the surface of the Moon, the total operational time for the Krechet-94 would be as high as fifty-two hours.[128]

---

126. U. S. Central Intelligence Agency, "National Intelligence Estimate 11-1-69: The Soviet Space Program," Washington, DC, June 19, 1969, p. 14, as declassified in 1997 by the CIA Historical Review Program.

127. Semenov, ed., *Raketno-Kosmicheskaya Korporatsiya*, p. 573; Kamanin, "A Goal Worth Working for," no. 47; Kamanin, "A Goal Worth Working for," no. 49; Afanasyev, "N1: Absolutely Secret"; Afanasyev, "Unknown Spacecraft."

128. Kamanin, "A Goal Worth Working for," no. 48; Kamanin, "A Goal Worth Working for," no. 49.

Presumably because of the results of the IMI tests, Mishin was unable to meet the March 1968 deadline for launch, informally delaying the attempt to May. Military units evidently did not completely master all operations related to the work of the huge emergency rescue system on top of the NI. To add to the problems, work was disrupted on the booster in April by the death of two men during ground tests.[129] Oxygen systems on the support tower were also incomplete for a launch. At a meeting on April 22, Mishin targeted May 5 for another full-scale testing of the flightworthy NI-LI on the pad. The first launch article finally arrived at its pad on May 7, 1968. The launch was set for late May, despite concerns over the state of the booster engines, which were in less than perfect condition and only barely within the specified limits for testing.

The original payload for booster no. 3L had apparently been a 7K-LIE spacecraft equipped to test firings of the Blok D stage. At some point in 1968, the spacecraft was replaced by a dedicated circumlunar spacecraft re-equipped for flight in lunar orbit. In an example of the cross-pollination among the various lunar programs, this variant, known as the 7K-LIS, seems to have been left over from the short-lived plan to have the 7K-LI dock in Earth orbit with a Soyuz spacecraft prior to its circumlunar mission. The spacecraft was equipped with the Engine Orientation Complex (known as the "DOK") from the L3's Lunar Orbital Ship. The complex, having a mass of around 800 kilograms, was installed at the forward end of the 7K-LIS on its prominent support cone to carry out attitude control. Because there was no need for docking on the NI's launch, the engine complex did not have the active node of the Kontakt docking system. The DOK was manufactured by a new entrant to the Soviet space program, the Arsenal Machine Building Plant based in Leningrad, whose design bureau was headed by Chief Designer Petr A. Tyurin.[130] The first complete 7K-LIS vehicle was assembled in March 1968, in time for the planned NI launch in two months.

The launch was not to be. At some point during the prelaunch testing, technicians discovered cracks in the first stage, Blok A, which had evidently formed when the rocket was mated to its payload.[131] In such a condition, there was only one option: bring the booster back to the assembly-testing building and repair the cracks. The restoration took much longer than expected, introducing what would prove to be a fatal delay in the NI-L3 problem. Days turned to weeks, which eventually turned to months. It was not just the cracks on the NI, but also cumulative delays in the delivery of reliable equipment for ground operations, which was a significant factor in pushing back the deadline. In August, Mishin met with Ustinov and reported that subcontractors were continuing to break deadlines, that many electrical systems at the launch site did not meet specifications, and that there were many failures during ground testing. There was also a severe shortage of military personnel at Tyura-Tam for NI operations. Afanasyev and Mishin were looking at a best chance for launch in late 1968, yet *another* year behind schedule. The hopes of the Soviet Union in reaching the Moon before the Americans hopelessly sank into an intractable quagmire. By this time, NASA had already flown a second Saturn V booster and launched the first automated Lunar Module into Earth orbit.

129. Military workers for the NI were part of the Sixth Scientific-Testing and Experimental Directorate at Tyura-Tam. See Jacques Villain, ed., *Baïkonour: la porte des étoiles* (Paris: Armand Colin, 1994), p. 73.

130. That the original payload for booster no. 3L was a 7K-LIE is noted in Semenov, ed., *Raketno-Kosmicheskaya Korporatsiya*, p. 573. The manufacture of the DOK-DKP for the 7K-LIS was probably the first venture for the Arsenal Machine Building Plant in the Soviet space program. Later, in 1969–70, the organization took on "design escorting" for the US naval reconnaissance satellite system originally developed by TsKBM under General Designer V. N. Chelomey. See M. Tarasenko, "The Scientific Program of the KB 'Arsenal'" (English title), *Novosti kosmonavtiki* 6 (March 11–24, 1996): 47–48; Dmitriy Litovkin, "Space Projects of 'Arsenal'" (English title), *Krasnaya zvezda*, January 13, 1996.

131. Afanasyev, "NI: Absolutely Secret."

Unlike the NI, the Saturn V used a high-performance cryogenic upper stage fueled by liquid hydrogen and liquid oxygen. Throughout 1968, as the race slowly slipped through their hands, many Soviet designers clearly realized that although the NI had arrived as a real quantity on the launch pad at Tyura-Tam, it had much room for improvement, specifically in its use of propellants. An increased payload would allow engineers to amend one of the weakest elements of the NI-L3 plan and increase the crew size from two to three. The late Korolev had persistently tried to create a liquid hydrogen engine development program in the early 1960s, and the effort was finally producing results by 1967–68 with the establishment of a modest production base as well as the first static tests of actual engines.

The model with the best prospects, which began static tests in 1967, was the 11D56 engine with a thrust of seven and a half tons, a creation of the Design Bureau of Chemical Machine Building under Chief Designer Isayev based in Kaliningrad. Two other engines, the 11D54 and 11D57, built by the Saturn Design Bureau under Chief Designer Lyulka, were also approaching the ground testing stage by 1968. A fourth engine, a derivative of the NI's NK-15V motor, was the most powerful of the lot: it was a 200-ton-thrust engine proposed by NI engine architect Chief Designer Kuznetsov. This engine was, however, far behind in its development curve than the others. Possible applications of the Kuznetsov engine on future variants of the NI were discussed only in January 1968. Each of the four engines had a specific application in a modernized NI:

- The NK-15V would replace the current engines in Blok B (stage II).
- The 11D54 would replace the current engines in Blok V (stage III).
- The 11D57 in the new Blok S would replace the current Blok G (stage IV).
- The 11D56 in the new Blok R would replace the current Blok D (stage V).[132]

Perhaps it was the success of the Saturn V or perhaps it was Isayev and Lyulka's progress in developing the engines, but the Soviet liquid hydrogen–liquid oxygen rocket engine program seems to have interested a most unlikely party at this time. After years of vociferously opposing such engine applications in space rocket boosters, in early 1967, Chief Designer Glushko suddenly emerged with an idea for a 200- to 250-ton liquid hydrogen–liquid oxygen engine. The idea was evidently discussed at a ministerial level in January 1968, but by this time, Mishin was not interested in Glushko's reconciliatory gesture.

Proposals for the four engines from Isayev, Lyulka, and Kuznetsov allowed Mishin to table realistic modifications of the NI in 1967–68. In May 1968, The chief designer had one of his aides prepare a letter to Minister Afanasyev proposing three modifications of the NI—designated the NIF-V2, the NIF-V3, and the NIF-V4—each distinguished by the particular liquid hydrogen stage it used. The NIF-V2 would use a new second stage, the NIF-V3 would use a new third stage, and so on.[133] In August 1968, an "expert commission" consisting of

---

132. Semenov, ed., *Raketno-Kosmicheskaya Korporatsiya*, p. 262.

133. The aide was V. K. Bezverbiy. TsKBEM engineers had begun work on modernized variants of the NI prior to Korolev's death. Korolev had signed a "technical account" on November 9, 1965, that described four primary versions of the NI: the NIU (a variant with better mass characteristics and more reliable engines), the NIF (a model with improved engines on the first and second stages), and the NIM (two radically improved versions with new engines on all three stages). Each of these would also have subvariants, depending on their use of high-performance liquid hydrogen–liquid oxygen engines on the second or third, or both, stages. Their designations included the letter "V" to denote the Russian word for hydrogen ("vodorod") and a number to denote the stage application. These subvariants were the NIU-V3, the NIF-V3, the NIM-V3 (two different versions), the NIF-V2/V3, and the NIM-V2/V3. Lifting capability stretched from ninety-five tons on the NIU to 230 tons on the NIM-V-2/V3. See B. V. Raushenbakh, ed., *S. P. Korolev i ego delo: svet i teni v istorii kosmonavtiki: izbrannyye trudy i dokumenty* (Moscow: Nauka, 1998), pp. 632–33.

representatives from various other organizations examined the NIF-V3 and NIF-V4 concepts, evidently giving a positive recommendation to both. The latter version, the NIF-V4, was discussed at the Central Committee level the same month, although a formal decision on development was not forthcoming at the time. In their pursuit of high-performance engines, TsKBEM engineers considered many other proposals, including redesigning the Blok D fifth stage for liquid hydrogen, uprating the current first- and second-stage engines for higher thrust and reusability, upper stage nuclear rocket engines, and even combined liquid/air-compressed engines working on liquid hydrogen for the first stage of the N1.[134]

As the preparations for the first N1 launch at last began to pick up, space officials finally addressed a most critical, but often-postponed issue: a training program for cosmonauts for the L3 lunar landing program. In contrast to NASA astronauts who had been involved in lunar operations training for several years already, the Soviets were typically behind on the curve. Air Force Aide Kamanin had agreed on an initial list of six men on September 2, 1966, to prepare for the lunar landing.[135] Unfortunately for the cosmonauts, they did not do much training; by the end of 1967, there were still no L3 simulators available at the Cosmonaut Training Center. Kamanin claims in his journals that much of this delay in the delivery of simulators had to do with TsKBEM's continuous redesign of the L3 complex, which made it impossible for the prime contractor of the simulators, the Specialized Experimental Design Bureau at the M. M. Gromov Flight-Research Institute, to produce them. Another obstacle was what Kamanin calls the "ideology" of the L3 complex. In the fall of 1966, official documents specified that unlike previous Soviet piloted spacecraft, the L3 would afford cosmonauts a significant degree of control over the course of a mission.[136] In a year, Mishin's engineers had backed away from this requirement, falling back on Korolev's old adage about having them serve only as passengers. Thus, from the point of view of TsKBEM, L3 cosmonauts could manage with a compressed training program. In their view, civilian engineers from the design bureau would be the best candidates for lunar landing flights.

The issue of L3 simulators and the cosmonaut training program finally came to a head in December 1967 during several meetings between Air Force and TsKBEM representatives. The former were particularly surprised to find that Mishin had canceled contracts for two simulators: a turbo-flier and a V-10 helicopter with LK controls. Mishin's unilateral actions seem to have seriously raised the wrath of many officials, who were increasingly tiring of the chief designer's somewhat abrasive ways. Eventually by December 15, two deputy chief designers at TsKBEM, Tregub and Tsybin, agreed in principle to a new list of twenty cosmonauts, consisting of ten civilian engineers and ten military officers under Air Force command. TsKBEM and Air Force officials also came to a preliminary agreement on a list of simulators needed for the landing.[137]

---

134. The last concept is mentioned in Semenov, ed., *Raketno-Kosmicheskaya Korporatsiya*, p. 279.

135. The men were Yu. A. Gagarin, V. V. Gorbatko, Ye. V. Khrunov, A. A. Leonov, A. G. Nikolayev, and V. A. Shatalov. See N. Kamanin, "A Goal Worth Working for" (English title), *Vozdushniy transport* 45 (1993): 8–9.

136. Kamanin, "A Goal Worth Working for," no. 47.

137. The civilians were K. P. Feoktistov, G. M. Grechko, V. N. Kubasov, O. G. Makarov, V. P. Nikitskiy, V. I. Sevastyanov, N. N. Rukavishnikov, V. N. Volkov, V. I. Yazdovskiy, and A. S. Yeliseyev. The military officers were V. F. Bykovskiy, A. V. Filipchenko, V. V. Gorbatko, Ye. V. Khrunov, A. P. Kuklin, A. A. Leonov, A. G. Nikolayev, G. S. Shonin, V. A. Voloshin, and B. V. Volynov. By December 26, Nikitskiy and Voloshin had been replaced by V. Ye. Bugrov and P. I. Klimuk, respectively, although the latter two did not effectively join the group until February 1968. See Kamanin, "A Goal Worth Working for," no. 48. Note that TsKBEM had evidently established its own group of cosmonauts for the L3 program earlier on August 18, 1967. These six cosmonauts were S. N. Anokhin, V. Ye. Bugrov, G. A. Dolgopolov, V. P. Nikitskiy, V. I. Patsayev, and V. A. Yazdovskiy. See I. A. Marinin and S. Kh. Shamsutdinov, "Soviet Programs for Lunar Flights."

The eighteen-member L3 group, commanded by the ubiquitous Aleksey A. Leonov, finally began preliminary training in January 1968, later joined by two others the following month. On March 13, Air Force Commander-in-Chief Marshal Konstantin A. Vershinin signed off on a two-and-a-half-year-long training program for these men. At the time, the first L3 missions in Earth orbit were set for late 1968. The first lunar landing, under normal circumstances, was expected in the 1970–71 period, although most designers desperately still clung to the hope of carrying out the mission by late 1969. The shift to 1970–71 was evidence of a marked but subtle feeling among most Soviet space officials that it would be all but impossible for NASA to fulfill Kennedy's goal of landing an American on the Moon before the end of the decade. This belief was not without validity. By March 1968, NASA had still to recover from the Apollo 1 tragedy and was months away from flying a piloted Apollo spacecraft in Earth orbit, let alone in lunar orbit. Many Soviet officials believed that it would take a miracle to successfully carry out a sequential series of completely successful piloted Apollo missions in the perhaps fourteen months leading to a first landing. In many ways, the Soviets were viewing American capabilities through the prism of their own record. Failures were simply an accepted part of testing systems in space for the Soviets. In a diary entry in March 1968, Kamanin wrote:

> It took us three extra years to build the N1 and the L3, which let the United States take the lead. The Americans have already carried out the first test flight of a lunar space-craft, and in 1969 they plan to perform five manned flights under the Apollo program. It is worth noting that there are bottlenecks in the American program—I mean the use of liquid hydrogen as fuel for the second and third stages of the Saturn V and of pure oxygen inside the Apollo. So far hydrogen has been successfully "working" for the United States, but it may throw them back as was the case with oxygen which let them down, causing the death of three astronauts in January of last year.[138]

But the Soviets did not count on the fact that Apollo was one of the most thoroughly ground-tested programs in the history of the U.S. space program. They could not and did not anticipate that Apollo would fail to fail.

---

138. Kamanin, "A Goal Worth Working for," no. 49, p. 8.

CHAPTER FIFTEEN
FINAL LAP
TO THE MOON

Through the ten years after Sputnik, two powerful nations engaged in a competition whose underpinnings had as much to do with ideology as it did with strategic power. Space was, of course, only one component of this race, and some would argue less important in its immediate ramifications than the ideological and often bloody confrontations played out all across the world. But when John F. Kennedy's singular pronouncement in 1961 changed the tenor of the space race from one of the grander conquest of space to the less encompassing and more specific reach for the Moon, the meaning of space also changed. For a brief period in the 1960s, for most people, space exploration did not immediately bring to mind images of communications satellites, weather pictures, interplanetary probes, or even military fortifications. It was the Moon that caught the eye—the Moon, always mystical in nature, but now imbued with earthly concerns and earthly rivalries. For many, he who would reach the Moon first would not lay claim to the Moon, but rather Earth itself. As such, the last gasp to the finish line from September 1968 to July 1969 was as remarkable as anything ever seen before in the history of space exploration.

Return to Flight

As the summer gave way to the fall in 1968, the record of the Soviet piloted circumlunar program was dismal. Original plans were to carry out four automated lunar flights before flying cosmonauts around the Moon. In the four attempts since late 1967, there had been three complete failures and one partial success, the deep space mission of Zond 4 in March 1968. To add insult to injury, another L1 spacecraft had been destroyed during ground preparations for a launch in July 1968, delaying flight plans by several months. The first of the three remaining 7K-L1 spacecraft arrived at the Baykonur Cosmodrome to inaugurate a new series of attempts beginning with the lunar launch window in September 1968. The pace and results of ground preparations would determine the possibility of launching L1 missions in the October, November, and December windows.

L1 State Commission Chairman Tyulin, accompanied by Kamanin and a number of L1 cosmonauts, including Bykovskiy and Popovich, arrived at Tyura-Tam on September 10, 1968, for the launch, set for just after midnight on September 15. Kamanin appointed Bykovskiy, one of the leading contenders to command the first lunar mission, to be in charge of controlling preparations for the new launch. As the most experienced Soviet cosmonaut, he had recently, on July 11, been appointed commander of the cosmonaut detachment.[1] On the morning of

---

1. The "real" designation of the cosmonaut detachment was military unit no. 26266. See Sergey A. Voevodin, VSA053, October 23, 1994, an electronic newsletter, available at NASA History Office, NASA Headquarters, Washington, DC, file on cosmonauts.

September 13, there were reports from representatives of the search-and-rescue services for the L1 spacecraft. Resources were evidently very limited at the backup site in the Indian Ocean, primarily as a result of financial constraints; the State Planning Organ, responsible for budget appropriations, had recently cut monies for the service by half. If the spacecraft splashed down in the Indian Ocean, it would be during night time on September 21, making the recovery even more difficult with the limited resources at hand, especially because the L1 descent apparatus had no light beacon. Later in the day, the L1 State Commission met at a new three-story building at site 81 near the Proton launching pad. Deputy Chief Designers Trufanov and Shabarov, responsible for the booster and spacecraft, respectively, confirmed that all was ready for a successful circumlunar flight.[2]

The 7K-L1 spacecraft no. 9 carried a most interesting assortment of biological payloads to allow doctors to prepare for a piloted circumlunar mission. The central component of the payload was a set of two Steppe tortoises (*Testudo horsfieldi* Gray), each with a mass of 0.34–0.4 kilograms. As part of the experiment, there were two other tortoises in the control group and four more that were left untouched. Soviet doctors picked tortoises over other animals because they did not need complex systems for "security" and also "the method of fixing them on board spacecraft [could] be stringent."[3] The two flight tortoises were placed in the spacecraft on September 2, at which time their food supply was terminated. Physicians would study the deprivation of food until the recovery of the spaceship, to study the pathomorphological and histochemical changes in the animals over the course of several weeks. Apart from tortoises, spacecraft no. 9 also carried hundreds of *drosophila* eggs of the Domodedovo-32 line, air-dried cells of wheat, barley, pea, pine, carrot, and tomatoes, a flowering plant of *Tradescantia paludosa*, three different strains of chlorella, and a culture of lysogenic bacteria.[4]

The launch was perfect. The Proton booster lifted off just 0.07 seconds late, at 0042 hours, 10.77 seconds Moscow Time on September 15, 1968. With the Moon suspended squarely above the pad, the rocket gained speed as it sped into the night sky. At an altitude of 160 kilometers, the third stage switched off as planned, letting the booster coast up. After an agonizing 251-second interval, Blok D switched on as planned and fired for a nominal 108 seconds to insert the stack into a perfect Earth orbit of 191 by 219 kilometers. After a circuit around Earth, about sixty-seven minutes after launch, Blok D fired successfully a second time to impart sufficient velocity to its payload to send it toward the Moon. After the translunar-injection maneuver, the Soviet press finally announced the launch, designating the mission Zond 5. It was the first time in the circumlunar program that a spacecraft had been successfully sent toward the Moon.

While the initial results from the flight were encouraging, as it progressed, there were some malfunctions that threatened to destroy any hope of a complete success. During the outbound flight to the Moon, ground controllers at the main flight control center at Yevpatoriya discovered that the 100K stellar attitude control sensor had failed. Later diagnosis showed that the failure was a result of a contamination of the sensor's optical surface from residue released by

2.   N. Kamanin, "A Goal Worth Working for" (English title), *Vozdushniy transport* 50 (1993): 10–11.

3.   N. A. Gaidamakin, G. P. Parfenov, V. G. Petrukhin, V. V. Antinov, P. P. Saksonov, and A. V. Smirnova, "Pathomorphological and Histochemical Changes in the Organs of Tortoises Carried on Board the Spacecraft Zond 5" (English title), *Kosmichesiye issledovaniya* 7 (November–December 1969): 931–39.

4.   O. G. Gazenko, V. V. Antipov, and G. P. Parfenov, "Results of Biological Investigations Undertaken on the Zond-5, Zond-6, and Zond-7 Stations" (English title), *Kosmicheskiye issledovaniya* 9 (July–August 1971): 601–09. The payload was evidently carried in a 150-kilogram cone and also included instruments for the study of radiation, primary cosmic rays, the composition of the solar atmosphere, and photometry of several stars. See Christian Lardier, *L'Astronautique Soviétique* (Paris: Armand Colin, 1992), p. 161.

the heat given off from the interior coating. With one sensor malfunctioning, positioning the vehicle for mid-course corrections became a difficult proposition. Upon hearing news of the failure, Chief Designer Mishin and State Commission Chairman Tyulin flew to Yevpatoriya from the Baykonur Cosmodrome to direct compensatory measures, joining a group of cosmonauts, including Bykovskiy and Popovich, who were already at the center. On the morning of September 17, controllers were able to use the less accurate solar and Earth orientation sensors to maneuver the spacecraft successfully to carry out the first mid-course correction, sufficient to make the vehicle circle the Moon and head directly toward the Earth. At the time of the firing, at 0611 hours Moscow Time, Zond 5 was at a distance of 325,000 kilometers from Earth.[5]

The spacecraft circled around the far side of the Moon at a distance of 1,960 kilometers from the surface and was flung onto a return trajectory toward Earth. Special cameras took high-quality photographs of Earth from a distance of 90,000 kilometers, which were, in fact, the first complete pictures of Earth from the Moon, three months before Apollo astronauts returned with similar photographs. On the night of September 19–20, the British astronomical observatory at Jodrell Bank monitored transmissions from Zond 5 and picked up a Russian voice calling out instrument values from the spacecraft.[6] At the time, observers believed that the voice was prerecorded, but more than likely, cosmonauts, including Popovich at Yevpatoriya, were playing the role of a real crew by transmitting their reports via the spacecraft.

Zond 5's journey back was a difficult and challenging ordeal for ground controllers. To the alarm of the flight control team, the 101K Earth sensor also failed at the time. The problem was later traced back to incorrect procedures during the spacecraft's preparation at the technical complex. There was evidently an error in the operational documentation that caused the sensor to fall out of coordination with the mechanical operation of the spacecraft's main omni-directional antenna. To make matters worse, the three-axis stabilization platform spuriously switched off the guided reentry system. With all these failures, there was little hope that the spacecraft could carry out a guided reentry onto Soviet territory because that would require a highly precise attitude during the firing of the main engine. Engineers instead focused on bringing the vehicle back on a ballistic trajectory into the Indian Ocean using the remaining 99K solar sensor in conjunction with the smaller attitude control thrusters. Over the course of twenty hours, controllers at Yevpatoriya fed a series of singular commands to "swing" the ship from one side to the other, so that the resulting thrusts of the two engines would fire in the direction of Earth. After alternately turning on the small thrusters on each side of the vehicle, the ship gathered enough velocity and hit a tiny thin corridor in Earth's atmosphere for a ballistic reentry into the Indian Ocean.[7]

Tensions were high at both control centers, the primary one at Yevpatoriya and the supporting one located at the Ministry of General Machine Building's Coordination-Computation Center at TsNIIMash, next door to Mishin's design bureau. A number of high-level officials, including Georgiy N. Pashkov, a Deputy Chairman of the Military-Industrial Commission, and Maj. General Andrey G. Karas, the Commander of the Central Directorate of Space Assets, were present for the reentry at the center. Air Force representative Kamanin, who was also present, summarized the possible fate of Zond 5·as controllers watched their terminals:

---

5. Kamanin, "A Goal Worth Working for"; Yu. P. Semenov, ed., *Raketno-Kosmicheskaya Korporatsiya "Energiya" imeni S. P. Koroleva* (Korolev: RKK Energiya, named after S. P. Korolev, 1996), pp. 243–44; V. P. Glushko, ed., *Kosmonavtika entsiklopediya* (Moscow: Sovetskaya entsiklopediya, 1985), p. 130.

6. Kenneth Gatland, *Robot Explorers* (London: MacMillan, 1972), p. 141. Some reports also suggest that a second mid-course correction was effected on the return trip. See Glushko, ed., *Kosmonavtika entsiklopediya*, p. 130.

7. Semenov, ed., *Raketno-Kosmicheskaya Korporatsiya*, pp. 244, 354. Note that there were two 99K solar sensors on the ship. One of them had failed to turn on, leaving a single solar sensor available for use.

The spacecraft, according to estimates, should enter the atmosphere at an angle of 5–6 degrees to the local horizon. Even minus one degree in the reentry angle would mean that Earth's atmosphere would fail to "catch" the spacecraft. Even one degree would increase the g-load by 10–16 units above the estimated 30–40 units, and a greater angle would be dangerous not only for the crew, but may also destroy the spacecraft. In other words, the spacecraft should fly over 800,000 kilometers along the Earth-Moon-Earth route at a speed of 11 kilometers per second and hit the zone ("funnel") of safe entry 13 kilometers in diameter. Such high precision can be compared only to that of hitting a one-kopek coin from a 600 meter distance.[8]

To the credit of the resourceful ground controllers at Yevpatoriya, the ship slipped perfectly through its intended corridor into Earth's atmosphere. Within three minutes of the splashdown at 1908 hours Moscow Time on September 21, the commander of the search-and-rescue service, Air Force Maj. General Kutasin, reported that Zond 5 had landed 105 kilometers from the nearest Soviet ship in the Indian Ocean.[9] The first flight of a spacecraft to the Moon and back had lasted six days, eighteen hours, and twenty-sour minutes.

The rescue of the Zond 5 descent apparatus was complicated not only by the nighttime conditions but by the presence of some uninvited guests. U.S. Navy vessels were in the area at the time, evidently to observe the recovery process and to collect information on the Zond spacecraft. The lingering U.S. ships caused undue anxiety back at Yevpatoriya, especially for "flight director" Pavel A. Agadzhanov, the chief of the Chief Operations and Control Group, who did not want to compromise the secrecy of the landing. It took the *Borovichi*, an Academy of Sciences ship equipped with radio direction finders and powerful searchlights, several hours to find the capsule in the rough seas. Rescuers then lifted the 2,046-kilogram capsule onto the ship's deck and covered it with a large tarpaulin. The American ships left within minutes of having observed the recovery.[10] After recovery, an oceanography ship, the *Vasiliy Golovnin*, carried the spacecraft to Bombay on October 3, where it was packed into a container to hide its appearance. Officials drove the capsule to the airport, from where it was flown directly to Moscow on an An-12 aircraft. Through it all, the tortoises survived their ordeal, despite enduring a rough sea landing.[11] The descent apparatus, including the animals, arrived in Moscow on October 7; four days later, doctors were able to finally begin their medical analysis.[12]

---

8.   Kamanin, "A Goal Worth Working for," p. 10. Others present at the Coordination-Computation Center included K. P. Feoktistov (Department Deputy Chief, TsKBEM) and A. G. Mrykin (First Deputy Director, TsNIIMash).

9.   The exact location of the landing was 32°38' S by 65°33' E.

10.   B. A. Pokrovskiy, *Kosmos nachinayetsya na zemlye* (Moscow: Patriot, 1996), pp. 283–84. Curiously, the CIA in its report on the recovery of Zond 5 stated: "The spacecraft splashed down late on 21 September after completing a seven-day flight around the Moon. Soviet recovery ships were unable to locate the vehicle for some ten hours, and it was another three hours—mid-morning—before they recovered it. A U.S. destroyer observed this first Soviet water recovery at close range." See Peter Pesavento, "Two Weeks That Killed the Soviet Dream," *New Scientist* (December 18, 1993): 29–32.

11.   *Soviet Space Programs, 1966–70: Goals and Purposes, Organization, Resources, Facilities and Hardware, Manned and Unmanned Flight Programs, Bioastronautics, Civil and Military Applications, Projections of Future Plans, Attitudes Toward International Cooperation and Space Law*, prepared for the Committee on Aeronautical and Space Sciences, U.S. Senate, 92d Cong., 1st sess. (Washington, DC: U.S. Government Printing Office, December 1971), p. 242.

12.   Gaidamakin, et al., "Pathomorphological and Histochemical Changes." According to their analysis: "The effects of space flight, in conjunction with starvation, produced changes mainly of atrophy type in the organs of the animals. . . ." In addition, "Starvation at the space center (of tortoises of a control group) led to less pronounced atrophy of the tissue. Comparison of the changes which occurred in the test and control animals indicates that the main structural changes in the tortoises were caused by starvation and to a lesser degree by the action of the flight factors."

The Zond 5 mission, despite its attendant flaws, was the first unequivocal success in the LI program. It allowed Tyulin and Mishin to seriously plan on flying a crew on a circumlunar mission in January 1969, contingent upon two more successful LI flights. By the time Zond 5 splashed down in the Indian Ocean, there were three lunar launch windows left before 1969— in October, November, and December. Based on the pace of preparations, Mishin hoped to fly LI spacecraft no. 12 in November and spacecraft no. 13 in December. The ship and cosmonauts for a piloted flight would be ready in January. Such a schedule would still fulfill the original mandate of flying four robotic spacecraft before a crewed attempt.

Crews for the piloted mission had nearly completed their training program by this time, with a final spurt during the Zond 5 flight, when some of the LI cosmonauts trained at Feodosiya. On September 27, Kamanin and Mishin agreed to three final crews for the first circumlunar mission. With any luck, one of these crews would make history as the first humans to fly from Earth to the Moon. The crews were:

- Crew 1: Aleksey A. Leonov and Oleg G. Makarov
- Crew 2: Valeriy F. Bykovskiy and Nikolay N. Rukavishnikov
- Crew 3: Pavel R. Popovich and Vitaliy I. Sevastyanov

All three crews were judged to be equally prepared for the flight, although it seems that Kamanin had favored the Bykovskiy crew as the primary candidates for the first outbound mission. As with all other Soviet piloted missions, a final decision on the issue was expected at the State Commission meetings prior to launch. Each of the three crews also had a single understudy—Anatoliy P. Kuklin, Petr I. Klimuk, and Valeriy A. Voloshin, respectively. The three backup cosmonauts were trained and ready to step into either the commander's or flight engineer's position in case a primary crewmember was indisposed.[13]

The nine men training for a circumlunar mission were not the only cosmonauts preparing for spaceflight in the fall of 1968. By August 1968, trainees Beregovoy, Volynov, and Shatalov had completed training for the first piloted Soyuz mission since the Soyuz 1 tragedy more than a year before. In the autumn of 1968, Ivan I. Utkin, the chair of the subcommission investigating the accident, finally declared the Soyuz landing system completely ready for piloted flight.[14] Less by plan than by coincidence, Chief Designer Mishin set the "return to flight" Soyuz mission in time for the fifty-first anniversary of the Great October Revolution. The flight plan was for one cosmonaut in an active Soyuz to link up with a passive automated Soyuz. The two ships would remain docked for a few hours before separating and carrying out independent missions. The conservative rendezvous and docking flight would then open the way for the long-delayed EVA transfer attempt. There was one major difference on this mission from the previous "rehearsal" docking missions of Kosmos-186/188 and Kosmos-212/213; in this case, engineers decided to launch the passive instead of the active vehicle first. The older profile was clearly more suited for simulating operations in lunar orbit when the active LOK would await the passive LK after it had lifted off from the Moon. The Soviets themselves have never revealed the reasons for this unusual switch. Perhaps it was dictated by engineering concerns over checking the operation of the *Igla* rendezvous radar system before committing to a piloted mission. Less likely, but certainly possible, it may have been TsKBEM's attempt at rehearsing an Earth-orbit rendezvous for a lunar landing mission in case the N1 was not deemed safe for carrying cosmonauts into orbit. Such a prospect was, in fact, given serious consideration throughout 1968–69.

---

13.   Kamanin, "A Goal Worth Working for." On September 24, three days before the final decision, Kamanin was leaning toward the following crews: A. A. Leonov/A. F. Voronov, V. F. Bykovskiy/N. N. Rukavishnikov, and P. R. Popovich/O. G. Makarov. Obviously, this crew composition was modified by September 27.

14.   Semenov, ed., *Raketno-Kosmicheskaya Korporatsiya*, p. 183.

This meeting of the State Commission occurred prior to the Soyuz 3 mission in October 1968. Sitting at left are Commission Chairman Kerim Kerimov and Chief Designer Vasiliy Mishin. Standing next to Mishin is Air Force Aide Nikolay Kamanin. Standing next to Kamanin from left to right are cosmonauts Georgiy Berogovoy (primary), Vladimir Shatalov, and Boris Volynov. Sitting on the extreme right is Marshal Sergey Rudenko, a Deputy Commander-in-Chief of the Soviet Air Force. (copyright Christian Lardier)

The Soviet political leadership was particularly anxious to resume space missions after the long gap, particularly because of NASA's well-publicized launch of Apollo 7 on October 11, 1968. It was the first crewed U.S. spaceflight since the Apollo 1 fire in January 1967. A few days after the Apollo 7 launch, Mishin met with Communist Party General Secretary Brezhnev to brief him on the state of various projects at TsKBEM, including the N1-L3, Soyuz, and RT-2 ICBM programs. Mishin also spoke to Minister of General Machine Building Afanasyev by telephone after arriving at the launch site. The two Soyuz missions were set for mid-October 1968, but there were numerous malfunctions during prelaunch testing, which prompted Afanasyev to order Mishin to delay the launches. On October 23, the day after the Apollo 7 crew's splashdown, the State Commission for Soyuz met at the Baykonur Cosmodrome to discuss preparations for the Soviet launches. Kamanin presented cosmonaut Beregovoy as the primary candidate, with Shatalov and Volynov as his backups. There seems to have been some serious doubt as to Beregovoy's qualifications for the flight. He had failed his prelaunch examination, receiving a "2" ("bad") out of a possible "5" ("excellent"). Instead of flying his backup Shatalov, Air Force officials organized a second examination, in which Beregovoy managed to get "4" ("good").[15] All three men—Beregovoy, Shatalov, and Volynov—had trained for the Voskhod 3 flight in 1966, whose cancellation had been one of Mishin's first actions after his official appointment as chief designer. Another issue at the meeting was what to call the first automated 7K-OK vehicle in the press—that is, whether to give it a nondescript "Kosmos" designation to hide its true mission or to bestow it with the Soyuz moniker. Commission

15.   I. Izvekov and I. Afanasyev, "How From a Failure Was 'Forged' the Next Victory" (English title), *Novosti kosmonavtiki* 23/24 (1998): 64–66.

members agreed to call the spacecraft Soyuz 2, but to announce it only after the launch of Beregovoy with Soyuz 3.

The 7K-OK spacecraft no. 11 lifted off successfully from site 1 at the Baykonur Cosmodrome at noon on October 25, 1968. The initial orbital parameters were 185 by 224 kilometers at a 51.7-degree inclination. All systems aboard the automated Soyuz spaceship seemed to be working without fault, but conservatism crept into the proceedings. Chief Designer Mnatsakanyan of the Moscow-based Scientific-Research Institute for Precision Instruments recalls that on the night of the first launch, thirteen members from the Chief Operations and Control Group at Yevpatoriya sent a telegram to him at the Tyura-Tam control center to drop the idea of docking on the mission and simply try a two-part rendezvous—first to thirty kilometers and then down to 100–200 meters. The abrupt change in plans was evidently motivated by a lack of confidence in the Igla radar system, whose chief architect was Mnatsakanyan. By his own account, the chief designer had no one to consult, and he unilaterally decided to reject their recommendation, taking full responsibility for the decision.[16]

The following day at 1134 hours Moscow Time, as the target vehicle passed over the launch site, the 7K-OK spacecraft no. 10 lifted off with Colonel Georgiy T. Beregovoy aboard. It was the first-ever piloted launch from site 31, the second launch complex at the Baykonur Cosmodrome built for launch vehicles derived from the old R-7 ICBM. At forty-seven years old, Beregovoy was the oldest person to venture into space at the time. His initial orbital parameters were 205 by 225 kilometers also at a 51.7-degree inclination. Soon after the launch, the Soviet press announced Beregovoy's mission as *Soyuz 3* and the target as *Soyuz 2*.

On Soyuz 3's first orbit, ground controllers switched the Igla rendezvous system into operation, bringing the vehicle to a distance of only 200 meters from the Soyuz 2 target after at least two orbital corrections. At that point, as external TV cameras beamed down images to Earth, test pilot Beregovoy took over manual control to bring his spacecraft in for a docking. As he closed into a range of forty to fifty meters, his spaceship automatically banked 180 degrees from the target despite his best attempts to compensate for the guidance system.[17] After the sudden failure, the two ships moved apart while several senior officials, including Minister Afanasyev, Academician Keldysh, Col. General Kamanin, Space Assets Commander Maj. General Karas, and Chief Designer Mishin, flew to Yevpatoriya from the launch site. There was evidently some controversy on whether the docking failure was the result of an Igla system failure. Mnatsakanyan insisted that his system worked flawlessly and that:

> the cosmonaut had been confused by the light beacons [on the target spacecraft], and thereby [had maneuvered his spacecraft in such a way] that a certain angle had been formed between the antennas of the [two] ships, causing the [active] ship to "turn away" to one side.[18]

Later analysis confirmed Mnatsakanyan's hunch and clearly pointed to pilot error as the primary reason for the failure. Once the Igla system had brought Soyuz 3 to within 200 meters of Soyuz 2, Beregovoy took over manual control. At that point, the two ships were still not aligned perfectly. However, instead of gingerly stabilizing his ship along a direct axis to the target, Beregovoy used a stronger firing to put his spacecraft into a completely incorrect orientation relative to the target. The passive Soyuz 2's radar system, sensing the improper deviation, automatically turned its nose away from Soyuz 3 to prevent an incorrect docking.

16.   Yu. A. Mozzhorin, et al., eds., *Dorogi v kosmos: II* (Moscow: MAI, 1992), p. 35.
17.   Semenov, ed., *Raketno-Kosmicheskaya Korporatsiya*, p. 190.
18.   Mozzhorin, *Dorogi v kosmos: II*, p. 35.

Beregovoy, not sensing the real problem, completed a fly-around, and then tried to approach the target a second time. The same thing happened again. In the process, he practically exhausted all the propellant remaining for orientation. Because there was barely enough propellant remaining for reentry only, any further docking attempts had to be called off.[19]

After the initial rendezvous, Beregovoy retreated from Soyuz 2, and throughout the remainder of the day, the ships drifted 565 kilometers apart. At the end of his work day, on Soyuz 3's fifth orbit, the cosmonaut moved into the spheroid living compartment at the forward end of the ship and began his sleep period.[20]

On October 27, after waking up, Beregovoy exercised for about twenty-five minutes before beginning his day's activities. Perhaps taking a cue from the recent live transmissions from the Apollo 7 spaceship, the State Commission allowed Beregovoy to "host" a TV performance later that day that was beamed down to Soviet television, providing the public their first view of the interior of the Soyuz spaceship. Viewers saw the cosmonaut wearing a woolen training suit and a white helmet with earphones as he spoke of the comfort afforded by the new spaceship. The following morning, the automated Soyuz 2 spacecraft separated into its component parts, and despite a malfunctioning astro-orientation sensor, the descent apparatus carried out a successful guided reentry, landing at 1056 hours Moscow Time near the target region in Kazakhstan. The parachute system worked without fault. On October 28, Beregovoy devoted his time to a modest suite of scientific and Earth observation experiments. He carried out:

> observations of the stellar sky, the earth, and other heavenly bodies; detected the storm centers of typhoons and cyclones on the earth's surface; made reports to earth on fires in forests and jungles; studied the brightness of the earth's surface; photographed its cloud cover and snow cover; and photographed its horizon in daylight and twilight.[21]

This last experiment involved taking photographs using photometrically marked black-and-white film with orange-colored light filters.[22]

After midday, Beregovoy performed a second TV transmission for public benefit, pointing out instrumentation within the vehicle. One orbital maneuver the same day on the thirty-sixth orbit changed his orbit to 199 by 244 kilometers. His fourth working day began on October 29 at 0345 hours Moscow Time, and it culminated with his third public TV broadcast, during which he gave viewers a look through the portholes in the Soyuz. There were evidently no anomalies during the flight, and the cosmonaut worked without interruption on his experimental observations. He maintained a good appetite throughout the mission and did not display any sign of disorientation, although he later admitted that it took him about twelve hours to get fully used to the weightless state.

Soyuz-3's reentry program was the source of great anxiety at the control centers, not the least because it was the first piloted return to Earth since Komarov's tragic death. After an initial aborted attempt, Beregovoy fired his main engine for 145 seconds over the Atlantic

---

19.   Izvekov and Afanasyev, "How From a Failure Was 'Forged' the Next Victory." As a comparison, during the twenty minutes of the automatic portion of the rendezvous, Soyuz 3 used only thirty kilograms of propellant. In the ensuing two minutes, Beregovoy used up forty kilograms, after which there were only eight to ten kilograms remaining, sufficient for only one reentry attempt.

20.   Evgeny Riabchikov, Russians in Space (Moscow: Novosti Press Publishing House, 1971), p. 244; Peter Smolders, Soviets in Space (New York: Taplinger Publishing Co., 1973), p. 163.

21.   Riabchikov, Russians in Space, p. 245.

22.   G. V. Rozenberg and A. B. Sandomirsky, "Altitude Variation of the Scattering Coefficient from Spaceship Soyuz 3 Measurements and Aerosol Stratification," in K. Ya. Kondratyev, M. J. Rycroft, and C. Sagan, eds., Cospar: Space Research XI: Volume I (Berlin: Akademie-Verlag, 1971), pp. 633–38.

Ocean to brake from orbit on the morning of October 30. Flying over Africa and then the Caspian Sea, the descent apparatus successfully carried out a guided reentry landing at 1025 hours Moscow Time near Karaganda in Kazakhstan. Luckily for Beregovoy, a blizzard at the landing area had passed by morning time, and the cosmonaut landed safely on a snow-covered steppe, welcomed by a bewildered local boy on a donkey.[23] During a three-day, twenty-two-hour, fifty-minute, forty-five-second mission, Beregovoy had circled the Earth sixty-four times. While his flight may not have been completely successful, the Soyuz 2/3 mission was a significant boost to the confidence of engineers working on the program. Almost every single automated system aboard the Soyuz 3 spacecraft, including the Igla rendezvous system, the life support systems, the main engine, the attitude control sensors, and the parachute landing system, worked flawlessly. Beregovoy's postflight report on October 31 to the State Commission was illuminating. He recalled that payload fairing jettisoning was "unpleasant." Once in orbit, there were problems with the viewports: the right viewport was fogged up from the exterior, and there was dust between the glasses of the viewports. In general, Beregovoy reported that there was a lot of dust in the descent apparatus. Most critically, he reported that the manual control during the approach to Soyuz 2 was "too sensitive," implying that the "human automation" dynamics had room for improvement. When asked later by the press whether his age had made it difficult for him to be chosen for the mission, Beregovoy replied that his height (180 centimeters) had been more of a problem than anything else.[24]

Crew-rating the Soyuz spacecraft was critically important for the future of the Soviet space program, but for immediate purposes, the focus was on the Moon—in particular, the L1 circumlunar program. Delays in the preparation of the next flight-ready L1 vehicle had forced Mishin to skip the October 13–15 lunar launch window, thus shifting the launch into November. With rumors on the possibility of an Apollo lunar-orbital mission circulating in the Western press, Soviet public spokespersons suddenly found themselves in a difficult position. As a result, throughout October and November, Soviet officials expressed often contradictory positions on their policy on the "race to the Moon." On October 14, Academician Sedov, representing the Soviet Union at the 19th Congress of the International Astronautical Federation in New York, in a clear obfuscation of the truth, stated that "the question of sending astronauts to the Moon at this time is not an item on our agenda. The exploration of the Moon is possible, but is not a priority."[25] Then, as if to contradict himself, he added that "the program for the exploration of the Moon depends upon the success [of the Zond] experiments. Since the experiments may have various results, it is not possible at this time to be positive about lunar landings."[26]

The press conference for the Soyuz 3 mission, held on November 5, was also an interesting exercise in public relations. Despite hesitance on talking about lunar plans, Academician Keldysh was forced by the numerous questions from journalists to finally concede that the Soyuz spacecraft was not designed for a flight around the Moon. He strongly implied that the Soviets were not planning a piloted flight around the Moon in the near future. It was the first step on the slow and painful road for the Soviets in their cover-up of the piloted lunar programs. After years of vociferously voicing opinions in favor of crewed lunar operations, Soviet spokespersons were all of a sudden caught in a web of confusion, having to emphasize that they were not interested in the Moon while confirming as such, often in the very same sentence. Keldysh, for example, added at the Soyuz 3 press conference that before cosmonauts

23. Riabchikov, *Russians in Space*, pp. 243–44; Smolders, *Soviets in Space*, pp. 165–66.
24. Reginald Turnill, *The Observer's Book of Manned Spaceflight* (London: Frederick Warne & Co., 1975), p. 132.
25. "Russian Denies Moon Race Is On," *New York Times*, October 15, 1968, p. 48.
26. *Ibid.*

actually carried out a lunar landing, a complete mission from liftoff to lunar landing and return to Earth would be carried out automatically. The Soyuz 3 mission itself was the subject of a lie; when a journalist asked Beregovoy why he had not docked with the Soyuz 2 spaceship, the cosmonaut replied calmly, "That was not on the program."[27] No doubt, he was only saying what his "handlers" had asked him to say. As if to confirm that the Soviets were finally backing away from any public association with the Moon, Academician Sedov emphatically announced during a visit to the University of Tennessee Space Institute on November 7 that the "U.S.S.R. would not conduct manned lunar operations within the following six months."[28]

Apollo Versus Zond

In this penultimate lap toward the Moon, the tenor of the competition between the Soviet Union and the United States dramatically changed in the late fall of 1968 with the fast pace of events in the Apollo program. U.S. space officials had been carefully watching Soviet accomplishments throughout the year for hints of their ambitions toward the Moon. Circumlunar missions had been raised in classified CIA briefs as early as April 1967, and it was no surprise to U.S. observers when Zond 5 successfully carried out its flight exactly as predicted. The CIA, in a top-secret "National Intelligence Estimate" on the Soviet space program dating from April 1968, claimed that the Soviets might attempt a piloted circumlunar mission by "the last half of 1968."[29] One senior NASA astronaut, Frank Borman, recalls that in early August, news of the Soviet deadline of late 1968 had trickled down from the CIA to NASA, prompting NASA officials to establish a more ambitious timetable for Apollo.[30] In the alphabetical sequence of Apollo missions, the "C" mission (Apollo 7) in Earth orbit was to be followed by the "D" mission (Apollo 8), the first flight of the combined Command and Service Module with the Lunar Module, also in Earth orbit. The "E" mission (Apollo 9) would then be a Lunar Module test in high-Earth orbit.

In early August 1968, George M. Low, the Deputy Director of NASA's Manned Spacecraft Center in Houston, ordered his staff to work on a plan to eliminate the "E" mission in favor of the much more ambitious "C-prime" flight—one in which an Apollo Command and Service Module launched on a Saturn V would go directly to lunar orbit. It was a decision laden with risks. It would only be the third launch of the Saturn V booster, and the risks of a lunar-orbital mission would be exponentially more than one in Earth orbit. But based on their analysis, Low and Air Force General Samuel C. Phillips, Apollo program manager at NASA Headquarters, were willing to commit. As NASA historian Roger D. Launius accurately observed in retrospect:

> The advantages of this could be important, both in technical and scientific knowledge gained as well as in a public demonstration of what the United States could achieve. So far Apollo had been all promise; now the delivery was about to begin.[31]

27.   "Soyuz 3 Moon Trip Called Unlikely," *New York Times*, November 6, 1968, p. 44; *Soviet Space Programs, 1966–70*, p. 370.

28.   NASA Science and Technology Division, *Astronautics and Aeronautics, 1968: Chronology of Science, Technology, and Policy* (Washington, DC: NASA Special Publication (SP)-4010, 1969), p. 267.

29.   U.S. Central Intelligence Agency, "National Intelligence Estimate 11-1-67: The Soviet Space Program," Washington, DC, April 4, 1968, p. 2, as declassified in 1997 by the CIA Historical Review Program.

30.   Frank Borman and R. J. Sterling, *Countdown: An Autobiography* (New York: William Morrow, 1988), p. 189.

31.   Roger Launius, *NASA: A History of the U.S. Civil Space Program* (Malabar, FL: Krieger Publishing Co., 1994), pp. 89–90.

By mid-August, the Manned Spacecraft Center received clearance from NASA Headquarters on the new plan; a final decision was still contingent upon the success of the initial piloted Apollo mission in Earth orbit, then slated for October 1968. If Apollo 7 was an unequivocal success, NASA would move ahead to the lunar-orbital Apollo 8 in December.[32]

On October 11, 1968, NASA launched Apollo 7 into Earth orbit with three astronauts. After a highly successful eleven-day flight, the crew splashed down safely in the Pacific Ocean. NASA management's case for lunar orbit in December was further bolstered by the outstanding achievement of Zond 5, which had successfully circled the Moon and splashed down in the Indian Ocean. There was little doubt among independent observers that the Soviets were targeting the Moon for a piloted circumlunar flight, possibly for their lunar launch window, also in December 1968. On November 11, Phillips composed a final memorandum on launching Apollo 8 to lunar orbit, and Acting NASA Administrator Thomas O. Paine announced it publicly a day later.[33]

By early November, the Soviets were still planning two more automated L1 missions, one in mid-November and one in early December, to be followed by a piloted launch in January. The question begs itself: Once the Apollo 8 announcement was made public by NASA, did Soviet officials consider skipping one of the precursor flights and moving the piloted launch to December? The Soviets had a significant advantage. To have the best lighting conditions for potential lunar landing sites for future missions, NASA officials had set the Apollo 8 launch window for December 21, 1968. Because of differences in trajectories, the circumlunar launch window for a Soviet launch from central Asia would be earlier in the month, around December 8–10. Thus, launching cosmonauts to the Moon in December would guarantee a first-place finish at a time when the rivalry between the two space programs was approaching a climactic finish. But contrary to a plethora of speculation in the West, there was, in fact, no real plan for a December 1968 piloted launch to preempt Apollo 8.[34]

Cosmonauts, chief designers, and military officials arrived at Tyura-Tam in early November to direct the preparations for the launch of the 7K-L1 spacecraft no. 12. The launch went off without incident at 2211 hours, 31 seconds Moscow Time on November 10, 1968. Within sixty-seven minutes of the launch, the Blok D upper stage successfully fired to boost the spacecraft, named *Zond 6* by the Soviet press, toward the Moon. As soon as the spacecraft was on its way to the Moon, controllers discovered that an antenna boom had not deployed, effectively preventing operation of the stellar attitude control sensor mounted on the boom. Despite the problem, ground controllers managed to command the vehicle to perform its first mid-course correction at a distance of 246,000 kilometers from Earth on the morning of November 12 using a backup stellar attitude control sensor that used the Sun and Sirius as fixed points. Flying what seemed to be a perfect flight, Zond 6 flew around the far side of the Moon two days later at a closest distance of 2,420 kilometers.

A camera on the spacecraft took high-resolution black-and-white photographs of the Moon from distances of 11,000 and 3,300 kilometers. The first session was intended for filming the lighted surface of the Moon for measuring its photometric characteristics and determining its amount

32. For a discussion of the switch to "C-prime," see also William David Compton, *Where No Man Has Gone Before: A History of Lunar Exploration Missions* (Washington, DC: NASA SP-4214, 1989), pp. 132–33; Donald K. "Deke" Slayton with Michael Cassutt, *Deke! U.S. Manned Space: From Mercury to the Shuttle* (New York: Forge, 1994), pp. 213–16.

33. This memorandum of General Samuel is reproduced in full as "Reading No. 15: NASA Decides to Make a Circumlunar Apollo Flight," in Launius, *NASA: A History of the U.S. Civil Space Program*, pp. 207–10.

34. Col. General N. P. Kamanin's diaries confirm as such. On November 9, 1968, he wrote: "We have two more test launches to complete the program of preparing a piloted flight around the Moon." These two test launches were planned for mid-November and early December 1968. Later on November 10, 1968, Kamanin wrote that "our flight around the Moon with a crew on board is scheduled for the first half of 1969." See N. Kamanin, "I Feel Sorry for Our Guys" (English title), *Vozdushniy transport* 12 (1993): 11.

and form. The closer shots enabled large-scale photography for photometric measurements and the mapping of hidden portions of the Moon. The camera used panchromatic film and had a focal length of 400 millimeters; it produced frame sizes of thirteen by eighteen centimeters. Stereo imaging was made possible by the angles of some of the images. The photographs covered areas of the Moon both visible from Earth and on the far side. Apart from the camera, Zond 6 also carried a photo-emulsion detector to record the paths of cosmic rays, as well as another device to measure micrometeoroid impacts.[35] The spacecraft also carried biological specimens, although the Soviets have never provided any details. These possibly included tortoises, drosophila, Tradescantia plants, bulbs of the Allium series, dried wheat germs, various strains of chlorella, B. coli, and other samples. Explicit mention was only made of air-dried cells of wheat, barley, peas, pine, carrots, and tomatoes.[36]

After the spacecraft circled the Moon, controllers had to refine the trajectory of Zond 6 sufficiently to allow it to perform a guided reentry into Earth's atmosphere and land on Soviet territory instead of the Indian Ocean. The first correction was successfully accomplished on the morning of November 16 at a distance of 236,000 kilometers from Earth. It looked as if everything was on track for a perfect mission until sometime the same day when ground controllers detected a disastrous problem: the air pressure within the descent apparatus had dropped from a normal level of 760 mm Hg down to 380 mm, indicating a compromise of the spacecraft's integrity.[37] There was also an associated drop in temperature within the hydrogen peroxide tanks for reentry attitude control. Despite the partial depressurization, later found to be the result of a faulty rubber gasket, the critical systems on the ship remained operational, and the controllers were able to carry out the third and final mid-course correction, just eight and a half hours prior to reentry at a distance of 120,000 kilometers from Earth on the morning of November 17. Zond 6 separated into its two component modules prior to reentry, and at 1658 hours Moscow Time the same day, the descent apparatus entered its tiny entry corridor into Earth's atmosphere at a velocity of 11.2 kilometers per second. Passing through its 9,000-kilometer-long reentry corridor, it skipped out of the atmosphere, having reduced velocity down to 7.6 kilometers per second, and began a second reentry that further lowered velocity to only 200 meters per second. Throughout the reentry, engines on the descent apparatus automatically fired to vary roll control so as to change lift force and reduce g-loads. Unlike its predecessor, the Zond 6 descent apparatus was subjected to a maximum of four to seven g's.[38] The complex reentry was a remarkable demonstration of the precision of the L1 reentry profile.

The guided reentry may have been successful, but the depressurization problem was a failure difficult to ignore. During part of the descent, pressure in the descent apparatus reduced further down to only twenty-five millimeters, certainly killing any biological payloads on board. No doubt, a crew within the ship would have experienced the same fate. The near-total depressurization caused the gamma-ray altimeter of the descent apparatus to issue a false command to release the single parachute system, whose container was also depressurized, at an altitude

35.   Glushko, ed., Kosmonavtika entsiklopediya, p. 130; Soviet Space Programs, 1966–70, p. 243. Note that more recent Russian accounts state that the lunar photography was carried out at distances of 8,000 and 2,600 kilometers. See Semenov, ed., Raketno-Kosmicheskaya Korporatsiya, p. 245. The pictures had a resolution of fifty lines per millimeter.

36.   One Soviet source implies at several points that the biological payloads for Zond 6 were almost, but not completely, identical to Zond 5. See Gazenko, Antipov, and Parfenov, "Results of Biological Investigations."

37.   Semenov, ed., Raketno-Kosmicheskaya Korporatsiya, p. 245.

38.   Glushko, ed., Kosmonavtika entsiklopediya, p. 130; G. V. Petrovich, ed., The Soviet Encyclopaedia of Space Flight (Moscow: Mir Publishers, 1969), pp. 513–14; I. B. Afanasyev, "Unknown Spacecraft (From the History of the Soviet Space Program)" (English title), Novoye v zhizni. Nauke, tekhnike: Seriya kosmonavtika, astronomiya no. 12 (December 1991): 1–64.

of 5,300 meters above the ground instead of much later. Without a parachute, the ship simply plummeted down to the ground and smashed into pieces. Remarkably, the impact occurred only sixteen kilometers from the Proton launch pad at the Baykonur Cosmodrome, where Zond 6 had lifted off just six days and nineteen hours previously.[39]

What lay ahead for rescuers was yet another situation fraught with danger. The crushed descent apparatus clearly had a lot of valuable materials, including the in-flight data recorder as well as exposed film from the Zond 6 camera, which possibly could have survived the crash. On the other hand, the capsule contained ten kilograms of TNT, whose condition was unknown and which would pose a threat to any recovery operation. Groups from TsKBEM and the Scientific-Research Institute for Automated Devices arrived at the site on the day of the crash, followed by Deputy Chief Designer Bushuyev the following day, November 18. The plan was to extract all available recoverables from the broken chassis of the spacecraft with manual tools, but without striking any blows to the ship. It was a long, step-by-step, and arduous process, but rescuers eventually dismantled the explosive system and handed it over to an Air Force team, which later blew it up in a nearby steppe. For their demanding work, Chief Designer Mishin personally ordered commendations for all rescuers. A cursory inspection of the remains of the descent apparatus showed that the parachute system had indeed been jettisoned; moreover, the main undeployed antenna boom had remained attached to the capsule through reentry instead of being automatically discarded prior to entry into the atmosphere, although this did not affect the success of its guided reentry. Among the items recovered intact from the wreckage was the exposed film from the Zond 6 camera. Beautiful pictures of both Earth and the Moon were later published in the journals, serving to confirm Soviet assertions that everything about the flight had been successful. While all the biological specimens had been killed, Soviet scientists were able to glean information from some of the seedlings on board.[40]

Following the Zond 6 crash, Mishin postponed any plans for a piloted L1 mission in the near future; the dreams of Soviet engineers and scientists of circling the Moon prior to the United States also went up in smoke. It was the final and ignominious end of three years of intensive work—work plagued by unprecedented delays and failures. It was not a pretty picture for the Soviets in November 1968. Given the results of Zond 6, an automated launch would have to be skipped for the December launch window. The next available window was in January 1969. If and only if that mission was completely successful, officials could hope for a piloted circumlunar mission for the next window, perhaps in March or April 1969. Kamanin wrote in his diary on November 26, 1968:

> I have to admit that we are haunted by U.S. intentions to send three astronauts on board Apollo 8 around the Moon in December. Three of our unpiloted L1 spacecraft have returned to Earth at the second cosmic velocity, two of them having flown around the Moon. We know everything about the Earth-Moon-Earth route, but we still don't think it is possible to send people on that route.[41]

---

39.   Semenov, ed., *Raketno-Kosmicheskaya Korporatsiya*, p. 245; Afanasyev, "Unknown Spacecraft." The parachute system was evidently discarded at the time that the "frontal shield" for the descent apparatus was jettisoned. See Major I. Kuznetsov, "The Flight That Did Not Occur" (English title), *Aviatsiya i kosmonavtika* no. 8 (August 1990): 44–45. In the final conclusion on the Zond 6 failures, TsKBEM engineers believed that two problems—the drop in temperature in the hydrogen peroxide tank to minus five degrees Centigrade and the capsule depressurization—were related events. After the temperature drop on the night of November 14, engineers had attempted to heat the tank by facing it toward the Sun. The excess heat evidently affected the weak seal of the main hatch and led to slow decompression.

40.   Semenov, ed., *Raketno-Kosmicheskaya Korporatsiya*, p. 246; Gazenko, Antipov, and Parfenov, "Results of Biological Investigations."

41.   Kamanin, "I Feel Sorry for Our Guys," no. 12.

Ironically, it was the same day that the Soviet press for the first time explicitly connected the Zond circumlunar flights to a *piloted* space project. A journalist wrote in *Soviet News*, "[The] space station Zond 6, like Zond 4 and Zond 5, was launched in order to improve the automatic functioning of manned spaceship which will be sent to the Moon."[42] It was a particularly curious time for such an admission, especially because the LI program was at its nadir then, with little prospect of a piloted mission in the near future.

The impending launch of Apollo 8 on December 21 raised the ante of the space race to a dramatic level, especially in the public forum. Many mainstream Western publications reported that the Soviets were planning to go ahead with a piloted circumlunar launch on December 8.[43] Early in December, the popular magazine *Newsweek* quoted "U.S. sources" claiming that the Soviets would "default because of unspecified technical problems with their Zond spaceship."[44] A week later, the same magazine asserted that:

> Intelligence sources confirm that the Soviet Union was ready but unable to send a manned mission to the moon earlier this month when the launch window was open. Unspecified technical difficulties developed in the Zond spacecraft. In the past week, the Soviet space tracking and recovery ships in the Indian Ocean have dispersed or returned to port.[45]

These rumors contributed to a veritable cottage industry of stories that the Soviets had prepared a booster and that cosmonauts had been ready on the launch pad going through a countdown, which had been canceled at the last moment. The evidence, however, suggests that there was no such attempt, nor were there plans for such a launch, at least on the part of senior officials and designers. The cosmonauts training for the LI, however, apparently had other ideas.

Civilian cosmonaut Sevastyanov, an engineer on one of the three crews training for the circumlunar mission, recalls that the LI group of six cosmonauts wrote a letter directly to the Politburo asking for permission to fly to the Moon in December. They argued that despite all the failures on Zond 5 and Zond 6, the presence of a crew aboard the ship would make a flight more safe. Their proposed mission would begin with a launch on December 9, with sufficient time to beat Apollo 8. According to Sevastyanov, despite the absence of permission from higher officials, the cosmonauts flew to Tyura-Tam during the first days of December and were there for more than a week. The Proton booster and the 7K-LI spacecraft no. 13 were ready in the assembly-testing building, apparently the same articles that had been planned for a robotic flight in December before the Zond 6 failure. With zero support from most space officials, the cosmonauts never received permission to fly.[46] Given the inordinate levels of confusing information concerning Soviet space history, Sevastyanov's account is probably purely apocryphal. As evidenced by

---

42. Phillip S. Clark, "Topics Connected With the Soviet Manned Lunar Programme," *Journal of the British Interplanetary Society* 40 (May 1987): 235–39. See also Donald C. Winston, "Soviets Admit Zond 6 Manned Capability," *Aviation Week & Space Technology*, December 2, 1968, pp. 18–19. A prominent article in the official Soviet government newspaper *Izvestiya* also highlighted the possibility of a Soviet piloted circumlunar mission as a result of the "success" of Zond 6. See "Russians Cite Readiness for Manned Lunar Flights," *New York Times*, November 26, 1968, p. 4.

43. See, for example, "Radiating Confidence," *Aviation Week & Space Technology*, December 2, 1968, p. 15.

44. "Soviet Moon Shot Postponed?," *Newsweek* (December 16, 1968): 24.

45. "Soviet Moon Shot That Fizzled," *Newsweek* (December 30, 1968): 11. See also "Cosmos 260 Launched," *New York Times*, December 18, 1968, p. 35.

46. I. A. Marinin and S. Kh. Shamsutdinov, "Soviet Programs for Lunar Flights" (English title), *Zemlya i uselennaya* no. 4 (July–August 1993): 62–69; S. Shamsutdinov and I. Marinin, "Flights Which Never Happened: The Lunar Program" (English title), *Aviatsiya i kosmonavtika* no. 2 (February 1993): 30–31.

Kamanin's personal journals, the cosmonaut overseer was not even at Tyura-Tam on December 8, instead spending the day at the Cosmonaut Training Center in Moscow overseeing minor bureaucratic issues unrelated to the lunar program.[47] True or not, Sevastyanov's story adds to the mythology of the Soviet space program, growing ever more richer and imaginative year by year.

As the Apollo 8 launch grew closer, Soviet spokespersons for the space program began their efforts to neutralize what was threatening to become a public relations disaster. In a propaganda offensive that would last a year, Soviet officials engaged in a complete about-turn, backing away from their insistent statements of years before. Veteran cosmonaut Titov, on a trip to Bulgaria, told journalists the day before the Apollo 8 launch, "It is not important to mankind who will reach the Moon first and when he will reach it—in 1969 or 1970."[48] But matter it did. When Apollo 8 lifted from Cape Kennedy on December 21, 1968, the eyes of world were upon the three astronauts, Colonel Frank Borman, Captain James A. Lovell, Jr., and Lt. Colonel William A. Anders, who were embarking on a journey as important as any in history—to leave the bonds of Earth and head out into deep space. For many Soviets, it was a bittersweet day. Kamanin wrote in his diary:

The flight of Apollo 8 to the Moon is an event of worldwide and historic proportions. This is a time for festivities for everyone in the world. But for us, the holiday is darkened with the realization of lost opportunities and with sadness that today the men flying to the Moon are not named Valeriy Bykovskiy, Pavel Popovich, nor Aleksey Leonov, but rather Frank Borman, James Lovell, and William Anders.[49]

The Apollo 8 Command Module splashed down in the Pacific Ocean on December 27, 1968, after a mission successful beyond the best of hopes, during which the crew had circled the Moon ten times. After years of uncertainty and a lack of self-confidence, the United States had convincingly taken a dramatic lead over its only competitor. The time for payback had arrived for both countries. For the United States, it was payback for excellent management, high levels of funding, and a state-level commitment; for the Soviet Union, it was precisely the opposite. In their meager responses to Apollo 8, Soviet spokespersons weakly defended their positions. Academician Sedov, still referred to as the "father of the Sputnik," told Italian journalists a day after the Apollo 8 splashdown that the Soviets had not been competing in a race to orbit or land on the Moon. Referring to Apollo 8, he added:

There does not exist at present a similar project in our program. In the near future we will not send a man around the moon. We start from the principle that certain problems can be resolved with the use of automatic soundings. I believe that in the next 10 years vehicles without men on board will be the first source of knowledge for the examination of celestial bodies less near to us. To this end we are perfecting our techniques.[50]

Automation was a big theme in Soviet public statements throughout 1969. The topic was prominent at a meeting of the Military-Industrial Commission on December 30, 1968, to discuss

---

47. N. Kamanin, "For Him, Living Meant Flying" (English title), *Vozdushniy transport* 18–19 (1994): 12. Even more damaging for the Sevastyanov story, Kamanin on December 5–6 had spent the days off at his dacha resting and "clearing paths of snow."
48. *Soviet Space Programs, 1966–70*, p. 371.
49. Lev Kamanin, "From the Earth to the Moon and Back" (English title), *Poisk* 12 (July 1989): 7–8. See also Abe Dane, "The Moon Mission That Wasn't," *Popular Mechanics* (March 1990): 38–39.
50. "Soviet Space Aide Denies Moon Race," *Washington Post*, December 29, 1968, p. A4.

a response to Apollo 8. Grasping at straws, commission members decided to move ahead with one possible glimmer of light at the time: the Ye-8-5 robot spacecraft capable of recovering soil samples from the surface of the Moon. Kamanin had a cynical view of the exercise, writing:

> They cannot possibly get into their heads the very simple thought that it is impossible to answer the piloted flight of Apollo 8 with a flight of an automatic machine . . . any automatic machine cannot possibly be a satisfactory answer. Only landing people on the Moon and successfully recovering them on Earth would serve as an answer to the triumph of Apollo 8. But we are not ready for an expedition to the Moon, in the best case we will be ready for such a flight in about 2–3 years.[51]

As with many other lunar projects at the time, there was much still unknown about the Ye-8-5; engineers at Chief Designer Babakin's design bureau had not even built a complete model of the spaceship by the end of the year. Regardless, the Central Committee and the USSR Council of Ministers issued a new decree, no. 19-10, on January 8, 1969, titled "On the Work Plans for Research of the Moon, Venus, and Mars by Automatic Stations."[52] The decree evidently called for the acceleration of various automated programs, including the Ye-8-5 robot. It was the first clear response to Apollo 8, and it established a new direction in Soviet space policy that would remain entrenched for many years to come. Handed their biggest defeat yet, officials now went about neutralizing the effects of the Apollo victory by claiming that the Soviet Union had never intended to reach the Moon. It was clearly much easier to change history when the details of that history were originally obscured or hidden beyond recognition.

Transfer in Orbit

When the Soviets were finally ready to carry out their long-delayed docking and EVA Soyuz mission, it was already an anachronism. Originally, Korolev had conceived such flights as means to master rendezvous, docking, EVA, long-duration missions, and other complex operations in Earth orbit to provide expertise for future piloted lunar excursions. It would serve in much the same capacity as Gemini did for Apollo in the U.S. space program. To extend the analogy, by the time the Soviets were ready to fly *their* Gemini, the United States was already flying Apollo. In fact, much of the technology used on the Soyuz was different from that on the L3. For example, cosmonauts would use the *Yastreb* EVA suits on Soyuz unlike the *Orlan* and *Krechet-94* on the L3. The Soyuz used the Igla rendezvous radar system, while the L3 used Kontakt. The actual docking contraptions were completely different, and the launch vehicles had no common elements. Still, it was an important step in moving slowly to piloted lunar operations by providing crucial experience to ground controllers, cosmonauts, and designers in performing complex operations in Earth orbit.

Rumors about the mission were bolstered in November 1968 when Mishin, under cover as the anonymous "Chief Designer," spoke to Soviet journalists about the assembly of two Soyuz spacecraft in orbit.[53] Preparations for the missions culminated in a meeting at Tyura-Tam of the Soyuz State Commission on January 11, 1969. Kamanin presented the two primary and two backup crews to the commission for final approval.[54] Like no other crew before, the four

---

51. N. Kamanin, "I Feel Sorry for Our Guys" (English title), *Vozdushniy transport* 13 (1993): 8–9.
52. *Ibid.*
53. Donald C. Winston, "Soyuz Series Aims for Orbital Platform," *Aviation Week & Space Technology*, November 18, 1968, pp. 121–23.
54. Kamanin, "I Feel Sorry for Our Guys," no. 13.

members of the primary crew, all rookies, each had distinctive backgrounds, breaking tradition with earlier Soviet cosmonauts. The commander of the active spacecraft was Vladimir Aleksandrovich Shatalov, the first of a new generation of Soviet cosmonauts to fly into space. Born in Petropavlovsk in Kazakhstan on December 8, 1927, he had graduated with distinction from the Red Banner Air Force Academy with honors in 1956. When training to become a test pilot in the early 1960s, Shatalov applied for admission into the ranks of cosmonauts at a time when the Air Force was expanding its pool base from young, inexperienced pilots to accomplished engineering-test pilots with graduate degrees. It seems that Shatalov had been the top ranked in the group of fifteen military officers selected in January 1963. The light-haired and powerfully built man had plenty of experience preparing for space missions. He would have flown on one of the later Voskhod missions in 1966 had the program not been canceled. He

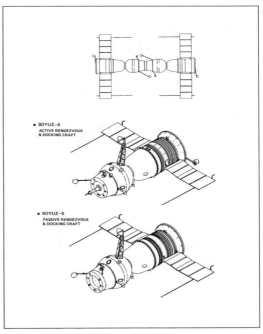

Soviet cosmonauts performed the first docking of two piloted spacecraft in orbit during the Soyuz 4/5 mission in January 1969. (copyright R. F. Gibbons)

had also served as ground communicator for the Voskhod, Voskhod 2, and Soyuz 1 flights.[55]

The passive vehicle crew consisted of Volynov, Yeliseyev, and Khrunov. Commander Volynov had served as a backup crewmember for a number of Vostok, Voskhod, and Soyuz missions, and might have have commanded Voskhod 3, had it not been canceled only two weeks prior to liftoff. He would also have the honor of being the first Russian Jew to fly into space, a distinction that would posit him in many difficult situations in the future. Both Volynov and Khrunov had joined the cosmonaut detachment in 1960 as part of the famous "Gagarin group," although both had to wait almost nine years for their first chance to fly in space. The self-effacing Khrunov, like Volynov, had also served in important backup positions, including for cosmonaut Leonov during his historic first spacewalk on Voskhod 2 in 1965. The final member of the crew, Yeliseyev, was the first of the new civilian group from TsKBEM, whose candidacy had been pushed so hard by Korolev and then Mishin. On this mission, Khrunov and Yeliseyev would carry out the actual EVA transfer from one Soyuz to another—the mission they had been trained to perform in 1967 on Komarov's ill-fated flight.

On January 13, 1969, Shatalov boarded his ship for the first Soyuz launch, which was set for 1300 hours Moscow Time. Given the fact that Shatalov's home telephone number also ended in "13" and that he was slated to be the thirteenth Soviet cosmonaut, many were a

55.   Rex Hall, "Soviet Air Force Cosmonauts," in Michael Cassutt, ed., *Who's Who in Space: The International Space Year Edition* (New York: Macmillan, 1992), p. 261.

little apprehensive about the launch. Fortunately for the superstitious, nine minutes prior to liftoff, the countdown abruptly stopped.[56] There was evidently a failure in a hydraulic system on Blok A of the booster; the State Commission postponed the launch to the following day.[57] It was another freezing day on January 14 when launch operations began for a second launch attempt at pad 31. Witnesses recall the entire launch area being covered with a thick layer of snow. This time, there were no problems. Lt. Colonel Vladimir A. Shatalov, forty-one years old at the time, lifted off at 1032 hours Moscow Time on January 14 in his Soyuz spaceship, vehicle no. 12. The initial orbital parameters of the ship, named *Soyuz 4*, were 173 by 225.3 kilometers with a 51.72-degree inclination. During his initial hours in orbit, Shatalov manually fired the main Soyuz engine on the fifth orbit, about six hours after launch, to change parameters to 207 by 237 kilometers, sharpening his approach trajectory in wait for the target vehicle. He also hosted a television session, which was broadcast to Moscow TV, clearly showing two extra but empty seats in his spaceship, thus arousing speculation that there would be a linkup of some kind in the following days.[58]

The next day, January 15, the 7K-OK spaceship no. 13 lifted off precisely on time at 1005 hours Moscow Time with its three-cosmonaut crew of thirty-four-year-old Lt. Colonel Boris V. Volynov (Commander), thirty-four-year-old civilian Aleksey S. Yeliseyev (Flight Engineer), and thirty-five-year-old Lt. Colonel Yevgeniy V. Khrunov (Research Engineer). The initial orbital parameters of the now-named *Soyuz 5* were 198.7 by 230.2 kilometers at a 51.69-degree inclination. As soon as Soyuz 5 was in orbit, both spacecraft immediately began their program of approach toward each other. In contrast to the original plans for the mission, which envisioned a docking on the very first orbit of the passive ship, the maneuvers were carried out in a much leisurely pace over the period of a day. Volynov on Soyuz 5 fired his main engine on his fifth orbit to change the orbit to 211 by 253 kilometers, thus moving closer to Shatalov's chosen orbit. After a second maneuver by Shatalov on the morning of January 16 on his thirty-second orbit, ground controllers switched on the Igla system at 1037 hours Moscow Time. Through the next half hour, the radar system brought the two vehicles to a distance of only 100 meters. Shatalov later vividly described the program from then on:

> At this point, I went over to manual control, and Boris Volynov did the same. The problem was to make sure that the docking units of both spacecraft were properly oriented toward each other. Throughout this time I was manually controlling the appropriate thrusters. With the control stick on the left-hand side I regulated the craft's linear velocity—slowing it down or speeding it up—and damped out the lateral velocity. When we were over the shores of Africa—some seven or eight thousand kilometers from the borders of the Soviet Union—we approached to within [forty meters] of each other and started to hover. At this range, Boris Volynov and I performed several maneuvers.[59]

As he closed in on Soyuz 5, there were some problems, including erroneous signals from the docking control and contact lights, that were apparently related to the spurious activation of the control and diagnostics system on Soyuz 4. At a ginger twenty-five centimeters per second, the two spacecraft hard-docked at 1120 hours to Volynov's exclamation of "Welcome!"[60]

---

56.   M. F. Rebrov, *Kosmicheskiye katastrofy: Russkiye sensatsii* (Moscow: IzdAT, 1993), pp. 41–42.

57.   The launch problem may have occurred because of ground operator error. K. P. Feoktistov recalls that there were "incorrect actions of ground control" in inputting settings. See *Russian Space History, Sale 6516* (New York: Sotheby's, 1993), description for Lot 57.

58.   Smolders, *Soviets in Space*, p. 170.

59.   Riabchikov, *Russians in Space*, p. 256. With his right hand, Shatalov exercised roll control.

60.   *Ibid.*, p. 257; Smolders, *Soviets in Space*, pp. 171-172; *Russian Space History*, Lot 57.

One cosmonaut on the passive ship was rumored to have been much more excited. Unconfirmed reports suggest that at the moment of docking, when the pin was inserted in the cone of Soyuz 5, one of the crewmembers on the latter ship shouted out "We're being raped! We're being raped!" While initial TV broadcasts of the segment carried the exclamations intact, all later replays omitted the offending words.[61]

After docking, it seems that the two vehicles had suffered excessive rotations because of the problems with the diagnostics system but settled down sufficiently for the cosmonauts to begin preparing for the crew transfer. Somewhat overextending its claims, the Soviet press dramatically announced that the link up of Soyuz 4 and Soyuz 5, a combined mass of 12,924 kilograms, as "the world's first experimental space station."[62] The complex did, however, have a common power system during the docked duration by means of a plug-and-sockets system on the docking nodes. On the thirty-fifth orbit of Soyuz 4, Khrunov and Yeliseyev began their preparations for their transfer EVA by entering the living compartment of Soyuz

Soyuz 4 Commander Vladimir Shatalov displays how Soyuz 4 and Soyuz 5 docked in Earth orbit in January 1969. It was the first-time that two piloted spacecraft docked to each other in space. (files of Peter Gorin)

5 and unstowing two Yastreb suits from a side cupboard. Commander Volynov assisted them during the procedure, which proved to be relatively difficult with three men in the cramped confines of the module. Each suit had a self-contained backpack attached to one of their legs instead of their waists, as was the case on the earlier Yastreb versions for the abandoned Soyuz 1/2 mission. Both cosmonauts were, however, tethered safely to the spacecraft via umbilicals, which carried lines for communications and health telemetry. In a ceremonial move, Soyuz 5, launched a day after Soyuz 4, had carried into orbit a bunch of mail addressed to Shatalov, as well as a number of newspaper articles on the Soyuz 4 launch. The letters were not only from his family, but also from Minister Afanasyev, Chief Designer Mishin, State Commission Chairman Kerimov, Col. General Kamanin, and others. During the transfer, Khrunov and Yeliseyev were to carry the mail and media materials, presumably in their pockets, in addition to a camera.[63]

After the suits were tested and pressurized, Volynov bid the two cosmonauts goodbye and retreated back into the descent apparatus and shut the intermediary hatch between the two modules before commanding the living compartment to depressurize. Khrunov then opened up the outer hatch of the living compartment on Soyuz 4's thirty-fifth orbit and poked his head out cautiously. After Volynov's final permission to egress, Khrunov moved his body out of the spacecraft, briefly getting entangled in his safety cord. The combined complex was over South America at the time. Khrunov recalled later:

> I emerged from the spacecraft without difficulty, and looked around. I was amazed by the marvelous, magnificent spectacle of two spacecraft linked together high above the earth. I could make out every tiny detail on their surfaces. They glittered brilliantly as

61. James E. Oberg, *Red Star in Orbit* (New York: Random House, 1981), pp. 98–99.
62. Petrovich, *The Soviet Encyclopaedia of Space Flight,* p. 388.
63. Riabchikov, *Russians in Space,* p. 258; Smolders, *Soviets in Space,* p. 172.

*they reflected the sunlight. Right in front of my eyes was Soyuz-4, looking very much like
an aircraft. The big, long spacecraft was like a fuselage, and the solar panels were like
wings.*[64]

Yeliseyev followed after Khrunov, letting the latter lead in EVA activities. Khrunov crawled
toward the docking unit of Soyuz 5 and removed a TV camera from a support and turned off
its power supply. Before exiting the spacecraft, Yeliseyev had forgotten to fasten a still-photo
camera to his suit. The instrument floated out into space, depriving the Soviets of high-quality
photographs of the historic event. Among their modest activities during the excursion, the two
men also "made observations of the Earth's horizon, [and] checked the operation of the attitude-
control jets."[65] Khrunov, followed by Yeliseyev, then moved over to the living compartment of
Soyuz 4, opened its hatch, and crawled in. They were received by a welcome note from Shatalov,
who was at the time in the spaceship's descent apparatus. After the pressurization of the living
compartment, the hatches between the two modules were opened, and Shatalov embraced his
comrades, treating them to a toast of black currant juice instead of the customary vodka, which
was prohibited aboard the spacecraft. The entire episode had lasted one hour, although the two
cosmonauts had been out of the spacecraft for thirty-seven minutes.

Wasting little time, the two commanders, Shatalov and Volynov, began immediately
to prepare for undocking. At 1554 hours, just four hours and thirty-four minutes after docking,
the two spacecraft separated and went on their own ways, Soyuz 4 now with three cosmonauts
and Soyuz 5 with one. They had been joined together for three orbits. In continuing
their independent missions, the crews carried out a number of scientific experiments, which
included the use of a new stellar-navigation sextant, the operation of the RSS-1 spectrograph for
geophysical studies, and the testing of instrumentation for medical and biological experiments.
Earth observational experiments included observing and photographing terrestrial cloud cover,
storm formations, snow and ice cover, and various geological structures. One set of activities
included astronomical investigations, such as observing the astral sky during both day and
night, photographing the night sky in a direction opposite the Sun, and studying the initial
stages of the development of comet tails. The RSS-1, on Soyuz 5, was used for a spectrophotometry experiment on Earth's twilight aureole over a spectral range of 400–650 nanometers on
the second and fifteenth orbits from a mean altitude of 240 kilometers. Khrunov also carried
out experiments related to the passage of radio waves through the ionosphere. Finally, both
Soyuz 4 and Soyuz 5 carried special targets on the exterior for measurements of tritium and
helium-3. Each target consisted of a package of fourteen plates made from one sheet of aluminum.[66]

Soyuz 4 was the first to return from orbit. On January 17, Shatalov, Yeliseyev, and Khrunov
carried out a guided reentry, landing at 0953 hours Moscow Time, forty kilometers northwest
of the town of Karaganda in Kazakhstan. The mission had lasted two days, twenty-three hours,
twenty minutes, and 47 seconds. Volynov, now alone, had a much more difficult time, facing
perhaps the most dramatic and dangerous reentry in the history of the Soviet space program.
During the early morning of January 18, in preparation for his reentry around midday, Volynov
reported that all systems were fine aboard the ship. At 1020 hours, he passed over the Gulf of

64.   Riabchikov, *Russians in Space*, p. 259.
65.   *Ibid.*, pp. 259–60.
66.   I. A. Alimova, V. O. Naidenov, B. S. Boltenkov, and V. N. Gartmanov, "Measurement of Tritium and
Helium-3 in Aluminum Targets" (English title), *Kosmicheskiye issledovaniya* 9 (January–February 1971): 149–51;
K. Ya. Kondratyev, A. A. Buznikov, B. V. Vinogradov, V. N. Volkov, V. V. Gorbatko, and O. I. Smotky,
"Spectrophotometry of the Earth From Manned Spacecraft," in Kondratyev, Rycroft, and Sagan, eds., *Cospar:
Volume I*, pp. 619–32; Smolders, *Soviets in Space*, p. 175; Lardier, *L'Astronautique Soviétique*, pp. 187–88; G. S.
Narimanov, ed., *Ot kosmicheskikh korabley - k orbitalnym stantsiyam* (Moscow: Mashinostroyeniye, 1971),
pp. 65–66.

Guinea near Africa before firing the S5.35 engine for the predetermined period. Six seconds after the termination of retrofire, Volynov heard the pyrocartridges triggering to separate the three major modules of the spacecraft: the living compartment, the descent apparatus, and the instrument-aggregate compartment. As he looked through the viewport, he noticed something deadly wrong: he could clearly see the antennas attached to the solar arrays on the cylindrical instrument-aggregate compartment, meaning that the section, also known as the service module, had not separated from the descent apparatus. While similar failures had occurred on early Vostok and Voskhod flights, it posed a much greater threat on Soyuz because of the relatively huge size of the module. Volynov immediately reported in code to ground controllers about his predicament. Most simply believed that Volynov had little chance to live.[67]

The descent apparatus tumbled in somersaults as it remained attached to the three-ton service module and began its long journey through the atmosphere. Turning over and over, with the thermal shield unexposed to the heat because it was still covered by the service module, the heat began to affect unprotected portions of the descent apparatus. Smoke began to appear within the capsule as the light heat insulation began to burn. Normally, during a reentry, hydrogen peroxide jets would fire during this period to guide the capsule to provide lift and reduce thermal and gravitational stresses. In this case, Volynov noticed that his instrument panel indicated that the valves for the thrusters were open, but there had been no firings. All the propellant had been used up at the initiation of retrofire, when the computer had tried in vain to correct the spaceship's incorrect attitude.

Volynov recalls that he was sure that only a few minutes separated him from death. The normally unflappable cosmonaut considered saying goodbye to his relatives, but instead decided to hurriedly save all the recorded materials on the docking procedure by ripping the important pages from the log book, rolling them up tightly, and sticking them into the middle of the book. Then, amid the cauldron around him, he calmly began to speak into a tape recorder, describing all the details of his experience to assist in identifying the reasons for the failure. Through it all, there were terrifying moments. Once, there was a sharp clap, indicating that the propellant tanks of the service module had blown apart with such force that the crew hatch was forced inwards and then upwards like the bottom of a tin can. Plummeting through a ballistic trajectory, he realized that the service module had finally disintegrated and he had survived. His relief soon turned to anxiety when the parachute system triggered at an altitude of ten kilometers. The straps on the main parachute began to twist, preventing them from unfurling properly. For the second time in minutes, he was convinced of his end. Remarkably, the braids of the parachute began to untwist slowly; by the time that the descent apparatus landed with its soft-landing engines, it was sufficient to ensure Volynov's safety, although the landing was so hard that the roots of his teeth in his upper jaw were broken off. It was only the specially built shock-absorbing seat that saved him from broken bones and more serious injuries.[68]

The Soyuz 5 descent apparatus landed 600 kilometers from its originally intended landing site, 200 kilometers southwest of Kustanay. TASS only announced that "the flight took place successfully, a unique experiment was conducted, and the vehicle touched down in the designated area." Volynov landed at 1108 hours Moscow Time on January 18, after a three-day, fifty-four-minute, fifteen-second mission. In their investigation of the Soyuz 5 reentry, TsKBEM engineers found that the connection locks between the descent apparatus and the instrument-aggregate compartment had failed to release. The two modules finally separated from each other when the intermediary transfer compartment, carrying hydrogen peroxide tanks for the attitude control thrusters, exploded. Despite the dangerous situations, the designers were

67. Mikhail Rebrov, "A Difficult Re-Entry From Orbit" (English title), *Krasnaya zvezda*, April 27, 1996, p. 5.
68. *Ibid.*

extremely pleased with the performance of the descent apparatus, which had withstood temperatures and stresses far above nominal during the reentry and specifically ensured the safety of the crew in a sudden switch from a guided to a ballistic reentry. The mission also confirmed the correctness in using an advanced titanium frame for the descent apparatus, as well as the propitiousness of countless design and statistical tests to ensure the stability of the capsule with any angle of attack.[69]

In spite of the near catastrophe at the end of the flight, the Soyuz 4/5 mission was a landmark flight in the Soviet space program. It was not only the first docking of two piloted spacecraft in space and the first transfer of a crew in orbit from one spacecraft to another, but also the first completely successful piloted space mission in the post-Korolev era. While the mission had been accomplished nearly two years late, the complexity of the flight indicated a certain maturity in Soviet space operations from the almost primitive Voskhod missions during Korolev's last years. Still, compared to the U.S. space program, it was a poor match. NASA astronauts had accomplished the first docking in space as early as March 1966 on Gemini VIII. Even the Soviets themselves had already accomplished automated docking twice in orbit. But after the humiliating defeat of Apollo 8, the Soviet leadership was willing to take anything remotely successful as a godsend. What was at best an interesting and moderately complex operation in Earth orbit was made out to be the most dramatic step in the exploration of space. At the subsequent press conference for the Soyuz 4/5 cosmonauts, the Soviets made much of the fact that the docked complex had been the world's first "experimental orbital station." In one of the few interesting moments of the presentation, cosmonaut Khrunov let out that "in the design of our spacesuits certain aspects of Leonov's suit were taken into consideration. Our experiences on this flight may well contribute to the designs of a moon suit."[70]

There was a bizarre postscript to the Soyuz 4/5 mission. On January 22, a number of famous cosmonauts, including Nikolayev, Tereshkova, Leonov, and Beregovoy, were being driven to the Kremlin for an awards reception in the back of a Zil limousine. As they entered the gates of the Kremlin, a man in a hat and dark glasses stepped from the shadows with a gun in each hand and began firing at the limousine with the cosmonauts. He managed to fatally wound the driver. Leonov remembers:

> I looked down and saw two bullet holes on each side of my coat where the bullets passed through. A fifth bullet passed so close to my face I could feel it go by. This man was shooting at me, thinking that I was Brezhnev. He was angry because he had been conscripted into the army. When it was over, Brezhnev took me aside and told me: "Those bullets were not meant for you, Aleksei. They were meant for me, and for that, I apologize."[71]

The man, a young army lieutenant named Ilyin, was apprehended, and later spent twenty years in a special prison.

Dazed and Confused

More than any other U.S. space achievement of the 1960s, the flight of Apollo 8 froze the Soviet space industry into a kind of collective shock. Nothing the Soviet Union was capable of doing in December 1968 could have been neutralized the worldwide accolades for the impressive achievement of Borman, Lovell, and Anders. If the Communist Party was only too eager to

---

69.   Semenov, ed., *Raketno-Kosmicheskaya Korporatsiya*, p. 184.
70.   Smolders, *Soviets in Space*, p. 176.
71.   Thomas O'Toole, "The Man Who Didn't Walk on the Moon," *New York Times Magazine*, July 17, 1994, pp. 26–29.

use space achievements as a means to sell the virtues of socialism in the early 1960s, now Soviet officials were almost embarrassed by it. In this backdrop, senior Soviet space officials convened in January 1969 to discuss not only an adequate response to the U.S. space program, but also to talk in general about the larger direction of their entire piloted space effort.

The first meeting, presided over by Minister of General Machine Building Sergey A. Afanasyev, was held on January 10 amid the cold snowy weather at Tyura-Tam, just a few days prior to the Soyuz 4/5 launches. Among those present were all the members of the Council of Chief Designers involved in lunar programs, as well as deputy chief designers and department heads from many design bureaus and institutes.[72] Afanasyev was aghast. He asked the distinguished assemblage, perhaps, rhetorically, "How can we get out of this mess?!" The primary questions at hand were:

- How should the success of Apollo 8 be neutralized in the short term?
- What should be done with the L1 circumlunar program now that its importance had been all but neutralized by Apollo 8?
- How should the L3 landing project proceed, and was there any way the USSR could beat an American landing?
- How should the N1 be modified to improve its capabilities for the future of the Soviet space program?[73]

On the first point, the Party and government had just passed a resolution accelerating the Ye-8-5 sample returner project. In a compensatory measure to allay public opinion, many senior Soviet government officials were shifting their thinking to automation. Kamanin emphasized as such in his diary entry for January 20, 1969, lamenting that:

> in the Academy of Sciences and in the industry there is a very strong mood for the use of robots and against the active development of piloted flights. This aspiration is supported by the Central Committee, the [Military-Industrial Commission], and the [Strategic] Missile Forces.[74]

Boris A. Stroganov, one of Serbin's deputies in the innards of the Central Committee's Defense Department, proposed that all parties should assist the Lavochkin design bureau to quickly accomplish its task of completing a sample return mission before an Apollo landing. If Soviet officials publicly touted the value of automated lunar exploration, then privately most knew that it was a poor substitute at best. The majority of participants at the meeting vocally supported piloted exploration. In fact, Afanasyev asked the attending chief designers whether a thirty-day-long Soyuz mission could be mounted in the near future to boost Soviet claims as a leading space power.

On the issue of the circumlunar L1 project, opinions were divided. Some, such as Babakin, Ryazanskiy, and Chertok, supported moving on to piloted missions regardless of the success of Apollo 8, while others, such as Mishin's deputies Kozlov and Kryukov, argued for only further automated launches. Yuriy A. Mozzhorin, the powerful Director of TsNIIMash (formerly NII-88), openly voiced a means to "save" the L1 program. Because the Soviet Union had declared that it had a space program as accomplished as the American one, simply continuing

---

72. Among those present were S. A. Afanasyev (MOM), G. N. Babakin (GSMZ Lavochkin), V. K. Bezverbiy (TsKBEM), B. Ye. Chertok (TsKBEM), G. I. Degtyarenko (TsKBEM), B. A. Dorofeyev (TsKBEM), V. P. Finogeyev (NII AP), A. G. Iosifyan (VNII EM), D. I. Kozlov (TsKBEM Kuybyshev Branch), S. S. Kryukov (TsKBEM), V. P Mishin (TsKBEM), A. S. Mnatsakanyan (NII TP), Yu. A. Mozzhorin (TsNIIMash), S. O. Okhapkin (TsKBEM), N. A. Pilyugin (NII AP), M. S. Ryazanskiy (NII Priborostroyeniya), and B. A. Stroganov (TsK KPSS Defense Industries Department).

73. Interview, Peter Gorin by the author, November 18, 1997.

74. Kamanin, "I Feel Sorry for Our Guys," no. 13.

the L1 program would not do. Instead, he suggested giving the project a "scientific flavor," as if to suggest that the Soviet Union had higher goals than simply competition. It was in fact exactly such a tack that official Soviet spokespersons took in the coming months as the USSR half-heartedly continued the circumlunar project in its automated variant. Plans for piloted missions were indefinitely postponed in March 1969, while the remaining 7K-L1 spacecraft were prepared for use only in robotic mode.

As for the N1-L3 program, some, such as Chertok and Mishin's principal aide for new projects, Vitaliy K. Bezverbiy, admitted openly for the first time what was privately beyond debate for over a year: that the Soviet Union could no longer overtake the United States in a landing of humans on the Moon. There was, however, overwhelming support for reconfiguring the N1-L3 program so as to use two launch vehicles to assemble a lunar complex in Earth orbit, instead of the one planned for several years. Participants considered two separate options: one using the current variants of the N1 and one using advanced and uprated versions. The first option, supported by Kryukov, Mozzhorin, Pilyugin, and Ryazanskiy, among others, was motivated primarily by the poor rated performance characteristics of the first four flight models of the N1, vehicle nos. 3L, 4L, 5L, and 6L; none of them were capable of lifting the ninety-five tons required for a bare-bones L3 lunar mission. Thus, two launches would ensure that all the components of the L3 complex would reach orbit. It must have been particularly demoralizing to hear Chief Designer Pilyugin state that engineers were not sure they could make the ninety-five-ton mass limit for the L3 complex, *even if* the N1 could lift such a payload into Earth orbit. His Deputy Vladlen P. Finogeyev reminded everyone that because the L3 design had been redrafted three times in the last few years, there was not even an LOK or an LK spaceship in any shape or form existing anywhere.

The second option—using uprated N1s—was attractive because it would enable engineers to expand the landing crew size from one cosmonaut to two—a crucial issue that factored into the discourse on the safety of cosmonauts on the Moon. Among the variants considered at the time were the N1F-V3 and N1F-V4, with liquid hydrogen stages in the third and fourth stages, respectively. The most favored option seems to have been the use of the two N1F-V4s to launch a huge lunar complex into Earth orbit, called the L5, which would allow four to five cosmonauts to spend up to two months on the surface of the Moon. In the end, nothing was decided. It seems to have been a meeting to air the "dirty laundry," a catharsis of sorts. Perhaps the most pointed comments were from TsKBEM Deputy Chief Designer Chertok who, during his speech, very accurately observed that the Soviet space program had less resources than the U.S. program and yet was spending its money with even less rationality. It was a dead-on observation on the poverty, not only of money, but also of management, in the Soviet space program in the 1960s.[75]

Major consultative meetings of the Council of Chief Designers were set for late January 1969, and in preparation, Mishin met with many leading officials through the month to discuss various aspects of the piloted lunar program. On January 24, he examined both the current N1-L3 effort as well as possible modernized variants. One of the issues at hand was the possibility of eliminating the testing of the T1K, T2K, and L1E Earth orbital test beds to reduce the amount of work. In addition, once again, there was some discussion on the complicated LK plus LK$_R$ (backup lander) plus Ye-8 (rover) profile planned for the L3. He also drew up preliminary documents on inviting other organizations—namely the S. A. Lavochkin State Union Machine Building Plant—to manufacture the payload block for the proposed N1F-V3 rocket.[76] Problems with the LOK and the LK had also cropped up. Both spacecraft were still overweight, the former by five kilograms. As an example of the lengths to which the Soviets worked on "shaving

---

75.   Gorin interview, November 18, 1997.
76.   The payload block of the N1F-V3 consisted of Blok G, Blok D, the transfer fairing, and Blok I.

off" mass from the lander, engineers proposed eliminating an eighteen-kilogram visor and a sighting instrument from the LK.

The following day, January 25, Mishin met with Chief Designer Pilyugin of the Scientific-Research Institute of Automation and Instrument Building, one of Korolev's old associates from the 1940s who now presided over the development of most control and guidance systems for Soviet spacecraft. The meeting was important because, for the first time, there was serious discussion of using Mars to neutralize the success of Apollo. The two chief designers discussed a three-step Mars exploration program:

- Mars '73—a robotic vehicle to Mars for sample return (on the N1)
- Mars '75—a piloted satellite of Mars (on the N1F-V3)
- Mars '77—a piloted landing on Mars using an N1 with nuclear rocket engines

In the meantime, Pilyugin suggested continuing the current N1-L3 program, but in a two-launch scheme, both with and without the Ye-8 rovers. He suggested that to reduce extraneous work, Soviet designers should focus on creating a single modernized version of the N1, the N1F-V3. Perhaps prompted by the discussion with Pilyugin, Mishin brought up the issue of Mars at an internal meeting on January 26, at which he considered the possibility of inviting the Ministry of Medium Machine Building to develop nuclear power sources for Martian spacecraft.

These discussions culminated in widely attended and important meetings of the Council of Chief Designers on January 26 and 27, 1969. Apart from the usual chief designers and their deputies, a number of important scientists from the Academy of Sciences and representatives from the military were also present.[77] Academician Keldysh set the meeting off with the admonition that there was no hope of carrying out the N1-L3 program as it then stood. Instead, he believed that designers should focus on improving the capabilities of the N1 with liquid hydrogen upper stages and carrying out the three-step Mars exploration program, with missions in 1973, 1975, and 1977–80. Although most at the meeting agreed that Mars should be the next goal for the Soviet space program, there was little support to completely abandon the Moon. For the Mars expedition, most of the attendees supported the creation of the uprated N1 booster, the N1F-V3, which would have a new third stage equipped with Chief Designer Lyulka's 11D54 liquid hydrogen–liquid oxygen engine. One attendee, Viktor I. Shcheulov, an officer in the Strategic Missile Forces, cautioned that liquid hydrogen stages would not be ready for use until 1971 at the earliest.

Shcheulov made one of the more prophetic statements at the meeting. He believed that the creation of Earth-orbital space stations would smooth the effect of recent U.S. successes in space. TsKBEM had, for many years, explored various conceptions of space stations, one of them being a huge complex in orbit called the Multirole Space Base-Station (better known simply as the "MKBS"), which would allow for the replacement of crews on board, thus establishing a permanent piloted presence in space. The space station option, while not as attractive as Mars, was slowly beginning to emerge at the time as a possible alternative long-range goal for the Soviet space program. In January 1969, with the recent success of Apollo 8 in mind, there was, however, more of an interest in the Moon and Mars, and this clearly influenced the formation of a post-1968 space policy for the Soviets. At a meeting on January 29 for his senior staff at the design bureau, Mishin brought up the issue of the Moon, Mars, and Earth-orbital stations.[78] Most of the designers agreed on a two-prong long-range program:

---

77. Among those present were K. D. Bushuyev (TsKBEM), A. G. Iosifyan (VNII EM), M. V. Keldysh (AN SSSR), M. S Khitrik (NII AP), G. P. Melnikov (NII-4), V. P. Mishin (TsKBEM), A. S. Mnatsakanyan (NII TP), Yu. A. Mozzhorin (TsNIIMash), G. N. Pashkov (VPK), M. S. Ryazanskiy (NII Priborostroyeniya), G. I. Severin (KB Zvezda), V. I. Shcheulov (TsUKOS), and G. I. Voronin (KB Nauka).

78. Among those present were V. K. Bezverbiy, B. Ye. Chertok, K. P. Feoktistov, V. P. Legostayev, S. O. Okhapkin, V. N. Pravetskiy, I. I. Raykov, and Ye. F. Ryazanov. All were from TsKBEM.

- The development of the MKBS in Earth orbit, whose design would be based on old designs for the Heavy Interplanetary Ship dating from the Korolev days
- The use of the MKBS to mount a Mars expedition

Much of the discussion was focused on the development of closed-cycle life support systems to ensure survival over a period of two to three years in space, as well as nuclear-electric power sources for such advanced missions.[79] The MKBS would also be used for defense goals.

The general consensus from the meetings was that the Soviet Union should continue work intensively on the NI-L3 program, now as part of a dual-launch Earth-orbit rendezvous/lunar-orbit rendezvous profile, but at the same time begin planning for the *coup de grace*—a progressively sophisticated Mars landing program over the next decade, which promised to bring the prestige of the Soviet space program out of its current doldrums. The Mars program would use components of the large Earth-orbital station, the MKBS, which would also be dedicated to defense purposes. The somewhat diffuse and perhaps hasty response to the success of Apollo 8 was not confined to the restricted corridors of the Soviet space establishment. Academician Keldysh, in a statement to Moscow Radio on January 24, hinted at the uncertain prospects for the future of the Soviet space program. Putting a bright face on the recent Soyuz 4/5 success, he spoke clearly about new directions: the establishment of permanent orbital stations and the accomplishment of interplanetary flights. Speaking of the Zond spacecraft and its capability to carry cosmonauts around the Moon, he added that such a flight should not be expected in the next two or three weeks. In closing, he said simply that piloted lunar operations "depends somewhat on our further considerations as to what we shall do with automatic apparatus and with manned ones."[80]

The NI in Flight

It is ironic that at precisely the time when the Soviets were having second thoughts about the Moon, a number of their lunar projects approached the flight testing stage, making 1969 one of the busiest years for lunar-related space launches in the history of the Soviet space program. The armada was inaugurated by a launch during January that punctuated the intensive high-level discussions on the Moon program. Prompted by TsNIIMash Director Mozzhorin's suggestion to continue robotic LI launches with a "scientific" tenor, it seems that Minister Afanasyev had sanctioned further launches in the beleaguered program, beginning with one in January 1969. Ironically, a number of the LI cosmonauts, including Bykovskiy, were on hand at Tyura-Tam to view the launch, no doubt fully aware that their chances of ever flying around the Moon had abruptly dropped dramatically.

The 7K-LI vehicle, spacecraft no. 13, was the same article that was to have been launched in early December on a robotic circumnavigation of the Moon, but was stood down because of the catastrophic crash of Zond 6.[81] The Proton booster lifted off successfully at 0414 hours, 36 seconds Moscow Time on January 20, 1969. After first-stage cutoff, the second stage began firing, but at T+501 seconds, the booster began to fall. After several minutes, controllers reported to State Commission Chairman Tyulin at the command center at site 2 that search-and-rescue services had detected the LI spacecraft, saved by the emergency rescue system, southeast of Irkutsk near the border with Mongolia. It took about four hours for analysts to produce a preliminary accident

---

79. The Soyuz Moscow Machine Building Plant (formerly OKB-300), headed by Chief Designer S. K. Tumanskiy, had developed nuclear power sources capable of producing ten kilowatts by 1969. He expected to create more powerful units—at fifty kilowatts by 1970 and up to 2,500 kilowatts by 1975.

80. *Soviet Space Programs, 1966–70*, p. 372.

81. Interestingly, in early January 1969, a cable network on spacecraft no. 13 had been severed. Engineers opted to replace part of the network with parts from the already-flown spacecraft no. 7, which had been launched in April 1968 and recovered after a launch failure.

report. One of the four engines of the second stage had shut down abruptly, twenty-five seconds prior to the planned cutoff point. At this point, the third stage could have easily fired to compensate and inserted the payload into orbit, but a diagnostics computer on the booster, as soon as it had detected the engine failure, aborted the mission and fired the emergency rescue system for the L1 spacecraft.[82] Thus was lost 100 million rubles and another chance to fly to the Moon.[83] It was the fourth launch failure in the circumlunar program out of only nine launch attempts, illustrating that one of the weakest links in the project was the UR-500K Proton booster, designed and built by a branch of Chelomey's design bureau.

With that inauspicious beginning, engineers moved on to more ambitious prospects. In February 1969, both Babakin's first Ye-8 lunar rover and Mishin's first N1 rocket were ready for liftoff. In fact, in what was certainly not a coincidence, their launches were timed a day apart. The specially made 7K-L1S lunar spacecraft would arrive in lunar orbit and attempt to photograph the Ye-8 rover on the surface. Since the original May 1968 launch date, engineers had spent months mired in a frustrating delay. Although the first flight-rated N1 vehicle, booster no. 3L, was completely ready for launch and the basic construction of the first launch pad had been finished by the end of 1967, problems with many pad-booster interface systems forced launch date postponements for weeks and then months.[84] On September 18, 1968, Afanasyev had presided over a meeting of the State Commission for the N1 at Tyura-Tam at site 112 near the N1 pads. Approximately 100 chief designers, deputies, Strategic Missile Forces and Air Force officers, and government officials were present during the five-hour meeting. The participants noted that three different deadlines stipulated by Central Committee decrees had not been met. At the time, the 1M1 mock-up was on the completed pad at site 110P with a functional payload undergoing fueling tests to allow service teams to train and gain experience for actual launches. Kamanin, who attended the meeting, recalled, "There have been lots of drawbacks, improper quality of work and plain bungling—in particular there was an accident with a bulldozer cutting the main power supply of the launch pad."[85] At the meeting, Afanasyev scheduled the first launch for late November 1968 and the second one for February 1969.

---

82.   Kamanin, "I Feel Sorry for Our Guys," no. 13.
83.   Although Russian sources suggest that this launch was an attempt at a circumlunar flight, the launch date for the mission would seem to indicate that it may have been a deep space mission, much like Zond 4. Richard Flagg's analysis of L1 launch windows during 1968–70 suggests that Zond circumlunar launches were only attempted when trajectories could be flown that were close to coplanar with the Moon's orbit to minimize the effects of the Moon's gravity. If a 7K-L1 craft approached the Moon from a transfer orbit with a large angle relative to the Moon's orbital plane, then the force of lunar gravity would have changed the plane of the orbit, deflecting the craft from the required return trajectory. In such a situation, the 7K-L1 spacecraft's main engine would not have been powerful enough to effect mid-course corrections to return the vehicle on an Earth-bound trajectory. An additional scientific requirement was for the far side of the Moon to be illuminated during the mission to carry out surface photography. In examining the launch windows of the 7K-L1, Flagg observed that there were no circumlunar launch windows that satisfied these criteria from January 1969 to July 1969. However, lunar age, declination phase, and opening angle were close to permitting a 7K-L1 launch during January 7–9, 1969, although those parameters "were definitely outside those . . . as defined by the successful Zond flights." See *Soviet Space Programs, 1981–87: Space Science Applications, Military Space Programs, Administration, Resource Burden, and Master Log of Spaceflights*, prepared for the Committee on Commerce, Science, and Transportation, U.S. Senate, 101st Congress, 1st sess. (Washington, DC: U.S. Government Printing Office, April 1989), pp. 384, 386. Curiously, on January 3–4, 1969, news of a Soviet launch failure that was apparently detected by NORAD was "leaked" to the Western press. The failure was said to have occurred because of a second-stage malfunction. See Gatland, *Robot Explorers*, p. 144.
84.   Boris Arkadyevich Dorofeyev, "History of the Development of the N1-L3 Moon Program," presented at the 10th International Symposium on the History of Astronautics and Aeronautics, Moscow State University, Moscow, Russia, June 20–27, 1995.
85.   Kamanin, "A Goal Worth Working for." There was a complete fueling exercise involving the N1 stack in October 1968 that lasted ten days. See Maj.-Gen. Valery Aleksandrovich Menshikov, "The Toilers of the Cosmodrome: The Test Personnel of Baykonur" (English title), *Aviatsiya i kosmonavtika* no. 1 (January 1993): 39–41. The flight version of the N1, booster 3L, was moved to the pad in November 1968, but it was replaced briefly by the 1M1 again.

Engineers were unable to remedy the interface problems until December 1968, forcing another two- to three-month delay. A few days before the Soyuz 4/5 missions, on January 9, 1969, amid discussions about a post–Apollo 8 strategy, Afanasyev convened another meeting of the N1 State Commission. It was unusual for a minister to head a State Commission, and Afanasyev's appointment to the position underlines the importance with which space program head Ustinov viewed the N1 rocket program. After hearing a number of reports, Afanasyev set the launch date for the first N1 as February 18, 1969, within the launch window for a lunar-orbital flight. The proceedings were interrupted by an alarming report from Baykonur Cosmodrome Commander Maj. General Aleksandr A. Kurushin, who refused to agree to a launch of the rocket because of many "deficiencies" in both the ground equipment and the rocket itself. After pressure from most of the members of the State Commission, including Afanasyev and Mishin, as well as Party Central Committee representa-

The first N1 rocket being brought to the pad for launch in February 1969. (files of Asif Siddiqi)

tive Stroganov, Kurushin backed down and promised to have these "deficiencies" removed by the slated launch date.[86] Needless to say, Kurushin's initial outburst did little to instill confidence in a success.

The final prelaunch cycle for the first N1 launch began in mid-January 1969. The twenty-eight-day program involved 2,300 people from dozens of different organizations and fifty tank wagons for liquid oxygen fueling of the rocket.[87] The majority of the site workers were Army conscripts, who, as one participant recalls, had come from backgrounds unrelated to the space program:

> The test officers at the time were principally 35–40 years old, without higher education and came from all over. Tankers and artillerymen, pilots and sailors, combat engineers and chemists—in short, it would be easier to list who was not there—were encountered among them.[88]

The men completed their job on time. On February 3, booster no. 3L was slowly moved from the assembly-testing building to the launch pad on a special crawler-transporter. At the pad itself, the giant booster was lifted to a vertical position and held up by a sixteen-meter support ring with forty-eight explosive bolts at the base of the first stage. The mass of the booster and its L3S payload was exactly 2,772,103 kilograms. By the time of its first launch, models off the first-stage engines for the rocket had accumulated over 100,000 seconds of test operating time on the ground.[89]

---

86.   Kamanin, "I Feel Sorry for Our Guys," no. 13.
87.   J. Villain, "A Brief History of Baykonur," presented at the 45th Congress of the International Astronautical Federation, IAA-94-IAA.2.1.614, Jerusalem, Israel, October 9–14, 1994.
88.   Menshikov, "The Toilers of the Cosmodrome," p. 40.
89.   The L3S designation was confusingly applied to the payload for the first two N1s, which consisted of Blok G, Blok D, and the 7K-L1S lunar orbiter.

The assault on the Moon in February 1969 began with the launch of the first Ye-8 lunar rover. A Proton booster lifted off successfully at 0948 hours Moscow Time on February 19 with its payload, Ye-8 vehicle no. 201 and its translunar-injection Blok D stage. As Babakin's engineers watched the rising rocket, just over fifty-one seconds after launch, the payload abruptly fell apart, and the booster eventually exploded. The debris from the accident, including portions of the lunar rover, fell fifteen kilometers from the launch site. A later investigation found that the source of the problem had been a new payload fairing designed and built specifically for the rover payload. Aerodynamic vibrations during passage through maximum dynamic pressure tore the shroud off at its weakest tension points. The debris tore into the lower stages of the rocket, resulting in a massive explosion at T+54 seconds. Despite an intensive search of the debris area, engineers were unable to find the Polonium-210 radioactive isotope in the rover payload designed for heating the spacecraft on the Moon. Unconfirmed rumor has it that soldiers at Tyura-Tam discovered the isotope

This photograph was taken moments before the launch of the first N1, booster no. 3L, in February 1969. (copyright Quest)

package and used it to heat their barracks during the bitter winter of 1968–69.[90] With two failures out of two Proton launch attempts in the year, space officials turned their attention to the long-awaited first launch of the N1 rocket.

The launch was originally set for February 20, but it was delayed to the afternoon of February 21 because of poor weather conditions at the launch site.[91] Boris A. Dorofeyev, Mishin's deputy for testing the N1, directed all the launch preparations; he would perform the same on-site technical direction carried out by the late Leonid A. Voskresenskiy back in the 1940s and 1950s. Before the launch, a senior engineer ceremoniously broke a bottle of champagne on the main body of the N1's launch transporter.[92] It was a clear and cold day at the Baykonur Cosmodrome, and prelaunch operations proceeded without delays. Almost four years late, the most powerful rocket ever built by humans fired its engines precisely on time at 1218 hours, 7 seconds Moscow Time on February 21, 1969. The thirty first-stage engines generated a total of approximately 4,590 tons of thrust, and within thirteen seconds, the N1 soared off the pad and headed out into the skies with its L3S payload. Deputy Chief Designer Chertok vividly described the launch of this monster:

90.   Konstantin Lantratov, "Anniversaries: 25 Years From Lunokhod-1" (English title), *Novosti kosmonavtiki* 24 (November 19–December 2, 1995): 70–79; O. A. Sokolov, "The Race to the Moon: A Look from Baykonur," presented at the 45th Congress of the International Astronautical Federation, IAA-94-2.1.610; N. G. Babakin, A. N. Banketov, and V. N. Smorkalov, *G. N. Babakin: zhizn i deyatelnost* (Moscow: Adamant, 1996), p. 57; Kamanin, "I Feel Sorry for Our Guys," no. 13; I. Lisov, "Launch and Flight of the 'Mars-96' Station" English title), *Novosti kosmonavtiki* 22–23 (October 21–November 17, 1996): 48.
91.   Igor Afanasyev, "N1: Absolutely Secret" (English title), *Kryla rodiny* no. 9 (September 1993): 13–16.
92.   Vad. Pikul, "The History of Technology: How We Conceded the Moon: A Look by One of the Participants of the N1 Drama at the Reasons Behind It" (English title), *Izobretatel i ratsionalizator* no. 8 (August 1990): 20–21.

Even if you have attended our Soyuz launches dozens of times, you can't help being excited. But the image of an N1 launch is quite incomparable. All the surrounding area shakes, there is a storm of fire, and a person would have to be insensitive and immoral to be able to remain calm at such moments. You really want to help the rocket: "Go on, go up, take off."[93]

And go it did, despite the fact that between three and ten seconds of ignition, the Engine Operation Control (KORD) system erroneously shut down two first-stage engines. All seemed well until T+70 seconds, when the KORD system abruptly shut down *all* the engines of the first stage, well before planned engine cutoff. This let the behemoth fly upward to an altitude of twenty-seven kilometers and then gradually descend on a trajectory that led to impact about fifty kilometers from the launch site. The emergency rescue system was activated after engine cutoff, and the descent apparatus of the 7K-L1S spacecraft landed without incident thirty-two to thirty-five kilometers from the pad area.[94]

Because it was the first launch attempt of a booster whose first stage had not been tested on the ground, engineers were not unduly discouraged by the failure, although the timing of the loss, as NASA was gearing to land on the Moon, perhaps lent a disheartening tenor to the recovery operation. Military-Industrial Commission Chairman Smirnov was apparently satisfied with the performance of the rocket, and Mishin himself reassured his engineers that "this is normal for a first launch."[95] Official historians of Mishin's design bureau were more specific:

Despite the accident, this launch confirmed the correctness of the selected dynamic scheme, the dynamics of the launch, the control processes of the

These movie stills pieced together show the launch of the first N1 Moon rocket in February 1969. *(copyright VideoCosmos Co., via Don Pealer/Quest)*

93. Sergey Leskov, "How We Didn't Get to the Moon" (English title), *Izvestiya*, August 18, 1989, p. 3.

94. Marinin and Shamsutdinov, "Soviet Programs for Lunar Flights"; Kamanin, "I Feel Sorry for Our Guys," no. 13.

95. M. Rebrov, "But Things Were Like That—Top Secret: The Painful Fortune of the N1 Project" (English title), *Krasnaya Zvezda*, January 13, 1990, p. 4; Pikul, "The History of Technology."

[booster] with the aid of coordinated engine thrusts, and allowed the receipt of experimental data on the loads on the [booster] and its precision, the influence of acoustical loads on the rocket and the launch system and [on its] operational characteristics in realistic conditions.[96]

It was clear after the launch that during the forty-first second of flight, one of the thirty engines of the first stage had failed and ignited others around it. As designers gathered after the launch, Mishin seemed to believe that the failure was probably caused by a malfunction in the turbogenerators, which provided electric current for the booster. First Deputy Chief Designer of the All-Union Scientific-Research Institute for Electro-Mechanics Nikolay N. Sheremetyevskiy recalls that Mishin squarely laid the blame on him before leaving the launch site. Later analysis of telemetry proved that Mishin was wrong. In fact, when the turbogenerators were recovered from the debris, both units were still in operating condition.[97]

Senior N1 engineers were able to report on the results of a preliminary investigation on the causes of the failure by March 11, 1969. The critical KORD system had clearly failed to meet the required standards for flight operation. As designers reported, the KORD system had not passed acoustical testing; an analysis of the reliability of the system had shown that KORD could not react to all possible conditions. As reconstructed from telemetry and an analysis of debris, 0.37 seconds prior to engine ignition, the KORD system shut down engine no. 12, and then by its logic, the opposite engine no. 24, although both were functioning without problem. Thus, by the time the rocket lifted off from the pad, twenty-eight of the thirty engines were firing; the remaining engines compensated fully for the absence of the two shutdown units and kept the booster aimed perfectly on a nominal trajectory. At T+5.5 seconds, excessive vibrations in the gas generator of engine no. 12 caused a line connected to a gas-pressure sending unit behind the turbine to rupture. The engine was beset by a second problem at T+23.3 seconds when, after the throttling down of thrust to reduce loads during maximum dynamic pressure, a two-millimeter-diameter pipe for measuring the fuel pressure in front of the engine's gas generator punctured. Consequently, "acid" gas with a temperature of 340 degrees Centigrade began mixing with the propellant, forming an extremely flammable solution. Eventually, at T+54.5 seconds, a fire broke out in the tail section of the first stage. Ground telemetry clearly showed a sharp rise in temperature at that point in engine nos. 3, 21, 22, 23, and 24. At T+68.67 seconds, the fire burned through the cable insulation, thus causing a short circuit in the 1,000-hertz direct-current and alternate-current circuits of the KORD system, which issued a command to shut down all the remaining twenty-eight engines of the first stage.[98]

96.   Semenov, ed., *Raketno-Kosmicheskaya Korporatsiya*, p. 257.
97.   Yu. A. Mozzhorin, et al., eds., *Dorogi v kosmos: I* (Moscow: MAI, 1992), p. 195.
98.   Afanasyev, "N1: Absolutely Secret"; Jeffrey M. Lenorovitz, "Trud Offering Liquid-Fueled Engines From N1 Moon Rocket Program," *Aviation Week & Space Technology*, March 30, 1992, pp. 21–22. An excerpt from the official accident investigation of the 3L launch is included in R. Dolgopyatov, B. Dorofeyev, and S. Kryukov, "At the Readers' Request: The N1 Project" (English title), *Aviatsiya i kosmonavtika* no. 9 (September 1992): 34–37. There are conflicting versions of the accident. One commonly quoted scenario is that at T+66 seconds, "the elevated vibrations caused by acoustical loads ruptured a line that feeds oxidizer to the gas generator of one of the [engines]; the leaking liquid oxygen started a fire in the aft section." See Afanasyev, "Unknown Spacecraft." The "elevated vibrations" arose because at T+65–66 seconds, the first-stage engines "throttled back to full power, but much stronger than expected causing strong vibration[s]. The oxidiser pipeline of one engine broke spilling liquid oxygen. The KORD control system was unable to shut the engine down quick enough and a fire broke out." See V. A. Lebedev, "The N1-L3 Programme," *Spaceflight* 34 (September 1992): 288–90; I. A. Marinin and S. Kh. Shamsutdinov, "Soviet Programs for Piloted Flight to the Moon" (English title), *Zemlya i vselennaya* no. 5 (September–October 1993): 77–85. The official history of the Korolev design bureau states that there was a failure in engine no. 2 because of high-frequency oscillations in its gas generator. As a result, a pressure carbine punctured, allowing the propellants to cause a fire in the tail end of the rocket. The fire disrupted the operation of the on-board cable network of the KORD system, which issued a command at T+68.7 seconds to shut off all the engines. See Semenov, ed., *Raketno-Kosmicheskaya Korporatsiya*, p. 257.

Overall, it was clear that the main problem for the booster was the lack of integrated ground testing for the first stage. In addition, there had been inadequate testing of the first-stage engines because of the absence of vibration stands. The space industry's leading research and development institution, TsNIIMash, recommended the introduction of a burn-monitoring system on the engines and stages prior to assembly as part of a flight model, but these recommendations were apparently rejected because of the lack of time and resources—a familiar reasoning offered throughout the 1960s.[99] Mishin and Kuznetsov introduced some cosmetic changes to the following flight models of the N1, including the deletion of the pressure sending unit and its pipe behind the turbine. The KORD system's main network was moved from the aft compartment into the intertank section. Additional improvements included adding new ventilation openings below the fuel pipeline covers to allow external air into the inside compartment.[100] Booster no. 4L was moved out of the queue of flights to allow for the cosmetic modifications as well as more substantive ones to improve lifting capacity. The next N1 launch would instead use booster no. 5L.

To the Finish Line

In March and May 1969, NASA performed two highly successful Apollo missions, Apollo 9 and Apollo 10, respectively, bringing the United States ever so closer to landing astronauts on the surface of the Moon. On Apollo 9, astronauts had thoroughly tested the Lunar Module in complex rendezvous and docking operations in Earth orbit. Such activities were repeated in lunar orbit on Apollo 10. In the Soviet canon, such missions would have been out of the question in 1969 because none of its lunar spacecraft were flightworthy: Chief Designer Yangel's engineers static-fired the important Blok Ye engine of the Soviet lunar lander for the first time only in February 1969.[101] Through the dampening enthusiasm, an increasingly small group of cosmonauts continued to train for lunar landings at both the Yu. A. Gagarin Cosmonaut Training Center or at the M. M. Gromov Flight-Research Institute, both located near Moscow. On March 28, 1969, veteran cosmonaut Bykovskiy was appointed the chief of the lunar department of the cosmonaut detachment.[102] By June 18, this department included only eight men out of the original group of approximately twenty-five from early 1968 who had trained for lunar landing missions. The eight included three Air Force officers training to land on the Moon—Valeriy F. Bykovskiy, Yevgeniy V. Khrunov, and Aleksey A. Leonov—and five others training to remain in lunar orbit during surface operations—Oleg G. Makarov, Viktor I. Patsayev, Nikolay N. Rukavishnikov, Anatoliy F. Voronov, and Aleksey S. Yeliseyev.[103]

The training was most challenging for the three preparing to land on the Moon. A dynamic simulator, built on the basis of an Mi-9 helicopter (itself modified from the Mi-8), allowed the cosmonauts to train for the actual landing phases. Having finished helicopter school, the trainees flew the helicopters to simulate worst-case scenarios for landing. Leonov recalls: "I

---

99. Rebrov, "But Things Were Like That." The article does not explicitly mention TsNIIMash, but rather "the head institute," which was usually a euphemism for TsNIIMash.

100. Afanasyev, "N1: Absolutely Secret"; Alexander Yasinsky, "The N-1 Rocket Programme," *Spaceflight* 35 (July 1993): 228–29; Marinin and Shamsutdinov, "Soviet Programs for Piloted Flight to the Moon."

101. V. Pappo-Korystin, V. Platonov, and V. Pashchenko, *Dneprovskiy raketno-kosmicheskiy tsentr* (Dnepropetrovsk: PO YuMZ/KBYu, 1994), p. 77. The tests were conducted at the giant testing facilities at Zagorsk belonging to NII KhimMash.

102. Voevodin, *VSA053*. The cosmonaut detachment as a whole was split up into different departments, including orbital space stations (headed by G. S. Shonin), spaceships (P. R. Popovich), air-space systems (G. S. Titov), and candidate cosmonauts (P. I. Belyayev).

103. Kamanin, "I Feel Sorry for Our Guys," no. 13.

made nine very difficult landings in that helicopter with the engines cut. Normally pilots don't do such landings because they usually end in a catastrophe, but we did it. We cosmonauts and pilots perfected the art."[104] They also took training courses at the M. M. Gromov Flight-Research Institute to master the ability to choose a landing site in the shortest time with minimal propellant reserves, while evaluating vertical velocity, to enable a survivable landing on the ground. After TsKBEM engineers had completed their preliminary landing simulations of the LK at the testing station at Zagorsk, the cosmonauts were invited to participate in landing trials on fake lunar landscape in specially built landing simulators at the Kiev Institute of Civil Aviation Engineers.

Cosmonaut Aleksey Leonov appears here in training for lunar landing approaches using a specially equipped helicopter. This photo dates from around 1969. (NASA photo)

The training eventually had cosmonauts wearing the Krechet-94 lunar suit in simulated lunar gravity. One of the fears among engineers was the possibility of the cosmonaut falling over on the surface and being unable to get up in the low gravity in the cumbersome lunar suit. To circumvent this problem, engineers came up with an ingenious solution consisting of a large hula-hoop–type ring that would be attached to the waist of the spacesuit before disembarking on the lunar surface. The larger part of the hoop was at the back side so as not to interfere with arm movements. The cosmonauts participated in sessions in special aircraft that simulated one-sixth gravity during which they "fell down" on their backs and simply rolled over and lifted themselves up. Another concern was depressurization after launch from the Moon. In a grueling exercise carried out in 1968, an Air Force captain dressed in a cumbersome pressure suit spent twelve torturous hours in an LOK cabin placed in a pressure chamber.[105]

The cosmonauts may have been engaged in intensive training to land on the Moon, but if the barometer of public statements from Soviet officials was any indication, the USSR was very confused about its next destination. Academician Blagonravov, the veneered doyen of Soviet space spokespersons, intimated in a statement reported by TASS on March 14, 1969, that there was still much work to be done before a Soviet lunar landing. Yet less than a month later on April 9, recently flown cosmonaut Shatalov told the Hungarian press that the Soviet Union would need "six, seven, and perhaps more months" of preparations before a landing on the Moon. He added with confidence that "who makes the better preparations will get to the Moon first, and it is our wish to do so."[106] L3 trainee Leonov was also unequivocal in his belief in the power of Soviet science:

104. "The Russian Right Stuff: The Dark Side of the Moon," NOVA television show, #1808, WGBH-TV, Boston, February 27, 1991. The M. M. Gromov Flight-Research Institute had put in a request for an Mi-4 helicopter to train for lunar landings as early as March 8, 1965. See N. P. Kamanin, Skrytiy kosmos: kniga vtoraya, 1964–1966gg (Moscow: Infortekst IF, 1997), p. 210.

105. Lardier, L'Astronautique Soviétique, p. 176; Afanasyev, "Unknown Spacecraft"; Kuznetsov, "The Flight That Did Not Occur"; S. Leskov, Kak my ne sletali na lunu (Moscow: Panorama, 1991), p. 12; V. M. Filin, Vospominaniya o lunnom korablye (Moscow: Kultura, 1992), pp. 60–61. The Air Force captain was Zhon Gridunov.

106. Soviet Space Programs, 1966–70, p. 372.

The Soviet Union is also making preparations for a manned flight to the Moon, like the Apollo program of the United States. The Soviet Union will be able to send men to the Moon this year or in 1970. We are confident that pieces of rocks picked from the surface of the Moon by Soviet cosmonauts will be put on display in the Soviet pavilion during the Japan World Exposition in Osaka in 1970.[107]

Leonov's somewhat misplaced confidence was astonishing because it came quite possibly at the utmost nadir of the Soviet space program in the 1960s. Removed from actual decision making within the Soviet space program, the cosmonauts were in general prone to more dramatic and often outlandish statements than older officials at conferences. However, even the cosmonauts must have surely known that there would be no Soviet cosmonaut on the Moon in 1969 or indeed in 1970.

The mainstream of Soviet public pronouncements was, however, turning to Earth-orbital space stations as the "mother lode" of future operations. Following the intensive high-level discussions in January 1969, the Soviets persistently began to emphasize two major directions: automated lunar exploration and permanent space stations in Earth orbit à la Tsiolkovskiy. Statements from academicians, anonymous chief designers, cosmonauts, and official radio commentators proliferated into the new Soviet propaganda offensive even before an American had set foot on the Moon.[108] A third option, piloted Martian missions, would be emphasized in the future as the technology became available. These statements were the first in a long series in 1969 to bombard the Western media with the idea that the Soviet space program was neither politically motivated (which is why the "race to the Moon" was unimportant) nor narrow (which was why Earth-orbital stations were being planned). These pronouncements were hard to counter because real Soviet intentions had always been cloaked in mystery. But the Soviets themselves were fully aware of this obfuscation of truth. Air Force Aide Kamanin wrote in his diary during the Apollo 10 mission of the "unrestrained lying" by Soviet officials on the issue of Soviet intentions with respect to the Moon. He added bitterly, "We have come to the end to drink the bitter chalice of our failure and be witnesses to the distinguished triumph of the U.S.A. in the conquest of the Moon."[109]

For Soviet government and Communist Party leaders, the impending humiliation was a hard pill to swallow. In early April 1969, Communist Party General Secretary Brezhnev invited Vasiliy P. Mishin to report on the work of the Soviet piloted space program during his three-and-a-half-year tenure as chief designer of the leading Soviet space enterprise. Mishin painstakingly explained the root reasons for the poor showing of the Soviet program in comparison to Apollo—all symptoms evident to any high-level space official in the Soviet Union. There was the institutional disarray in the organization of the space industry. Although there were many multi-profile design bureaus, there were severe shortages of subcontractor institutions. The production plants were badly organized with poor quality control, and each plant handled too many different production lines and was not specialized enough. Most tellingly perhaps, Mishin also touched on ideological reasons; he emoted on the lack of material incentives among workers in fulfilling plant orders of experimental models of articles.[110]

Among the four major points discussed at the meeting was an agreement to limit the L1 circumlunar project to only further automated flights, thus unequivocally terminating any hopes of cosmonauts flying around the Moon in the near future. As far as the N1-L3 program,

107. *Ibid.*, p. 374.
108. For several statements from the period, see *ibid.*, pp. 373–75.
109. Kamanin, "I Feel Sorry for Our Guys," no. 13.
110. Gorin interview, November 18, 1997.

Mishin could only report that a piloted landing would be preceded by a complete robotic mission, including landing and takeoff from the Moon. Future N1 missions would include the docking of spaceships in Earth orbit using liquid hydrogen stages, such as Blok S, before embarking on the voyage to the Moon. Repeating a mantra that had been uttered dozens of times by both Korolev and Mishin, the latter asked for more funding to pursue liquid hydrogen research, which, despite the best efforts of many, had enjoyed only lukewarm support from the government. Mishin's two final proposals to Brezhnev involved the creation of new generations of space weapons for ballistic missile defense using the N1 as a launch vehicle, and advanced flights to the Moon, Mars, Venus, and the outer planets. All of these were in the future. As far as the race to the Moon was concerned, there would be little to show from the Soviet side in 1969.

One of the more common stories proliferating in the Western media during the summer of 1969 was that the Soviets would do something spectacular before the first Apollo landing mission, Apollo 11. After the unqualified success of Apollo 10 in May 1969, NASA was looking at a lunar landing flight in July, with the ideal launch date being July 16. The question was: Could the Soviets do something to preempt the climax of the greatest American adventure of the 1960s? Nothing that the Soviets had accomplished in 1968 or 1969 had indicated that they had even a modicum of capability to attempt a full-scale lunar landing. Evidence now suggests that in June 1969, Chief Designer Mishin's most optimistic timetable for a first Soviet lunar landing was "by the end of 1970."[111] Wernher von Braun claimed in early June that it was still possible for the USSR to reach the Moon before the United States if the Apollo 11 mission was delayed, and he strongly believed that the Soviets would undertake piloted lunar flight in the "latter part of 1969" using a giant booster.[112] The CIA clearly had less confidence in Soviet capabilities than von Braun. In a top-secret "National Intelligence Estimate" issued a month before the launch of Apollo 11, the CIA predicted that "we estimate that a [Soviet] manned lunar landing is not likely to occur before 1972 although late 1971 cannot be ruled out."[113] But von Braun also referred to the most widely discussed scenario: that in the few remaining weeks leading up to the launch of Apollo 11, a robotic spacecraft would scoop up some soil and bring it back to Earth.

Prompted by Apollo 8, the Soviet Communist Party and government had decreed in January 1969 to accelerate their robotic lunar exploration program. Chief Designer Babakin's engineers had done an outstanding job of producing at least five flight models of the Ye-8-5 sample return spacecraft by the summer of 1969 in sufficient time to beat Apollo 11. Apart from the fact that the Ye-8 class series of heavy lunar probes had not been tested in space even once, the engineers had to address another possible problematic issue: the poor performance of the UR-500K Proton booster. By the end of April 1969, four consecutive launches of the rocket had failed to deposit their payloads into Earth orbit, let alone into deep space.[114] Of the total thirteen launches of the three-stage UR-500K variant (most with a fourth stage), seven had been unequivocal failures. In this context, the State Commission for the L1 circumlunar program met on May 29, 1969, to address "the Proton factor." While none of the failures pointed to errors in design, they did not exonerate quality control procedures during manufacturing. Designers Chelomey, Glushko, and Konopatov promised State Commission Chairman Tyulin that the next booster would not fail, but confidence was at a high premium at that point.[115] Perfect operation of the Proton booster

111. Kamanin, "I Feel Sorry for Our Guys," no. 13. Kamanin mentions this in his diary entry for June 19, 1969.

112. NASA Science and Technology Division, *Astronautics and Aeronautics, 1969: Chronology of Science, Technology, and Policy* (Washington, DC: NASA SP-4014, 1970), p. 170.

113. U. S. Central Intelligence Agency, "National Intelligence Estimate 11-1-69: The Soviet Space Program," Washington, DC, June 19, 1969, p. 20, as declassified in 1997 by the CIA Historical Review Program.

114. Of the four launches, one carried a 7K-L1 (in January), one carried a Ye-8 lunar rover (in February), and two carried M-69 Mars probes (in March and April).

115. Kamanin, "I Feel Sorry for Our Guys," no. 13.

was particularly critical at the time, not because of its use in the now-dying piloted circumlunar program, but because the Proton was to launch the Ye-8-5 lunar scooper to the Moon.

The confluence of activity in both the Soviet and U.S. space programs during the summer of 1969 was unprecedented. Babakin's lunar scooper had two chances to fly to the Moon, in the June and July lunar launch windows. At the same time, Mishin was almost ready to bring the second flight model of the N1 rocket to the launch pad. If the attempt was successful, the rocket would send the 7K-L1S spacecraft on an ambitious fully automated lunar-orbital flight, followed by the vessel's return to Earth. NASA would, of course, launch perhaps the most important mission in the history of American efforts to explore space. The race was now in its final lap.

Ye-8-5 spacecraft no. 402 was launched from Tyura-Tam on June 14, 1969, to reclaim some glory for the Soviet space program. If all went well, a sample of lunar soil would be back on Soviet territory in a little more than eleven days. Unfortunately, the spate of Proton failures did not abate. After the third stage had completed firing, the fourth Blok D stage was to fire to insert the payload into Earth orbit. Because of a disruption of an on-board circuit, the control system failed, preventing the Blok D engine from firing. The payload instead traced an arc that deposited it into the Pacific Ocean.[116] The odds were decreasing day by day now. Babakin still had four more scoopers left, and one could be launched in the second week of July 1969 for a repeat attempt. After five straight launch failures of the Proton, engineers and officials could be forgiven for harboring a pessimistic attitude on the chances of success.

The focus of the race to the Moon now shifted to the N1 rocket. By early April, based on the pace of preparations, Mishin had set May 30 as the date for transporting the next flight-ready N1, booster no. 5L, from the assembly-testing building to the launch pad. The launch would be during the lunar launch window in June, on June 13–15, 1969. The preparations for the launch were far more speedy than usual. One participant recalls:

> The first launch of the N1 (article 3L) aroused dual feelings among those contributing to the events: on the one side [Central Committee] Secretary D. F. Ustinov demanded acceleration of the launch of the "fifth—article 5L." The commotion at the plant rose extraordinarily. The issuing of the complete equipment and nodes for the assembly of the N1 managed to be on the current schedule, the fulfillment of which was overseen personally by the Deputy Minister of Defense of the country. On the other side, people from a multitude of commissions proposed some highly practical [modifications]. . . . However, the events unfolded so fast that many of the conceived measures for the 5L rocket simply physically did not have the time to be "spread out."[117]

The inevitable delays in the schedule meant that Mishin rescheduled the launch of the rocket from the June launch window to the one in July, just three weeks before Apollo 11. It would be a truly extraordinary few weeks in July, with plans for the launch of the second N1, the second Ye-8-5 lunar scooper, and, of course, Apollo 11.

The launch of N1 booster no. 5L was set for the night of July 3, 1969. The day before, there were rumors from unofficial sources in Moscow that something spectacular was imminent, but all these reports predicted a sample return mission on or about July 10.[118] Given the level of activity at the Baykonur Cosmodrome, it is testament to the power of the Soviet shroud

116. *Ibid.*; K. Lantratov, "The 'Late' Lunar Soil" (English title), *Novosti kosmonavtiki* 15 (July 16–29, 1994): 41–43; Sokolov, "The Race to the Moon."

117. Mikhail Rudenko, "Four Steps From the Moon" (English title), *Moskovskaya pravda,* July 19, 1994, p. 10. The quote is from Vadim Pikul.

118. *Soviet Space Programs, 1966–70,* p. 374; NASA Science and Technology Division, *Astronautics and Aeronautics, 1969,* p. 195.

Two N1 Moon rockets appear on the pads at Tyura-Tam in early July 1969. In the foreground is booster number 5L with a functional payload for a lunar-orbiting mission. In the background is the 1M1 ground test mock-up of the N1 for rehearsing parallel launch operations. (files of Asif Siddiqi)

This is the spectacular night launch of the second N1 booster in July 1969. Within seconds, the rocket collapsed back onto the pad, destroying the entire pad area in a massive explosion.
(copyright VideoCosmos Co., via Don Pealer/Quest)

the Baykonur Cosmodrome, it is testament to the power of the Soviet shroud of secrecy that, without exception, there was not a single leak to the Western media on any impending launch of a giant booster from Soviet central Asia. The hubbub at Tyura-Tam was unlike anything seen in recent memory. Ministers, deputy ministers, chief designers, senior military officers, and cosmonauts had all flown in for the launch—a final gasp for the sinking hopes of the Soviet reach for the Moon. Valeriy A. Menshikov, then a young lieutenant in the Strategic Missile Forces, who was duty officer at site 112 near the N1 pads, later provided one of the best personal accounts of that fateful night:

> There were hundreds of vehicles on the roads with soldiers, officers and civilians. They bore combat banners, documents and various materiel. The dust and heat, the roar of the automobile engines, the human chaos, the congestion and traffic jams, the hoarse shouts of the traffic-control personnel—all of this was reminiscent of frames from movies of the first months of the [Second World] war. The only thing missing were German dive bombers.[119]

As night fell, Menshikov ordered the launch site group to assemble and then led them away from the rocket to a bunker close to the N1 pad at site 110P to await the launch. Like most observers, lunar cosmonauts Leonov, Makarov, and Rukavishnikov witnessed the launch from a distance of six to seven kilometers. Prelaunch operations began at 0600 hours Moscow Time on

119. Menshikov, "The Toilers of the Cosmodrome," p. 40.

the morning of July 3 and continued through the day. By 1540 hours, personnel had begun fueling the first three stages, a procedure that was completed within one hour and fifty minutes. Fueling of the L3S payload block began in the early evening at 1900 hours. There were evidently no serious anomalies during the ensuing countdown as the clocked ticked closer to midnight.

The N1 ignited to life at exactly 2318 hours, 32 seconds Moscow Time on July 3 (it was after midnight on July 4 at Tyura-Tam). Menshikov remembers the experience vividly:

> We were all looking in the direction of the launch, where the hundred-meter pyramid of the rocket was being readied to be hurled into space. Ignition, the flash of flame from the engines, and the rocket slowly rose on a column of flame. And suddenly, at the place where it had just been, a bright fireball. Not one of us understood anything at first. A terrible purple-black mushroom cloud, so familiar from the pictures from the textbook on weapons of mass destruction. The steppe began to rock and the air began to shake, and all of the soldiers and officers froze.[120]

Rukavishnikov's remembrance is almost surreal: he could see the booster double over in an explosion on the pad, but there was no sound. Those few seconds of "deathly silence" lasted an eternity until the full roar of the launch and the ensuing explosion reached the viewing stands.[121] The young Lieutenant Menshikov adds:

> Only in the trench did I understand the sense of the expression "your heart in your mouth." Something quite improbable was being created all around—the steppe was trembling like a vibration test jig, thundering, rumbling, whistling, gnashing—all mixed together in some terrible, seemingly unending cacophony. The trench proved to be so shallow and unreliable that one wanted to burrow into the sand so as not to hear this nightmare . . . the thick wave from the explosion passed over us, sweeping away and leveling everything. Behind it came hot metal raining down from above. Pieces of the rocket were thrown ten kilometers away, and large windows were shattered in structures 40 kilometers away. A 400 kilogram spherical tank landed on the roof of the installation and testing wing, seven kilometers from the launch pad.[122]

By some estimates, the strength of the explosion was close to 250 tons of TNT—not a nuclear explosion, but certainly the most powerful explosion ever in the history of rocketry. The booster had lifted off to a height of 200 meters before falling over and exploding on the launch pad itself, about twenty-three seconds after launch. The emergency rescue system fired in the nick of time, at T+14.5 seconds, to shoot the descent apparatus of the payload two kilometers from the pad, thus saving it from destruction. Remarkably, no doubt because of the stringent safety precautions, there were no fatalities or injuries, although the physical devastation was phenomenal. When the first teams arrived near the pad in the early-morning hours of July 4, there was only carnage left behind:

> We arrived at the fueling station and were horrified—the windows and doors were smashed out, the iron entrance gate was askew, the equipment was scattered about

---

120. *Ibid.*
121. Rudenko, "Four Steps From the Moon," p. 10.
122. Menshikov, "The Toilers of the Cosmodrome," p. 40.

with the light of dawn and was turned to stone—the steppe was literally strewn with dead animals and birds. Where so many of them came from and how they appeared in such quantities at the station I still do not understand.[123]

By 0800 hours the morning of July 4, Minister of General Machine Building Afanasyev had convened a meeting of the State Commission and began the long process of determining the reasons behind the disaster by looking at films of the launch and analyzing telemetry. Afanasyev also telephoned Brezhnev and Kosygin, the latter of whom was particularly dissatisfied with the results. Perhaps most sobering of all was Chief Designer Barmin's assessment on the destruction of the launch area. The right launch pad at site 110P was completely destroyed; the explosive force also displaced the 145-meter-tall service tower from its rails and destroyed all the special ground equipment of the launch installation, including a lightning arrester. The top two and a half floors of the five-story underground pad support structure had collapsed.[124] The left launch pad at site 110L had remained unscathed. A second N1 had in fact been mounted at the pad during the failed launch presumably to rehearse dual launches planned for later in the lunar program. Barmin believed that restoration of the destroyed complex would be faster and cheaper than building a completely new one.

To pursue an investigation of the accident, Afanasyev created a commission headed by Chief Designer Mishin; this commission consisted of seven subcommissions for particular areas of the N1 rocket.[125] The stress of the previous few months of relentless work seem to have taken their toll on the fifty-two-year-old Mishin; at a meeting three days after the disaster, he suffered serious heart trouble, although he was apparently back at work very soon after. Beginning on July 4 and continuing through the waning weeks of July, the commission focused on malfunctions in the KORD engine control system. It was immediately clear after the accident that at least five engines had been turned off within one second of ignition. According to early data, KORD turned off *all* engines save one, engine no. 18, about ten seconds into the mission. Engineers also detected early on a short circuit in an oxygen line in the area of two other engines, nos. 8 and 9. But the question remained: Why had KORD shut the engines down in the first place? By July 11, a researcher from the P. I. Baranov Central Institute of Aviation Motor Building was able to report that perhaps a foreign object had entered an NK-15 engine's oxygen pump, causing a cascade of failures. By the time of Mishin's visit to Kuybyshev on July 16 for

---

123. *Ibid.*, p. 40. A U.S. CORONA photo-reconnaissance satellite photographed the aftereffects of the pad explosion by early August 1969. One such picture, taken on CORONA mission 1107 during pass 169 on August 3, 1969, has been published. See the back cover of *Quest* 4(2) (Summer 1995). CORONA information was probably the primary basis for a description of the N1 launch failure in a top-secret CIA document from March 1970. See U.S. Central Intelligence Agency, "National Intelligence Estimate 11-1-69: The Soviet Space Program," Washington, D.C., March 26, 1970, p. 1, as declassified in 1997 by the CIA Historical Review Program. News of the disaster eventually leaked out into the open media. The first revelations emerged on November 17, 1969, simultaneously in Great Britain and the United States. See Stuart Auerbach, "Soviet Moon Rocket Exploded in Test," *Washington Post*, November 18, 1969, p. A1; "Soviets Suffer Setbacks in Space," *Aviation Week & Space Technology*, November 17, 1969, pp. 26–27; "Disaster at Tyuratam," *Time*, November 28, 1969, p. 27. Curiously, the February 1969 launch attempt was never detected by Western intelligence, although they apparently did expect a launch in early 1969. See "Countdown for Biggest Rocket Yet," *Newsweek*, February 24, 1969, p. 28; Donald C. Winston, "Soviet Space May Include Large Booster Test," *Aviation Week & Space Technology*, March 10, 1969, pp. 132–33.

124. Rudenko, "Four Steps From the Moon." Note that another source says that "all six underground levels of the launch structure were destroyed by the explosion." See Marinin and Shamsutdinov, "Soviet Programs for Piloted Flight to the Moon."

125. The subcommissions were headed by N. D. Kuznetsov (engines), G. I. Degtyarenko (temperatures and loads), A. G. Iosifyan (electrical supply), V. P. Finogeyev (guidance and control systems), Ye. V. Shabarov (launch escape system), B. A. Dorofeyev (specialty unknown), and Kupavin and Dorofeyev (KORD).

discussions with engine Chief Designer Kuznetsov as well as his own First Deputy Kozlov, there were four likely reasons for the accident out of a possible seven at the beginning of the investigation.

The search for the causes of the disaster would continue on for many months, but the damage inflicted not only on hardware but also on the spirits of Soviet engineers on the night of July 3, 1969, was irreparable. Kamanin wrote in his diary the day after the accident:

Yesterday the second attempt to launch the NI rocket into space was undertaken. I was convinced that the rocket would not fly, but somewhere in the depth of my soul there glimmered some hope for success. We are desperate for a success, especially now, when the Americans intend in a few days to land people on the Moon, and when the American astronaut Frank Borman is our guest. But all such hopes were dispelled by the powerful explosion of the rocket five seconds after the "launch" command . . . on its first time, the rocket flew 23 kilometers, and did not cause harm to the launch platform and launch site. This time it fell two kilometers [sic] from the pad and caused huge damage to the launch site. This failure has put us back another one to one and a half years.[126]

Soviet Ambassador to the United States Anatoliy Dobrynin had indeed invited Apollo 8 astronaut Colonel Frank Borman for a nine-day visit to the USSR. Although Borman and his family were not considered official guests of the Soviet government, it was the first visit of an American astronaut to the country. On the night of July 4, 1969, Borman was present at the U.S. embassy's reception to celebrate Independence Day. The timing could not have been worse for the Soviets. Instead of being feted by reporters on a new success in space, Soviet cosmonauts were on hand, less than twenty-four hours after the catastrophe at Tyura-Tam, glum and reticent. When asked about the possibility of a Soviet lunar scooper timed to fly before Apollo 11, Beregovoy, Feoktistov, and Titov declined to confirm or deny the rumors.[127] The following day, Borman visited the Gagarin Cosmonaut Training Center, where he was received by the newly appointed Commander-in-Chief of the Soviet Air Force Marshal Pavel S. Kutakhov and Col. General Nikolay P. Kamanin.[128] The many cosmonauts attending the function could only watch in damaged pride as the NASA astronaut gave an impressive slide show of his recent flight to the Moon.

Through their despair, the Soviets had one final gasp left: a flight of the Ye-8-5 sample return spacecraft during the July launch window. If it succeeded, the mission would vindicate their recent abrupt emphasis on automation versus piloted flight. Even more dramatic would be a success for the scooper if Apollo 11 failed. Such a scenario, no doubt given consideration during those desperate weeks in early July, would have, in one fell swoop, eliminated all the failures, explosions, and delays of the year so far.

Chief Designer Babakin's engineers prepared his spacecraft, Ye-8-5 vehicle no. 401, for launch at the same time that workers were scouring the remains of the NI at Tyura-Tam. There were problems with the mass of the spacecraft right up until the final days before launch. Engineers calculated that the ascent stage of the robot, called the RYe-85, had a mass of 513.3 kilograms instead of the allotted 512 kilograms. After much soul searching, Babakin ordered the deletion of one of two 1.28-kilogram radio transmitters on the ship, leaving the primary one with no backup. It was a gutsy move, underlining the risks inherent in the mission in general. The launch itself was a blessing. After five straight failures of the Proton launch vehicle, the rocket lifted off on time at 0554 hours, 41 seconds Moscow Time on July 13, 1969; precious payload was deposited on a perfect trajectory heading for the Moon. The Soviet press, announcing the

126. Kamanin, "I Feel Sorry for Our Guys," no. 13.
127. James F. Clarity, "Top Soviet Aides Observe the 4th," *New York Times*, July 5, 1969.
128. Riabchikov, *Russians in Space*, pp. 265–66.

mission as *Luna 15*, merely stated that the spacecraft would study circumlunar space, the Moon's gravitational field, and the chemical composition of lunar rocks, and would carry out surface photography.[129]

The world's eyes and ears, however, were not on the Soviet spacecraft, but on the three American men who set off for the Moon on July 16, just three days after the launch of Luna 15. For a brief moment, Apollo 11 astronauts Neil A. Armstrong, Michael Collins, and Edwin E. Aldrin, Jr., represented not only NASA and not just simply the United States, but, in the justifiably hyperbolic language of the day, all humanity itself. But there was also a more earthly aspect of the mission, too: they carried the baton on the last lap of the "space race," inaugurated by the Soviet Sputnik twelve years previously. This more political dimension had gradually receded from the foreground as it seemed that the Soviets had, for reasons unclear, relinquished their claim to answer President Kennedy's challenge. For Soviet space engineers, however, the "space race" as a living artifact was far more imposing in 1969 than to their counterparts across the ocean. Their last hopes were pinned on Luna 15 much more than anyone would care to admit at the time.

The responsibility of directing the Luna 15 mission fell on the shoulders of First Deputy Minister of General Machine Building Georgiy A. Tyulin, the fifty-four-year-old retired artillery general whose career in the missile and space industry had now spanned more than twenty-five years. Tyulin, as chair of Luna 15's State Commission, ran into trouble with the spacecraft after only one day of flight. Controllers detected unusually high temperatures in the propellant tanks of the S5.61 engine, which would be used for takeoff from the lunar surface after the collection of the lunar sample. With the specter of a possible explosion of the entire engine complex en route to the Moon, Tyulin assembled all the senior program engineers, including Chief Designer Babakin. After a quick analysis, some participants proposed a seat-of-the-pants method of turning the spacecraft in such a way as to keep the suspect tank in the Sun's shadow at all times. Despite some acrimonious exchanges and stiff resistance from engineers, Tyulin sided with trying the unorthodox procedure; telemetry later showed that the tank temperature stabilized at acceptable levels.[130]

Luna 15 fired its main engine to enter lunar orbit at 1300 hours Moscow Time on July 17. Engineers planned two major orbital corrections prior to landing on the Moon. The first (KIII) on July 18 was to bring the spacecraft's perigee to sixteen plus or minus four kilometers altitude. If the altitude was too high, then there would be insufficient propellant to brake the ship down to the surface, and if it was too low, then there would not be enough time to slow the vehicle down for a survivable landing. The second correction (KIV) on July 19 would determine the longitude of the ascending node to posit the ship over the precise landing corridor. The State Commission did not, however, anticipate the ruggedness of the lunar surface, and the altimeter showed wildly varying readings for the projected landing area. Controllers instead spent three to four days carefully analyzing incoming data. Over twenty to twenty-two communications sessions per day, engineers laid the groundwork for carrying out corrections, built a support system of coordinates, established thrust orientation vectors, and carried out trajectory measurements over consecutive orbits. Two carefully prepared maneuvers were carried out at 1608 hours on July 19 and at 1716 hours on July 20, the latter putting the spacecraft into the planned 110- by sixteen-kilometer orbit at a retrograde inclination of 127 degrees.[131]

129. *Soviet Space Programs, 1966–70*, pp. 196; Babakin, Banketov, and Smorkalov, *G. N. Babakin*, p. 64.
130. Babakin, Banketov, and Smorkalov, *G. N. Babakin*, pp. 60–62.
131. *Ibid.*, pp. 62–63; Lantratov, "The 'Late' Lunar Soil." The orbit after the first correction was 221 by ninety-five kilometers.

Some members of the State Commission for the Luna sample return spacecraft are shown in a photo from 1970 at Simferepol. Sitting in the foreground from left to right are Commission Chairman Georgiy Tyulin, Chief Designer Georgiy Babakin, and Minister of General Machine Building Sergey Afanasyev. The tall figure standing at the back on the right is Yuriy Koptev, the current director of the Russian Space Agency who was an engineer at the Lavochkin design bureau at the time. Sitting in the second row at left is Academician Boris Petrov, one of the principal international spokespersons for the Soviet space program. (copyright Asif Siddiqi)

The Western press closely followed the mission of Luna 15. Kenneth Gatland, a respected British journalist who hosted the Apollo 11 broadcasts for British television, recalled:

> *Even as the Apollo 11 programme was on the air, and we sat before the cameras discussing how Neil Armstrong and Edwin Aldrin would land, the Russian robot was maneuvering in orbit. There was even the suggestion from one scientist that Russia might be preparing to set down on the Moon a mooncraft capable of rescuing the Americans if, by some accident, they were stranded on the Moon! To the delight of the TV producers, the drama was kept up until the last.*[132]

Even NASA, busy as it was with Apollo 11, managed to join in the drama of the race. The Apollo 11 crewmembers were kept apprised of the progress of Luna 15. There was also some concern that Luna 15's orbit might, in an unlikely situation, interfere with that of Apollo 11.

---

132. Gatland, *Robot Explorers*, p. 145.

Astronaut Borman played a critical role in passing on detailed orbital information on Luna 15 from the Academy of Sciences to the White House, which evidently laid to rest any fears the Apollo flight control might have had back in Houston.[133]

To Western observers, the closeness of the race in lunar orbit was without precedent. A little less than six hours after Luna 15's second and final orbital correction, the Apollo 11 Lunar Module began its voyage toward the lunar surface. After a thrill-laden descent, the two astronauts, Armstrong and Aldrin, safely put down the ungainly looking lander onto the lunar surface at 2017 hours GMT on July 20. In Moscow, it was 2317 hours, close to midnight. Luna 15, meanwhile, was still in orbit, as controllers pored over their data. Originally, their plan was to put down the robot less than two hours after Apollo 11. The delays in mapping out a correct trajectory for Luna 15, however, took their toll. Unsure of the terrain below, Tyulin delayed the landing a full eighteen hours, awaiting a final and unanimous affirmative from his engineers. During this no doubt demoralizing period, Neil A. Armstrong exited the Lunar Module and set foot on the surface of the Moon.

As a mesmerized world watched the ghostly images of human beings walking on another celestial body, Luna 15 became a footnote to history. Tyulin's State Commission finally commanded the robot to fire its descent engine at 1847 hours Moscow Time on July 21, a little more than two hours prior to the planned liftoff of Armstrong and Aldrin from the Moon. It was the spacecraft's fifty-second orbit around the Moon. Controllers impatiently followed the signals from Luna 15 as it descended swiftly to the lunar surface. Landing would be six minutes after the beginning of powered descent. To the collective shock of all those present, transmissions abruptly ceased four minutes after deorbit, at an altimeter reading of three kilometers.[134] Later analysis showed that the spacecraft had unexpectedly hit the side of a mountain at a velocity of 480 kilometers per hour. The impact point was at 12° N, 60° E in Mare Crisium. The Soviet news agency TASS characteristically announced that Luna 15's research program had been completed and the spacecraft had "reached" the Moon in the "preset" area.[135] There was one small irony to the whole mission. Even if there had not been a critical eighteen-hour delay in attempting a landing, and even if Luna 15 had landed, collected a soil sample, and safely returned to Earth, its small return capsule would have touched down on Soviet territory two hours and four minutes *after* the splashdown of Apollo 11.[136] The race had, in fact, been over before it had begun.

Armstrong and Aldrin, meanwhile, lifted off successfully, and with crewmember Collins, headed back to Earth, splashing down safely in the Pacific Ocean on July 24, 1969, concluding one of the most dramatic voyages of exploration in the history of humankind. Outside the USSR, Soviet officials were unusually magnanimous in their praise of this incredible feat, but within the country, to their own citizens, they were less than generous. By the end of the 1960s, official Soviet doctrine had showed a marked positive evaluation and reportage of American space achievements, but Apollo 11, given its paramount importance as a defining moment of the space race, was an anomaly. Many within the space industry, including TsNIIMash Director Mozzhorin, were themselves responsible for deemphasizing the importance of Apollo 11, perhaps partly to hide their own shortcomings. In the *glasnost* days of reevaluating the black holes of Soviet history, one Soviet journalist wrote with undisguised vitriol:

133. Slayton and Cassutt, *Deke!*, p. 240; *Soviet Space Programs, 1966–70*, p. 196.
134. Babakin, Banketov, and Smorkalov, *G. N. Babakin*, pp. 63–64; Lantratov, "The 'Late' Lunar Soil."
135. *Soviet Space Programs, 1966–70*, p. 197. Rebrov, "But Things Were Like That."
136. Lantratov, "The 'Late' Lunar Soil."

The task of Mozzhorin's group consisted of misinforming the public and concealing from the people the blunders and the real state of our affairs in space. But the deception became obvious when, on July 21, 1969 . . . Neil Armstrong became the first earthling to set foot on the surface of the Moon and planted the American flag. Our deceitful propaganda, supervised then by M. A. Suslov (now one of Moscow's boulevards has been named after him), was forced to show this historical event on our television screens during a volleyball match between two local teams.[137]

The news itself was not accompanied by any TV footage, merely a dry news report. Actual video of the landing was evidently restricted to a select group within the Soviet Union, including the chief designers.

With the final and ignominious end to the "race to the Moon," the uncertainty of the numerous pronouncements of the last eight months disappeared, replaced by two clear and consistent themes: the Soviet objective to the explore the Moon by automated means and the longstanding goal of establishing piloted orbital space stations in Earth orbit. Implicit, of course, in both these themes was the claim that the Soviet Union had never planned to send humans to the Moon because its program had always been geared more toward scientifically productive rather than politically motivated objectives. Academician Blagonravov claimed on Moscow Radio on July 21 that the only advantage of sending cosmonauts to the Moon was to provide freer choice in picking up Moon rocks. He emphasized that the space programs of the two superpowers had moved at about the same pace but along parallel paths.[138] Even if he knew of the existence of the N1-L3 program, he would have been committing treason against the state had he stated that the Soviet Union had indeed tried to race the Americans to the Moon. Later in the month, in another statement, he added that Soyuz spacecraft would be converted into "modules of orbital space laboratories designed for research in lengthy flight."[139]

Salvaging the wreckage of the Soviet piloted space program was not an easy task. Discussions in early 1969 had given focus to three possible future tracks:

- A piloted Mars mission
- Improved lunar landing missions
- Earth-orbital space stations

Publicly, Soviet spokespersons focused only on the third item. Academician Sedov, for example, on a visit to Japan in late August 1969, claimed that a new type of "spacecraft" would be used to put a large space station into Earth orbit. There was, he said, no necessity in sending humans to the Moon because automated lunar probes could return soil back to Earth.[140] The decision to move ahead with space stations was, however, fraught with much more internal acrimony than Sedov's statement would suggest. The three major possibilities available to the Soviets in the post–Apollo 11 climate raised not only the hope of restoring prestige to a rudderless Soviet space program, but also gave rise to yet more acrimony among the major players in the industry. The lessons of losing the Moon race had, it seems, not been learned very well.

---

137. German Nazarov, "You Cannot Paper Space With Rubles: How to Save Billions" (English title), *Molodaya gvardiya* no. 4 (April 1990): 192–207.

138. *Soviet Space Programs, 1966–70*, p. 375.

139. NASA Science and Technology Division, *Astronautics and Aeronautics, 1969*, p. 256.

140. *Soviet Space Programs, 1966–70*, p. 376. Disingenuously, Sedov added that a soft-landing had not been a goal for Luna 15; the spacecraft had been "sent to study the Moon from lunar orbit."

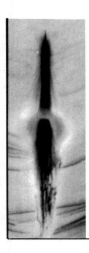

Chapter Sixteen
Options

Conventional wisdom would suggest that after such a fatal blow as the triumphant landing of American astronauts on the Moon, the Soviet Union would simply fall back into a period of conservatism, characterized more by self-appraisal rather than any further grand gestures at competition. But Soviet officials, from the highest arbiter of the Soviet space program, Dmitriy F. Ustinov, down to the lowest engineers, differed in one key respect to their American counterparts. For the Soviets, the race to the Moon might have been over, but the less specific "space race" was not. Ironically, it was, in fact, the American space program that entered an uncertain period of soul searching as it sought to define a direction in the post-Apollo frontier—a direction that for the first time was not determined exclusively by Cold War competition with the Soviet Union. The Soviets, on the other hand, continued to propose, define, and implement newer programs, which harked back to political imperatives of the Kennedy-Khrushchev era. If the Americans had beaten the Soviets to the Moon, then the Soviets would beat them to Mars. If the Americans were going to build a space station in Earth orbit, then the Soviets would build one sooner. While Soviet motivations in late 1969 were a little more complex than such simplistic rhetoric, by and large, the Soviet space program did not abandon the space race in 1969. In fact, its piloted lunar programs continued to serve as a major force in policy, years after Neil A. Armstrong stepped on the Moon in July 1969.

Rummaging Through the Wreckage

Much of the activity in the Soviet program during the latter part of 1969 resulted more from inertia rather than any new goals. As policy planners gradually sought to establish clear directions for the overall effort, space vehicles intended for flight earlier in the decade were finally ready for launch. With little to lose after Apollo 11, Ustinov, Smirnov, and Afanasyev allowed some token launches in the piloted lunar program, which on superficial examination seem to make little sense. The first such mission was a circumlunar flight of the 7K-L1 spacecraft in the late summer of 1969. Although the piloted component of the circumlunar program had been officially suspended in March 1969, Chief Designer Mishin continued flights of the trouble-prone spacecraft in the hope of flying crews on board at some uncertain time in the future. Carrying out a simple automated circumlunar mission less than a month after Apollo 11 might indicate a disregard for public perceptions of the Soviet space program, but the timing of the launch was apparently more of a coincidence than anything else. The Soviets did, however, go to great lengths to play down news of the mission.

As with previous L1 launches, cosmonauts were present at the Baykonur Cosmodrome, although this time they were involved to a greater degree in flight operations. Leonov and

Makarov trained to acquire skills of "controlling the [descent apparatus] as operators" in preparation for a piloted flight.[1] The 7K-L1 spacecraft, vehicle no. 11, had been the last model manufactured for automated flight and contained mannequins. The ship was, however, redesigned for piloted flight with powered control panels and blocks removed from the switches. The spacecraft lifted off from the Baykonur Cosmodrome at 0248 hours, 6 seconds Moscow Time on August 8, 1969, and successfully headed toward the Moon an hour later. Called *Zond 7* by the Soviet press, the ship, like its predecessors, carried a menagerie of living specimens, including four Steppe male tortoises, which were part of a group of thirty tortoises selected for a biological study.[2] The spacecraft was said to have been equipped with improved instrumentation, although few details were provided. After a mid-course correction at a distance of 250,000 kilometers from Earth on August 9, the ship circled the far side of the Moon at a range of 1,200 kilometers two days later. The only anomaly on the flight was a communications problem—the main parabolic antenna failed to unfurl because of a jam in the securing cables—although this did not prevent the accomplishment of any of the main flight objectives.[3]

For the first time on a Zond mission, the on-board camera took color photographs. The first session took place on August 8 when the camera took pictures of Earth at a distance of 70,000 kilometers, clearly showing a large part of the globe, including Asia, Africa, and the Middle East. Three days later on August 11, there were two further sessions. The first ten-minute run was at a distance of 10,000 kilometers when the ship was closing in on the Moon; it covered the western side of the Ocean of Storms and nearby heavily cratered areas. An hour later, the spacecraft took a further series of photographs showing far side features from a range of 2,000 kilometers. Several of these spectacular shots were reminiscent of those taken by Apollo astronauts, with Earth majestically setting over the Moon's horizon. Although the Moon generally tends to look gray, scientists hoped that color photos from different angles might reveal differences in its microstructure. Apart from photography, the spacecraft also performed "measurements of the physical characteristics of circumlunar space as well as technical experiments for developing motion controlling systems with the onboard [computer], astro-orientation systems, deep space communications apparatus, and other onboard systems."[4]

The Zond 7 spacecraft flew back to Earth without incident, once again flying over the South Pole and then moving north over the Indian Ocean. It entered the correct corridor on August 14, lost velocity, skipped out, and then reentered again for a perfect, aerodynamically controlled reentry onto Soviet territory. Parachutes deployed at an altitude of seven and a half kilometers and soft-landing engines fired a meter above the ground for a faultless touchdown south of Kustanay in Kazakhstan, just fifty kilometers from the intended target point. The mission had lasted six days, eighteen hours, and twenty-five minutes. Two years late, TsKBEM finally accomplished a fully successful 7K-L1 circumlunar mission. It was, of course, too late for

1.   Yu. P. Semenov, ed., *Raketno-Kosmicheskaya Korporatsiya "Energiya" imeni S. P. Koroleva* (Korolev: RKK Energiya, named after S. P. Korolev, 1996), p. 246.

2.   O. G. Gazenko, V. V. Antipov, and G. P. Parfenov, "Results of Biological Investigations Undertaken on the Zond-5, Zond-6, and Zond-7 Stations" (English title), *Kosmicheskiye issledovaniya* 9 (July–August 1971): 601–09. Other specimens included air-dried cells of wheat, barley, peas, pines, carrots, tomatoes, mustard bulbs of common onion, one strain of single-celled chlorella algae, and a culture of lysogenic bacteria.

3.   V. P. Glushko, ed., *Kosmonavtika entsiklopediya* (Moscow: Sovetskaya entsiklopediya, 1985), p. 130; Semenov, ed., *Raketno-Kosmicheskaya Korporatsiya*, p. 246.

4.   Glushko, ed., *Kosmonavtika entsiklopediya*, p. 130; *Soviet Space Programs, 1966–70: Goals and Purposes, Organization, Resources, Facilities and Hardware, Manned and Unmanned Flight Programs, Bioastronautics, Civil and Military Applications, Projections of Future Plans, Attitudes Toward International Cooperation and Space Law*, prepared for the Committee on Aeronautical and Space Sciences, U.S. Senate, 92d Cong., 1st sess. (Washington, DC: U.S. Government Printing Office, December 1971), p. 244; Kenneth Gatland, *Robot Explorers* (London: MacMillan, 1972), p. 150.

politicians to extract any mileage from the resounding success of Zond 7, coming as it did less than a month after the American lunar landing. But the conclusion of the mission did raise the possibility of moving ahead to piloted missions on the L1 spacecraft. At a meeting of the L1 State Commission on September 19, 1969, the members discussed such an option. The Air Force Commander-in-Chief's Aide for Space Col. General Kamanin recalled that "the success of Zond-7 . . . gave some encouragement to Mishin, Tyulin, and Afanasyev who were gradually recovering from the shock caused by the failure of the N1 and the brilliant Apollo missions."[5] The State Commission tentatively decided to make use of the three remaining 7K-L1 spacecraft still left on the ground. The first would be launched in early December 1969 on an automated flight followed by the second in April 1970, perhaps carrying the first Soviet cosmonauts around the Moon. While Mishin and State Commission Chairman Tyulin may have wished for such, the forces against piloted L1 missions were too overwhelming. There was little to be gained politically from a piloted L1 mission at this point. Both Brezhnev and Ustinov had more or less decided on the program's termination in the spring of 1969, and the plans to launch a crew in April 1970 eventually died a quiet death. By the end of 1969, the piloted portion of the UR-500K-L1 project was irrevocably over, and while Mishin had plans to fly the remaining unflown vehicles, these were redirected toward primarily technological goals.

The dilemma facing Soviet space planners in the direct aftermath of the Apollo landing was how to respond in the immediate months. What kind of a piloted mission could be mounted in the waning months of 1969 that would not underline the weak position of the USSR in comparison to the United States in the exploration of space? In the landmark January 1969 meetings after Apollo 8, Minister of General Machine Building Afanasyev had suggested a thirty-day Soyuz mission in Earth orbit. A month later, Soyuz State Commission Chairman Maj. General Kerimov emerged with a more modest seven-day Earth-orbital flight of two cosmonauts in a Soyuz ship. Space program chief Ustinov wanted more, telling the commission that a seven-day mission was too "thinnish" and that "it should be thick."[6] Kamanin, on February 11, underlined the confusion in how to proceed with the Soyuz program, writing in his diary:

> We have reached a fully absurd [situation]: there is not one man in this country who would be able to say what the next flight into space will be. Ustinov does not know this, Keldysh, Smirnov, and Mishin do not know this—generally no one knows! All my attempts to obtain from the state the composition of plans for piloted space flights lead nowhere: there are no such plans, and it is most unlikely that there will be.[7]

Originally, prior to the Soyuz 4/5 docking-and-EVA mission in January 1969, Mishin had had plans to fly repeat Earth-orbital flights of the 7K-OK Soyuz spacecraft, but equipped with the Kontakt rendezvous radar system earmarked for the lunar version of the Soyuz instead of the less advanced Igla. While Kontakt was not ready for flight at the time, the Soyuz 4/5 repeat mission plans offered an answer on how to formulate a response to Apollo. By late February, Mishin's idea was to launch three 7K-OK Soyuz spaceships into Earth orbit, two of which would dock automatically with each other, while the third would hover at 300 to 400 meters range by means of manual control and take photographs of the experiment.[8] Although a poor

5. N. Kamanin, "I Feel Sorry for Our Guys" (English title), Vozdushniy transport 14 (1993): 11.
6. N. Kamanin, "I Feel Sorry for Our Guys" (English title), Vozdushniy transport 13 (1993): 8–9.
7. Ibid.
8. Dmitriy Payson, "Eternal 'Soyuz'—Today Marks the 25th Anniversary of the First Docking in Orbit" (English title), Nezavisimaya gazeta, January 15, 1994, p. 6; Christian Lardier, L'Astronautique Soviétique (Paris: Armand Colin, 1992), p. 188.

match for a lunar mission, such a flight would not only demonstrate the capacity of the Soviet space program to perform complex operations in space, but also provide a long overdue public relations extravaganza from the potentially spectacular photographs. On a purely technical level, the flight would also allow engineers to perfect rendezvous and docking operations and control multiple vehicles in orbit in preparation for future space station missions.

By April 1, 1969, Mishin had a short-range plan for the 7K-OK Soyuz program:

Missions	Dates
Soyuz 6, 7, and 8	Triple flight in August 1969
Soyuz 9 and 10	Docking flight in October 1969
Soyuz 11 and 12	Docking flight in February 1970

Apart from rendezvous and docking, the triple joint mission would have other important elements. A special unit named the *Vulkan* ("Volcano") was installed on Soyuz 6 (spacecraft no. 14) to allow its crew to carry out a complex series of welding operations in conditions of microgravity and vacuum. The Ye. O. Paton Institute for Electro-Welding based at Kiev had developed the unit on a contract handed out during the Korolev era. Cosmonaut Fartushniy, a scientist from the institute, had been slated to fly the Vulkan unit into space, but by April, he had been moved from Soyuz 6 to Soyuz 11, evidently because of mass constraints when the crew size was reduced from three to two.[9] Additional instrumentation on Soyuz 6 included the *Svinets* apparatus, a military experiment for detecting and identifying the plumes from ICBM launches. The triple ship experiment would have a record seven cosmonauts flying in space simultaneously, most of whom had been training in various capacities on the piloted circum-lunar and landing projects during the previous two years.[10] The two final docking missions—Soyuz 9/Soyuz 10 and Soyuz 11/Soyuz 12—would include at least one very long-duration mission to reclaim the absolute endurance record for a space mission, held for almost four years by NASA's Gemini VII mission. These four missions would also use the long-delayed Kontakt rendezvous system.[11]

Mishin discussed these plans with Ustinov during a meeting on June 7, 1969, but the possibility of carrying out the triple Soyuz mission quickly gained a new urgency after the second catastrophic blow to the Soviet space program in eight months, the Apollo 11 landing. Once the inevitable delays crept into the ambitious Soyuz plan, Soviet space program leaders began to get cold feet. In late September, less than two weeks before the projected launches, Chief Designer Mishin met again with Ustinov to discuss preparations for the triple mission. Mishin noted in his personal office notes that "there is a fear in taking decisions."[12] Ustinov forbade Mishin to begin propellant loading of the boosters and spaceships, despite the latter's protest to adhere to the original program. Ustinov told Mishin that the final decision to proceed with

9. Fartushniy was to have flown in the third seat on Soyuz 6. See Rex Hall, "Soviet Civilian Cosmonauts," in Michael Cassutt, ed., *Who's Who in Space: The International Space Year Edition* (New York: Macmillan, 1992), pp. 290–91.

10. On April 7, 1969, the planned crews for the three ships were G. S. Shonin/V. N. Kubasov (Soyuz 6), A. V. Filipchenko/V. N. Volkov/V. V. Gorbatko (Soyuz 7), and A. G. Nikolayev/V. I. Sevastyanov (Soyuz 8). The back-up crewmembers were A. P. Kuklin, G. M. Grechko, and P. I. Kolodin. The crews began training for the missions on April 10, 1969.

11. The crews for the last four missions were (on April 7, 1969) Ye. V. Khrunov/A. S. Yeliseyev (Soyuz 9), A. P. Kuklin/G. M. Grechko (Soyuz 10), V. A. Shatalov/V. G. Fartushniy (Soyuz 11), and G. S. Shonin/V. A. Yazdovskiy/V. I. Patsayev (Soyuz 12).

12. Interview, Peter Gorin by the author, November 18, 1997.

the launches would be discussed at the Politburo level, an unusual state of events for a space launch. It is quite likely that Soviet leaders such as Brezhnev and Kosygin were extremely sensitive to the possibility of a catastrophic failure in the Soviet space program so soon after Apollo 11; such a mission would also once again raise the question of the direction of the Soviet space program. How were officials to answer to the obvious comparisons with Apollo?

On September 29, Mishin spoke with Ustinov, Smirnov, and Afanasyev. The chief designer had already received permission to begin fueling the first Soyuz, spacecraft no. 14, but was still awaiting approval to move ahead with prelaunch preparations for vehicle nos. 15 and 16. The Politburo met a day later and finally granted permission to carry out the triple flight. The mission would be touted as a major step in the creation of Earth-orbital stations, the "true calling" of the Soviet space program. The activity leading up to the launches was further intensified by major changes in the crew complement of the three Soyuz vehicles. Originally, the third Soyuz—the active vehicle during the docking exercise—would have been crewed by cosmonauts Nikolayev and Sevastyanov. Colonel Nikolayev, the veteran from the Vostok days, would also serve as the overall commander of all seven cosmonauts in space. Unfortunately for him, he had performed poorly during a preparatory exam in late July 1969.[13] Perhaps expecting an improvement in his abilities, planners continued to maintain the original crew complements until September 17, when Mishin and Kamanin agreed to replace the Nikolayev-Sevastyanov crew with a new two-cosmonaut crew fresh off their own recent spaceflights: Shatalov and Yeliseyev. Shatalov, of course, had the distinction of being the only Soviet cosmonaut who had actually carried out a docking in space, and his inclusion in the crew for the third Soyuz was probably a boon to confidence. A final decision on the crew replacement was taken in early October, after all the primary and backup cosmonauts for the three ships had arrived at the Baykonur Cosmodrome.[14]

Apart from the uniqueness of having three Soyuz ships in orbit at the same time, the joint flight would also mark a significant expansion of Soviet communications capabilities. Transmissions were normally limited to flight over the Soviet landmass or with a small flotilla of modest seafaring vessels under the control of the Department of Naval Expeditionary Work under the Academy of Sciences since 1967. That same year, the Soviets began the construction of the first of a new generation of vastly improved tracking ships. The first of these, with a displacement of 17,850 tons, was the *Kosmonaut Vladimir Komarov*, a Poltava-class dry cargo vessel that was converted to its new role at Leningrad in 1967. The 121-strong crew and 118-member science team were three and seven times larger, respectively, than predecessors such as the *Dolinsk*. The prominent features of the *Kosmonaut Vladimir Komarov* were the unusual hull sponsons and the massive plastic radomes, which enclosed huge antenna arrays for tracking and communications. For the Soyuz program, the ship would serve as one node of a communications bridge, from the Soyuz spacecraft, to the *Kosmonaut Vladimir Komarov*, to Molniya-1 satellites in Earth orbit, to the NIP-16 Flight Control Center at Yevpatoriya. The ship's first active role during a piloted mission had been on the Soyuz 4/5 docking flight, although it had provided support during the circumlunar Zond 5 mission when it had been stationed at Havana.

---

13.   Gordon Hooper and Bert Vis, "Meeting the Space Explorers: Vitali Sevastyanov," *Spaceflight News* (January 1991): 34–36.

14.   I. Marinin, "Russia. The Extraordinary Incidents of the 'Vulkans'" (English title), *Novosti kosmonavtiki* 17 (August 12–15, 1996): 22–25. Nikolayev and Sevastyanov, meanwhile, were consigned to serving as backups for the mission. The backups were A. G. Nikolayev/G. M. Grechko (Soyuz 6), A. G. Nikolayev/G. M. Grechko/P. I. Kolodin (Soyuz 7), and A. G. Nikolayev/V. I. Sevastyanov (Soyuz 8). Note that the original backup crews were different and included at various points as crew commander both A. P. Kuklin (who was dropped because of health problems in July 1969) and Ye. V. Khrunov (who was penalized in July 1969 for being involved in a hit-and-run automobile accident during which he had not come to the aid of the victims).

704

As the *Kosmonaut Vladimir Komarov* entered duty in August 1967, even larger vessels were on the drawing board—ones capable of controlling both Earth-orbital and deep space missions.[15] All of these served to significantly expand communications-link times for piloted missions.

The architect behind much of the radio-tracking and communications equipment on these ships was Chief Designer Mikhail S. Ryazanskiy of the Scientific-Research Institute for Radio Instrument Building (formerly NII-885). One of the original members of Korolev's old Council of Chief Designers from the 1940s, he also had a very interesting career. Obsessed with building radios since he was a child, in the late 1920s, Ryazanskiy became a radio technician and a leading member of the Young Communist League at Nizhniy Novgorod (or Gorkiy). It was there that he came under the suspicion of the Soviet secret police, having been accused of destroying important equipment. Incriminating evidence that his grandfather had been a priest, an "unacceptable" heritage for any Communist Party member at the time, bolstered the absurd charges. With the support of many of his coworkers, a possible death sentence

Chief Designer Ryazanskiy was one of the six original members of the Council of Chief Designers. His organization, originally called NII-885, was responsible for all radio-control guidance systems for Soviet ballistic missiles and spacecraft. (files of Peter Gorin)

was commuted to one month's hard labor. Rising through the ranks, Ryazanskiy eventually made important contributions to the Soviet wartime effort in radio and radar technology before joining the Moscow-based NII-885 as a chief designer in 1946 after the A-4 recovery operations in Germany.[16] Along with Korolev, Glushko, Pilyugin, Barmin, and Kuznetsov, Ryazanskiy completed the original Council of Chief Designers.

Ryazanskiy's career as a chief designer was briefly interrupted in January 1951 when he was appointed the chief *engineer* of NII-88—a position superior to Korolev at the time. The turned tables do not seem to have disrupted their own personal relationships. Ryazanskiy was promoted out of the missile design business to an administrative position in 1952 as chief of the Seventh Chief Directorate of the Ministry of Armaments under Ustinov, but in less than two years, he returned to his chief designer spot at NII-885, saying that "administrative work is not for me."[17] Back at the institute, life was not easy for Ryazanskiy. Secret police mastermind Lavrentiy P. Beriya had a particularly strong dislike for the chief designer because of his father's

15.   B. A. Pokrovskiy, *Kosmos nachinayetsya na zemlye* (Moscow: Patriot, 1996), pp. 347–48; *Soviet Space Programs: 1976–80: Supporting Vehicles and Launch Vehicles, Political Goals and Purposes, International Cooperation in Space, Administration, Resource Burden, Future Outlook*, prepared for the Committee on Commerce, Science, and Transportation, U.S. Senate, 97th Congress, 2d sess. (Washington, DC: U.S. Government Printing Office, December 1982), p. 127. During the Soyuz 4/5 mission, the *Kosmonaut Vladimir Komarov* had served in conjunction with two other older vessels, the *Morzhovets* and the *Nevel*. See G. S. Narimanov, ed., *Ot kosmicheskikh korabley - k orbitalnym stantsiyam* (Moscow: Mashinostroyeniye, 1971), p. 57.

16.   Col. M. Rebrov, "The Whiteness of Martian Seas . . .: Pages From the Life of the Chief Designer of Radio Control Devices" (English title), *Krasnaya zvezda*, March 11, 1989, p. 4.

17.   *Ibid.*

political leanings in the 1930s. Several people from NII-885 were, in fact, arrested in 1952–53 by Beriya's henchmen, while Ryazanskiy himself was charged with withholding evidence. His fate and possibly his life were saved by the deaths of Stalin and Beriya in 1953. Later, Ryazanskiy was instrumental in choosing the site of the Baykonur Cosmodrome, an action that would prompt Korolev to often grumble: "Mikhail is to blame for everything. He chose this God-forsaken hole. . . ."[18] The final ignominy Ryazanskiy had to face was in 1961, when all the original members of the Council of Chief Designers received their second Hero of Socialist Labor award—all except Ryazanskiy. As rumor had it, Ryazanskiy had been witness to one of Brezhnev's drinking binges around 1960. When the latter had offered the chief designer a cognac, Ryazanskiy disgustedly refused his offer. Brezhnev remembered this event when the awards were handed out for Gagarin's flight. Ryazanskiy's name was crossed off of the list and substituted with that of Brezhnev. At the time of the triple Soyuz mission, Ryazanskiy was sixty years old.

Troika

The first 11A511 booster with its Soyuz payload was moved to the pad at site 31 at Tyura-Tam on the morning of October 8 to begin its prelaunch processes. It would be an intensely active period for ground personnel; over a period of three consecutive days, Strategic Missile Forces troops would launch three different Soyuz stacks into orbit. Each spacecraft would remain in orbit for five days, all three overlapping for the middle three days. News about an impending Soviet space spectacular evidently leaked out of Moscow, with some press reports, on October 9, predicting the launch of three Soyuz spaceships that might be used for "building an orbital station."[19]

7K-OK spacecraft no. 14 lifted off on time at 1410 hours Moscow Time on October 11, 1969, with two rookie cosmonauts Lt. Colonel Georgiy S. Shonin (the commander) and civilian Valeriy N. Kubasov (the flight engineer), both thirty-four years old at the time. The spacecraft, named *Soyuz 6*, which was not equipped with a docking probe but did carry the small Vulkan apparatus in its living compartment, entered an initial orbit of 186.2 by 222.8 kilometers inclined at 51.68 degrees. It had been almost ten months since the last Soviet piloted mission. Among the objectives announced by the Soviet media were perfecting spacecraft control systems, testing navigational devices, carrying out Earth resources photography, investigating atmospheric phenomena, performing biomedical research, and experimenting with welding in vacuum and weightlessness.[20] It seems that the cosmonauts did not do much during their first day in orbit apart from a main engine firing on the fourth orbit at 2008 hours to change orbital parameters. Some minor activity on the fourteenth orbit involved Shonin carrying out navigational exercises using the astro-orientation system and automatic stellar sensor. Kubasov, meanwhile, tried out a new sextant, the SMK-4, whose measurements were compared with computations on the ground to verify the accuracy of the instrument. Kubasov later took photographs of the low-lying Caspian Sea coast and the Volga delta, forests in Central Russia, and cloud formations.[21]

Mounting rumors of more Soyuz launches were confirmed the following day, when 7K-OK spacecraft no. 15 lifted off from site 1 at Tyura-Tam at 1345 hours with not two, but three rookie

---

18. *Ibid.*

19. *Soviet Space Programs, 1966–70*, p. 377.

20. G. I. Petrov, ed., *Conquest of Outer Space in the USSR, 1967–70* (New Delhi: Amerind Publishing Co., 1973), pp. 117–18.

21. Peter Smolders, *Soviets in Space* (New York: Taplinger Publishing Co., 1973), p. 179.

cosmonauts. The ship, named *Soyuz 7*, entered an initial orbit of 207.4 by 225.9 kilometers at a 51.68-degree inclination to the equator. Aboard were Lt. Colonel Anatoliy V. Filipchenko (the commander), civilian Vladislav N. Volkov (the flight engineer), and Lt. Colonel Viktor V. Gorbatko (the research engineer). Filipchenko was forty-one at the time, while Volkov was thirty-three and Gorbatko was thirty-four. TASS announced the goals of the mission as including maneuvering in orbit, navigational investigations jointly with Soyuz 6 in group flight," and scientific research consisting of the observation of celestial bodies and Earth's horizon, the determination of the actual brightness of stars, and measurements of illumination by the Sun.[22] Naturally, there was no mention that the ship was equipped with a passive docking mechanism, nor that the spacecraft was to dock with a third Soyuz.

Preparations for the launch of 7K-OK spacecraft no. 16 had begun immediately after the launch of Soyuz 6 from the pad at site 31. Within two hours of launch, the new booster-payload stack was moved to the pad to begin *its* prelaunch operations. Once the two cosmonauts were settled into the descent apparatus of the spacecraft, Commander Shatalov ran into a minor problem while tightening the wheel on the hatch lock between the two Soyuz modules when one of its three spokes cracked under excess pressure. The crew reluctantly reported the problem to ground control, who advised that as long as pressure integrity was maintained, the problem would not hinder a timely launch.[23] Thus, within twenty-four hours of the launch of Soyuz 7, Strategic Missile Forces personnel launched the third Soyuz spacecraft in three days. The launch was at 1319 hours Moscow Time on October 13, 1969. Veteran cosmonauts Colonel Vladimir A. Shatalov (the commander), who was forty-one, and civilian Aleksey S. Yeliseyev (the flight engineer), who was thirty-five, entered an initial orbit of 204.5 by 223.7 kilometers at a 51.68-degree inclination. TASS announced that the new ship, named Soyuz 8, would carry out complex scientific observations with Soyuz 6 and Soyuz 7, including group flight and the even more general "joint orbital maneuvering to solve a number of problems connected with manned space flights."[24] TASS also reported that Shatalov would be in overall command of the three ships. Both he and Yeliseyev had the distinction of holding the record for the shortest turnaround for space missions, having flown in space less than ten months earlier.

Initially, after Soyuz 8 entered orbit, the three spacecraft carried out independent flight focused on their own experiments program, although several orbital corrections by all three ships on October 13 and 14 seemed to have been preliminary maneuvers to allow for the eventual intersection of their orbits. In general, the experiments program in orbit was divided up. The Soyuz 6 crew carried out biomedical research (such as inner ear tests) and Earth photography. The Soyuz 7 crew performed photography of Earth and stellar objects in differing spectral bands. The Soyuz 8 crew focused on research on the polarization of sunlight reflected by the atmosphere. Biomedical experiments included using "functional probes" and individual and group psychological tests to assess working capacity in orbit. Earth photography focused on the development of cyclones and the movement of storm fronts. The Soyuz 7 cosmonauts, in particular, conducted detailed remote-sensing exercises, including the study of geological areas to detect reserves of mineral raw materials. Soyuz 8 Flight Engineer Yeliseyev, like his compatriot Kubasov on Soyuz 6, also used a new SMK-4 sextant to determine orbital elements independently of help from ground stations. One major experiment involved the determination of reflective properties of forests, deserts, and other areas of Earth's surface. The crews remained in regular contact with each other and for the first time jointly used the Molniya-1 satellite

---

22. Petrov, *Conquest of Outer Space*, p. 123.
23. M. F. Rebrov, *Kosmicheskiye katastrofy: Russkiye sensatsii* (Moscow: IzdAT, 1993), pp. 43–44.
24. Petrov, *Conquest of Outer Space*, p. 129.

Here are the seven cosmonauts of the Soyuz 6/7/8 mission. Sitting from left to right are Valeriy Kubasov, Georgiy Shonin, Vladimir Shatalov, and Aleksey Yeliseyev. Standing from left to right are Viktor Gorbatko, Anatoliy Filipchenko, and Vladislav Volkov. (files of Peter Gorin)

system and the *Kosmonavt Vladimir Komarov*.[25] A military component of the Soyuz 6 mission was the *Fakel* ("Torch") experiment for visually detecting the launch plumes of ballistic missiles from orbit.[26] Evidently using the Svinets apparatus, Shonin later reported that he could clearly see special light projectors on ground targets and that the measurement of background illumination was not difficult. On three occasions on October 12, R-16 ICBMs were launched from Tyura-Tam while Soyuz 6 passed over the launch range. All the launches were at night, limiting the applicability of the experiment. It is unlikely that the Svinets instrument would have been capable of detecting launches during daytime.

25.   Older ships, such as the *Bezhitsa, Borovichi, Dolinsk, Kegostrov, Morzhovets, Nevel,* and *Ristna,* were also used for communications. See Evgeny Riabchikov, *Russians in Space* (Moscow: Novosti Press Publishing House, 1971), p. 273. For the general experiments program, see Smolders, *Soviets in Space,* pp. 181, 184; Riabchikov, *Russians in Space,* pp. 273–74; Kenneth Gatland, *Manned Spacecraft* (New York: Macmillan, 1976), pp. 143–45; Lardier, *L'Astronautique Soviétique,* p. 188; Narimanov, *Ot kosmicheskikh korabley,* p. 72.

26.   "In Memory of Cosmonaut G. S. Shonin" (English title), *Novosti kosmonavtiki* 7(March 24–April 6, 1997): 25–27.

By October 14, the three spacecraft were in a common orbit of roughly 200 by 225 kilometers at a 51.7-degree inclination. As planned, the Soyuz 7 and Soyuz 8 spacecraft approached each other to within a distance of 500 meters, while Soyuz 6 watched nearby. Docking between Soyuz 7 and Soyuz 8 had been planned to be semi-automatic, with the Igla system bringing the two ships to a distance of 100 meters of each other, after which Shatalov would take over manual control. As backup cosmonaut Sevastyanov recalled later, the ships did not come closer than 500 meters of each other:

> There was a mistake during the preliminary stage of the docking and the [Igla] radio system didn't work [on Soyuz 8]—it didn't give the information on where the second spacecraft was. They tried to use an optical channel, but at that time they didn't have a special laser device for measuring the distance, and they had no possibility to measure the distance between the two spacecraft.[27]

The "optical channels" were evidently bright light signals on the ships used at range distances of 1,500 meters and 500 meters. In two attempts to close in on Soyuz 7 manually from those distances, an increasingly stressed Shatalov on Soyuz 8 found it too difficult to measure the relative distance to the passive spacecraft while the ships were in Earth's shadow. The cosmonauts' frustrations were exacerbated by on-board indicators showing that the Igla system was completely operational. Recent reports indicate that one or more of the ships may also have been inserted into the wrong orbit, further complicating matters.[28] Because of the malfunctioning Igla system, the Soyuz 8 cosmonauts were unable to move close enough to Soyuz 7 to transfer to manual control and dock. As a last desperate move, ground control decided to try and maintain station-keeping between the two ships using only ballistics data transmitted from the ground. The docking attempt was rescheduled for the following day, October 15. Unfortunately, without the use of the Igla system, the cosmonauts were unable to bring the ships closer than 1,700 meters. The third ship, Soyuz 6, which did not carry the Igla system, was unable to independently complete any close approaches to the other two spacecraft.

That the mission was a complete mess was underlined in a U.S. intelligence report, which was declassified in 1997. The CIA wrote:

> The five rendezvous attempts made during the mission were all unsuccessful for several different reasons. The first failed because the automatic rendezvous system [that is, Igla] would not indicate radar lock-on between Soyuz 7 and 8. Two orbits later the first manual rendezvous attempt was made but it was broken off after Soyuz 8 used more than the authorized amount of attitude-control propellant. A second manual attempt, made the next day, failed because Soyuz 8 did not properly control its lateral velocity relative to Soyuz 7. The attempt by Soyuz 6 to carry out a cosmonaut-controlled rendezvous with the other two spacecraft failed because of insufficient time to correct for a three kilometer out-of-plane separation between it and the other vehicles. The final manual attempt at rendezvous and docking between Soyuz 7 and 8 was poorly timed and the vehicles could not establish the correct interval and relative velocity between them required for a docking operation before they entered the earth's shadow.[29]

27.   Hooper and Vis, "Meeting the Space Explorers: Vitali Sevastyanov," p. 36.
28.   "In Memory of Cosmonaut G. S. Shonin."
29.   U.S. Central Intelligence Agency, "National Intelligence Estimate 11-1-71: The Soviet Space Program," Washington, DC, July 1, 1971, p. 29, as declassified in 1997 by the CIA Historical Review Program.

According to official Soviet data, during three days of jointly coordinated flight, the ships completed thirty-one orbital maneuvers. Using Soyuz 7 as a target vehicle, Soyuz 6 and Soyuz 8 completed three and four close rendezvous, respectively. On two occasions, the approaches were simultaneous—that is, all three vehicles were in very close proximity for a total of four hours and twenty-four minutes of "co-orbiting." Soyuz 7 and Soyuz 8, meanwhile, spent as much as thirty-four hours and nineteen minutes "co-orbiting" with each other.[30] During these rendezvous exercises:

> The crews made observations of the other spaceships, took photographs, and used movie cameras to determine the visibility of objects at various distances. They also investigated the possibility of exchanging information by means of light indexes and visual optical devices.[31]

The exchanging of information was probably related to military experiments. A former CIA official later recounted that:

> The cosmonauts experimented with methods of communicating with each other and used light sources that could not be monitored by normal electronic intelligence listening devices. They also conducted experiments to determine the visibility of objects at various distances from their spaceships, which among other things is the type of information used by military planners for designing equipment for photographing and inspecting hostile satellites.[32]

No pictures taken during the mission have ever been published by the Soviet or Russian press in the thirty years since the mission. With the disappointments of the several failures behind them, Chief Designer Mishin had the unfortunate task of telephoning both Brezhnev and Ustinov to inform them of the situation.

It was on October 16 that cosmonauts Shonin and Kubasov on Soyuz 6 prepared for one of the main goals of the entire experiment, the welding exercise with the Vulkan unit. The instrument itself was a squat green cylinder resembling "a round refrigerator" with a mass of about fifty kilograms, installed in the living compartment of Soyuz 6. The object consisted of two sections, one of which contained various instruments and power sources, measuring and converter devices, and communications and automation equipment in a pressurized nitrogen atmosphere. The other section contained the welding devices. Scientists at the Paton Institute had painstakingly designed the unit based on extensive tests in vacuum chambers and on parabolic weightless flights in aircraft. On their seventy-seventh orbit, the Soyuz-6 cosmonauts shut the hatch between the descent apparatus and the living compartment and depressurized the latter module. Flight Engineer Kubasov, using remote-control switches, then turned on the welding unit, initiating three different methods. The system first performed a low-pressure compressed arc welding. This was followed by an attempt at electron beam welding. The final method was arc welding using a consumable electrode. The actual welding was performed using an electron gun with samples of titanium, aluminum alloys, and stainless steel. All the welding was automated, and the only major role of the crew was to turn on the system and recover the samples. Kubasov was, however, able to follow the work of the unit with a special

30.   Lardier, *L'Astronautique Soviétique*, p. 188.
31.   Riabchikov, *Russians in Space*, p. 273.
32.   Peter N. James, *Soviet Conquest From Space* (New Rochelle, NY: Arlington House Publishers, 1974), p. 116.

indicator panel in the descent apparatus, while data were also directly transmitted to ground stations.[33] Academician Paton later glowingly reported that:

> *The experiment in welding in orbit had opened a new page in the exploration of space. An engineering procedure involving the heating and melting of metal has been performed in space for the first time. The age of space metallurgy has dawned.*[34]

While much was made of the fact that welding would be a requisite for future orbital assembly operations in space, the Vulkan experiment was, in fact, a near catastrophe for the Soyuz 6 crew. Soviet authorities revealed twenty-one years later that "the welding experiment which was supposed to be carried out on one of the ships, ended unsuccessfully. They almost burned a hole in the ship."[35] During one of the three methods tested, possibly the low-pressure compressed arc, the Vulkan unit evidently incorrectly aimed a beam and melted the internal wall of the living compartment. The cosmonauts were apparently unaware of the danger during the experiment, and they only discovered the damage once the living compartment was repressurized to recover the samples of the experiment.[36]

Soyuz 6 returned to Earth almost as soon as the Vulkan exercise was over. The two cosmonauts landed at 1252 hours Moscow Time on October 16, 1969, in the frozen and barren steppes of Kazakhstan, 180 kilometers northwest of the town of Karaganda. Their mission had lasted four days, twenty-two hours, forty-two minutes, and forty-seven seconds. It was chilly cold with a powerful wind at the landing site, and despite landing twenty kilometers from the intended landing point, rescue services were able to reach the cosmonauts relatively quickly.

The Soyuz 7 and Soyuz 8 cosmonauts continued their missions in Earth orbit. The remainder of the mission was uneventful except for a malfunction on Soyuz 7 on October 17. One of three cosmonauts accidentally activated the automatic landing system display in the descent apparatus. The unit was supposed to turn on automatically at an altitude of eleven kilometers after reentry for use during the parachute descent. Because the display was to be used on the last leg of the mission, there was no provision to turn it off in orbit. Some ground controllers were concerned that if the display remained continuously turned on for more than a day, there might be a possibility of failure during descent.[37] With little to do to rectify the situation, the crew continued to orbit Earth with the system left active. The Soyuz 7 and Soyuz 8 crews carried out the perfunctory medical experiments and Earth photography exercises during the remainder of their missions before preparing to return to Earth. Soyuz 7 cosmonauts Volkov and Gorbatko, in particular, carried out complex spectrophotometry and photography of the twilight aureole of Earth, its clouds, and its underlying surface using the handheld RSS-2 spectrograph. The experiment was carried out on the spacecraft's eighty-seventh orbit over northeast Africa from an altitude of 218 kilometers. An earlier session on October 13 over the Arabian penin-

33. *Soviet Space Programs, 1966–70*, p. 237; Lardier, *L'Astronautique Soviétique*, p. 188; Gatland, *Manned Spacecraft*, p. 143; Riabchikov, *Russians in Space*, p. 274; Narimanov, *Ot kosmicheskikh korabley*, p. 73.

34. Riabchikov, *Russians in Space*, pp. 274–75.

35. German Nazarov, "You Cannot Paper Space With Rubles: How to Save Billions" (English title), *Molodaya gvardiya* no. 4 (April 1990): 192–207.

36. The inference that it was the low-pressure compressed arc that caused the problem is based on the premise that the Soviets at the time touted the success of the other two methods, but refrained from doing so for the compressed arc test. See Gatland, *Manned Spacecraft*, pp. 143–45, for positive evaluations of arc welding, and see *Soviet Space Programs, 1966–70*, p. 237, for the same for electron beam welding. See also "In Memory of Cosmonaut G. S. Shonin."

37. L. N. Kamanin, "Removing the Cosmetic Retouching: N. Kamanin—From His Journal Entries for 1970" (English title), *Sovetskaya kultura*, July 14, 1990, p. 15.

sula was coordinated with ground observatories and two specially equipped Li-2 scientific aircraft flying at altitudes of 2.7 kilometers.[38]

The three-cosmonaut Soyuz 7 crew returned to Earth without incident, landing safely 155 kilometers northwest of Karaganda at 1226 hours Moscow Time on October 17, 1969, almost exactly a day after Soyuz 6. Their mission had lasted four days, twenty-two hours, forty minutes, and twenty-three seconds. The weather was worse this time, with stinging cold winds of snow and sleet as well as low visibility. Soyuz 8 crewmembers Shatalov and Yeliseyev settled down a day later at 1210 hours Moscow Time on October 18, 145 kilometers north of Karaganda in a raging blizzard. The last crew had completed a mission lasting four days, twenty-two hours, fifty minutes, and forty-nine seconds. The triple Soyuz flight was over.

As much as the flight bewildered Western observers with its meandering nature and lack of docking, Soviet spokespersons went on the offensive after all three ships had touched down. They had had little to celebrate during the year, and the modest achievements of Soyuz 6, Soyuz 7, and Soyuz 8 would have to do. The cosmonauts' return to Moscow was made into a celebratory event of national proportions. As bands played and salutary guns fired, Communist Party and government leaders and thousands of Muscovites welcomed the seven men. At the ceremonial reception at the Kremlin Palace of Congresses, all the cosmonauts were awarded, like their predecessors, the title "Hero of the Soviet Union."[39] This occurred, despite the obvious failure to achieve the primary goal of the mission—the docking between Soyuz 7 and Soyuz 8—which was, of course, not announced as such. All Soviet press reports of the time clearly put forward the notion that docking had not been planned for the flight. As for the failure, the cosmonauts were exonerated of any wrongdoing during the mission. A thorough investigation that took three months proved that the failure in the Igla system had been caused by errors in ground preparations. When the Scientific-Research Institute for Precision Instruments had tested Igla on the ground for pressurization, engineers had used a 95-percent helium mixture. Investigators later discovered that this particular mixture harmed the radio components and thermostats of the flight units. After two more instruments from the same institute had failed in orbit by the end of 1969, engineers changed the mixture to either inert gases or a 5-percent helium solution.[40]

The postflight period for the triple Soyuz mission was particularly important because of the insistent and precise nature of Soviet statements on orbital stations. It finally seemed that the apparent confusion of the earlier part of the year on future prospects for the Soviet space program was finally over. Academician Sedov, the man who had made the infamous announcement on the launch of a Soviet satellite during the International Geophysical Year in 1955, told reporters in Peru in late October 1969 that the Soviet Union had never announced that it would send men to the Moon.[41] Fortunately for Sedov, no one bothered to read to him his pronouncements on the topic from earlier in the decade. Perhaps the most important public policy statement by a top Soviet figure emerged amid the celebrations for the Soyuz 6/7/8 mission. In a speech on October 22 at the Kremlin Palace of Congresses that retrospectively proved to be as important for the Soviet space program as Kennedy's speech in 1961 was for the United States, First Secretary of the Central Committee of the Communist Party Leonid I. Brezhnev made no bones about the "true direction" for the future Soviet cosmonaut:

38.   K. Ya. Kondratyev, A. A. Buznikov, B. V. Vinogradov, V. N. Volkov, V. V. Gorbatko, and O. I. Smotky, "Spectrophotometry of the Earth From Manned Spacecraft," in K. Ya. Kondratyev, M. J. Rycroft, and C. Sagan, eds., *Cospar: Space Research XI: Volume I* (Berlin: Akademie-Verlag, 1971), pp. 619–32.

39.   Riabchikov, *Russians in Space*, p. 276.

40.   Yu. A. Mozzhorin, *et al.*, eds., *Dorogi v kosmos: II* (Moscow: MAI, 1992), pp. 35–36.

41.   *Soviet Space Programs, 1966–70*, p. 378.

Our country has an extensive space program, drawn up for many years. We are going our own way; we are moving consistently and purposefully. Soviet cosmonautics is solving problems of increasing complexity. . . . Our way to the conquest of space is the way of solving vital, fundamental tasks, basic problems of science and technology. . . . Our science has approached the creation of long-term orbital stations and laboratories as the decisive means to an extensive conquest of space. Soviet science regards the creation of orbital stations with changeable crews as the main road for man into space. They can become cosmodromes in space, launching platforms for flights to other planets. Major scientific laboratories can be created for the study of space technology, biology, medicine, geophysics, astronomy, and astrophysics.[42]

He added a second thread—that of a Soviet space program working purely for improving the welfare of Soviet citizens: "Space for the good of people, space for the good of science, space for the good of the national economy. Such in brief, is the substance of the Soviet space program—its philosophical credo."[43] The implication was clear: while Americans were chasing the Moon with Apollo, an empty, politically motivated enterprise, Soviet cosmonauts were doing their all for the advancement of science and ultimately for the benefit of humankind. From the moment Brezhnev finished his speech, it was clear to most participants in the Soviet space program that the age of the space station had begun—an era that ultimately led to the *Mir* space station.

At a postflight press conference for the Soyuz 6/7/8 mission on November 4, Academy of Sciences President Keldysh stressed that Soviet efforts in space would focus on the creation of the first permanent orbital space station. The timeframe would "certainly be within ten years, and [probably] less than five years . . . literally in the nearest future."[44] On October 24, Keldysh told the Swedish press that "we no longer have any scheduled plans for manned lunar flights."[45] Commentators through the end of the year also repeatedly stressed the importance of cost in future planning, suggesting that automatic exploration of the Moon was far cheaper than piloted exploration. The suggestion was that the high cost of space exploration had forced a redirection in the overall effort.[46] All this worked to neutralize the success of Apollo. In one of the more bold pronouncements of the period, *The New York Times* claimed in a page-one story in late 1969 that:

according to some observers in Washington and some American scientists, the Russians may never have had a high-priority goal and timetable for a lunar landing in the same sense as the Apollo project's commitment to land men on the Moon in this decade.[47]

---

42. This excerpt from his speech is a slightly modified version of that published in *ibid.*, p. 378. Some corrections have been added based on the excerpts in James F. Clarity, "Brezhnev Says Soviet is Following the 'Main Road' in Space," *New York Times*, October 23, 1969, p. 20.

43. Riabchikov, *Russians in Space*, p. 278.

44. Bernard Gwertzman, "Soviet Expert Predicts Space Station in 5 Years," *New York Times*, November 5, 1969, p. 16. There was an amusing exchange at the press conference that was not reported in the West. Upon being asked by a U.S. reporter whether the Soviets were preparing to send a man to the Moon, Keldysh replied confusingly, "I think the Moon has to be sent to the man." The audience burst into laughter, but it took Keldysh a long time to realize why the audience was laughing. It was only when Shatalov prompted him that the academician tried to correct himself, but he did it so clumsily that there was more laughter. See Kamanin, "I Feel Sorry for Our Guys," no. 14.

45. John Noble Wilford, "Soviet Apparently Drops Plan to Put Men on Moon," *New York Times*, October 26, 1969, pp. 1, 11.

46. See, for example, Bernard Gwertzman, "Soviet Curbs Space Work for Economy," *New York Times*, December 31, 1969.

47. Wilford, "Soviet Apparently Drops Plan to Put Men on Moon," p. 1.

From an outside perspective, the direction of the Soviet space program seemed simple. While the Soviets may have been looking to compete with Apollo in the early 1960s, they abandoned that goal early, perhaps around 1964–65, and had then focused only on the development of an Earth-orbital space station. For almost twenty years, this would indeed be the dominant paradigm in understanding Soviet motives during the 1960s and 1970s.[48]

If Westerners proved to be easier to convince of Soviet intentions, the USSR's own citizens proved less gullible. A Moscow-based journalist, recalling the Brezhnev speech, wrote with sarcasm in 1990:

> *Orbital stations at that time did not represent an end itself, but a political response. Following the spectacular lunar landing by Neil Armstrong and Edwin Aldrin in July 1969, Brezhnev was obliged to come up with an alternative space project to save face, as well as the badly tarnished myth of Soviet superiority in space. He was told about an alternative. Brezhnev mentioned the U.S. success in reaching the moon and said that "we are following a different course, which is consistent and purposeful." Designers, cosmonauts, and thousands of other people probably laughed up their sleeves, knowing full well that the General Secretary was lying.*[49]

USSR Academy of Sciences President Mstislav Keldysh's scientific, managerial, and advisory contributions to the Soviet space program were matched by only a few individuals during the Soviet era. Keldysh also had the distinction of being one of the few high-ranking individuals in the space program whose identity was public knowledge. (files of Peter Gorin)

Brezhnev's pronouncements notwithstanding, in reassessing the trajectory of the Soviet piloted space program in 1969, a few questions come to mind. Did the Soviets really abandon their piloted lunar program in 1969? In other words, was the space station option put forward as a substitute or a complement to the lunar program? Why space stations? As with most policy issues in the Soviet space program, the answers to these questions are not simple, nor can they be isolated from the myriad of programs and proposals dating from the Korolev era.

The Space Station Arrives

By the late spring of 1969, Soviet space officials had already decided on three options available for a suitable response to Apollo, prompted by the stunning success of Apollo 8 in December 1968. These options were a piloted mission to Mars, the modification of the N1-L3

---

48. See, for example, Nicholas Daniloff, *The Kremlin and the Cosmos* (New York: Alfred A. Knopf, 1972), pp. 153, 164; William H. Schauer, *The Politics of Space: A Comparison of the Soviet and American Space Programs* (New York: Holmes and Meier, 1976), pp. 164–78;
49. Leonard Nikishin, "Inside the Moon Race," *Moscow News*, April 11, 1990, p. 15.

program for extended visits to the Moon, and the creation of Earth-orbital stations. Although Brezhnev's speech served to move the third option into the forefront, the Soviet space establishment did not give up the other two options in late 1969. In fact, if funding was any indication, money for the N1-L3 piloted lunar program reached a peak in appropriations for 1970, about $1.8 billion, a year after Apollo 11.[50] While there was certainly a state commitment for the lunar landing program well past 1969, as well as a modicum of interest in the Mars project, the space station program seems to have offered the quickest return. Ustinov, Smirnov, and Afanasyev needed something big, perhaps as early as 1970. Neither the N1-L3 nor any proposed Mars expedition would be ready by then. Space stations were seen as an acceptable alternative.

As with most Soviet space projects of the period, there was another external factor. The U.S. Department of Defense had forged ahead with the Manned Orbiting Laboratory for the latter part of the 1960s, but that program had been canceled in May 1969. On the civilian side, NASA had been studying space station options almost since its birth in 1958, and in 1965 these studies evolved into the Apollo Applications Program—a project that would make maximal use of Apollo hardware to build a modest space station in Earth's orbit. In July 1969, NASA selected a final design for the project, a "dry workshop" based on an upper stage of the Saturn V booster. A month later, the space agency "definitized" a contract with McDonnell Douglas to build the station, renamed Skylab in February 1970.[51] The station was expected to be ready for launch by mid-1972. Afraid of losing another race in space, Ustinov did not want to react with too little too late.[52]

In some ways, the space station option was one hoisted upon Soviet space engineers. Many in the upper echelons of the Soviet space industry, having invested almost ten years on the N1-L3 lunar program, were reluctant to see it consigned to second place behind some hastily put together space station program. TsKBEM Deputy Department Chief and veteran cosmonaut Feoktistov hinted later at the discord brewing within the design bureau.

In the 1960s it was clear to us engineers that the most important development for manned flights would be the creation of orbital space stations, but the administration was against it. Mishin, the Design Bureau Chief, was totally opposed to this. He thought that it was important to carry on with the Moon program. Everything else was nonsense and not worth doing.[53]

The debate over the space station versus the Moon program split the design bureau into opposing factions, and in a few years, this small fracture in unity would ultimately lead to cataclysmic consequences. But even as early as 1969, the "pro space station" group had powerful supporters in highly placed positions and managed to pull the right strings. Feoktistov later described how his faction managed to influence the content of Brezhnev's famous October 1969 speech:

50. V. P. Mishin, "Why Didn't We Fly to the Moon?" (English title), *Znaniye: tekhnike: seriya kosmonavtika, astronomiya* no. 12 (December, 1990): 3–43. The amount in Soviet currency, according to Mishin, was 600 million rubles. The total appropriations for the N1-L3 program up to January 1, 1971, was 2.9 billion rubles, or roughly $8.7 billion.

51. Linda Neuman Ezell, *NASA Historical Data Book, Volume III: Programs and Projects 1969–1978* (Washington, DC: NASA Special Publication (SP)-4012, 1988), pp. 98–100; Roger D. Launius, *NASA: A History of the U.S. Civil Space Program* (Malabar, FL: Krieger Publishing Co., 1994), pp. 97–98.

52. The Skylab option as a rationale for the Soviet space station program is mentioned in Semenov, ed., *Raketno-Kosmicheskaya Korporatsiya*, p. 264.

53. "The Russian Right Stuff: The Dark Side of the Moon," *NOVA* television show, #1808, WGBH-TV, Boston, February 27, 1991.

We didn't know how to get the bosses to change their minds, but some well-wishers in the Party Central Committee cunningly inserted a passage into Brezhnev's speech saying that orbital stations promised the right way forward.[54]

While the identity of the "well-wishers" remain undisclosed, one of them was probably Dmitriy F. Ustinov, who, unhappy about the results of the lunar program, apparently wanted some immediate results from an aimless space program.[55] He also had his own reputation to protect. As the secretary of the Central Committee for Defense Industries and Space, he was directly responsible for the Soviet space program. When his boss Brezhnev announced the space station as the "main road into space," it cemented the pro space station faction's position. The N1-L3 program would, of course, continue, as would work on a Mars project, but results from the new option were expected in 1970 or 1971.

Since the early 1960s, the late Korolev had tasked engineers at his design bureau to explore the possibility of designing what was generically called the Heavy Orbital Station (TOS). Reportedly nicknamed Zvezda, work on the proposal continued throughout the 1960s with neither official sanction nor much financial support.[56] Diverted by more pressing programs such as Soyuz and eventually the N1-L3 effort, it seems that Korolev had viewed the idea as one left for fruition during the 1970s.

A special subdivision of the Korolev design bureau studied several different variants of the TOS during the 1960s, from relatively small designs to giant space stations. One small space station design consisted of three floors: the living quarters, a controlling compartment, and an airlock chamber. One end of the station had a multiple docking adapter for four visiting Soyuz-type spacecraft. In this variant, the TOS was six meters in length, just under three meters in diameter, and cylindrical in shape, with the floors akin to "slices" along the longitudinal axis. A mock-up of the station was built in assembly shop no. 444 at the Experimental Machine Building Plant at Kaliningrad, the very same site where workers assembled Soyuz ships.[57] Another similar conception, also apparently built, had four floors. The floors were for lockers and "cupboards," for a crew compartment with a kitchen and toilets, for a laboratory and a control post, and for a multiple docking unit for five visiting spacecraft. The docking unit would also serve as an airlock adapter for performing EVAs.[58] By 1969, as space stations began to assume a more crucial role in the future of the Soviet space program, a group at TsKBEM began work on a much more ambitious version of the TOS, a 100-ton behemoth to be launched into Earth orbit by the N1 rocket. The station proper was a cylinder twenty meters long and six meters in diameter. Four Soyuz spacecraft could dock at a special multiple docking section at one end of the station, each node angled at thirty degrees to the main axis of the vehicle, giving the entire station the look of an arrow with feathers.[59] None of the TOS conceptions went beyond exploratory studies. As one Soviet space historian later recalled, "Korolev assumed that he would be able to realize [the] notion of a manned station, but he was so overloaded with other work, he wasn't able to do it."[60]

54. *Ibid.*
55. Kamanin suggests that it was Ustinov, Smirnov, and Keldysh who were instrumental in "putting these words into [Brezhnev's] mouth." See Kamanin, "I Feel Sorry for Our Guys," no. 14.
56. The designation Zvezda is from Yaroslav Golovanov, *Korolev: fakty i mify* (Moscow: Nauka, 1994), p. 768.
57. V. M. Petrakov, "Soviet Orbital Stations," *Journal of the British Interplanetary Society* 47 (September 1994): 363–72.
58. C. Wachtel, "The Chief Designers of the Soviet Space Program," *Journal of the British Interplanetary Society* 38 (December 1985): 561–63.
59. Semenov, ed., *Raketno-Kosmicheskaya Korporatsiya*, p. 278.
60. I. B. Afanasyev, "Unknown Spacecraft (From the History of the Soviet Space Program)" (English title), *Novoye v zhizni. Nauke, tekhnike: Seriya kosmonavtika, astronomiya* no. 12 (December 1991): 1–64.

The 100-ton variant of the TOS, dating from 1969, may have been a part of a much larger conceptual design that had slowly evolved at TsKBEM throughout the late 1960s. Around 1965, Korolev had approved exploratory studies of an integrated large modular space station in Earth orbit, very much similar to the ideas of Tsiolkovskiy and Oberth from the early part of the century. Designated the Multirole Space Base-Station (MKBS), it would be part of the larger Multirole Orbital Complex (MOK). Korolev had evidently entrusted this early work on the MOK to First Deputy Mishin, who continued to pursue the topic once he had become chief designer after Korolev's death.[61] Work on the MKBS involved not only the main design bureau, but also TsKBEM's branch at Kuybyshev under First Deputy Chief Designer Kozlov. Discussions during the post–Apollo 8 period had focused on the MOK/MKBS as a possible vehicle for responding to the success of Apollo. Some officials at the time suggested integrating defense goals into the effort, perhaps to elicit some interest from the Ministry of Defense to fund the endeavor. In early August 1969, soon after the Apollo 11 mission, Ustinov had expressly ordered Mishin to accelerate work on the MKBS.

While the MOK/MKBS was an attractive long-term option, it suffered from the same limitations in time as piloted Mars missions and an expanded lunar landing project: the earliest possible flight would not be until the mid-1970s at best. Keeping the MOK/MKBS as a future proposition, Ustinov instead turned his attention to existing hardware to bring his space station idea to a realistic conclusion. At the end of 1969, the Soviets had two modest space station programs in progress, although neither had any actual hardware to fly in space. Both were primarily military in nature, and they were products of two different design bureaus. The smaller of the two was TsKBEM's Soyuz-VI station, consisting of the OB-VI block, which was about the size of a Soyuz spacecraft, and a ferry vehicle, the 7K-S, a variant of the basic 7K-OK Soyuz modified for internal crew transfer into the OB-VI block. Under Deputy Chief Designer Okhapkin's control, the design bureau had already issued the complete design documentation for facilitating a program of experimental work on the station.[62] Early plans to launch the Soyuz-VI in 1969, however, proved to be too optimistic. Given Mishin's lukewarm support for creating the OB-VI, it was not surprising that delivery dates for flight-ready articles had been pushed back into 1970. Mishin was much more supportive of the 7K-S Soyuz ferry, arguing at many meetings in 1969 that the Ministry of Defense increase funding support for the project. Touted as an improved and more reliable version of the trouble-prone Soyuz, he believed that it was important that the 7K-S be introduced into service as quickly as possible.

Going through the list of options, Ustinov was not particularly enthused by the Soyuz-VI as an appropriate response to Apollo. What the Soviet image needed was something more substantial, something more "thick." And Ustinov found his "thick" solution not in Mishin's hands, but in the empire of General Designer Vladimir N. Chelomey. Since about 1966, Chelomey's TsKBM had been engaged in the development of the Almaz space station complex, aiming provisionally to launch the first completed product into orbit by the 100th birthday of V. I. Lenin on April 22, 1970.[63] For the most part, progress on the project had been steady. By late 1969, work on the actual hull of the station and certain service systems was on schedule, although there were major delays in some of the internal instrumentation. As of 1970, Chelomey's engineers had built the hulls of eight test stand units and two flightworthy vehicles. At the same time, ground testing of the control system, solar panels, and some of the

61. The MKBS and the MOK were mentioned at a meeting in late 1965 to discuss changes in the 1966–70 five-year plan for space exploration. See V. Denisov, "The Last Lesson" (English title), *Aviatsiya i kosmonavtika* no. 12 (December 1991): 40–43. Korolev prepared notes on a tactical-technical requirement for the MKBS on September 30, 1963.

62. Semenov, ed., *Raketno-Kosmicheskaya Korporatsiya*, p. 211.

63. Lardier, *L'Astronautique Soviétique*, p. 189.

station's other components was under way.[64] A small group of cosmonauts had been training for the Almaz program from as early as December 1967, and by 1969, four crews had been formed for the first flights to the station, headed by veteran commanders Belyayev, Popovich, Volynov, and Gorbatko.[65] Even more impressive, taking a page from Mishin's book, Chelomey had dozens of civilian engineers from his organization screened for cosmonaut training. Three of those passed tests and began further training in 1969 in anticipation of the formal selection process by the State Interdepartmental Commission, the body with the final word on selecting cosmonauts in the Soviet Union.[66]

The Almaz option was ideal for Ustinov's push to get a space station into orbit as soon as possible—ideal except for two major problems. First, there was the lag in developing and testing the Almaz's "auxiliary" systems, such as control and guidance systems, power supplies, and so forth. There were conflicts with the military in sharing instrumentation on the station, which also contributed to delays in configuring and delivering on-board systems. Chelomey was trying his best, but he expected the problems with the systems to put a wrench in the works and delay a launch to early 1972 or late 1971 at best.[67] Second, Ustinov despised Chelomey. Having opposed Chelomey's plans at critical junctures throughout the 1960s, it would put Ustinov in an awkward position if, of all people, it was Chelomey who would chalk up a victory for the Soviet space program.

In late 1969, Ustinov began wholeheartedly supporting an unthinkable, but typically brilliant solution: why not have Mishin's design bureau use one of the almost-finished Almaz units, complete it with instrumentation from the Soyuz, and then launch it into space, all within one year?[68] There is still some confusion on the source of this idea. Some attribute it to Ustinov and some to a group of Mishin's subordinates at his design bureau. One common story is that three leading deputy chief designers at TsKBEM—Bushuyev, Chertok, and Okhapkin—in alliance with three important department chiefs—Feoktistov, Kryukov, and Raushenbakh—approached Ustinov with a proposal to use elements of the Almaz orbital station re-equipped with the auxiliary systems that had already been tested in orbit on the Soyuz spacecraft. In addition, they would build a delivery vehicle, a modified Soyuz named the 7K-T, specifically to serve as a ferry to and from the station. According to Bushuyev and the others, a preliminary analysis had evidently showed that the idea was not only feasible but could be fulfilled in the shortest time.[69] According to one source, Mishin, who wanted to maintain the N1-L3 lunar program as the primary focus of his organization, was bypassed in these initial discussions in late 1969, being on holiday at Kislovodsk at the time. Possibly, this was not a coincidence, and Mishin's deputies may have taken advantage of the chief designer's absence to solidify the

---

64.   Afanasyev, "Unknown Spacecraft"; Vladimir Polyachenko, "The 'Pep' of Almaz" (English title), *Krylya rodiny* no. 1 (January 1992): 18–19..

65.   Polyachenko, "The 'Pep' of Almaz." Another source suggests that there were three commanders in 1969: P. R. Popovich, V. D. Shcheglov, and O. A. Yakovlev. See E-mail correspondence, Sergey Voevodin to the author, January 30, 1997.

66.   The three candidates from TsKBM were A. A. Grechanik, V. G. Makrushin, and D. A. Yuyukov. See Nina Chugunova, "Chelomey's Cosmonauts: Why There Are No Crews From NPO Mashinostroyeniya in Outer Space" (English title), *Ogonek* 4–5 (January 1993): 24–29.

67.   Petrakov, "Soviet Orbital Stations"; Chugunova, "Chelomey's Cosmonauts"; Semenov, ed., *Raketno-Kosmicheskaya Korporatsiya*, p. 264.

68.   Dmitriy Payson, "Without the 'Secret' Stamp: *Salyut*' and Star Wars" (English title), *Rossiyskiye vesti*, November 21, 1992, p. 4.

69.   Semenov, ed., *Raketno-Kosmicheskaya Korporatsiya*, p. 264. In one source, the idea for using the Almaz as a basis for the new station is attributed to Mishin himself, but given later events, this is extremely unlikely. See S. A. Zhiltsov, ed., *Gosudarstvennyy kosmicheskiy nauchno-proizvodstvennyy tsentr imeni M. V. Khrunicheva* (Moscow: RUSSLIT, 1997), p. 74.

"pro space station" contingent within the design bureau. Ustinov was clearly supportive of the idea, not the least because it would be a big blow to Chelomey's indefatigable ambitions. As the ball started rolling on the idea, Chelomey was acutely aware that it was Ustinov who was the main sponsor to this latest blow against his empire.[70] At a meeting of TsKBEM senior staff on January 3, 1970, Ustinov offered his complete backing and ordered the preparation of a formal Communist Party and government decree on the matter.[71]

It may have been a brilliant idea for Ustinov, but implementing the concept proved to be a little more difficult. Ustinov did not want to deal directly with Chelomey's central organization, and thus he invited a subsidiary of Chelomey's design bureau, his Fili Branch, to the preliminary discussions with Mishin. This cooperation between two unlikely partners was, in fact, stipulated in Ustinov's initial order to Mishin to:

- Have the space station ready in a year to a year and a half
- Make maximal use of ready instrumentation from the Soyuz spacecraft
- Arrange with the chief of TsKBM's Fili Branch, Viktor N. Bugayskiy, concerning the participation of that branch in the new program[72]

TsKBM's Fili Branch had a long and distinguished history in the Soviet aviation, rocketry, and space industries. In the 1950s, it had been an independent design bureau (OKB-23), headed by the famous Chief Designer Myasishchev, and had built some of the most famous long-range bombers for the Soviet Air Force. Among its more ambitious, albeit unrealized, achievements was the conceptualization of one of the Soviet Union's first spaceplanes, the M-48, as well as an intercontinental cruise missile, the Buran. After it was subordinated to Chelomey's design bureau in 1960 as Branch No. 1, the organization slowly shifted its design focus to ICBMs and space launch vehicles. Under Chelomey's general leadership, the branch created the UR-200 ICBM (later canceled), the UR-100 ICBM, and the UR-500 (Proton) launch vehicle.[73] All of these rockets were manufactured at the massive M. V. Khrunichev Machine Building Plant, collocated with the Fili Branch in Moscow.

Detailed discussions on the cooperation between TsKBEM and TsKBM's Fili Branch took place in January 1970 at Bakovskiy near Moscow, where Mishin was on holiday at the time. Ustinov evidently presided over the negotiations, which were attended not only by Mishin and Bugayskiy, but also the director of the Khrunichev Plant, Mikhail I. Ryzhikh. It was then that "basic questions were solved about the joint work of the three organizations in the development and creation of the orbital station."[74] There were also exchange visits among the three entities. On January 4, Mishin visited the Khrunichev Plant, while the following day Bugayskiy and Ryzhikh returned the favor by visiting Mishin's design bureau at Kaliningrad. Ustinov completely excluded Chelomey from the negotiations, despite the fact his First Deputy, Bugayskiy, was an essential participant in the talks. The discussions culminated with a decree (no. 105-41) of the Central Committee and the USSR Council of Ministers dated February 9,

70. See Chugunova, "Chelomey's Cosmonauts," for Chelomey's reaction upon hearing of the idea and his suspicions of Ustinov.
71. The preparation of the decree was entrusted to A. I. Tsarev (VPK), K. A. Kerimov (MOM), and K. D. Bushuyev (TsKBEM).
72. Petrakov, "Soviet Orbital Stations."
73. Zhiltsov, ed., *Gosudarstvennyy kosmicheskiy*, pp. 56–65. By 1970, it had already begun the development of two modifications of the UR-100 ICBM, designated the UR-100M and the UR-100K. Note that the UR-500 Proton had begun development as an ICBM-with-orbital-weapons delivery system.
74. Petrakov, "Soviet Orbital Stations."

1970, which called for the development of a new space station complex, the DOS-7K.[75] "DOS" stood for "Long-Duration Orbital Station" and represented the station proper, while the 7K denoted the Soyuz ferry vehicle. In later years, it would publicly be known first as *Salyut* and later as *Mir*. Apart from formally approving the project, the decree also stipulated the transfer of an already manufactured hull of Chelomey's Almaz station to the hands of Mishin's engineers. The latter, in cooperation with people under Bugayskiy and Ryzhikh, would reequip the Almaz to create the DOS vehicle.[76]

By the time that the Soviet leadership issued a formal decree on the DOS, the leaders of the relevant organizations had already shuffled their priorities to bring a high priority to the program. By late December 1969, Bugayskiy's Fili Branch had established a group of "lead designers" for the orbital station project headed by Vladimir V. Pallo, which included veterans of the group that had designed the Proton booster.[77] At Mishin's design bureau, the senior staff had proposed the appointment of thirty-four-year-old Yuriy P. Semenov as the "lead designer" of the DOS-7K complex, a position that gave him direct design control over the project. Semenov had served in the same capacity since May 1967 for the L1 circumlunar project, a remarkable distinction for such a young man. A clearly competent engineer, it was rumored that his rapid rise was owed in part to the fact that he was the son-in-law of Politburo member Andrey P. Kirilenko.[78] On February 4, Mishin handed out assignments on the DOS-7K project. As one would expect, most of the key assignments went to those who had proposed the project in the first place, including Bushuyev, Chertok, and Feoktistov.[79]

In Soviet terms, the pace and acceleration of the project were remarkable. By December 31, 1969, literally in the course of a few days, TsKBEM engineers prepared a document, "Basic Provisions for an Orbital Station," which was the precise origin of the DOS-7K design. In February 1970, the design bureau's Department No. 241 issued the technical plan for the DOS, with which the leadership of TsKBM's Fili Branch concurred. In early March, a group of engineers from TsKBEM, TsKBM's Fili Branch, and the Khrunichev Plant met for the first time to discuss the project and agreed on the basic requirements and direction of work.[80] The distribution of labor among the three enterprises laid the foundation for a cooperation that

75.   *Ibid.;* Semenov, ed., *Raketno-Kosmicheskaya Korporatsiya,* p. 267; Zhiltsov, ed., *Gosudarstvennyy kosmicheskiy,* p. 75.

76.   A subsequent decree (no. 57ss) of the Ministry of General Machine Building (MOM) dated February 16, 1970, also specified more details of each side's participation in the project. See Zhiltsov, ed., *Gosudarstvennyy kosmicheskiy,* p. 75.

77.   *Ibid.;* G. Amiryants, "Ivensen's 'Chayka'" (English title), *Aviatsiya i kosmonavtika* no. 4 (April 1990): 36–38; Andrey Tarasov, "Space Science of the Future: Selection of Paths and Orbits" (English title), *Pravda,* May 17, 1990, p. 3.

78.   Semenov's official appointment as lead designer of the DOS-7K complex, dated January 20, 1970, has been reproduced in full in Semenov, ed., *Raketno-Kosmicheskaya Korporatsiya,* p. 265–66. For a biography of Semenov, see K. Lantratov, "Yu. P. Semenov (on 60 Years)" (English title), *Novosti kosmonavtiki* 6 (April 9–22, 1995): 54–55. For the Kirilenko connection, see Roald Z. Sagdeev, *The Making of a Soviet Scientist: My Adventures in Nuclear Fusion and Space From Stalin to Star Wars* (New York: John Wiley & Sons, 1993), p. 180.

79.   The main assignments were: Yu. P. Semenov (lead designer for the DOS-7K complex), K. D. Bushuyev (chief of DOS-7K development), K. P. Feoktistov (deputy chief of DOS-7K development), P. V. Tsybin (lead designer for the 7K ship), L. A. Gorshkov (lead designer for the DOS orbital block), B. Ye. Chertok (chief of the guidance system), B. V. Raushenbakh (deputy chief of the guidance system), I. Ye. Yurasov (deputy chief of the guidance system), Ya. I. Tregub (chief of flight tests), V. I. Zelenshchikov (deputy chief of flight tests), and A. P. Abramov (chief of the ground complex, technical position, and fueling equipment).

80.   Semenov, ed., *Raketno-Kosmicheskaya Korporatsiya,* p. 266. Among those present were: Ye. A. Bashkin, E. K. Demchenko, K. P. Feoktistov, L. A. Gorshkov, A. A. Nesterenko, and Yu. P. Semenov from TsKBEM; V. N. Bugayskiy, G. D. Dermichev, Ya. B. Nodelman, and V. V. Pallo from TsKBM's Fili Branch; and B. G. Britkov, Ye. M. Kupryakov, M. P. Parfenov, and A. I. Tsimmerman from the Khrunichev Plant.

existed among the same three entities into the 1990s in the design, development, testing, and launch of the *Mir* space station and its various add-on modules. Never before had the Soviet space industry engaged in such a cooperative project that was primarily civilian in nature.

Mishin's TsKBEM worked on the overall design of the station, supplied almost all the complete systems, developed new systems for the station, ensured the launch and return of station crews, and had control over flights. It also manufactured the basic systems of the station and carried out preflight testing of the fully built station. Bugayskiy's TsKBM Fili Branch developed the layout of the station, carried out modeling, developed a small portion of the systems, issued the design documentation, supervised the manufacturing at the plant, and participated in the preparation of the station at the launch site. The Khrunichev Plant had already manufactured the pressure hull, manufactured new ones at its Building 160, and carried out the full assembly of the product.[81]

As soon as the official government decree was issued, the leading architects for the DOS—Bushuyev, Feoktistov, and Semenov—developed a simplified initial concept for the station, which was then delivered to Bugayskiy's team. At the basic level, the designers introduced four major modifications to Chelomey's Almaz station to turn it into the DOS:

- A new transitional compartment with a passive docking node, which forced a redesign of the forward bulkhead
- A truncated airlock compartment at the rear of the station with deletion of the associated passive docking node
- A new aggregate compartment at the rear of the station with a much smaller diameter than the rest of the station, which would contain the main engines
- New large solar panels installed like wings on the transitional and aggregate compartments (the old Almaz panels would be deleted)[82]

These initial changes to the Almaz station design were incorporated into a special wooden mock-up of the station built to specifications at the Fili Branch. More difficult was the actual appropriation of the several complete Almaz models, which Chelomey naturally was reluctant to give up. In March 1970, DOS lead designer Semenov for the first time met with Chelomey at the latter's offices in Reutov. The meeting was long and did not go very well; the proud Chelomey evidently gave Semenov an earful. The younger man invoked the recently passed Central Committee and Council of Ministers decree, but Chelomey refused to give in. It was only after personal intervention by Minister of General Machine Building Afanasyev that the matter was resolved. Chelomey capitulated and handed over four already-built hulls of the Almaz station to Mishin's engineers.[83] Ultimately, eight station hulls, associated equipment, and documentation were transferred to the DOS program. All of this was done via Chelomey's Fili Branch—that is, without going through the general designer. One of Chelomey's deputies recalled:

> The TsKBM Branch was instructed to hand over all blueprints related to the TsKBEM project. Chelomey's Deputy at the Branch implemented the order, having made the diazotype copies of our drawings, and he had not even wiped out our signatures from the developed drawings related to the DOS . . . which he handed over.[84]

---

81.   *Ibid.*, p. 268; Zhiltsov, ed., *Gosudarstvennyy kosmicheskiy*, p. 74.
82.   Zhiltsov, ed., *Gosudarstvennyy kosmicheskiy*, p. 75; Petrakov, "Soviet Orbital Stations."
83.   Semenov, ed., *Raketno-Kosmicheskaya Korporatsiya*, p. 267. Another source says that the MOM order dating from February 16, 1970, stipulated that six Almaz stations were to be turned over from Chelomey to Mishin. See K. Lantratov, "Dmitriy Kozlov's 'Zvezda'" (English title), *Novosti kosmonavtiki* 6 (March 10–23, 1997): 74–80.
84.   Polyachenko, "The 'Pep' of Almaz," p. 19.

The convoluted story behind the genesis of the DOS could have been the brainchild of an author intent on confusing readers, a maze of abrupt turns, shifting alliances, and ultimately betrayal. No one could have possibly predicted such an outcome. Chelomey was ordered to hand over all his Almaz materials to Mishin, while at the same time, one of Chelomey's own branches was ordered to cooperate with Mishin on the project. And all this happened when both Chelomey and Mishin opposed the idea. For Chelomey, this was a blow of proportions comparable to the immediate post-Khrushchev period when the bottom fell out of so many of his programs. After that near catastrophe, he saw one after another of his piloted space projects disappear. Although he had a fairly strong automated space program, he staked all his hopes to claim some of the glory of the piloted space effort on Almaz. But his Almaz was near death. He was consoled by the fact that Ustinov was not singularly powerful enough to completely kill the military Almaz. Although it would be delayed, perhaps as much as two or three years, Ministry of Defense support ensured that eventually Chelomey would see his coveted Almaz fly in space.

Viktor Bugayskiy was the chief of the Chelomey design bureau's Branch No. 1 at Fili in Moscow. A veteran of the Ilyushin design bureau, Bugayskiy was primarily responsible for the serial production of Chelomey's many ballistic missiles and spacecraft. He was one of the principal architects of the first Salyut space station in 1970–71. (files of Peter Gorin)

Bugayskiy was put in an awkward position. He had had a distinguished career working as a deputy to renowned Soviet aircraft designer Sergey V. Ilyushin at OKB-240, where he led work on the famous Il-2 during World War II. He joined Chelomey's design bureau in 1960 to direct the plant production of the P-5 naval missile. The two men evidently had "excellent relations" with each other: while Chelomey had the creative vision, Bugayskiy knew how to work at the plant level, converting that vision into reality.[85] When, in 1960, Chelomey inherited the Fili Branch, he put Bugayskiy in charge. Throughout the 1960s, Bugayskiy was officially Chelomey's First Deputy, and thus ultimately responsible to him and no one else. But torn between Ustinov's whims and Chelomey's rank, he became a consistent supporter of the DOS despite heavy criticism from his boss. Chelomey was unable to dismiss Bugayskiy. With the help of the Ministry of Defense, Chelomey did manage to pass through an order limiting the number of employees at the Fili Branch who could work on the DOS. Opinions within the branch were divided—some supporting Chelomey, others Bugayskiy.[86] It was a remarkably discordant management situation. For his part, Chief Designer Mishin had been adamantly opposed to the DOS decision, believing it to be a diversion from the N1-L3 program. Writing twenty years later, his opinions apparently had not changed:

> The decision made no sense to me (and it still makes no sense to me now), inasmuch as the work on the Almaz orbital station was being done at the same time that work was being done on [the DOS]. . . . It would have been wiser to combine the efforts of both OKBs to develop a unified orbital station and to entrust that work to . . . Chelomey's firm, which

85. Telephone interview, Sergey Nikitich Khrushchev by the author, October 10, 1996; V. M. Petrakov, "From the History of Development and Creation of Carrier-Rockets in the USSR" (English title), in *Trudy XXup chteniy, posvyashchennykh razrabotke nauchnogo nasladeniya i razvitiyu idey K. E. Tsiolkouskogo* (Moscow: RAN, 1994), p. 170.

86. Petrakov, "Soviet Orbital Stations."

had long been working on that area. Such a decision would have relieved the burden being carried by our OKB substantially and would have given us the opportunity to concentrate our efforts on the work on the NI-L3 program.[87]

He added:

The decision could not help but complicate our relations with V. N. Chelomey, which were already strained because of the transfer to us (while Sergey Pavlovich [Korolev] was still alive) of subsequent work on the circumlunar flight.[88]

It was one of those rare instances when Chelomey and Mishin actually agreed on something, but their combined might could not stop the newest space station program. The manufacture of the first DOS flight article began at the Khrunichev Plant in February 1970, the first in a line of space vehicles that would ultimately lead to the *Mir* space station.

Eighteen Days

The Almaz was not the only casualty of the DOS decision. Concurrent with the decision to proceed with the DOS, on February 9, 1970, all work on the Soyuz-VI small military orbital station was terminated. Given the capabilities of the DOS, Ustinov believed that there was no rational need to have two space station programs at TsKBEM. The cancellation of Soyuz-VI was opposed by certain individuals in the military who had been patiently waiting for more than five years for a military version of the Soyuz, seeing each program neutralized one after the other. There was one bright spot in the otherwise dismal state of piloted military programs: while Minister Afanasyev canceled work on the OB-VI station portion of Soyuz-VI, he allowed work to continue on the 7K-S transport ship of the complex because he considered it "promising and having many improved characteristics compared to the [basic] 7K-OK [Soyuz]."[89] The 7K-S, with improved avionics, communications, safety, and capability characteristics over the basic Soyuz, would serve as the basis for autonomous military research Soyuz spacecraft in the 7K-S-I and 7K-S-II variants. A third version would serve as a ferry spacecraft to future DOS stations in Earth orbit. Mishin's interest in pursuing the 7K-S variant meant that funding for it was increased significantly by mid-1970, although progress was evidently slow because of a lack of facilities at the design bureau's plant. A first piloted flight was not expected until 1972–73.

The first DOS mission was scheduled for early 1971 at best. To fill the gap between piloted flights, Mishin had plans to conduct two Soyuz missions during 1970, each comprising two 7K-OK spacecraft that would dock with each other using the lunar Kontakt rendezvous radar system. One of these missions would also include a twenty-day long-duration flight of two cosmonauts in Earth orbit. By late December 1969, it was clear that the Kontakt system would not be ready for the 100th birthday of Lenin in April 1970, the target date for the first docking mission. Instead, Mishin formulated a plan to launch a single 7K-OK, spacecraft no. 17, with two cosmonauts on the twenty-day flight in April 1970.[90] In January 1970, the Military-Industrial Commission issued a formal decree for an eighteen-day flight, with the length of

87. Mishin, "Why Didn't We Fly to the Moon?"
88. *Ibid.*
89. Lantratov, "Dmitriy Kozlov's 'Zvezda'"; Semenov, ed., *Raketno-Kosmicheskaya Korporatsiya*, p. 211. The cosmonaut group for the Soyuz-VI program finally disbanded in August 1970.
90. There were apparently at least three other options to celebrate the April 1970 deadline, including one using the 7K-OK to dock with an Almaz Orbital Piloted Station (OPS) and another using the 7K-S to dock with an Almaz OPS.

duration determined by the safety reserves aboard the relatively cramped Soyuz spaceship. Such a flight would break the fourteen-day record set by the two Gemini VII astronauts almost five years earlier. This eighteen-day flight would then be followed by the Kontakt docking mission, perhaps as early as August 1970.

Six cosmonauts had begun training for the long-duration mission by November 1969, including primary contenders Nikolayev and Sevastyanov, who had lost their chance to fly on Soyuz 8 earlier in the year because of poor preflight preparations.[91] Insufficient training of the crew was also evidently a factor in postponing the new mission from early April to late May 1970. Apart from the purely physiological goals of monitoring the effects of prolonged microgravity, the two cosmonauts were also to reperform some of the rendezvous maneuvers tried in vain during the triple-Soyuz flight in late 1969. Their Soyuz ship would carry a new computer, named the Spacecraft Analogical Machine, to allow rendezvous in orbit with an imaginary target.[92] The computer was capable of locating targets at a range of thirty to fifty kilometers and of providing input on subsequent maneuvers. Throughout early 1970, the cosmonauts training for the flight performed extensive full-length flight simulations at the Gagarin Cosmonaut Training Center at Zvezdnyy gorodok to prepare for the mission. These were carried out to establish a "proper balance between reserve capacity of the air regenerative system and the metabolic processes of the crew."[93] Simulations included complete eighteen-day missions with ground crews matching the exact schedule planned for the mission. The cosmonauts used new state-of-the-art biomedical monitoring equipment as well as improved waste disposal systems.

On May 20, 1970, the Soviet Strategic Missile Forces launched a Zenit-4 reconnaissance satellite into orbit from site 31 at Tyura-Tam. Named *Kosmos-345* by the Soviet press, the satellite was launched from the same pad that was set aside for use for the long-duration flight. Because of extremely high winds at the launch site, up to and above twenty meters per second, there was some damage when the plumes from the rocket exhaust singed the launch trusses and cables of the pad structure. Pad personnel assured the Soyuz State Commission that repairs would be finished prior to the planned launch on May 31. Subsequent problems during ground testing of the 7K-OK vehicle at Tyura-Tam put that target date in question. During the integrated testing of the ship, engineers detected intermittent currents in its electrical system, measuring as much as sixty volts, instead of the nominal thirty-eight volts. Unusually, most of the members of the twenty-person State Commission had not arrived at the Baykonur Cosmodrome by this time. Air Force Aide Kamanin noted in his diary on May 22: "The attitude toward the preparations for the prolonged space flight, beginning with the highest leaders and ending with the rank-and-file workers, is mostly nonchalant."[94]

There was somewhat of a minor crisis on the evening of May 25, when Kamanin discovered primary crew Commander Nikolayev smoking a cigarette in direct violation of orders not to do so at the Baykonur Cosmodrome. Later, Sevastyanov also admitted that he had also been smoking contrary to medical orders. Kamanin was aghast, especially given that Nikolayev had been caught doing the same thing the previous December and had promised to quit smoking. The general noted with frustration that:

> If I had learned of this a month ago, I would have been against allowing Nikolayev and Sevastyanov to fly, but now, when there are only a few days left until the launch, and

---

91. The other cosmonauts in training by April 1970 were A. V. Filipchenko, G. M. Grechko, V. G. Lazarev, and V. I. Yazdovskiy.

92. Hooper and Vis, "Meeting the Space Explorers: Vitali Sevastyanov."

93. "Big Booster Paces Soviet Manned Flights," *Aviation Week & Space Technology*, July 6, 1970, p. 18.

94. Kamanin, "Removing the Cosmetic Retouching."

Nikolayev's crew has already been confirmed in fact as the primary crew in the Party's Central Committee and the government, it is impossible to raise the matter of replacing the cosmonauts with their backups.[95]

In the meantime, Minister of General Machine Building Afanasyev telephoned Mishin at Tyura-Tam that the Politburo had just discussed the impending flight. They had recommended that the press communiqués regarding the mission be low key, without all the pomp associated with past Soyuz missions.

On the evening of May 31, the complete State Commission met to formally approve the launch date and time of the launch, set for exactly midnight local time on June 1. At a subsequent press conference, Nikolayev and Sevastyanov were forbidden to talk about the main feature of the flight, its record-breaking length, and instead uttered the usual generalities. There seems to have been some tension between factions in the State Commission over the issue of length, a latent conflict that did not abate through the following weeks. Some, like Kamanin, were adamant that the length be limited to eighteen days, while others, like Mishin, were hoping for a possible extension to twenty days. On the afternoon of launch day, Kamanin tried to pre-empt any conflicts on the issue by explicitly forbidding either cosmonaut from asking for an extension of the flight over eighteen days once they were in space. Kamanin's concern was that any extension would severely strain the capabilities of the old Soyuz spacecraft and perhaps put the lives of the crew in jeopardy.

Throughout the day, Strategic Missile Forces personnel carried out all prelaunch procedures on time. The cosmonauts arrived at the pad a little over two hours prior to launch. Without further ado, the Soyuz spaceship lifted off precisely on time at 2200 hours Moscow Time on June 1, 1970, with forty-year-old Colonel Andrian G. Nikolayev as the commander and thirty-four-year-old civilian Vitaliy I. Sevastyanov as the flight engineer. The spaceship, named *Soyuz 9*, entered an initial orbit of 208 by 220.6 kilometers at a 51.7-degree inclination. For Nikolayev, it was his second spaceflight, having flown in space eight years before in 1962 as the pilot of Vostok 3. Sevastyanov was the fourth civilian engineer from TsKBEM to fly in space. NASA astronaut Neil A. Armstrong, the first human to set foot on the Moon, was on an official visit to the Soviet Union at the time. On the night of the launch, at the Cosmonaut Training Center near Moscow, he was clearly surprised when his host, cosmonaut Maj. General Beregovoy, turned on the TV to view film of the Soyuz 9 launch. Beregovoy reportedly told Armstrong, "This is in your honor."[96]

On their first day in space, the Soyuz 9 crew carried out two orbital maneuvers—the first on the fourteenth orbit to 213 by 267 kilometers and the second on the seventeenth orbit to 247 by 266 kilometers—sufficient enough to prevent orbital decay without additional maneuvers.[97] These maneuvers may have also been related to the mock rendezvous with an imaginary target. The two men began their extensive scientific experiments program by the end of the their first orbit. Within the first three to four days in orbit, ground controllers were already finding out that they would have to plan future long-duration missions differently. For example, the cosmonauts reported that they required nearly fifty minutes to complete their set of physical exercises, whereas they managed to do them in a half-hour during preflight training.

On June 4, most of the members of the State Commission, including Chairman Kerimov, Minister Afanasyev, Chief Designer Mishin, and Commander of Space Assets Karas, left Tyura-Tam for Moscow. In charge at the control point at the launch site were Col. General Kamanin and TsKBEM Deputy Chief Designer Yakov I. Tregub. During the latter part of the day,

---

95. *Ibid.*
96. Smolders, *Soviets in Space*, p. 186; Riabchikov, *Russians in Space*, p. 277.
97. Petrov, *Conquest of Outer Space*, pp. 171–73.

there was some alarm when ground readings showed that because of intermittent operation of the solar arrays' automatic equipment, the storage buffer batteries were showing higher levels of charge than normal. On the forty-seventh orbit, Sevastyanov reported that although the solar arrays had been turned off, the current in the batteries was twenty-six amperes, clearly indicating a malfunction in the control switch for the panels. During the previous two days of flight, the crew had to turn off the solar arrays manually more than twelve times, close to the limit of fifteen times the operation could be repeated.[98] One reason for the excess power was beyond the control of the ground or the crew. On this flight, the duration of "nighttime" was only forty seconds instead of the dozens of minutes on earlier Soyuz missions. Because the orbit of the current mission was such that the ship's orbit was nearly parallel with the terminator, the solar arrays were generating a nearly continuous stream of electric current. To compensate, the flight control team ordered the crew to turn the ship around at a rate of a half degree per second to turn the arrays away from the Sun. The solar panel switching system began operating normally the following day, indicating that either Sevastyanov had reported incorrect readings the previous day or that it had been a "self-repairing" problem.

A week into the mission, already the longest Soviet space mission, all systems seemed to be nominal. The cosmonauts reported that they felt significantly better on the sixth day than on the first two to three days of the flight. There were again murmurs of talk about extending the flight to twenty days, but such prognostication proved too premature at this point. One of the few negative indicators of the crew's health was the reduced consumption of drinking water (one liter per day) and oxygen (seventeen liters per day), indicating some fatigue. On June 10, Nikolayev and Sevastyanov had their first day off, and they spent time playing a game of chess with Kamanin and veteran cosmonaut Gorbatko on the ground. The players advanced their pieces twenty-five times over three orbits before agreeing to a draw.[99] The crew displayed the first real signs of fatigue and decrease in working efficiency on their twelfth day in orbit. Kamanin wrote in his diary that:

Nikolayev and Sevastyanov look somewhat puffy, and listlessness and irritability can be sensed in their actions. After talking things over with the cosmonauts, we decided to shorten significantly for the subsequent days of the flight the volume of experiments and to increase the rest periods.[100]

The activities of the Soyuz 9 crew in space were fairly intensive for such a relatively small spacecraft, with working days lasting on average between fourteen and sixteen hours. Both exercised twice a day in the living compartment with an expansion device that required an exertion tension of ten kilograms. On occasion, they wore a special suit named *Pinguin* ("Penguin") to simulate some of the effects of Earth's gravity. They assessed their condition before and after each exercise regime, recording arterial pressure, pulse, respiration, and contrast sensitivity of their eyes. The average daily calorific content for each cosmonaut was about 2,600 kilocalories. For the first time, a Soviet piloted spaceship carried a food heater, which allowed the crew not only to heat up their food, but also to get a fresh cup of coffee in the "morning." The men could not take baths in the ship, but they used wet and dry towels for rubdowns twice a day for personal hygiene. They were allowed a change of underwear once

---

98.   Kamanin, "Removing the Cosmetic Retouching." The limit of fifteen times was because every time the cosmonauts turned the switch off, hydrogen accumulated in the instrument compartment. With increasing amounts of hazardous hydrogen in the module, the controllers would have had to cut the flight short after eight days in space.

99.   *Ibid.*; Riabchikov, *Russians in Space*, p. 280.

100.   Kamanin, "Removing the Cosmetic Retouching."

a week. On this first space mission lasting more than two weeks, the cosmonauts maintained only intermittent contact with their families. On the birthday of Nikolayev's daughter Elena, she came to the Flight Control Center with her mother, former cosmonaut Valentina V. Nikolayeva-Tereshkova, to talk to her father via both video and audio.[101]

The actual scientific experimentation consisted of fifty experiments in various categories. On their fourth day in space, the crew used a new stellar sensor to calculate the orbital parameters and geographical latitude of the point above which the ship was flying, relative to the position of a selected star above the horizon, Vega in the Lyra constellation. The cosmonauts carried out this experiment, complicated by the motion and drift of Soyuz 9, over a period of two complete orbits without any communications with the ground as they manually maintained attitude and measured drift of the ship's gyroscopes. Other navigational exercises involved the use of the SMK-6 sextant, used in combination with solar and stellar sensors and an optical device in the spacecraft. On their fourteenth day, the cosmonauts explored the possibility of checking orientation with less "popular" stars, such as Arcturus, Deneb, and others, in conjunction with ground reference points on Earth, including lakes and mountains in Africa and South America. All these experiments led to precise determination of orbital elements to refine future rendezvous exercises.[102]

As usual, Earth photography comprised a large part of their work time and resulted in 1,000 pictures by the end of the mission. These included a special experiment on June 13 on Soyuz 9's 189th orbit. The crew investigated weather formations in the atmosphere and western portion of the Indian Ocean as part of an integrated exercise that included a Meteor-1 satellite at an altitude of 600 kilometers, the Soyuz 9 vehicle at 240 kilometers, and sounding balloons launched from the scientific research vessel *Akademik Shirshov* of the USSR Hydrometeorological Service located in the Indian Ocean. Less intensive observations included those of a large tropical storm in the Indian Ocean on their fifth day and forest fires in Africa near Lake Chad the day after. On the thirteenth day, the crew used both black-and-white and multispectral color film to identify different kinds of rock and soil on Earth, the moisture content of glaciers, the location of shoals of fish, and timber reserves. They studied aerosol particles in the atmosphere by observing twilight glow and carried out spectrographic measurements of the horizon to enhance definition of the horizon for navigational purposes. They also used the RSS-2 handheld spectrograph to make 200 spectrophotometric measurements of natural formations in different parts of the world. The same type of instrument had been used on Soyuz 7 the previous year. On day seventeen, they performed some brief photography of the Moon.[103]

Biomedical tests comprised a major part of their activities. On their ninth day in space, they reported that they were collecting air samples of their breathing before and after exercise to study the ration of oxygen and carbon dioxide. On day thirteen, Sevastyanov carried out a test of his mental capabilities by performing a simulated set of commands that had been preprogrammed into the on-board computer. His results would be compared to his performance before the flight on the same test. Nonhuman studies included those related to the micro and macro genesis of plants, the division of chlorella cells, the propagation of bacterial cultures in liquid media, and the development of insects in weightlessness.

As they were winding down their experiments program, there were some minor problems. At the scheduled beginning of their communications session on June 15, ground controllers were unable to wake up the crew despite three minutes of increasingly frantic calls. Both men

101. Riabchikov, *Russians in Space*, p. 280; Narimanov, *Ot kosmicheskikh korabley*, pp. 77–80; Smolders, *Soviets in Space*, p. 192; *Soviet Space Programs, 1966–70*, p. 238.

102. Narimanov, *Ot kosmicheskikh korabley*, pp. 81–83.

103. Riabchikov, *Russians in Space*, pp. 280–81; *Soviet Space Programs, 1966–70*, p. 239; K. Y. Kondratyev, et al., "Some Results of Spectrophotometry of Natural Formations From the Manned Spaceship Soyuz-9" (English title), *Kosmicheskiye issledovaniya* 10 (March–April 1972): 245–54.

apologized for sleeping through their wake-up time, but Sevastyanov, groggy from having been woken up, inadvertently switched on the button for the automatic landing system display when attempting to switch on the cabin light. It was an exact repeat of the situation on Soyuz 7, when the system, designed to operate after reentry at an altitude of eleven kilometers, remained turned on in space through the rest of the mission. Later the same day, the crew altered their orbit a third time, by firing their engine on orbit number 208. The following day, there was further anxiety when one of the batteries of the telemetry system failed, dropping out telemetry for a number of important parameters on the ship's systems. Both Chief Designers Mishin and Ryazanskiy, the latter responsible for the offending component, assured the State Commission that this was not a threat to continued flight.

There was an expanded meeting of the State Commission on June 16, when Mishin casually asked ballistics experts what the orbital parameters would be on the twentieth day of flight, clearly implying that he was interested in extending the mission from the planned eighteen days. The issue over mission length, a common conflict during many Soviet piloted space missions of the era, spilt out in the open during lunch the same day, when Mishin and Kamanin went head to head against each other. According to Kamanin:

> Mishin did not hold back and asked me, am I of a mind to fight? Knowing what he was driving at, I responded that, for the time being, I see no reasons for shortening the flight program. I did not begin to talk about the fact that members of the landing commission from the industry—[Chief Designers] Severin, Tkachev, and Darevskiy—had urgently requested that I not permit an increase in the duration of the Soyuz-9's flight beyond the [eighteen-day] program.[104]

From Soyuz 9 Commander Nikolayev's reports, it was clear that while food rations could be extended to twenty days, it would be difficult at best, and probably not worth the risk. The issue was finally resolved at a meeting of the inner circle of the State Commission on June 16. Both State Commission Chairman Kerimov and Mishin were clearly under political pressure to extend the flight to twenty days. Mishin's suggestion for an extension was, however, not taken lightly by the other attendees. Five men came out against Mishin. In frustration, Mishin turned on Ministry of Health representative Yevgeniy I. Vorobyev, responsible for dietary needs, accusing him of not providing enough food for twenty days. The final decision was to perform the landing on June 19, after eighteen days. Kamanin noted in his diary: "V. P. Mishin and K. A. Kerimov, having promised the high command in Moscow that they would carry the flight out to 20 days, will now have to concur with our decision."[105]

The last two days in orbit were relatively quiet for both the crew and ground controllers. On the morning of June 17, Kerimov, Mishin, and Kamanin congratulated the two cosmonauts on officially exceeding the record set by Gemini VII in 1965, thus reclaiming for the Soviet Union the absolute endurance record for a spaceflight. A day later, the State Commission approved a plan to land Soyuz 9 on its 287th orbit. In case of a possible ballistic reentry, the commission stationed a contingent of recovery forces, including amphibious craft, three

---

104. Kamanin, "Removing the Cosmetic Retouching." The chief designers were G. I. Severin of KB Zvezda (for spacesuits), N. A. Lobanov of NII AU (for parachutes), and S. G. Darevskiy of SOKB LII (for ground simulators and avionics).

105. *Ibid.* The other members of the inner circle were P. A. Agadzhanov (Deputy Chief of TsKIK and also Chief of GOGU), B. Ye. Chertok (Deputy Chief Designer of TsKBEM), N. P. Kamanin (Air Force Commander-in-Chief's Aide for Space), K. A. Kerimov (Chief of the Third Chief Directorate, MOM), Ya. I. Tregub (Deputy Chief Designer of TsKBEM), and Ye. I. Vorobyev (Chief of the Third Directorate, Ministry of Health).

helicopters, five sea launchers, and fifteen fishing vessels, in the Aral Sea. Nikolayev and Sevastyanov's journey back to Earth began on the afternoon of June 19. At least 150 people, including Minister Afanasyev, were present at the Flight Control Center at Yevpatoriya to observe the proceedings. Air Defense Forces radars tracked the capsule from an altitude of eight-three kilometers all the way down to parachute deployment. The whole crowd at the center burst into applause upon hearing Nikolayev's radioed message on a safe landing. Because of the precision of the landing, two helicopters were able to film the descending capsule and landed almost simultaneously with the cosmonauts. The Soyuz 9 ship landed seventy-five kilometers west of Karaganda at 1459 hours Moscow Time after a flight lasting seventeen days, sixteen hours, fifty-eight minutes, and fifty-five seconds. For the first time in more than four years, the Soviet Union held the absolute record for the longest piloted spaceflight.

When ground crews reached the cosmonauts, they found that the cosmonauts were unable to get out of the ship themselves and had to carry them out. After much discussion and dissent on the issue, Military-Industrial Commission Chairman Smirnov finally decided to cancel the immediate flight of the crew to Moscow's Vnukovo Airport. Instead, the cosmonauts remained at Karaganda for a day and arrived in Moscow on June 20 at Chkalovskaya Airfield. The plan was to escort the cosmonauts to the Cosmonaut Training Center for a press conference, but once Kamanin entered the aircraft to talk to the crew, these plans were changed. He wrote in his journal:

> When I entered the aircraft's cabin, Sevastyanov was sitting on the sofa, while Nikolayev was at a small table. I knew they were having a hard time enduring the return to the ground, but I had not counted on seeing them in such a sorry state. Pale, puffy, apathetic, without the spark of vitality in their eyes—they gave the impression of completely emaciated, sick people.[106]

The crew was eventually escorted off the plane by cosmonauts Shatalov and Yeliseyev, although both had said earlier that they could walk by themselves. In a weak voice, Nikolayev, the more debilitated of the two, gave a very brief speech about fulfilling their mission and being ready for another one. He and Sevastyanov were then put into cars and sent to the care of an Air Force medical support group at Zvezdnyy gorodok.

Over the period of the next few days, it was increasingly clear that part of the reason for the very poor shape of the Soyuz 9 crew was the slow spin of the spacecraft throughout the mission. The spinning also produced a weak field of artificial gravity, which affected the clarity of results of several experiments aboard the ship. Nikolayev and Sevastyanov spent several days in quarantine, not only to protect their weak bodies from infections, but also, as it turns out, because of the discovery of a mutation of two microbes not occurring on Earth that were found in their metabolic systems. For five days after their return, the microbes spread very rapidly but then died from the effects of gravity. During this period, the two cosmonauts were fed through a safe bio-interface system.[107] Briefing sessions were held during their confinement with engineers, physicians, and other scientists. One journalist wrote: "They were pale, and their faces furrowed with wrinkles. They tried to carry on a lively conversation and even make jokes; but they tired rapidly, and there were frequent lapses."[108] For the first four or five nights, they slept

---

106. Ibid.
107. Gordon R. Hooper, The Soviet Cosmonaut Team: Volume 2: Cosmonaut Biographies (Lowestoft, UK: GRH Publications, 1990), p. 264; Thomas O'Toole, "Soviet Union Still Trails U.S. in Space," Washington Post, June 17, 1973, pp. A1, A8.
108. Riabchikov, Russians in Space, p. 282.

fitfully, and the feelings of "acceleration" did not disappear until five or six days after landing. All ill symptoms finally disappeared eleven to thirteen days after landing. The men were sent off on short postflight vacation soon after.

The poor state of Nikolayev and Sevastyanov prompted a spate of debate over the issue of long-duration spaceflight. At one large postflight meeting at the Cosmonaut Training Center, two opposing factions expressed their views. Some believed that subsequent space missions should not exceed eighteen days by more than one or two days, and if the crew returned well after that, future missions could be extended conservatively. Other doctors argued that much longer missions were possible, but only with preventative measures such as medicine and exercising.[109] The debate over this issue to a significant degree affected plans for both the Almaz and DOS missions, with Soviet space officials looking to artificial gravity for very long missions on the Multirole Base-Station. Regardless of the condition of the cosmonauts, the Soyuz 9 mission was a landmark success for the Soyuz space program, precisely because it was the first fully unqualified success since the Soyuz 4/5 mission more than a year before.

Still Aiming for the Moon

It has been customary for Western observers of the Soviet space program to assume that the Soyuz 9 mission was the turning point for those involved the program—a signpost indicating their progression from quitting their piloted lunar program to creating Earth-orbital stations. This impression, partly supported by many official Soviet statements, has not been borne out by recent revelations. Even after Apollo 8, Apollo 11, and Apollo 12, the Soviets continued their vigorous search for successes on the Moon. When, in January 1969, Soviet space officials decided to move ahead with three different thematic directions—Earth-orbital stations, expanded lunar landings, and missions to Mars—all three were pursued for several years. Thus, in many ways, the story of the race to the Moon does not end in 1969—at least not for the Soviets. From both political and propaganda perspectives, future advanced lunar landings of cosmonauts offered a means to restore lost faith in the Soviet space program.

Much of the success of future lunar landings depended, of course, on the fate of the N1 rocket. The program had already been delayed by at least four years, and its record had been marred by two untimely failures in 1969. The investigation into the second failure in July 1969, which had destroyed one of the two available N1 pads at Tyura-Tam, was long and tedious. It took a full year before a formal report was ready on the accident, and even then there were multiple opinions on the cause of the accident within the investigation commission headed by Mishin. The reconstruction of the most probable chain of events was an exercise in detective work. A quarter of a second prior to liftoff, a metallic object, probably a portion of a steel diaphragm of a pressure oscillation sensor, had entered an oxidizer pump and caused engine number 8 of the first stage to explode. This disrupted the work of the on-board cabling network and damaged engines and telemetry instrumentation in the vicinity. As the lower part of the first stage was engulfed in fire, at T+0.6 second, the KORD system (for engine operation control) issued a command to shut down engine nos. 7, 8, 19, and 20. At T+8.76 seconds, it shut down engine no. 21 and its opposing engine no. 9. By T+10.15 seconds, all engines were shut down, except for engine no. 18, which continued to fire. The rocket, meanwhile, lifted up to a height of about 200 meters, and then it began to fall back vertically toward the launch pad, having been unable to turn on its nominal course because of the disruption of the cable network. The only operational engine gradually turned the rocket around its axis and, after

---

109. Yu. A. Mozzhorin, et al., eds., *Dorogi v kosmos: I* (Moscow: MAI, 1992), p. 64.

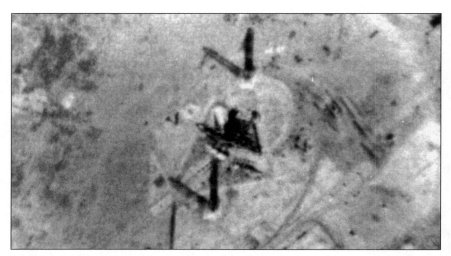

This remarkable photo of an NI booster on the still-intact pad at Tyura-Tam was taken from a U.S. CORONA photo-reconnaissance satellite on June 4, 1970. The three stages of the rocket, probably booster no. 6L, are clearly visible as is the associated pad structure. (copyright Charles P. Vick, KH-4B mission 1110-2, launched May 20, 1970, Frame A148)

a twenty-three-second flight, the booster fell almost broadside onto the launch pad and completely exploded. Earlier, at T+14.5 seconds, the emergency rescue system activated and shot off the descent apparatus of the 7K-L1S spacecraft.[110]

Mishin's commission had found in its investigation that during ground testing of the first stage's NK-15 engines, large metal objects (dozens of millimeters in diameter) had the propensity to get into the oxidizer pump, damaging the impellers and causing ignition and explosion of the pump. Small metal objects (chips, fillings, and so on) burning in the gas generator resulted in the destruction of the turbine vanes. Finally, nonmetallic objects (rubber, rags, and so on), which were fed into the inlet of the turbopump assembly, did not cause disruption of engine work. Booster 5L, which had exploded in July 1969, had been among the first batch of manufactured NIs, and thus it did not have filters for foreign objects installed in the inlets to the pumps. According to the program specifications, these filters were scheduled to be installed beginning with booster no. 8L—that is, on the fifth launch attempt of the NI.[111]

Mishin met with both Minister Afanasyev and Central Committee Secretary Ustinov in August 1969, explaining that the NI-L3 complex would still remain the primary system for researching the Moon. At a later meeting with Ustinov in September, Mishin was told that there would be a decision on the fate of the NI-L3 complex only after the causes of the July failure

---

110. Igor Afanasyev, "NI: Absolutely Secret" (English title), *Krylya rodiny* no. 9 (September 1993): 13–16; Semenov, ed., *Raketno-Kosmicheskaya Korporatsiya*, p. 257. The official report on the accident of booster 5L, in describing the initial cascade of events, states: "In the time interval between −0.2 second and +0.25 second, the following happened: (a) the rocket's airframe was subjected to pulsed loading; (b) there was a sharp rise in temperature in the vicinity of engine nos. 7, 8, and 9; (c) the telemetry equipment of engine nos. 8 and 9 failed." See R. Dolgopyatov, B. Dorofeyev, and S. Kryukov, "At the Readers' Request: The NI Project" (English title), *Aviatsiya i kosmonavtika* no. 9 (September 1992): 34–37.

111. Afanasyev, "NI: Absolutely Secret."

had been determined. The immediate plans after the July 1969 failure of booster 5L had been to perform full-scale one-way automated landings of the Lunar Ship (LK) on the Moon on N1 boosters 6L, 7L, and 8L. As the investigation into the disaster took longer and longer, these plans had to be shelved.

The fate of the N1 rocket itself seemed central to the future of the Soviet space program as a special governmental commission examined the program as a whole following the second accident. Coming at the nadir of the Soviet hopes in the "space race," the recommendations of the commission were positive in outlook: the commission believed that the N1 would be able to support all planned Soviet space projects for the subsequent ten to fifteen years.[112] In December 1969, after a review of the July catastrophe, the Commander-in-Chief of the Strategic Missile Forces, Marshal Nikolay I. Krylov, made his feelings known on the program. Traditionally an opponent of piloted space programs, Marshal Krylov wrote to Minister of General Machine Building Afanasyev that:

> The resulting analysis of the two failed launches of the N1-L3 complex, and also statistics from launches of other complicated rocket-space complexes show that the existing methods of developing rocket-space complexes do not ensure a high level of reliability upon entry into [flight-testing]. The existing methods of ground work on [rocket-space complexes], for the most part, are analogous to the methods of developing military missiles, which, as a rule, are considerably simpler than [rocket-space complexes] of the N1-L3 type. At the same time, the processes of [flight-testing] of military missiles differ by some tens of articles (from 20 to 60) to bring them up to a high level of reliability. In carrying out the [flight-design testing] of heavy [rocket-space complexes] the possibility of extended flight work is not feasible because of the great expenses of the rocket-carriers. In view of this, expedient changes in the volume and character of the ground work on these complexes up to the moment of entry to [flight-testing] should be introduced. In our opinion, new methods of ground work on heavy [rocket-space complexes] should include the basis for multi-use operations and [creation of] a large stock of resources of the complete system and equipment; preliminary firing tests of engines and rocket blocks without subsequent sorting out with the goal of discovering production defects and expirations of their working lives should also be carried out.[113]

The recommendations of Krylov, all clearly worthy of attention, were apparently taken into consideration in future planning for the program. One of the major changes during the 1969–70 period was reworking the procedural system by which engines for the first three stages of the N1 were selected for flight. The original method, known as KONRID, consisted of an efficiency control system in which a batch of six randomly selected engines were submitted for a flight article. Of these, two would be static tested on the ground. Depending on the results, the remaining four would then be consigned for the flight article. This meant that the actual engines used on the N1 were never tested prior to installation on the booster. Because the KONRID system had proved inadequate in the face of multiple engine failures on the first two launches, in July 1970, the Trud Design Bureau, under Chief Designer Kuznetsov, began using the old NK-15, NK-15V, and NK-21 engines of the first, second, and third stages of the booster to develop a new uprated set of three engines. According to the technical assignment issued by TsKBEM, these

---

112. Georgiy Stepanovich Vetrov, "Development of Heavy Launch Vehicles in the USSR," presented at the 10th International Symposium on the History of Astronautics and Aeronautics, Moscow State University, Moscow, Russia, June 20–27, 1995.

113. Semenov, ed., *Raketno-Kosmicheskaya Korporatsiya*, p. 257.

new engines would be capable of multiple firings, have much longer service lives, be delivered without reassembly after acceptance tests, and be tested on the ground prior to flight. Until these new engines were ready for flight, expected in late 1972, TsKBEM would use the older Kuznetsov engines.[114]

Apart from the engines, many other systems were reworked from 1969 to 1971. These included:

- Increasing the reliability of the oxidizer pumps (by increasing clearances and reducing the loads on bearings)
- Improving the quality of the manufacturing and assembly of the turbopump assembly
- Installing filters in front of the engine pumps to eliminate the entry of foreign objects
- Introducing the *Freon* fire extinguisher system
- Adding thermal protection elements into the instrumentation and cable system located in the tail section of the first stage
- Introducing blocking commands in the emergency engine shutdown system during the first fifty seconds of flight[115]

This ground model of the LK lunar lander is displayed at a museum in Kaliningrad (now Korolev) near Moscow. The four main landing legs are shown in retracted vertical position. The small spherical window at the upper center would have been used by the cosmonaut to observe the surface during landing. Sets of small attitude control engines can be seen near the top of the lander. (copyright Quest, via Luc van den Abeleen)

Furthermore, all piping in the N1's pneumo-hydraulic systems were still of the older flange pipe joint type. After the failure in July 1969, when engineers checked the already-manufactured and -tested units of another N1 booster, they found that many of the flange joints with fluorine plastic seals had leaked after long periods of storage. In July and August 1969, engineers decided to replace the flange joints with automated welded ones—an extensive redesign procedure that was performed by the Moscow-based NII TekhnoMash (formerly NITI-40) organization. Since 1970, all pipelines in Soviet launch vehicles have been joined during integration assembly by automated welding.[116]

The work on improving the characteristics of the recalcitrant booster was concurrent with continuing work on the L3 complex. Because of a continuous redesign process on the two

---

114. *Ibid.*, p. 258; Dolgopyatov, Dorofeyev, and Kryukov, "At the Readers' Request: The N1 Project"; Mozzhorin, *et al.*, eds., *Dorogi v kosmos: I*, p. 129; Igor Afanasyev, "N1: Absolutely Secret" (English title), *Krylya rodiny* no. 11 (November 1993): 4–5.

115. Semenov, ed., *Raketno-Kosmicheskaya Korporatsiya*, p. 258.

116. Oleg E. Ostrovsky and Valentin A. Kazakov, "Quality Provision for the Welded Joints of Pipelines for Space-Rocket Complexes," *Space Bulletin* 2(2) (1995): 9–11. NITI-40 was headed by O. Ye. Ostrovskiy during the late 1960s.

major components of the L3—the LK lander and the Lunar Orbital Ship (LOK)—neither component was ready for flight during 1969 and 1970. In the case of the LOK, sixteen ships had been originally ordered. Of these, by February 1970, seven had been manufactured, although only three were being ground-tested for future flight operations. As Mishin and his deputies stubbornly continued to pursue the old lunar landing plan, uncertainty in the mission profile continued to pervade the proceedings. In January 1970, six months after Apollo 11, engineers were still disagreeing about specifications of the Blok Ye engine for the LK lander. One of the major bottlenecks seemed to have been the components manufactured at the Arsenal Machine Building Plant in Leningrad. Engineers there faced many problems with tank production, thus missing deadlines for the delivery of the attitude control engines of the LK and the Engine Orientation Complex of the LOK. Consequently, there were changes in the powered descent profile of the lander, such as reducing the Blok D stage's deorbit operation time. In addition, they had still not adequately solved the question of mutual relationships among the LK, the LK_R (the reserve LK), and the Ye-8 rover. At a meeting of the TsKBEM leadership in May 1970, the prognosis was not good: although work on the N1 was proceeding relatively well, work on the L3 was, by far, in the worst condition at the design bureau, behind in its schedules than many other unrelated projects. Funding for the N1-L3 program in 1970 was evidently short by about 60 million rubles.

This is a close-up of the spherical living compartment of a ground model of the LOK lunar orbiter on display at a museum. Note the slightly asymmetrical shape of the module, significantly different from a standard Soyuz living compartment. The protrusion on the upper portion of the module is a porthole for a cosmonaut to directly observe rendezvous approaches with the LK lander. Attitude control thrusters, similar to the ones on the LK, are visible on the top left. The two small spherical protrusions on the top are propellant tanks for these engines. Curiously, the LOK living compartment seems to have been mounted on an L1 descent apparatus barely visible at the bottom of the photograph. (copyright Quest, via Luc van den Abeleen)

As far as the lunar landing itself, Mishin had informed Minister Afanasyev of a provisional schedule of N1 launches at a meeting in July 1970:

Date	Boosters	Missions	Engines
1970	6L, 7L	Automated lunar missions	Old engines
1971	8L, 9L, 10L	Automated lunar missions	Old engines
1972	11L, 12L, 13L	Piloted lunar-orbit missions	New engines
1973	14L, 15L, 16L	Two piloted lunar landings	New engines

The early automated flights would consist of robot variants of the LOK or LK or simply ballast, depending on what was available at the time. In the case of the LOK, the ships would carry special photographic equipment for imaging potential landing sites. Before an actual landing, it seems that Mishin had planned a fully automated lunar landing and return flight. The veracity of these projections depended to a great extent not only on the fortunes of the N1 but obviously on the flight rating of the L3 payload itself.

There was *some* good news in the L3 development program. Several vehicles were flown in 1970 that were directly part of the Soviet lunar exploration program. One of these was the 7K-L1E payload block, which consisted of a simplified 7K-L1 circumlunar vehicle, an experimental Blok D stage, and the payload fairing. The Blok D stage, the primary payload, was equipped with supplementary sensors for transmitting more complete information on the internal processes of the stage during firings in Earth orbit. The stage had special transparent "portholes" through which the internal volume of the tanks was illuminated. During the maneuvers in Earth orbit, special cameras would photograph the movement of propellants.[117] NASA had performed a very similar mission early on in the Apollo program during the mission of AS-203 in July 1966.[118] The first launch attempt of the 7K-L1E, spacecraft no. 1, had been at 1200 hours Moscow Time on November 28, 1969, on top of a three-stage Proton booster. Because of a third-stage failure, the payload never reached orbit. U.S. intelligence assets clearly monitored the telemetry from the attempted launch as pieces of the suspect stage inadvertently fell on Chinese territory about 200 kilometers north of Harbin.[119]

It was yet another in an unprecedented series of failures of the Proton booster in 1969. It took more than year to prepare a second L1E complex ready for launch. Spacecraft no. 2K was launched at 2000 hours Moscow Time on December 2, 1970. After reaching orbit, it was named *Kosmos-382*. Under the direction of TsKBEM Deputy Chief Designer Tregub, the Blok D stage was fired seven times in the course of six days in Earth orbit, simulating mid-course corrections, lunar orbit insertion, and powered descent from lunar orbit, thus rehearsing as closely as possible Blok D's nominal performance during an actual L3 lunar landing mission. All pertinent data on the stage's activities were transmitted successfully to Earth, adding significantly to confidence in the future use of Blok D.[120] During the mission, Western intelligence services were able to hear simulated voice transmissions from the spacecraft, prompting suggestions that the flight was related to a piloted project.[121]

Another L3-related precursor program was the flight of Earth-orbital versions of the LOK and the LK, designated the T1K and T2K, respectively. The original ambitious plans had been to fly these two spacecraft with crews on board to prove out both vehicles, much like the Apollo 9 mission flown a few months prior to the first lunar landing. Pressure from the Ministry of General Machine Building, in the form of financial restrictions, meant that Mishin had to completely eliminate flights of the T1K from the program; instead, the LOK would fly directly to the Moon on its first mission sometime in the future. The same fate probably would have befallen the T2K had it not been for intense pressure from Chief Designer Mikhail K. Yangel, whose organization, the Yuzhnoye Design Bureau (KB Yuzhnoye), created the main engine for the LK. Yangel's lobbying produced results, and Mishin was allowed to carry out three full flight tests of the T2K in Earth orbit in 1970–71—missions similar in many ways to the automated flight of the Lunar Module on Apollo 5.[122]

117. Semenov, ed., *Raketno-Kosmicheskaya Korporatsiya*, p. 228.

118. Roger E. Bilstein, *Stages to Saturn: A Technological History of the Apollo/Saturn Launch Vehicles* (Washington, DC: NASA SP-4206, 1996), pp. 338–40.

119. E-mail correspondence, Vladimir Agapov to the author, September 30, 1996. For a Western report on the launch failure, see "*Salyut* Elements Separate, Signals Lost," *Aviation Week & Space Technology*, April 30, 1973), p. 21.

120. Semenov, ed., *Raketno-Kosmicheskaya Korporatsiya*, p. 228.

121. See, for example, "Cosmos Changes," *Aviation Week & Space Technology*, December 21, 1970, p. 25; "Russian Moves," *Aviation Week & Space Technology*, May 3, 1971, p. 13.

122. V. Filin, "At the Request of the Reader: The N1-L3 Project" (English title), *Aviatsiya i kosmonavtika* no. 2 (February 1992): 40–41.

The T2K, while similar to the lunar version of the LK, was not identical to the latter. A number of systems necessary only for a real lunar landing were removed, while others necessary for testing were added. The most obvious difference was the omission on the T2K of the four landing legs comprising the Lunar Landing Unit with their stabilizing rocket engines. Engineers also removed the cosmonaut's ladder and two omnidirectional parabolic antennas on the rocket stage for deep space communications. As a substitute, a "weak" directional antenna was installed on the engine orientation compartment at the top of the lander. In addition, in place of the small suspended instrument compartment on the right side of the LK, designers added a large suspended compartment on the left side equipped with an ellipse-shaped cover. This compartment contained supplementary instrumentation for control and guidance, as well as an antenna system for radio control of the spaceship's maneuvers. The T2K also included an ionic orientation sensor instead of the standard adjusting and aiming sensors. On the pressurized cabin proper, there was also an additional telemetry antenna. The spaceship itself was equipped with a special control system capable of complete automated flight. The total launch mass of the T2K was around 5.7 tons, low enough to be launched by a modified variant of the Soyuz booster named the 11A511L. The rocket had an unusual "large-caliber" payload fairing to accommodate the spaceship.[123]

The flight program of the T2K was directed by yet another State Commission, this one headed by Maj. General Aleksandr A. Maksimov, the Deputy Commander of the Chief Directorate of Space Assets of the Strategic Missile Forces. One of Maksimov's more notable career duties had been service as the secretary of the State Commission for the R-7 ICBM and the early Sputnik launches during the 1950s. The T2K series would consist of three missions. The first flight would simulate a routine lunar landing, while the second and third would simulate potentially anomalous situations during a landing. About twenty primary systems would be monitored on each mission. In attendance for the first launch were Korolev's second wife Nina Ivanovna Koroleva and his daughter Natasha, who were at Leninsk for the opening of a new memorial in Korolev's honor.

The first T2K, vehicle no. 1, lifted on November 24, 1970, at 1400 hours Moscow Time and entered a 191- by 237-kilometer orbit inclined at 51.61 degrees. The spaceship was named Kosmos-379 by TASS; there was no hint that the flight had any relation to the piloted space program. After a thorough check of the on-board systems, at 0744 hours on November 25, controllers fired the main T2K engine under heavy throttling to simulate a landing on the lunar surface, including a "hover" phase. The resulting orbit was 192 by 1,210 kilometers at 51.65 degrees. Once again, controllers performed various checks of the T2K as it "rested on the Moon" for a day and a half. Finally, on November 27 at 1859 hours, the Lunar Landing Apparatus (the descent stage) was jettisoned, and the main engine fired once again, this time at maximum thrust, simulating a liftoff and entry into lunar orbit. After this maneuver, orbital parameters were highly elliptical: 177 by 14,041 kilometers at 51.72 degrees. The vehicle spent some subsequent time in stabilization mode to simulate maneuvers for rendezvous and docking with the LOK before the mission was declared a complete and unequivocal success.[124]

The momentum of this rare success in the piloted lunar program extended to the two remaining tests of the T2K. The second test was to simulate an aborted landing on the Moon. The spacecraft, named Kosmos-398 upon entering Earth orbit, was launched at 1514 hours Moscow Time on February 26, 1971. Initial orbital parameters were similar to the earlier ship: 191 by 258 kilometers at a 51.61-degree inclination. After two days in orbit, the Blok Ye main engine was fired at 0721 hours on February 28, simulating a landing attempt. After this, the descent stage was jettisoned, and the primary engine fired once more to insert the vehicle in its final orbit at 200 by 10,905 kilometers at a 51.59-degree inclination. Once again, the mission was flawless. The third and final test of the T2K was almost six months later. The ship, named

123. K. Lantratov, "The Fall From Orbit of the Last Soviet Lunar Ship" (English title), Novosti kosmonavtiki 25 (December 3–16, 1995): 32–36; Afanasyev, "Unknown Spacecraft."
124. Ibid.

Kosmos-434, was launched at 1250 hours Moscow Time on August 12, 1971, into an initial orbit of 189 by 267 kilometers at a 51.60-degree inclination. On this mission, the goal was slightly different: to use only the backup engine for liftoff "from the Moon," assuming that the primary one had failed. Less than a day after launch, at 0634 hours on August 13, the primary engine was fired, for the longest time on any of the missions, simulating a landing on the Moon. The new orbital parameters were 190 by 1,261 kilometers at the same inclination. Kosmos-434 remained static "on the Moon" for more than three days before using its reserve engine at 0840 hours on August 16 to fire into a new orbit of 186 by 11,804 kilometers at a 51.54-degree inclination. The second firing had been planned for an earlier time, but had to be delayed because of some minor technical problems, which did not detract from the completion of a successful mission.[125]

At the time of these apparently mysterious missions, Western observers closely monitored the orbital changes, concluding that the flights were part of a renewed Soviet effort to land cosmonauts on the Moon.[126] One of the more interesting postscripts to the T2K missions was the demise of Kosmos-434. In the summer of 1981, when the spacecraft was about to reenter, there were intermittent reports in the West that Kosmos-434 was a satellite with nuclear materials aboard, thus posing a threat to any people living over its descent track. The vehicle eventually reentered over Australian territory and fell harmlessly into the sea off the coast of China. To allay continuing fears, a spokesperson from the USSR Ministry of Foreign Affairs assured the Australian government on August 26, 1981, that the satellite did not carry any nuclear materials because it was "an experimental lunar cabin" with no "energy source."[127] Because "lunar cabin" was the term the Soviet press normally had used to describe the Apollo Lunar Module, the statement was a major landmark: it was the first official, albeit oblique, confirmation that the Soviet Union built hardware designed to land cosmonauts on the Moon.

The successful missions of the L1E and the T2K were significant morale boosters to the many thousands of engineers engaged in a program that had evinced few fruitful results. Firm commitments on a date for the lunar landing were fixed several times throughout 1970 and 1971. The original schedule produced by Mishin in July 1970, however, proved to be too optimistic. The launch date of the next N1 (booster 6L) was delayed primarily because of new concerns about discrete vibrations at launch. In addition, Mishin decided to begin using the new and improved Kuznetsov engines much earlier than planned (on booster 8L), requiring that rocket to be sent back to the plant for extensive redesign. The new schedule, truncated from before and prepared in September 1970, looked like this:

Date	Boosters	Missions	Engines
1971	6L, 7L	Automated lunar missions	Old engines
1972	8L, 9L, 10L	Automated and piloted lunar missions	New engines
1973	11L, 12L, 13L	Piloted lunar landings	New engines
1974	14L, 15L	Piloted lunar landings	New engines[128]

---

125. *Ibid.*

126. See, for example, Richard D. Lyons, "Experts Say Russia Plans Manned Landing on Moon," *New York Times*, September 6, 1971, pp. 1, 33; Donald C. Winston, "Soviets Prepare for Manned Moon Landing," *Aviation Week & Space Technology*, March 8, 1971, pp. 43–46; "Recent Cosmos Believed Advanced Hardware," *Aviation Week & Space Technology*, March 15, 1971, p. 18; "Russian Moves," *Aviation Week & Space Technology*, May 3, 1971, p. 13.

127. Robert Gillette, "Soviets Hint 'Experimental' Fallen Satellite Lost Its Way to Moon," *Washington Post*, August 30, 1981, p. A25.

128. The first piloted lunar landing would use a second booster, probably the unflown booster 4L, to launch the reserve LK known as the LK_R. See K. Lantratov, "Anniversaries: The 'Deceased' Lunar Plan" (English title), *Novosti kosmonavtiki* 14 (July 2–15, 1994): 60–61.

The final launches in 1974 would officially end the N1-L3 program, at the same time that more advanced lunar missions, still in the early stages of planning in 1970, would begin in 1974. Mishin personally briefed Soviet leader Brezhnev with this schedule at a meeting in October 1970. Even at this late date, Mishin continued to appeal to Brezhnev to commit to funding to build a full-scale static test stand for the N1 first stage, but these entreaties fell on deaf ears. If Mishin's promised schedule was met, however—and it seemed a fairly realistic assessment given the current pace of operations—then a Soviet cosmonaut would finally land on the Moon sometime in 1973, four years after Apollo 11.

The Scooper Comes Home

The untimely failure of Luna 15 during that historic week in July 1969 had not discouraged the design bureau at the S. A. Lavochkin State Union Machine Building Plant at Khimki in pursuing its primary objective of using the Ye-8-5 robotic spacecraft to recover lunar soil and bring it back to Earth. Although the unusually high stress of the summer of 1969 had evapo-rated, the pressure never completely disappeared. Because one of the new public doctrines of the Soviet space program was the automated exploration of the Moon, Chief Designer Georgiy N. Babakin had the dubious role of serving to fit the needs of the Soviet propaganda by delivering a successful sample return mission. The first attempts to do so after the July 1969 debacle were in the fall of 1969. On September 5, 1969, Maj. General Tyulin, the chair of the State Commission for the Ye-8 series of probes, reported that the central cause of the Luna 15 failure had still not been determined by engineers. Despite the gap in data, Tyulin opted to launch another scooper, the third in the series, on September 23.[129]

Ye-8-5 probe no. 403 was launched from site 81 at Tyura-Tam at 1700 hours Moscow Time on September 23, 1969. The Proton booster successfully inserted the payload into Earth orbit, but the Blok D translunar-injection stage failed to fire a second time to impart Earth escape velocity to the probe. Telemetry the following day indicated that a fuel injection valve had evidently become stuck during the first firing of Blok D to insert the payload into Earth orbit, and all the liquid oxygen had been sucked out before the second firing. Remaining as an inert payload in Earth orbit, the Soviet press quietly designated the satellite as *Kosmos-300* and promptly forgot about it. Ground controllers evidently attempted to control the descent of the probe for about four days, but the spacecraft eventually reentered harmlessly over the oceans. A second try came less than a month later. Ye-8-5 probe number 404 was launched on October 22 and successfully entered Earth orbit. After an hour, when the Blok D engine was timed to fire, the readings abruptly went off the scale, and communications were interrupted. For two hours, the flight control team attempted to regain communications, before finally receiving a report from the Kamchatka tracking station that not only had the probe not left Earth orbit, but that it had reentered and fallen in the ocean near Australia. This time, there was a failure in one of the radio-command blocks. Apparently a "minus" sign had not been removed from a program to command the guidance system for the firing.[130] The stranded probe was named *Kosmos-305*.

---

129. Kamanin, "I Feel Sorry for Our Guys," no. 13.

130. K. Lantratov, "The 'Late' Lunar Soil" (English title), *Novosti kosmonavtiki* 15 (July 16–29, 1994): 41–43; Kamanin, "I Feel Sorry for Our Guys," no. 14; O. A. Sokolov, "The Race to the Moon: A Look from Baikonur," presented at the 45th Congress of the International Astronautical Federation, IAA-94-2.1.610, Jerusalem, Israel, October 9–14, 1994. Note that the description of the failure given in the last source probably refers to Kosmos-305 and not Kosmos-300, as indicated by the author.

Trudging on, Babakin's engineers prepared the fifth sample returner, Ye-8-5 probe no. 405, for a launch in early 1970. The launch went off on February 6, 1970, but 126 seconds into the flight, the first stage exploded, destroying any hopes of a success.[131] Clearly, one of the bottlenecks in the program was the performance of the UR-500K Proton launch vehicle. Its record during 1967–70 had been perhaps one of the most dismal in the record of any launch vehicle developed by any spacefaring nation. Out of nineteen launches of the four-stage variant of the Proton booster up until February 1970, ten had completely failed to deposit their payloads into orbit, three had reached orbit but failed to send their payloads to escape velocity, and only the remaining six had been completely successful:

No.	Launch Date	Payload	Mission	Result
1	March 10, 1967	7K-L1/Zond	Kosmos-146	Success
2	April 8, 1967	7K-L1/Zond	Kosmos-154	Blok D failure
3	September 28, 1967	7K-L1/Zond		Stage I failure
4	November 22, 1967	7K-L1/Zond		Stage II failure
5	March 2, 1968	7K-L1/Zond	Zond 4	Success
6	April 23, 1968	7K-L1/Zond		Stage II failure
7	September 14, 1968	7K-L1/Zond	Zond 5	Success
8	November 10, 1968	7K-L1/Zond	Zond 6	Success
9	January 20, 1969	7K-L1/Zond		Stage II failure
10	February 19, 1969	Ye-8/Luna		Shroud failure
11	March 27, 1969	M-69/Mars		Stage III failure
12	April 2, 1969	M-69/Mars		Stage I failure
13	June 1, 1969	Ye-8-5/Luna		Blok D failure
14	July 13, 1969	Ye-8/Luna	Luna 15	Success
15	August 7, 1969	7K-L1/Zond	Zond 7	Success
16	September 23, 1969	Ye-8-5/Luna	Kosmos-300	Blok D failure
17	October 22, 1969	Ye-8-5/Luna	Kosmos-305	Blok D failure
18	November 11, 1969	7K-L1E/Kosmos		Stage III failure
19	February 6, 1970	Ye-8-5/Luna		Stage I failure

In fact, if there was any one reason why the coveted L1 circumlunar program had achieved success so late, it was Chelomey's Proton rocket. The failures were so glaring that after the secret February 1970 launch failure, some Western observers were claiming, correctly so, that the Proton was a severe bottleneck in Soviet space ambitions.[132]

Babakin was naturally concerned about the Proton's record. In March 1970, he met with Minister of General Machine Building Afanasyev and asked him to stipulate that Chelomey address the dismal record of the rocket and make necessary changes. For his part, Chelomey's design bureau undertook a short development program to requalify the booster, especially its first and third stages. As part of this effort, on August 18, 1970, at 0645 hours Moscow Time, TsKBM launched a three-stage UR-500K rocket on a suborbital mission to verify certain systems of the launch vehicle.[133] The flight, named 82EV, was evidently successful, as

---

131. N. G. Babakin, A. N. Banketov, and V. N. Smorkalov, *G. N. Babakin: zhizn i deyatelnost* (Moscow: Adamant, 1996), p. 66.

132. See, for example, Stewart Alsop, "Salt and Apollo 13," *Newsweek* (April 27, 1970): 112. According to the source, the Soviets had spent about $2 billion on the Proton program up to 1970.

133. Agapov correspondence, September 30, 1996; Babakin, Banketov, and Smorkalov, *G. N. Babakin*, p. 86.

space-related Proton launches finally resumed the following month after a long gap. In fact, the record of Proton flights following August 1970 showed a dramatic improvement, with failures becoming an occasional rarity.

The sixth scooper probe, Ye-8-5 no. 406, was launched at 1626 hours Moscow Time on September 12, 1970, into a parking orbit around Earth. First Deputy Minister of General Machine Building Tyulin once again served as chair of the State Commission. About seventy minutes after entering orbit, the Blok D stage fired to boost the payload toward the Moon. There was a short mid-course correction the following day before the spacecraft, named *Luna 16*, successfully flew into lunar orbit on September 17 using the 11D417 engine. The orbit was circular at an altitude of 110 kilometers at a 70-degree inclination to the lunar equator. There were two planned burns to adjust the orbit on September 18 and 19, the final firing leaving Luna 16 in a low elliptical orbit at fifteen by 106 kilometers at 71 degrees. The landing approach began as soon as the ship reached its low perigee. Unlike the Apollo Lunar Module, which followed a complex shallow approach to the landing site, the Ye-8-5 ship simply fired the main engine to cancel orbital velocity, causing a drop toward the surface, and then performed a final burn to ensure a soft-landing. The spacecraft's on-board control system fed attitude and altitude information into the internal gyro, and the ship's two side units were cast off just before commencement of the descent to the Moon. The engine fired for about 270 seconds, beginning at 1112 hours on September 20. The free-fall itself followed a preprogrammed instruction set modified by radar altimeter information on altitude and rate of descent. At a height of 600 meters, with the spaceship falling at a rate of 700 kilometers per hour, the on-board computer fired the main engine again. The engine cut off at twenty meters, prompting the two smaller engines to ignite to complete the descent. Luna 16 landed safely in the northeast portion of the Sea of Fertility about 100 kilometers east of the Webb crater. The landing velocity was nine kilometers per hour.[134]

Two cameras similar to the ones used on the earlier Ye-6 landers were installed on the main instrument section to swivel and return facsimile stereo images of the area between the ship's two landing pads to determine a precise spot for obtaining a sample. The spacecraft, however, landed in an area not illuminated by the Sun, and it is probable that the cameras were of no use. The hollow rotary/percussion bit, a hollow cylinder with cutters on the edge, was driven thirty-five centimeters into the surface for a seven-minute period to capture a small soil sample. During this phase, ground controllers were alarmed when telemetry information showed that soil resistance to the drill increased with depth, and then abruptly decreased, raising the possibility of a broken drill. Luckily, there was no damage, although Tyulin's team at Yevpatoriya terminated drilling at that point. The boom then lifted the sample to the open hatch of the small spherical return apparatus. Evidently, a significant amount of soil dropped out of the scooper during this upward movement. The total amount in the capsule was 105 grams. At 1043 hours on September 21, after more than a day on the surface, the ascent stage of Luna 16 fired its S5.61 engine to lift itself on a direct return trajectory to Earth. Roll thrusters provided spin control during the trip, ensuring proper thermal regulation. The remaining portion of Luna 16 continued to return data on local temperature and radiation conditions.

Straps holding the return apparatus to the ascent stage were severed at 0450 hours on September 24 at a distance of 48,000 kilometers from Earth. While the ascent stage burned up over Earth, the spherical capsule hit the atmosphere vertically at thirty degrees, traveling at eleven kilometers per second. As temperatures reached an incredible 10,000 degrees Centigrade, the capsule decelerated at up to 350 g's; a signal from a barometer commanded the

---

134. Andrew Wilson, *Solar System Log* (London: Jane's Publishing Co., 1987), p. 60–61. The exact landing coordinates were: 0° 41' S and 56° 18' E. See *Soviet Space Programs, 1966–70*, p. 198.

ejection of the top of the sphere at an altitude of fourteen and a half kilometers, thus unfurling the drogue parachute. The main parachute and four beacon antennas deployed at eleven kilometers. The capsule landed safely at 0826 hours, about eighty kilometers southeast of the town of Dzhezkazgan in Kazakhstan. After a trip to the Moon and back, the landing was only thirty kilometers from the projected target.[135]

This first recovery of soil from a planetary body by automated means was an outstanding accomplishment and a tribute to the ingenuity of Soviet engineering expertise. State Commission Chairman Tyulin recalled later that "the emotional strain on the State Commission and the technical leadership, as well as on all of participants of this unusual operation, was clearly noticeable, especially over the last 12 days," but that when the signals were received confirming a safe return of Luna 16, there was "boundless rejoicing" at the Flight Control Center.[136] Rescue teams located the capsule within minutes of the landing. They removed the soil container, which was flown to Moscow. There it was unsealed in a sterile chamber filled with inert helium. The analysis of the soil, performed by the V. I. Vernadskiy Institute of Geochemistry and Analytical Chemistry's Laboratory of Comparative Planetology, showed the composition of the dark powdery basalt material to be very similar to samples returned on Apollo 12 in November 1969. In June 1971, three grams of the Luna 16 sample were forwarded to NASA as part of a scientific agreement in exchange for three grams from the Apollo 11 samples and three grams from the Apollo 12 collection.[137]

The success of Luna 16 raised the inevitable comparisons with the Apollo program. Soviet commentators naturally made much of their recent accomplishment. Academician Boris N. Petrov told TASS on September 24 that automatic exploration cost one-twentieth to one-fiftieth as much as piloted space exploration. TsKBEM Department Deputy Chief Raushenbakh was more specific in his comparisons, suggesting on September 28 that the cost of the samples returned by Luna 16 were considerably less than those brought back by the Apollo missions.[138] While the two programs are difficult, if not impossible, to compare, it is a fact that the two Apollo missions up to that point had returned a far greater amount (sixty kilograms) of lunar rocks and soil than Luna 16 (0.105 kilograms). Based on the per capita cost of a kilogram of lunar soil from the Luna mission versus the Apollo missions, there is no doubt that the latter were far superior. But the amount of lunar soil returned is clearly poor measure of the true scientific value of a mission. In purely scientific terms, the U.S. astronauts conducted a wide array of experiments on the surface while Soviet controllers were extremely limited in their choice of research. The Apollo astronauts, for example, had a much greater ability to choose particular samples from a very large area compared to Luna 16. Finally, the costs of Apollo were associated with numerous intangible benefits—primarily associated with prestige—which clearly cannot be measured in the traditional sense. Luna 16 was certainly a remarkable technological accomplishment, but it was probably not, as Soviet officials of the day touted, a "cheaper and better" alternative to Apollo.

Luna 16 was followed, also in 1970, by another equally impressive achievement in the Soviet lunar exploration program: the flight of the Ye-8 lunar rover, which was named *Luna 17*. Incorporated into the L3 piloted lunar landing plan, the rover effort had by then assumed a life of its own. The first attempt to launch the mobile crawler had failed in February 1969. It was

135. Wilson, *Solar System Log*, pp. 61–63; Sokolov, "The Race to the Moon"; N. Kamanin, "I Feel Sorry for Our Guys" (English title), *Vozdushniy transport* 15 (1993): 12.

136. Mozzhorin, et al., eds., *Dorogi v kosmos: I*, p. 164.

137. Elizabeth K. Newton, *A Preliminary Study of the Soviet Civil Space Program, Volume 1: Organization and Operations* (Pasadena, CA: Jet Propulsion Laboratory, JPL D-7513, 1990), pp. 3–4, 33–34; Wilson, *Solar System Log*, p. 63.

138. *Soviet Space Programs, 1966–70*, p. 383.

almost two years before the second flight model, spacecraft no. 203, was ready for launch. At 1744 hours Moscow Time on November 10, 1970, a four-stage Proton lifted off and injected the spacecraft toward the Moon soon after. Following two mid-course corrections on November 12 and 14, Luna 17 entered orbit around the Moon on November 15. The parameters were eighty-five by 141 kilometers. The following day, the perigee was lowered to nineteen kilometers. The spacecraft deorbited and safely landed at 0646 hours, 50 seconds Moscow Time on November 17 in an ancient crater in the Sea of Rains. The landing profile was identical to that used on Luna 16. Two sets of ramps were lowered, and the five-person steering team at Yevpatoriya in Crimea commanded the strange-looking eight-wheeled robot down to the surface about an hour and a half after touchdown. Contact with the Ye-8 lunar rover, called *Lunokhod 1* in the Soviet press, was limited to about six hours a day when the Moon was above Earth's horizon.[139]

Among the scientific experiments aboard Lunokhod 1 was a penetrometer to test the soil's mechanical characteristics, which was used more than 500 times during the rover's sojourn. The *Rifma* x-ray fluorescence spectrometer, used about twenty-five times, was used to irradiate soil and record induced radiation to identify quantities of different elements. Adding a slightly international flavor to the mission, Lunokhod 1 also carried a three-and-a-half kilogram French-supplied instrument above the forward cameras, consisting of fourteen ten-centimeter silica glass prisms to bounce back pulses of ruby laser light fired from observatories in Crimea and France. Scientists first used this reflector on December 5 and 6, allowing the Earth-Moon distance to be measured down to an accuracy of thirty centimeters. Similar instruments, with less reflective capacity, were also carried on the Apollo landing flights. There was also an x-ray telescope and a gamma spectrometer on the spacecraft. Lunokhod 1, the first mobile vehicle to travel on the surface of another planetary body, had an initial design life of three lunar days (about twenty-one Earth days), but in fact operated for eleven lunar days (about seventy-seven Earth days). Tyulin's team commandeered the rover across 197 meters during the first lunar day, peaking on the fifth by covering 2,004 meters between March 7 and 20, 1971. Steering through the lunar landscape was evidently very difficult for the control team, primarily because of the six-second delay between the command and the execution of a maneuver.[140]

The crawler's remarkable journey came to an end at 1605 hours Moscow Time on September 14, 1971, when the last communications session was finished. The day after, TASS reported that the internal temperature of the rover had fallen because of decay of the nuclear heater during the night. For several days, controllers tried to reestablish contact with Lunokhod 1, but with no success, and all attempts to do so were terminated on October 4. Lavochkin Deputy Chief Designer Ivanovskiy, one of the principal architects of the mission, later recalled that the rover's internal batteries had been designed for only a certain number of cycles of charging and recharging, equivalent to three months. After exceeding their design lifetimes by almost eight months, the batteries simply gave up. Ultimately, the mission had been an outstanding success. Lunokhod 1 had covered an area of 80,000 square meters and taken 20,000 photographs and 206 panoramas of the lunar surface. During its 301-day, six-hour, and thirty-seven-minute mission, it had traveled 10,540 meters. It had crossed craters, climbed inclines, observed solar eclipses, and even found its way back to its mother stage in January 1971, taking one of the more impressive photos of the mission—a beautiful shot of Luna 17

---

139. Wilson, *Solar System Log*, p. 63; Konstantin Lantratov, "Anniversaries: 25 Years From Lunokhod-1" (English title), *Novosti kosmonavtiki* 24 (November 19–December 2, 1995): 70–79. The five-person team was part of a larger eleven-member team of N. Yeremenko and I. Fedorov (commanders), G. Latypov and V. Dovgan (drivers), K. Davidovskiy and V. Samal (navigators), L. Mosenzov and A. Kozhevnikov (engineers), V. Sapranov and N. Kozlitin (omnidirectional antenna operators), and V. Chubukin (reserve driver and operator). The landing coordinates for Luna 17 were 38° 17' N, 35° 00' W.

140. Wilson, *Solar System Log*, pp. 63–64; Lardier, *L'Astronautique Soviétique*, p. 269.

with its ramps lowered to the lunar surface. For ten months, it had withstood temperatures ranging from the intense cold of the lunar night (minus 150 degrees Centigrade) to the searing heat of the lunar day (over 100 degrees). Before losing contact, the controllers had managed to park the rover so that the laser reflectors remained in a usable position.[141]

Both the Luna 16 and Luna 17 missions were not only important scientific and technological achievements in their own right, but they also added weight to Soviet claims of the benefits of automated over piloted lunar exploration. It was only fitting that a third robotic lunar flight in 1970, the very last gasp of the L1 piloted circumlunar program, was sandwiched between the Luna 16 and Luna 17 missions. Although the circumlunar project had long since lost its political utility, there was still hardware remaining, specifically three flight-ready 7K-L1 vehicles. Piloted flights in the series had been suspended in the spring of 1969, but Mishin had doggedly pursued the idea of launching a crew regardless of the decisions from above. His view did have some rationale; the entire circumlunar system, the 7K-L1 vehicle, Blok D, and the Proton booster were, by mid-1970, ready for piloted flight. Such a mission, perhaps even multiple missions, would provide valuable experience in mounting more complex crewed lunar operations in the future. But the pressure not to do so was intense, and he eventually abandoned the idea. As a compromise, he was allowed to continue automated technological flights. Thus, in the fall of 1970, TsKBEM prepared one final L1 ship, spacecraft no. 14, to fly around the Moon.

The vehicle was launched at 2255 hours, 39 seconds Moscow Time on October 20, 1970, a month after the recovery of lunar soil samples by Luna 16. Following the standard checkout in parking orbit around Earth, the ship, called *Zond 8*, headed for the Moon. The flight trajectory of the spaceship differed with respect to earlier Zonds because, on this mission, engineers planned to use a different reentry profile—one in which the spacecraft would fly in over the Northern Hemisphere instead of the South Pole. Such a profile would allow ground stations on the contiguous Soviet territories to control most portions of the flight; in addition, the profile "was more advantageous in terms of power consumption and ensured a more precise splashdown."[142]

The day after launch, during the trip to the Moon, scientists at the Shternberg Astronomical Institute, at an observatory in the Zaylinskiy Altay, photographed the spacecraft against the stellar background, partly to confirm the accuracy of its trajectory. Photomultiplier tubes allowed identification of the ship, which was 328,000 kilometers from Earth at the time. Zond 8 itself photographed Earth on October 21. Besides cameras, the spacecraft carried unshielded aluminum foil "targets" similar to those on the Apollo solar wind collector packages. These were mounted on the outside of the descent apparatus to detect the isotropic composition of the solar wind. There was one mid-course correction at a distance of 250,000 kilometers on the following day, allowing the spacecraft to circle the Moon on October 24 at a minimum distance of 1,200 kilometers. The standard black-and-white and color photographs of the lunar surface were taken at distances of 9,500 and 1,500 kilometers. On the way home, there were two further mid-course corrections to sharpen its trajectory for the new reentry profile. Ground stations within the Soviet Union were able to control the dynamics of reentry as the Zond 8 descent apparatus flew over the North Pole during a ballistic reentry. It eventually splashed down 730 kilometers southeast of the Chagos Islands, in the Indian Ocean, at 1655 hours Moscow Time on October 27. The landing was only twenty-four kilometers from the intended target—and twelve kilometers from the nearest ship, a Soviet oceanographic vessel named the *Taman*, which picked up the capsule fifteen minutes later. Rescuers then transferred the vehicle to the *Semyon Chelyuskin*, which took it to Bombay, India, from where it was flown to Moscow.[143]

141. Lantratov, "Anniversaries: 25 Years From Lunokhod-1"; Wilson, *Solar System Log*, p. 64.

142. Mishin, "Why Didn't We Fly to the Moon?"

143. Glushko, ed., *Kosmonavtika entsiklopediya*, p. 130; Gatland, *Robot Explorers*, 150–51; Semenov, ed., *Raketno-Kosmicheskaya Korporatsiya*, p. 246; *Soviet Space Programs, 1966–70*, pp. 244–45; Joel Powell, "Research From Soviet Satellites," *Spaceflight* 25 (January 1983): 33–34. One source states that originally a guided reentry was planned, but because of a failure in an attitude control sensor, the vehicle performed a direct ballistic reentry. See Afanasyev, "Unknown Spacecraft."

The LI program was finally over. Started by the late Korolev in 1965, it was originally to have been a symbol of Soviet power during the celebrations for the fiftieth anniversary of the Great October Revolution in 1967. But after eleven launches and billions of rubles, the program receded into the background as an example of how politics, poor planning, a terrible launch vehicle, and bad luck could sabotage even the best of intentions. The results were, of course, not all bad. TsKBEM had performed two fully successful (Zond 7 and Zond 8) and two partially successful (Zond 5 and Zond 6) automated circumlunar missions. Much of the technology and expertise cultivated during the project were invaluable for the well-being of more ambitious efforts, such as the L3 landing program. An official history of the Zond program rightly notes a remarkable list of technical accomplishments from the project, but ultimately does not shirk from listing the most glaring omission: that no LI spacecraft was ever flown with a crew on board.[144] It is, however, undeniable that had the Soviets chosen to fly a crew around the Moon in 1970, they could have. TsKBEM still had two flightworthy vehicles remaining. But as Mishin noted twenty years later:

> . . . as a result of a decision by the higher authorities, the circumlunar flight by two cosmonauts in the UR-500K-LI program did not take place, despite the fact that the material base and the cosmonauts for the flight were ready. This decision resulted from the fact that the United States had already taken the lead from us in that direction. I feel that the decision was erroneous and that it did not take into consideration the opinion of the rank-and-file people and specialists who had labored heroically to execute the program. . . .[145]

The Zond program took its place in history as yet another Soviet space program that was unfulfilled in its dreams.

---

144. Semenov, ed., *Raketno-Kosmicheskaya Korporatsiya*, pp. 246–47.
145. Mishin, "Why Didn't We Fly to the Moon?"

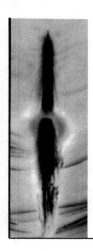

Chapter Seventeen
Dreams
Unfulfilled

Following the Soviet drive to reach the Moon during the Cold War is like chasing a trajectory that turns and twists at the least expected moments, often splintering into multidirectional paths, each road with its own story of triumph, tragedy, and irony. In 1969, the Soviet lunar program was at a crossroads and split into three distinct options for Soviet planners: the space station in Earth orbit, expanded lunar landing missions, and a Mars landing project. The Mars option was the most ambitious element of this triad, and the fact that it existed at all is testament to the often unrealistic ambitions of both space industry officials and the chief designers.

Aelita

One of the first Soviet-era science fiction novels was published in 1923. Authored by the well-known prose writer Aleksey N. Tolstoy, the novel was a narrative on the adventures of two Russian cosmonauts on the surface of Mars, a planet governed by a ruthless emperor. The novel, named _Aelita_ after its main character, the "Queen of Mars," was later turned into a movie of the same name, and it eventually became a widely popular film that was part of the cultural vernacular of the 1920s.[1] When the time came in 1969 to assign a cover name to the new Soviet Mars program, officials chose Aelita. Piloted expeditions to Mars had, of course, been part of exploratory studies in the Soviet Union well before 1969. Ten years earlier, a team under Maksimov at OKB-1 had begun research on the so-called Heavy Interplanetary Ship for flight around Mars and back. Another team, led by Feoktistov, studied a concept for landing a crew on Mars in a larger vehicle, also called the Heavy Interplanetary Ship. None of these studies had official sanction or funding from the Communist Party and government, but Chief Designer Korolev was sufficiently engrossed in the idea to assign a permanent team to study the problem. In the autumn of 1964, he established Department No. 92 under Ilya V. Lavrov to specifically study the prospects for a piloted Mars landing mission.[2]

As the N1-L3 program gathered steam during the mid-1960s, the work on the Heavy Interplanetary Ship moved ever so slowly to the sidelines. Still, Korolev managed to maintain his commitment to the idea and was particularly interested in the closed-loop life cycle systems that would be necessary for the long trip to Mars. Some of this research was carried out at the

1.   Peter A. Gorin, "Rising From a Cradle . . .: The Evolution of Public Perception of Spaceflight in Russian Prior to Sputnik," presented at the conference "Reconsidering Sputnik: 40 Years Since the Soviet Satellite," Washington, DC, September 30–October 1, 1997.
2.   Yu. P. Semenov, ed., _Raketno-Kosmicheskaya Korporatsiya "Energiya" imeni S. P. Koroleva_ (Korolev: RKK Energiya, named after S. P. Korolev, 1996), p. 168.

Physics Institute of the Siberian Department of the Academy of Sciences at Krasnoyarsk. Legend has it that two of the young scientists working on the problem once met with Korolev and offered him water regenerated by chlorella from human urine. The chief designer declined the offer, but remained very interested in the problem. The same institute designed a closed biosphere designated *Bios-1*, which was first tested by Iosif I. Gitelzon, a thirty-five-year-old medical doctor who had been one of the men who had met with Korolev.[3] Other organizations were also involved in the overall research. The Tomilino-based KB Zvezda designed one version of a system for the spacecraft in which food consisted of sublimated provisions based on two criteria: high nutrition value and low specific mass. A small hydroponic hothouse equipped with external solar concentrators would be used for additional nutrition.

Eventually by the late 1960s, presumably to optimize all work on Martian spacecraft, the two different Heavy Interplanetary Ship designs were unified into one, the Feoktistov proposal. A special ground test simulator for the ship was built after Korolev's death, and it was there on November 5, 1967, that three men—physician German A. Manovtsev (group leader), biologist Andrey N. Bozhko, and technician Boris N. Ulybyshev—entered the laboratory complex for a simulation of a long-duration piloted spaceflight. The team used water and oxygen regenerated from body waste, including urea, transpiration moisture, and exhaled carbon dioxide. For food, the researchers used freeze-dried food and green vegetables grown in a ground-based greenhouse. The greenhouse used simulated sunlight and ion-exchange resins saturated with nutrient substances instead of soil. Solid biological waste was simply removed from the cabin. They finally exited on November 5, 1968, a year after their entry.[4]

The work on the Heavy Interplanetary Ship slowed down after Korolev's death, but with the renewed interest in Mars after Apollo 8, these studies assumed an increased importance. Coincidentally or not, in 1969, TsKBEM issued an "experimental design" of a piloted Martian landing spacecraft, the most detailed technical description yet of such a vehicle. Spurred by the abrupt interest from Ustinov and Afanasyev to pursue a Mars project, engineers could be forgiven for hoping that this design would see the light. The ship, now called the Martian Expeditionary Complex (MEK), consisted of:

- An interplanetary orbital ship carrying the crew and primary on-board systems
- A Martian landing ship for landing on the surface of Mars
- A return apparatus for flight to Earth in which the crew would reenter Earth's atmosphere
- Powerful engine units with nuclear reactors and electric rocket engines

The basic requirements for the 1969 mission were to carry out a Mars landing during a 630-day (or 1.7-year) mission, with thirty days spent orbiting Mars. A total of six cosmonauts would be aboard the ship; three of them would spend at least five days on the surface. The primary propulsion system on the Martian ship would be electric rocket engines using nuclear power sources for the main part of the journey and liquid-propellant rocket engines for operations near Mars.[5]

3.   V. Nelyubin, "Three Flights to Mars. Soviet 'Cosmonauts' Made Them in the Early 1960's Without Leaving the Earth" (English title), *Komsomolskaya pravda*, January 15, 1992, p. 4. The other doctor who offered Korolev the recycled sample was I. Terskov.

4.   .G. V. Petrovich, ed., *The Soviet Encyclopaedia of Space Flight* (Moscow: Mir Publishers, 1969), pp. 376–77; Mikhail Rebrov, "Saga from the Archives of Document No. 23891 and an Unknown Space Project" (English title), *Krasnaya zvezda*, May 13, 1995, p. 6. Earlier simulations of ten to 120 days duration were carried out by 1964. See "For Future Space Flights: A Unique Experiment of Soviet Scientists" (English title), *Pravda*, September 11, 1964, p. 4.

5.   Semenov, ed., *Raketno-Kosmicheskaya Korporatsiya*, p. 280; Igor Afanasyev, "Piloted Flight to Mars . . . A Quarter Century Ago" (English title), *Vestnik vozdushniy flota* no. 7–8 (1996): 103–05.

This is a model of the pencil shaped MEK piloted Martian spacecraft offered in 1969 by Mishin's design bureau as part of the Aelita effort. At the base of the stand is another model evidently showing the Martian lander with its atmospheric braking aeroshell (shown in more detail in the bottom photo). On the right is a robotic Martian spacecraft. (copyright Mark Wade)

In Earth orbit, the MEK looked like a long needle. The 150-ton complex would be assembled in Earth orbit after two launches of a modified N1 booster. The first rocket would carry two components: the Martian Orbital Complex (MOK) and the Martian Landing Complex (MPK). The second N1 would carry a fully functioning low-thrust electric rocket engine powered by two nuclear reactors. Each reactor was installed on one extreme end of the complex and protected from other systems by a "shaded shield"; the cone-shaped propellant tanks for the electric rocket engines would provide additional protection to the crew from radiation from the reactors. The actual propulsion nozzles would be placed between the shade and the tanks. The complex would also have an extensible telescopic thermionic radiator for the energy sources, which would have a node to allow for docking and undocking to the MOK and MPK.

The MOK formed the main areas of living for the crew. From one end to the other, the complex had seven sequential sections: the instrument-aggregate compartment, the working compartment, the laboratory compartment, the biotechnology compartment, the living compartment, the "salon" compartment, and the orientation engine compartment. The MPK had an unfurlable aeroshell for aerodynamic braking into the Martian atmosphere. It was located behind the "shaded shield" of the main spacecraft. After separating from the main spacecraft complex in Martian orbit, it would discard its docking apparatus used for operations in Earth orbit and then use a liquid-propellant rocket engine to soft-land on the surface of the planet. The aeroshell encased a cylindrical "living compartment" linked to the main crew quarters via a hatch, as well as a two-stage ascent stage with a spherical cabin.

The MEK also contained the main crew return apparatus for returning the crew to Earth. The capsule was essentially a larger version of the "headlight-shaped" Soyuz descent apparatus with a lift-to-drag ratio of about 0.45, sufficient to significantly reduce g-levels upon terrestrial reentry. The capsule had a base diameter of 4.35 meters and a height of 3.15 meters.

The MOK and MPK would dock in Earth orbit with the electric rocket engine plus nuclear reactor payload. Docking would be followed by the ignition of the electric engines to begin its slow acceleration into ever larger spirals around Earth. After the complex cleared Earth's radiation belts, a Proton rocket would launch a 7K-L1 Zond-type spacecraft into Earth orbit with a crew. The Blok D fourth stage would accelerate the Zond to meet with the MEK in high orbit. Having entered the MEK, the crew would verify the operation of all systems on the complex with the option of abandoning the vehicle if there were serious problems. After reaching transplanetary velocity, the MEK would "shoot" out of Earth orbit in a trajectory toward the Red Planet. The electric engines would shut down at this point and stay in "cold storage."[6]

Calculations at the time had allowed engineers to compute the cumulative dose of radiation during periods of high solar activity that doctors believed would be acceptable for interplanetary crews. Based on these data, the crew of the MEK would stay in the special radiation shelter, which was in the form of a passage in the main instrument-equipment bay of the ship. The workload of the cosmonauts during both the outbound and inbound trips would be reduced as much as possible by making operations almost fully automated. Computers would deliver information on the spacecraft systems' operation based on an algorithm producing three values: "normal," "not normal," and "failure." The crew would be able to carry out any in-flight repair of the ship's radio and electronic equipment, designed to be easily accessible in the form of replaceable units. The effects of long-term gravity on the crew was still a potential unknown in 1969, and one option engineers seriously considered was the use of artificial gravity by rotating individual portions of the giant spacecraft around its axis. Research later proved that such rotations would be harmful to the body because of the appearance of "Coriolis" acceleration that distorted the human perception of gravity.[7]

The coast to Mars would take 150 days, after which the electric engines would start operating again to perform Mars orbit insertion. The MEK would take sixty-one days to brake into high orbit and a further twenty-four days to shift to low orbit. The crew would spend an additional week surveying possible landing sites for the MPK. Three of the six cosmonauts on board would then enter the lander and touch down on the surface. After about a weeklong mission on the surface, the ascent stage of the MPK would lift off and automatically rendezvous with the MOK. The crew would transfer from the former to the latter's living compartment, and the no-longer-needed lander would be discarded. A week later, the

---

6.   Afanasyev, "Piloted Flight to Mars."
7.   I. B. Afanasyev, "Unknown Spacecraft (From the History of the Soviet Space Program)" (English title). *Novoye v zhizni. Nauke, tekhnike: Seriya kosmonavtika, astronomiya* no. 12 (December 1991): 1–64.

crewmembers would begin their return trip in the MOK—seventeen days to escape Mars and another sixty-six days to gather velocity to reach Earth. During passive flight, the spaceship would pass as close to the Sun as possible, flying between the orbits of Venus and Mercury to accrue more velocity. Another seventeen days of active engine firing would lead to a second passive phase. Three days before reaching Earth, the electric rocket engines would be switched on again. The crewmembers would separate from the main MEK spacecraft in their return apparatus and land by parachute back on Earth with the results of their scientific experiments and Martian soil samples.[8]

Serious work on closed-cycle life support systems in support of the Mars program was carried out at the premises of the Moscow-based Institute for Biomedical Problems. In 1970, as part of the MEK project, scientists at the institute created a Scientific-Experimental Complex (NEK) for "special biomedical testing of prospective space life-support systems." The NEK consisted of three modules: one with a volume of 150 cubic meters, the second with a volume of 100 cubic meters, and the third, the aggregate compartment, with a volume of fifty cubic meters. Each module was connected with an airlock and had radio-television systems, anti-fire alarm systems, and extinguishers. Two of the modules had special areas for rest and athletic training. There was also a special kitchen for preparing food from sublimated products, as well as a doctor's area with a full complement of medication and instruments.[9]

One of the most intensive areas of focus in the design of the MEK was the nuclear energy source, not only to power the ship, but also to provide power to the electric rocket engines. In the 1960s, scientists and engineers at TsKBEM had engaged in research on creating a new class of slow-melting and high-temperature materials and new heat carriers—that is, new technologies for facilitating the creation of small-scale thermionic reactors. Several different complex test stands were built for testing methods, materials, and equipment at very high temperatures. Between 1965 and 1968, TsKBEM, together with the Physical-Power Institute at Obninsk, designed and manufactured a new thermionic reactor using fast neutrons. By 1970, they had created the new FS-1 critical test stand, essentially a reactor of zero power, to verify changes in the structure, geometry, composition, and configuration of the primary components of the nuclear-physical model of the thermionic reactor. Eight critical assemblies were made at the time, leading eventually to the creation of the 11B97 nuclear energy source.

Based on this research, for the MEK, TsKBEM engineers worked on a draft plan between 1966 and 1970 for nuclear energy units and electric rocket engines for the spacecraft and its launch vehicle. The power units and the rocket engines were created in single block (YaE-1 and YaE-1M) and triple block (YaE-2 and YaE-3) configurations, with each block consisting of one thermionic reactor. The performance characteristics were:

Unit	Power Output
YaE-1	2,500–3,000 kilowatts
YaE-1M	5,000 kilowatts
YaE-2	Three by 3,200 kilowatts
YaE-3	Three by 5,000 kilowatts

The spacecraft would have two low-thrust electric rocket engines of 6.2 and 9.5 kilograms thrust, respectively. Their specific impulses were remarkable, attesting to their high-performance

---

8.   Afanasyev, "Piloted Flight to Mars."
9.   Aleksandr Andryushkov, "The Secret Cosmodrome" (English title), *Krasnaya zvezda*, December 3, 1994, p. 6.

capabilities: 5,000 and 8,000 seconds, respectively. All of the materials on these energy units and associated engines were examined and approved by an expert commission of the Academy of Sciences under the leadership of Academicians Aleksandr P. Aleksandrov and Boris N. Petrov. The commission recommended further work to create the YaE-2 and YaE-3 units on an experimental basis.[10] Through early 1969, TsKBEM engineers were seriously considering using a *recoverable* nuclear reactor aboard the Mars spacecraft.

Work on the MEK was, of course, not isolated from the development of a suitable launch vehicle for sending the spacecraft to Mars. Conceptions of an uprated N1 to use for the mission remained in constant flux throughout 1969 as different models were proposed at different points. Early in the year, the most favored version was the N1F-V3, a technical proposal distinguished from the original N1 by the use of improved first and second stages and a completely new third stage using high-energy liquid hydrogen–liquid oxygen (LOX) as propellant. By March 1969, the most likely plan was to use two N1-derived boosters for launching components of the MEK into Earth orbit. One of the N1s would use the new liquid hydrogen-LOX Blok S on fourth stage. Based on this research, throughout April, there was intensive work on a radically improved variant, called the N1M. On May 28, 1969, Mishin signed the predraft plan for the N1M booster, designed specifically to carry out a Mars landing project. Among five projected variants of the N1M, three used liquid hydrogen-LOX engines on the second and third stages. The first stage would use thirty powerful 250-ton-thrust engines.[11]

Mishin's N1M-MEK Mars landing plan was not the only component of the new Mars offensive in 1969. When Soviet space leaders such as Smirnov and Afanasyev provisionally approved a Mars program to take the steam out of Apollo, there was a clear consensus that this would have to be a massive integrated project involving the major Soviet space design organizations. The official decree in support of the Aelita program was issued by the Ministry of General Machine Building in resolution no. 232 on June 30, 1969, two weeks before the flight of Apollo 11 to the Moon.[12] According to the order, the assigned chief designers were to deliver "materials" for the Aelita program by the third quarter of 1970. Participating in the effort was not only Mishin's TsKBEM, but also General Designer Chelomey's TsKBM and Chief Designer Yangel's KB Yuzhnoye. By August 1969, there were, in fact, three complete predraft plans for a Soviet Mars landing project, one each from the three design bureaus. The volume and scale of the work, however, seem to have discouraged even the most enthusiastic of participants. By the end of 1969, both Mishin and Yangel pulled out of Aelita, leaving it wholesale to Chelomey. Mishin clearly had a good reason: by the end of 1969, he was knee-deep in a new space station program. At the same time, his organization was involved in the flight testing of the troubled N1 and formulating variants of the N1 for improved piloted lunar landing missions for the early 1970s. It would simply be impossible to manage a Mars program concurrently.

10. Semenov, ed., *Raketno-Kosmicheskaya Korporatsiya*, pp. 411–12; A. Koroteyev, Yu. Demyanko, and Ye. Kuzmin, "From the History of Space Science: The Scientific-Research Institute of Reactive Propulsion" (English title), *Aviatsiya i kosmonavtika* no. 6 (November–December 1993): 39–41.

11. For mentions of the N1M in Russian documentation, see Semenov, ed., *Raketno-Kosmicheskaya Korporatsiya*, pp. 280, 412.

12. V. M. Petrakov, "Soviet Rockets for Space Apparatus," *Journal of the British Interplanetary Society* 49 (July 1996): 277–80; V. M. Petrakov, "From the History of the Development and Creation of Carrier-Rockets in the USSR" (English title), in *Trudy XXup chteniy, posvyashchennykh razrabotke nauchnogo nasladeniya i razvitiyu idey K. E. Tsiolkouskogo* (Moscow: RAN, 1994), p. 173; Sergey Khrushchev, *Nikita Khrushchev: krizisy i rakety: uzglyad iznutri: tom 2* (Moscow: Novosti, 1994), p. 526. One somewhat unreliable source states that the order came on June 30, 1968. See Mikhail Rudenko, "Designer Chelomey's Rocket Planes" (English title), *Vozdushniy transport* 52 (1995): 8–9.

Nothing is known about the Yangel offer, but given his previous record with piloted space projects, it is not surprising that he, too, did not participate in Aelita after 1969.[13]

Characteristically, Chelomey's offer for Aelita was far more ambitious than Mishin's NIM-MEK idea, bordering almost on fantasy. His own stab at a lunar landing project, the UR-700/LK-700 project, had died a slow death in early 1969, but it had provided a sound basis to consider more advanced concepts for the Mars effort. For Aelita, Chelomey used the UR-700 as a springboard and offered the even more gigantic UR-700M rocket—a launch vehicle so massive that it was quite possibly the most powerful booster ever seriously conceptualized anywhere in the world. The only comparable studies were NASA's Nova heavy-lift booster proposals dating from the early 1960s.[14]

By April 1969, General Designer Chelomey was looking at several different preliminary variants of the UR-700M, each with differing capabilities and configurations. The mass and performance characteristics were unprecedented:

Variant	Launch Mass	Payload to Earth Orbit	Propellants
1	4,820 tons	130–150 tons	Conventional
2	7,890 tons	230 tons	Conventional
3	7,890 tons	300 tons	Liquid hydrogen-LOX
4	Unavailable	350 tons	Nuclear
5	Unavailable	1,700 tons	Nuclear

Variants 1 and 2 differed by the composition of the number of strap-ons or engines. Among the missions being considered for these two versions were lunar landing expeditions lasting thirty days, automated flight to Mars and Venus with the landing of eleven-ton modules on the surface, and piloted landing expeditions to Mars with three cosmonauts on Mars for thirty days. In addition, on variant 2, two payload blocks of 230 tons would allow for the testing of a special nuclear rocket engine on an upper stage. Variant 3 would use high-performance cryogenic propellants, allowing for a landing on the Moon of six cosmonauts for missions lasting from one month to one year, piloted flight to Mars and Venus using atmospheric braking, and, with the use of two 230-ton payload blocks linked in Earth orbit, a landing expedition to Mars of four cosmonauts. Certainly, the most ambitious early conceptions were variants 4 and 5 using nuclear rocket engines on the third stage of the UR-700M. Engineers considered two different design schemes for the engine, one using a gas-phase reactor and the other a solid-phase reactor.[15]

These initial exploratory studies of the UR-700M led to two different layouts for the rocket. The first variant of the UR-700M was similar to the UR-700 in basic design—that is, it consisted of a core of three modules (the second stage) surrounded by three strap-on clusters (the first stage), each consisting of two modules. The change was in the engines: Chelomey substituted each RD-270 with four RD-253s. Thus, despite Chelomey's intensive criticisms of

13. In Petrakov, "Soviet Rockets for Space Apparatus," the author states that "Mishin, already heavily involved in manned spacecraft work for the space station programme, declined the work [on the Mars expedition], leaving the TsKBM in sole charge of Aelita."

14. For an overview of the Nova studies, see Keith J. Scala and Glen E. Swanson, "They Might Be . . . Giants," *Quest* 1(3) (Fall 1992): 12–27; Keith J. Scala and Glen E. Swanson, "They Might Be . . . Giants," *Quest* 2(1) (Spring 1993): 17–26; Keith J. Scala and Glen E. Swanson, "They Might Be . . . Giants," *Quest* 2(2) (Summer 1993): 4–20.

15. For another version of the UR-700M, see Christian Lardier, *L'Astronautique Soviétique* (Paris: Armand Colin, 1992), p. 252.

the NI booster for having too many engines on its first stage, he was now choosing that same option. One wonders if his criticisms of the NI abated with the emergence of the UR-700M. Chelomey's new rocket, with a total of thirty-six engines firing at liftoff, would develop about 5,400 tons of sea-level thrust. The rocket was topped off by a third stage of four modified RD-253s for altitude use, as in the UR-700. Its fourth stage was a bold new step in rocket design technology: Chelomey proposed the use of a nuclear rocket engine, the RD-410, a relatively unknown engine developed by Glushko's EnergoMash design bureau. Total Earth-to-orbit payload capability for this version of the UR-700M was 240 tons.[16]

Very little is known about the RD-410 engine, except that it had a thrust of seven tons. It is not clear whether the RD-410 engine was the same unit as the similarly designated RD-0410, also a nuclear rocket engine developed at the very same time as a cooperative effort between the Design Bureau of Chemical Automation (formerly OKB-154) and the Scientific-Research Institute of Thermal Processes. The RD-0410, with a thrust of just over three and a half tons, was a highly advanced engine, exceeding in its performance characteristics even concurrent American nuclear engine models. A stand for testing the engine was built beginning 1962 by KB Luch at a secret site about fifty kilometers southwest of Semipalatinsk-21. Testing began in 1971.[17]

The second variant of the UR-700M was truly a monster. Instead of the standard modules just over four meters in diameter so favored by Chelomey, engineers came up with a central core twelve and a half meters in diameter surrounded by four nine-meter-diameter blocks. The core (the second stage) would use twelve 600-ton-thrust engines, while the strap-ons (the first stage) would each use eight of the same engines. These engines, working on LOX and kerosene, would be developed by Glushko, who evidently had finally decided to abandon his boycott of LOX engines. A third stage, with a diameter of twelve and a half meters, would use six NK-35 engines, each with a thrust of 220 tons. These were new high-performance liquid hydrogen engines developed by Kuznetsov's KB Trud. Compared to the Saturn V's modest 130 tons, this behemoth would be capable of lifting 750 tons to Earth orbit. With a launch mass of 16,000 tons and a length of about 145 meters, this variant of the UR-700M was evidently the most preferred version for Aelita because it satisfied one of the main criteria of the plan—to use only a single docking (that is, assembly of a 1,500-ton complex) in Earth orbit to accomplish the Mars landing. Other requirements included simultaneous development of all the rocket engines, a "packet" layout for the booster, the use of multiple engines on each block, the possibility of manufacture of the giant in a major city, and extensive ground testing.[18]

Very little is known about the MK-700 Martian landing spacecraft conceptualized for the Aelita program. No doubt, the actual ship traced its lineage to the abandoned LK-700 lunar lander. The spacecraft looked roughly like a series of four truncated cones one on top of the other. The ship had a pair of large solar panels to provide power during the trip. As a whole, the development of the UR-700M rocket was assigned to TsKBM's Branch No. 1 at Fili, although Chelomey's main center at Reutov took the responsibility of developing the MK-700 piloted spaceship. Chelomey was enthusiastic about the entire effort, perhaps seeing in it a possibility to vindicate his various defeats in the space program at the hands of his enemies. Minister Afanasyev, a staunch supporter of Chelomey's, seems to have been the primary

16.   V. Karrask, O. Sokolov, and V. Shishov, "Known and Unknown Pages of the Russian Khrunichev Center's Space Activity," presented at the 47th Congress of the International Astronautical Federation, Beijing, China, October 7–11, 1996.

17.   Leonid Kvasnikov, Anatoliy Kostylev, and Vladimir Maksimovskiy, "Nuclear Rocket Engines" (English title), Vestnik vozdushniy flota no. 6 (November–December 1996): 53–55.

18.   Petrakov, "Soviet Rockets for Space Apparatus"; Petrakov, "From the History of the Development and Creation of Carrier-Rockets."

This is a close-up view of Chelomey's MK-700 Martian lander spacecraft that was proposed as part of the Aelita program to send Soviet cosmonauts to Mars in the late 1970s. In the background is a model of a variant of the Salyut space station. (copyright Mark Wade)

instigator in the government in favor of Chelomey's Aelita project. Remarkably, Chelomey delivered on his promise. In April 1970, he completed the predraft plan for the MK-700, and in October of the same year, he signed off on the predraft plan for the UR-700M rocket.[19] Somewhat unrealistically, Chelomey promised that he could bring the project to fruition within *three* years.

At the time, Aelita not only included the UR-700M-MK-700 project but also encompassed a larger Mars-directed offensive, including automated missions in 1971, 1973, and 1975, leading to a piloted landing between 1978 and 1980. One of the more interesting missions of this armada was the 5NM mission planned for launch on the N1 rocket. Also known as the Heavy Interplanetary Automatic Ship, the project was supervised by the Lavochkin Design Bureau under Chief Designer Babakin. Inspired by the success of the Luna 16 sample return, Babakin proposed using the 5NM spacecraft to recover a sample of Martian soil. The launch would occur in 1975, the ship would land on Mars in 1976, and then it would return to Earth in 1977.[20]

For all of the enthusiasm of Afanasyev and Chelomey for Aelita, the goals of the program could not be justified given the enormous amounts of expenditures involved. As a respected Russian space historian noted in 1991:

> . . . *even as the proposals for [the UR-700M-MK-700] program were being developed, it became clear that the impact of the first flight of a man to Mars on public opinion would be disproportionately small in comparison with the material expenses that would attend the flight.*[21]

Aelita really had no chance. By September 1970, the Military-Industrial Commission considered eliminating Aelita from its next five-year plan, 1971–75, but apparently after further discussions opted to include it. The participants were to produce a draft plan for the project in

19. Khrushchev, *Nikita Khrushchev: tom* 2, p. 526; Petrakov, "Soviet Rockets for Space Apparatus."
20. K. Lantratov, "To Mars!" (English title), *Novosti kosmonavtiki* 21 (October 7–20, 1996): 41–51.
21. Afanasyev, "Unknown Spacecraft," p. 40.

1972. The forces against the massive undertaking, however, proved to be too strong. Soon after Chelomey finished the predraft plans for his booster and spacecraft, his ambitious idea died a slow death. By the end of 1972, the Soviet piloted space program remained engrossed in both its space station and the lunar landing projects; cosmonauts on Mars would have to remain a dream of the future. Like most of Chelomey's other projects, disinterest from the Soviet leadership did not deter him from quietly pursuing his own ideas, and it is quite likely that some low level of work continued on the UR-700M and its MK-700 ship for several years into the early 1970s. Some components of Aelita escaped outright cancellation. In particular, Babakin's 5NM project remained a strong contender for approval. In general, however, the aborted Soviet Mars offensive of 1969–70 was a child of political circumstance. Born out of the shock of Apollo 8, it did not have sustainability to survive into less politically charged times of the space race.

A Month on the Moon

In July 1969, when the N1 rocket exploded on its pad at Tyura-Tam, no one could have guessed that it would take two years before another N1 launch took place. The investigation into the accident did not finish until July 1970; in fact, it was as late as May 1971 before one of the subcommissions submitted its findings, concluding that the explosion had occurred because of a problem in an oxygen sensor. By December 1970, Mishin was looking ahead to the next launch of booster no. 6L in January 1971, with booster no. 7L following in June. Throughout December 1970 and January 1971, the Council of Chief Designers in charge of the N1-L3 program met several times to discuss the prelaunch preparations.[22] There were still doubts on many technical issues that were not clarified to the satisfaction of several members, including such perennial problems as data on the pulsation pressure of the tanks and pipelines. On February 10, 1971, the technical leadership of TsKBEM met specifically to discuss the N1-L3 program, assessing the pace of preparations for future launches in the effort. The next N1 launch would carry a mass model of the L3 stack instead of the actual orbiter and lander—a decision most likely taken so as not to lose flight models of the Lunar Orbital Ship (LOK) and the Lunar Ship (LK) in case the N1 rocket failed to deposit its payload in orbit.

The assessment of preparations for the two subsequent launch stacks, boosters 7K and 8L, was mixed. While both N1 rockets were on schedule, there was still much uncertainty regarding their payload blocks, primarily because of delays in the delivery of components from subcontractors, in particular, the Arsenal Machine Building Plant, responsible for manufacturing the attitude control blocks for the orbiter and the lander. Booster 7L would carry an automated LOK and a mass model of the LK, while 8L would carry automated models of the LOK and LK. These two vehicles would carry out a fully automated lunar landing on the surface of the Moon. Booster 10L would carry the first piloted LOK in addition to an automated LK for a repeat of a robotic landing. The first piloted landing on the surface of the Moon was set for March 1973 on booster 11L.[23] Tests throughout 1971 continued for certifying the LOK for piloted flight.

---

22.   The Council of Chief Designers for the N1-L3 program in December 1970 were: V. P. Mishin (TsKBEM and council chief), A. P. Abramov (TsKBEM), V. P. Barmin (KB OM), A. G. Iosifyan (VNII EM), A. M. Isayev (KB KhimMash), I. I. Ivanov (KB Yuzhnoye), N. D. Kuznetsov (KB Trud), N. A. Lobanov (NII AU), A. M. Lyulka (KB Saturn), A. S. Mnatsakanyan (NII TP), N. A. Pilyugin (NII AP), M. S. Ryazanskiy (NII Priborostroyeniya), G. I. Severin (KB Zvezda), V. G. Stepanov (Turayevo Branch of MMZ Soyuz), G. I. Voronin (KB Nauka), and M. K. Yangel (KB Yuzhnoye). Additionally invited to participate in the proceedings were: P. A. Agadzhanov (TsKIK), A. Yu. Ishlinskiy (Institute of Mechanics AN SSSR), A. G. Karas (GUKOS), M. V. Keldysh (AN SSSR), V. Ya. Likhushin (NII TP), G. P. Melnikov (NII-4 Space Branch), Yu. A. Mozzhorin (TsNIIMash), B. N. Petrov (Interkosmos), G. I. Petrov (IKI AN SSSR), and V. P. Pukhov (NII KhimMash).
23.   N. Kamanin, "I Feel Sorry for Our Guys" (English title), *Vozdushniy transport* 15 (1993): 12.

These included water-landing tests and the verification of the giant launch escape tower for the L3 stack. The LOK would be the first Soviet piloted spacecraft whose primary landing target was the Indian Ocean. TsKBEM engineers also overcame major technical obstacles in building the first fuel cells for a Soviet piloted spacecraft, six years behind the United States. By 1971 and 1972, the engineers were ground-testing a four-and-a-half-kilowatt power supply unit called *Volna* for the LOK, which ran on alkaline and hydrogen-oxygen fuel cells with an efficiency rating of about 60 percent. Volna would provide electricity, oxygen, water for drinking, and support services during the lunar mission.[24]

Preparations for the next N1 launch were bogged down in technical delays, compounded by a lack of confidence in a success. It was only in late May 1971 that the rocket was finally moved to the second N1 launch pad at site 110L, the one that had remained intact after the catastrophic explosion during the summer of 1969.[25] Evidently, Chief Designer Barmin's engineers had not yet begun reconstructing the destroyed pad at site 110P. By December 1970, the consensus was to carry out the remaining N1 test launches from 110L while, at the same time, begin the construction of two completely *new* pads elsewhere at the launch range.

Booster 6L's payload consisted of mass mock-ups of the LOK and LK. There was no functional emergency rescue system on top of the L3 stack. The primary objective of the mission was to test, as simply as possible, the operation of the first three stages of the N1. No lunar operations were planned, and, in fact, the N1 was to be launched outside a convenient lunar launch window.[26] There was some minor drama during the searing hot days preceding the launch attempt. In an unexpected act of nature for arid Tyura-Tam, there was a violent rainstorm at the launch site while the N1 rocket was installed at the pad. For Kazakhstan, this was quite an anomaly, and engineers were very worried about the effect of rain on the N1 rocket's electronic circuitry. Some State Commission members proposed bringing the booster back to the assembly-testing building and then "drying" it, but Mishin was against this, apparently fearful that such a move would serve no purpose other than delaying the launch by days if not weeks. In the end, Mishin got his way.[27] The State Commission originally set the launch for June 20, but postponed it initially by two days to June 22. But there were more delays. Air Force representative Col. General Kamanin wrote in his diary on June 24 that:

> The launching of the N1 has again been put off. Now Mishin hopes to put it into space on June 27, but there are so many failures and malfunctions that this date may also prove unrealistic. General [Aleksandr G.] Karas [the commander of the Chief Directorate of Space Assets] called from the launch site today. He is dejected. The telemetry equipment on the N1 has given out, and there are other important malfunctions which may again delay the launching. This bad rocket is a great liability to our space program.[28]

The "bad rocket" was finally launched at 0215 hours, 8 seconds Moscow Time on June 27, 1971. As soon as the booster lifted off, telemetry on the ground indicated that the roll control system was behaving abnormally. There were unexpected gas-dynamic moments (eddies and countercurrents) at the base of the booster, which caused the N1 to roll around its axis. As the rate of roll increased steadily, by T+48 seconds, the large amount of torque began to destroy the second stage. Three seconds later, the KORD system shut down all the engines. As the

---

24.   Sergei Khudyakov, "Power Units Run on Fuel Cells," *Aerospace Journal* no. 6 (November–December 1996): 42–43.

25.   "Soviet Lunar Booster Souvenir," *Spaceflight* 34 (August 1992): 274.

26.   Afanasyev, "Unknown Spacecraft."

27.   Mikhail Rudenko, "Four Steps From the Moon" (English title), *Moskovskaya pravda*, July 19, 1994, p. 10.

28.   Kamanin, "I Feel Sorry for Our Guys," p. 12.

This photo captures the spectacular night launch of the third NI rocket in June 1971. Although the booster cleared the tower, unanticipated rotations around the main vertical axis of the vehicle led to yet another catastrophic failure. (copyright Quest)

rocket continued to break up in the air, it flew about twenty kilometers from the pad and hit the ground, creating a crater thirty meters wide and fifteen meters deep. Fragments of the rocket were scattered across an area of several kilometers.[29] While it was the first time that all of the NI's engines had fired together, the third failure in a row, not surprisingly, affected morale. Boris A. Dorofeyev, at the time the "lead designer" of the NI rocket, remembered that:

> such major accidents had a depressing effect on the personnel. But no one entertained the thought that the NI was doomed, or that its defects were of a chronic nature. People worked energetically, many asked to have their stay on the firing-range extended, and everyone felt that the rocket would "grow out of it," and that success was not far off.[30]

This bottom-up enthusiasm for a project that had spanned nearly a decade without any tangible results may sound irrational from a Western perspective, but the Russians were clearly in it for the long run. Despite three consecutive failures, many space officials continued to believe that the future of the Soviet space program depended on the NI rocket.

The NI may have become ensconced as a national, albeit secret, Soviet asset, but the L3 lunar landing program was in much bigger trouble by mid-1971. Already almost three years late, much of the technology and design approaches for the creation of the L3 complex were becoming outdated. Many continued to believe, with good reason, that flying the L3 to the Moon with

---

29. Afanasyev, "Unknown Spacecraft"; Igor Afanasyev, "NI: Absolutely Secret" (English title), *Krylya rodiny* no. 11 (November 1993): 4–5.

30. Sergey Leskov, "How We Didn't Get to the Moon" (English title), *Izvestiya*, August 18, 1989, p. 3.

the limited capabilities of the LOK and the LK would be a risk not worth taking. The early-1969 talks on a post-Apollo response had led to considerations of an improved lunar landing project with better characteristics, something that would not only guarantee the safety of a crew, but also be a significant improvement over any Apollo mission. In April 1969, Mishin told Brezhnev that his design bureau would work on a lunar project capable of sustaining a crew of three cosmonauts for extended periods on the Moon and equipped to travel long distances on the lunar surface. By September of the same year, Mishin was examining the preliminary documents on such a plan, and by January 1970, the proposal had a name, the L3M, which was a modified L3.

Engineers approached the formulation of the L3M with the weaknesses of the original L3 plan in mind—that is, what kind of improvements could be made given existing hardware and technology? Clearly, one of the vulnerable links in the chain was the docking of the LK with the LOK once the lander had lifted off from the lunar surface. Given Soviet weaknesses in microelectronics technology, the engineers had faced great difficulty in designing a completely automated rendezvous and docking system, as evidenced by the docking failures on Soyuz 2/3 and Soyuz 7/8. Lunar dockings would be even more complex given the poor knowledge of navigation conditions around the Moon and the difficulty of assisting the cosmonauts from Earth. One possibility was to design a very heavy launch vehicle capable of a direct ascent profile to the lunar surface, bypassing the need for rendezvous. Studies, however, proved that such a profile would require a very heavy launch vehicle well outside the capabilities of the N1 in the near future, even if augmented with high-energy stages. The other option was to launch huge components of the lunar ship separately and have them link up in lunar orbit. The extra mass afforded would allow the spacecraft to carry reliable rendezvous and docking instrumentation. It was the second option that the engineers decided to adopt for the L3M.[31]

The L3M lunar landing proposal depended on two major upgrades: an improvement of the N1 and a redesign of the L3. Upgrades to the N1 had, of course, been talked about for years. The most famous such modification was the N1M, proposed for use on TsKBEM's ambitious and abandoned Mars landing project. For the L3M plan, it seems that Mishin had returned to the old ideas of using high-performance liquid hydrogen-LOX upper stages for launching huge payload stacks into Earth orbit. Thus, the problem in many ways depended on the progress of developing these high-performance upper stages. Four such stages had been under development for several years, but only two—Blok S (to replace the current N1 stage IV) and Blok R (to replace stage V)—had achieved any modicum of success. Certainly, one of the major problems was the lack of support from higher authorities to finance such efforts. While NASA was already flying several excellent high-performance upper stages, Soviet engineers were still writing unanswered letters on the importance of such propellant combinations. Throughout 1969, work on Blok R and Blok S was given high priority, with Mishin meeting several times with Chief Designers Lyulka and Isayev, who were responsible for the two engines, respectively.

The results of testing of these two stages changed the design of the uprated N1 for the long-duration lunar expeditions. In February 1970, the most favored option was to launch three N1s, two with Lyulka's Blok S and one with a modified variant of Blok D, named Blok DM, which used conventional propellants. To optimize the work being done and also to unify disparate efforts, Mishin's engineers at the time emerged with a conception for a new upper stage, designated Blok S_R, with a fueled mass of 77.9 tons. This block was examined in two different versions, the first with one of Lyulka's 11D57 engines and the second with either two or four of Isayev's 11D56M engines.[32] In March 1970, the L3M proposal was narrowed down to an

---

31.   Afanasyev, "Unknown Spacecraft."
32.   Semenov, ed., *Raketno-Kosmicheskaya Korporatsiya*, p. 262. The 11D56M was a modernized variant of the older 11D56.

option with two sequential missions, each using a much more improved version of the LK lander called the LKM:

- One launch of the N1 with Blok S_R and the LKM for an automated lunar landing and return to Earth
- Two launches of the N1, one with a Blok S and one with a Blok S_R, which would link up in Earth orbit and take its piloted LKM on an extended visit to the surface of the Moon

While the L3M plan offered significant advantages over the original L3 profile, it was still by no means a certainty with respect to the Soviet space leadership. Throughout February and March 1970, there was much discussion on the preparation of an official governmental order from the Ministry of General Machine Building on the L3M proposal, but none seems to have been forthcoming at the time, evidently because of dissension among the chief designers on the details of the plan.

Mishin was clearly the primary sponsor of the L3M proposal, and it seems that he had trouble, at least initially, in gathering the necessary support to facilitate an official decree. Academy of Sciences President Keldysh offered lukewarm support, advising that TsKBEM first needed to perfect the old N1-L3 before moving on to the L3M. He cautioned that funding for all these proposals were limited and thus Mishin should reduce his requests to cater to the exigencies of the day. With wavering support, by the end of 1970, there was still no official word on the proposal. Meanwhile, Mishin, on his own initiative, had continued to focus work at his design bureau on L3M, simultaneously with all the work on the N1-L3 and of course the new space station program. This time, he did not want to make the same mistake of going with an "all-up" testing philosophy, which had, to an extent, crippled the N1 program. It was clear to him that one of the most important elements of the new L3M plan was the use of the high-energy upper stages. To fully test these out prior to their actual use on a lunar mission, Mishin's engineers, by April 1970, emerged with a proposal to develop a smaller version of the N1, called the N11. Such an idea, with the exact same designation, had been offered by Korolev in the early 1960s, but had never gotten off the ground because of a lack of funding. As envisioned at the time, the N11 was a three-stage rocket with:

Stage	Origin
I	Blok B (stage II) from the N1
II	Blok V (stage III) from the N1
III	Blok S_M (modification of the high-energy Blok S_R)

In October 1970, Mishin met with Brezhnev to brief the Soviet leader on the course of the piloted lunar landing program. One of the main topics of conversation was the creation of an improved N1-L3 complex—effectively the N1-L3M proposal—and the use of high-energy upper stages for both the N1 and the N11. TsKBEM was evidently short of money to build test stands for the stages.[33] As was typical of many other test programs, the N11 as a viable option did not last very long. Although it was discussed at the Military-Industrial Commission level in October 1970, by December, there were doubts about the feasibility of rapidly building the N11 in its current configuration. Eventually, it was completely abandoned. Mishin's engineers would have to make the direct jump from the N1 to its modified version with high-energy stages.

---

33. These stages included Blok S_1, Blok S_R, and Blok S_M. There were at least two conceptions of the N11: one the basic N11 and the other called the N11S.

The N I I may have been dropped from consideration, but Mishin was allowed to proceed with his important Blok S_R stage. In May 1971, a formal decision was taken on the development of Blok S_R replacing the various previous incarnations of high-energy upper stages, such as Blok R and Blok S. Blok S_R would be a universal upper stage for launching heavy space apparatus into geostationary Earth orbit and for sending heavy automatic stations on trajectories to the planets. Its primary job, however, would be to serve as the lunar-orbit insertion stage for the L3M mission. After an initial planning stage involving comparisons of different liquid hydrogen-LOX engine configurations for a single variant, engineers adopted Chief Designer Isayev's 11D56M engine for the stage because it would "ensure the best characteristics."[34] In the final analysis, Blok S_R was equipped with two of Isayev's 11D56M engines with a primary thrust regime of 15.08 tons and a medium thrust regime of eight tons. It was capable of being fired up to five times over a period of eleven days in a state of weightlessness and vacuum—that is, deep space. The performance characteristics for this first Soviet liquid hydrogen-LOX engine were remarkably high, comparing very favorably to NASA's Centaur RL-10 engine in terms of specific impulse. According to preliminary calculations, Blok S_R could deliver a mass of 23.8 tons (for a piloted ship) or 24.1 tons (for an automated ship) into lunar orbit, twenty tons to geostationary orbit around Earth, or 27.8 tons to a trans-Martian trajectory. The stage itself was sixteen and a half meters in length and just over five meters in diameter.[35]

The decision to select Isayev's 11D56M engine over Lyulka's 11D57 engine for Blok S_R had as much to do with technical considerations as it did with bureaucratic infighting. Lyulka's engine had run into serious technical trouble in 1970. By July, it was clear that its testing program was severely lagging, and by the end of the year, planners had all but given up on its use in the immediate future. The technical issues were compounded by interministerial jealousies. Lyulka's organization, the design bureau of the Saturn Plant, was part of the Ministry of Aviation Industry, and thus outside the "mainstream" of the Soviet space industry, which was part of the Ministry of General Machine Building. The latter's head, Minister Afanasyev, was evidently unwilling to have another chief designer from the aviation industry "interfere" in the N1-L3 program.[36] While Lyulka doggedly continued his work on Blok R, his engine was temporarily sidelined from the N1 program.

A specific technical design for Blok S_R enabled more precise definition of the N1-L3M proposal in 1971 because the capabilities and mission profile depended to a great extent on the performance characteristics of the upper stage. The main component of the plan was the N1F, an upgraded N1 that would incorporate improvements in each of its stages. The first three stages would use the new and better Kuznetsov engines capable of multiple firings, which were under development since 1970. Each replacement engine had the same thrust level as its predecessor. The fourth and fifth stages, strictly a part of the payload, would be replaced by a single high-energy upper stage, Blok S_R. The final configuration for the upgraded N1 was:

34. Semenov, ed., *Raketno-Kosmicheskaya Korporatsiya*, p. 262.
35. *Ibid.*
36. German Nazarov, "You Cannot Paper Space With Rubles: How to Save Billions" (English title), *Molodaya guardiya* no. 4 (April 1990): 192–207; Lardier, *L'Astronautique Soviétique*, pp. 174–75.

Stage	Stage Name	Engines	Thrust Levels
I	Blok A	30 X NK-33	30 X 154 tons (sea level)
II	Blok B	8 X NK-43	8 X 179 tons (vacuum)
III	Blok V	4 X NK-39	4 X 41 tons (vacuum)
IV (payload stage)	Blok S_R	2 X 11D56M	2 X 7.54 tons (vacuum)[37]

The L3M flight plan would use two of these NIFs to carry out the mission. The payload block for the first NI would consist of Blok S_R and a Blok DM stage with a total mass of 104 tons. The payload block for the second NI would consist of another Blok S_R and the actual lunar lander, the LKM, with a total mass of 103 tons.

Little has been revealed on the technical details of the LKM. In appearance, it looked like a greatly enlarged version of the smaller LK from the L3 project. The mass of the LKM—23.7 tons, which was about four times more than its predecessor—would seem to indicate a dramatic leap in abilities. The LKM had two distinct stages, the descent stage (or landing adapter) and ascent stage (the living compartment). The descent stage consisted of four long legs attached to a central framework, which included various systems. The nineteen-and-a-half-ton ascent stage was shaped like a huge cocoon consisting of two major portions, both within its external spherical hull: the descent apparatus and the instrument compartment.

The almost eight-and-a-half-ton descent apparatus was shaped somewhat like an enlarged Soyuz reentry capsule and installed on the upper portion within the cocoon. It was internally connected to the cylindrical instrument compartment in the lower portion of the cocoon. After launch and during flight, the cosmonauts would leave the descent apparatus and crawl into the instrument compartment to carry out all in-flight operations, including landing on the Moon. The instrument compartment afforded a large internal space with viewports to select an optimal landing site.[38] The main engine complex of the LKM was attached to the ascent stage and would be used several times throughout the mission. It included a primary and backup throttle-capable engine unit, both using storable hypergolic propellants; these may have been contracted to KB Yuzhnoye, which developed the LK main engine. The increased mass of the LKM over the earlier LK afforded significant upgrades in systems. One historian noted:

> The use of the "direct configuration" made it possible to equip the craft with a complicated system of more advanced radio gear for the precise and reliable performance of maneuvers connected with searching, meeting, and docking in lunar orbit. Such a larger LK would moreover have had greater freedom of maneuver close to the surface to select a landing site.[39]

---

37. Afanasyev, "Unknown Spacecraft"; I. Afanasyev, "The 'Lunar Theme' After NI-L3" (English title), *Aviatsiya i kosmonavtika* no. 2 (February 1993): 42–44; Jeffrey M. Lenorovitz, "Trud Offering Liquid-Fueled Engines From NI Moon Rocket Program," *Aviation Week & Space Technology*, March 30, 1992, pp. 21–22. An early version of Blok S_R was equipped with four 11D56M engines instead of two. See Semenov, ed., *Raketno-Kosmicheskaya Korporatsiya*, p. 262. The total mass of the NIF was 3,025 tons, and launch thrust (probably in vacuum) was 5,070 tons. The booster was capable of inserting 105 tons into Earth orbit, thirty-four tons to the Moon, and twenty-two tons into lunar orbit.

38. Afanasyev, "The 'Lunar Theme' After NI-L3." There was evidently another LKM configuration considered. This design resembled a Soyuz with its modules switched—that is, the descent apparatus on top of the living compartment. The crew would move from the former to the latter by means of "a crawlway-chute"—that is, not through a hatch in the heat shield. See Afanasyev, "Unknown Spacecraft."

39. Afanasyev, "The 'Lunar Theme' After NI-L3," p. 43.

Each NIF would launch its payload block toward the Moon using its own Blok S_R to accelerate to translunar injection. Near the Moon, the same stages would fire to put their respective payloads into lunar orbit. Once there, the two Blok S_R stages would be discarded, and the Blok DM stage would dock with the large LKM. If for some reason the docking failed, the cosmonauts could simply return to Earth in their LKM spacecraft without having to carry out any extraneous spacewalks. In case of a successful docking, Blok DM would decelerate the complex from lunar orbit and initiate a powered descent to the lunar surface. Much like the earlier L3 plan, after Blok DM's propellants were exhausted, the LKM would take over the remaining portion of the descent to the surface using its own engine. Depending on the size of the crew, the stay on the Moon would last from five (three cosmonauts) to fourteen days (two cosmonauts). After the entire surface exploration was over, the cosmonauts would lift off from the Moon in the living compartment, leaving behind the large descent stage on the surface. Once again, using its own engine, it would directly fire itself on a trans-Earth trajectory. Another less preferable option was to enter an intermediate orbit around the Moon before returning to Earth. Once near Earth, the living compartment would open up into two pieces, much like a clam, and release the actual descent apparatus containing the crew. After a controlled reentry into Earth's atmosphere, the cosmonauts would land either on Soviet territory or in the Indian Ocean.[40]

The NIF-L3M plan offered the hope of carrying out a series of impressive lunar landing missions. However, while the Communist Party, government, and industry had been lukewarm at best on the NI-L3, would they commit to the expanded version of it? The central question was obviously financing, and money was, in fact, one of the crucial issues in L3M planning within TsKBEM. There was also the question of what to do with the old L3 project. To resolve these issues, the Politburo signed an order on February 17, 1971, titled "On the Designation of an Expert Commission on NI-L3 Under the Chairmanship of M. V. Keldysh." The new Expert Commission would be tasked with three goals:

N1F (11A52F)
1972 (project)

The NIF variant of the basic NI rocket was intended to launch the advanced L3M complex for long-duration stays on the Moon. The program received preliminary approval by the Council of Chief Designers in May 1972. (copyright Peter Gorin)

---

40.   *Ibid.;* V. P. Mishin, "Why Didn't We Fly to the Moon?" (English title), *Znaniye: tekhnike: seriya kosmonavtika, astronomiya* no. 12 (December 1990): 3–43; Afanasyev, "Unknown Spacecraft."

- To evaluate the possibility of carrying out a lunar landing with one cosmonaut (that is, the L3)
- To evaluate the optimal program of work with regards to the Moon
- To evaluate prospective programs (that is, the L3M)

Col. General Kamanin, who was not a member of this Expert Commission, had some interesting comments on the body in his journal entry dated March 4, 1971:

> For several years I have argued (I made two special visits to the Central Committee of the Party and repeated visits to the Military-Industrial Commission) that the N1 rocket and the lunar L3 spacecraft were hopelessly outdated and that our Moon mission program should be drastically revised. Finally, the Central Committee and the Council of Ministers have appointed a commission, chaired by Keldysh, which has been given until May 1, 1971, to answer the question of what to do with the lunar complex and with the existing mission-to-the-Moon program. My answer would be most definitely that the N1 rocket and the L3 spacecraft should be scrapped, that Chelomey's UR-700 rocket should be modified and a new lunar probe designed with a view to sending the first mission to the Moon in 1974–75. Mishin and his supporters are afraid of such a prospect: they have staffed the panel with people who will toe their line. The most likely outcome will be that the Keldysh panel will recommend continued attempts to "cure" a bad rocket and an equally bad spacecraft.[41]

Kamanin was not entirely correct that the commission was staffed with people sympathetic with Mishin. Keldysh presided over six different subcommissions whose heads were chief designers and academicians from various branches of the aviation and missile industry, many of whose organizations had not participated in the N1-L3 program, nor had any vested interest in the project. Only five senior officials from TsKBEM were members of these subcommissions.[42]

At a meeting of the commission on May 31, Keldysh asked TsKBEM to prepare a formal proposal on the future of the N1-L3 program by June 15. Immediately, Mishin assembled his senior deputies, and through the ensuing days, there was much discussion on the issue. The preliminary plan was to follow through on piloted lunar exploration in three stages:

- Use the N1-L3 for piloted lunar-orbital flights with automatic landings of the LK
- Use the N1-L3 for a lunar landing, using both Earth-orbit rendezvous (to deliver the crew) and lunar-orbit rendezvous
- Use the N1F-L3M to link up elements in lunar orbit for an extended lunar landing and then directly return to Earth[43]

The pressure to completely abandon any thought of using the L3 for a piloted lunar landing was formidable. At a meeting in late July 1971, Minister of General Machine Building Afanasyev, Academy of Sciences President Keldysh, and Ministry Chief Directorate Chief

---

41. Kamanin, "I Feel Sorry for Our Guys," p. 12.

42. Among the subcommissions were Subcommission no. 1 chaired by MKB Fakel General Designer P. D. Grushin, Subcommission no. 3 chaired by TsKB Almaz General Designer B. V. Bunkin, Subcommission no. 4 chaired by USSR Academy of Sciences Vice President V. A. Kotelnikov, and Subcommission no. 5 chaired by Deputy Minister of Health A. I. Burnazyan. TsKBEM representatives on the subcommissions included S. O. Okhapkin and K. D. Bushuyev (on no. 1), M. V. Melnikov (on no. 2), B. Ye. Chertok (on no. 3), and Ya. I. Tregub (on both nos. 4 and 6).

43. There was an additional possibility within the second option: to carry out the N1-L3 mission with a single launch. This would, however, only be possible if engineers could increase the lifting capability of the N1, which was still not up to design levels by 1971.

Kerimov all agreed that the original NI-L3 complex should not be used for landing a cosmonaut on the Moon. They were even against using the NI-L3 for an *automatic* landing, as proposed by Mishin earlier. While debate over automated L3 landings continued, the original L3 piloted lunar landing plan received its final death knell at a meeting of the Keldysh commission on August 16, 1971. It had been almost exactly seven years to the day since the Soviet government had approved Korolev's L3 idea. The question of whether to use the remaining components of the L3 complex was left unresolved. At the same time, TsKBEM would commit its resources to perfecting the L3M plan. Mishin, in fact, signed the preliminary materials for a "prospective" lunar expedition—that is, the new L3M plan—on the same day as the Expert Commission's meeting. He was instructed to have the predraft plan ready by early 1972.

Throughout the latter part of 1971, Mishin's engineers continued evaluating various options for L3M.[44] This effort included freezing the design of the new descent apparatus with two new parachutes and reexamining the most optimal trajectories to and from the Moon—an exercise that evidently included studying data from the recently completed Apollo 15 mission. Support for the L3M option was growing at the time. Mishin later recalled:

> We finally managed to get technical tasking from the USSR Academy of Sciences for a lunar mission [that is, the L3M] with a list of problems that it was supposed to solve. It must be noted that no such specifications had ever been received from the Academy for the first version of the mission [that is, the L3].[45]

The Expert Commission's recommendation and the "technical tasking" of the Academy of Sciences were important factors in providing some much needed impetus to the NIF-L3M proposal. By the end of 1971, Mishin's engineers had evidently completed the detailed draft plan for the project. Even the all-powerful Military-Industrial Commission took an interest, issuing a decree on February 16, 1972, in support of further work on such a project. If obstinate opponents of the NI such as Chief Designer Glushko opposed the plan early in 1971, they all came around to the same point of view. On May 15, 1972, the Council of Chief Designers for the lunar program formally adopted the NIF-L3M plan, titled "Technical Proposals for the Creation of the NI-L3M Complex." Even Glushko signed the final document.[46]

In contrast to the utter chaos that had pervaded the birth of the NI-L3 in the early 1960s, this new project was not born out of jealous infighting among the chief designers, nor from external political imperatives. For the first time in a major Soviet space project, the pace was not dictated by what the United States was doing. This alone could have made the proposal worth pursuing, but the L3M also had excellent technical characteristics, well-planned schedules, and painstaking cost assessments to back it up. Mishin originally had planned for launches in the new program to begin simultaneously with the winding down of the original L3 project—that is, in 1974. According to preliminary plans in September 1970, there would be two launches in 1974 and four in 1975, the latter perhaps including actual piloted landings. By the time of the May 1972 decision, the timeframe was moved back by about two years, with launches

---

44. For example, options explored in late September 1971 included: (1) a two-launch scheme with docking in lunar orbit with either (a) a direct flight from the Moon to Earth or (b) with a second docking in lunar orbit and then returning to Earth; and (2) a two-launch scheme with docking in Earth orbit with either (a) a direct landing on the Moon and direct return to Earth or (b) a plan similar to the original NI-L3 in lunar orbit. One of the more interesting possibilities was using the new 7K-S Soyuz variant in lunar orbit.

45. Mishin, "Why Didn't We Fly to the Moon?"

46. *Ibid.*; Boris Arkadyevich Dorofeyev, "History of the Development of the NI-L3 Moon Program," presented at the 10th International Symposium on the History of Astronautics and Aeronautics, Moscow State University, Moscow, Russia, June 20–27, 1995.

beginning in 1976 and landings in 1977. These latter missions would include initial cosmonaut surface stays lasting fourteen days, leading up to full-fledged lunar surface missions lasting an unprecedented month on the Moon.[47]

What was even more astonishing about the L3M plan was that the Soviets did not stop there. There were even plans for permanent piloted bases on the lunar surface—plans that actually harked back to about 1965. Sometime before his death in January 1966, Korolev had discussed this idea with Chief Designer Barmin of the Design Bureau of General Machine Building. Although Barmin's main line of work was the design and development of ground launch complexes for Soviet missiles and launch vehicles, he was sufficiently interested in the topic to take on a modest subcontract from OKB-1 to explore the design of permanent lunar bases. These studies continued well after Korolev's death. Mishin's TsKBEM remained in overall charge of the research, but cooperated with Barmin's design bureau in formulating the goals of the base, the principles of construction, the stages of development, and the composition of scientific and special manufacturing equipment. Barmin's engineers also studied civil engineering methods, questions of life support systems and their maintenance, and power supply and radio communications systems.[48]

The overall effort was generically called the Long-Duration Lunar Base (DLB) and consisted of several different thematic directions with names such as *Kolumb* ("Columbia"), *Bolshoye koltso* ("Big Ring"), *Dal* ("Distance"), and *Osvoyeniye* ("Mastery"). Engineers designed a veritable menagerie of various insect-like vehicles for work on the lunar surface, including:

- Vehicles equipped with radio beacons (whose design was based on the Ye-8 descent stage), which would guide spacecraft down to specific landing sites
- Huge "closed" lunar rovers with pressurized compartments for crews to collect samples using long and jointed remote manipulator arms without leaving the comfort of their cabin
- Large utility vehicles for transporting vast amounts of raw materials across the lunar surface
- General crew mobiles capable of sustaining independent forays for days at a time
- Different automated rovers equipped with core-drilling manipulators built by Barmin's engineers for gathering soil samples

The L3M lander would serve as the initial transport vehicle to the lunar surface, and later N1s would bring the remaining assortment of rovers and beacons, many of which would be built by the Lavochkin Design Bureau—an appropriate choice given its experience in designing automated lunar and interplanetary probes. Long-term plans included mining the Moon for helium-3, hydrogen, oxygen, silicon, titanium, aluminum, and iron for various manufacturing and industrial processes. For the actual bases, Barmin considered different alternatives. In one conception, the cosmonauts would live underground to efficiently use sublunar heat-exchange processes. The actual production structures, landing sites for transport rockets from the Earth, and refueling stations would be located far away from these laboratories, but they would be connected via special tunnels, either by foot or by means of moving "strips" similar to those in airports today. Another option studied was to have residential and operational structures on the surface with dome-like protective coverings built from transparent material. The entire complex would contain perhaps three habitation modules, equipment for the production of oxygen and other gases, installations for the extraction and transportation of lunar materials, and a nuclear-type power plant.[49]

47. Mishin, "Why Didn't We Fly to the Moon?"
48. Yu. A. Mozzhorin, *et al.*, eds., *Dorogi v kosmos: I* (Moscow: MAI, 1992), p. 55.
49. Mikhail Rebrov, "Touching Upon the Legend of the DLB" (English title), *Krasnaya zvezda*, June 18, 1994, p. 6.

Small models of elements of the Long-Duration Lunar Base (DLB) are shown here in a museum display case. Note the models of the NI-L3 LK lunar lander at the left of the photograph. At least one of Babakin's Ye-8 descent stages is visible at center right. Several large mobile "crawlers" are placed at the top from left to right. (copyright Mark Wade)

All of these proposals, for both the L3M and the DLB, were, of course, restricted by the realities of the day, primarily financial ones. Neither had, by 1972, received formal approval from the Council of Ministers and the Central Committee. Sanction from the Soviet leadership would prove difficult, but given the multiple recommendations in favor of the L3M, it was not thought to be impossible. All would depend on the success of the remaining launches of the older NI model. Their success would be critical to convincing the Soviet leadership that the NI rocket project was an effort worth pursuing and funding for the long run. Awaiting a formal decision on the L3M, Mishin elected to doggedly continue to flight-rate the older LOK and LK spacecraft, because much of the technology from these vehicles would be used in upgraded form on the L3M. Thus, if his plans were approved, the Soviets would fly the remaining L3 hardware to the Moon by the mid-1970s, begin the advanced L3M missions by the late 1970s, and then slowly move to the DLBs by the first years of the following decade, possibly initiating the first colonization of the Moon.

Ironically, at the very same time that the Soviets were conceptualizing such grand projects, the U.S. civilian space program was suffering from post-Apollo malaise. In its early days, NASA had been well equipped to cope with repeated failures of its rockets and satellites, but in the aftermath of Apollo 11, it was unable to cope with success. Having been a single-issue agency, NASA leaders were facing the problem of using dwindling financial and human resources to create a tenable vision of the future. One of the most compelling components of this new vision was to create the means for "routine access to space"—that is, a shuttle vehicle that would service future space stations and haul scientific and applications satellites into space. Initially, in the fall of 1969, NASA had hoped that President Richard M. Nixon would approve an ambitious plan to build a space station, the Space Shuttle, and a piloted Mars project. As the financial realities sank in, this aggressive plan was reduced ultimately to just the Space Shuttle, which itself was redesigned several times to meet budget limitations, thus sacrificing much of its original *raison d'être*—that is, "routine access to space." In January 1972, Nixon met with NASA Administrator James C. Fletcher and issued a statement announcing the decision to "proceed at once with the development of an entirely new type of space transportation system designed to help transform the space frontier of the 1970s into familiar territory, easily accessible for human endeavor in the 1980s and

'90s."[50] Without a space station or a Mars mission, the United States was left with a plan of piloted space exploration that was lacking in a concrete vision, a means that had no end. At the same time, the Soviet Union was dramatically planning to up the ante, squarely targeting the Moon and building new space stations, including both the small DOS and the giant MKBS.

Building the *Salyut* Long-Duration Orbital Station

The Long-Duration Orbital Station, better known by its acronym "DOS," was designed, built, and tested over a remarkably short period of time. Not surprisingly, the mainframe of the station was identical to Chelomey's Almaz station—that is, roughly shaped like three cylinders of different diameters connected end on end. For the DOS (or product 17K), this design was augmented by a fourth cylinder. The total length in orbit was about 16 meters. From the station's forward end to its rear, the "cylinders" of the 18.9-ton station were the:

- Transfer compartment
- Working compartment (consisting of two of the "cylinders")
- Aggregate compartment

The transfer compartment at the forward end was equipped with the passive docking node for receiving Soyuz ferry vehicles. The length of this section, including the node, was three meters, and the diameter was two meters. This compartment primarily contained equipment for life support and thermal regulation. The major scientific component was the *Orion-1* ultraviolet telescope, which included a locked chamber for removable photo cassettes. Part of the telescope, designed by the Byurakan Astrophysical Observatory, jutted out of the compartment in a hemispherical depression embedded on the outside of the section. Other equipment included cameras and biological instrumentation. The short compartment also included an eighty-centimeter-diameter hatch for allowing crews to egress from the station for spacewalks. On the exterior of the compartment, there were two large solar panels fixed like bird wings, each with four paneled sections. With a wingspan of eleven meters, these were created on the basis of the solar panels on the 7K-OK Soyuz spacecraft. Other equipment on the exterior of the transfer compartment included the Igla rendezvous system antennas, lights for docking approaches, one of two external TV cameras, panels for heat regulation, ion sensors for the attitude control system, and panels for micrometeoroid detection.

Swimming from the Soyuz into the transfer compartment, a cosmonaut would open a second hatch and then enter the working compartment, the largest portion of the station. Its two cylinders of different diameters were connected via a conical transfer section. The smaller cylinder had a diameter of 2.9 meters and a length of 3.8 meters, while the measurements for the larger cylinder were 4.15 meters and 2.7 meters, respectively. The smaller diameter section contained the central command post for controlling the station with a control panel and on-board computers. The control system for the station was derived from the original 7K-OK Soyuz system, a measure adopted to eliminate extra effort. One movie camera and one still camera were installed on the "upper" wall of the section, allowing direct access to the outside. The small diameter area also contained a table for work and eating, facilities for heating food, drinking water, on-board documentation, a tape recorder, a library, a sketch album, and other items.

---

50.   John M. Logsdon, "The Evolution of U.S. Space Policy and Plans," in John M. Logsdon, gen. ed., with Linda J. Lear, Jannelle Warren-Findley, Ray A. Williamson, and Dwayne A. Day, *Exploring the Unknown: Selected Documents in the History of the U.S. Civil Space Program, Volume I: Organizing for Exploration* (Washington, DC: NASA Special Publication (SP)-4218), pp. 386–88; Roger D. Launius, *NASA: A History of the U.S. Civil Space Program* (Malabar, FL: Krieger Publishing Co., 1994), pp. 107–10.

Here is a model simulation of a 7K-T Soyuz spacecraft (left) docking with the early model of the DOS space station (Salyut). (NASA photo)

Moving aft, a cosmonaut would then enter the large diameter area. The primary equipment here was a large cone with its base on the "floor" and its apex almost to the "ceiling"; it was called the scientific apparatus compartment. The latter consisted of, among other instruments, the OST-1 two-meter-diameter solar telescope designed by the Crimean Astrophysical Observatory. At the rear of the large-diameter area, there were three posts for work on other scientific apparatus, which included the *Anna-3* gamma-ray telescope, the TEB telescope for studying charged particles in the upper atmosphere, the *Kalina* ("Viburnum") instrument, and the FEK-7A photo-emulsion camera. Other scientific instrumentation on the station included the RT-4 Roentgen telescope built by the Physical Institute of the Academy of Sciences, the ITSK infrared telescope-spectrometer, and the OD-4 optical visor instrument. The AFA-41/20 and AFA-M-31 cameras were for Earth resources surveys. Disciplines apart from Earth observation and astrophysics were also represented. There was a virtual menagerie of medical instrumentation with their own enigmatic names: *Polinom-2M, Levkoy-2M* ("Gillyflower"), *Tonometr, Rezeda-2M* ("Mignonette"), *Impuls* ("Impulse"), *Vertikal-M* ("Vertical"), *Plotnost* ("Density"), *Raduga* ("Rainbow"), and *Kreslo* ("Seat"). During exercises, the cosmonauts would wear the *Atlet* ("Athlete") suit, while at other times they would don the *Pingvin* ("Penguin") suit, which would force the crew to act against allowing the suit to assume its normal fetal position. In addition, the *Chibis* ("Lapwing") was a special "suit" designed to generate negative pressure on the lower body to reduce orthostatic intolerance during the return to Earth. Finally, there was also a special "antigravity" suit for the cosmonauts to wear before the end of the mission. The total mass of all scientific devices on the station was one and a half tons.

The large-diameter area also contained sleeping areas for the cosmonauts, physical trainers (including a stationary running track capable of ten kilometers per hour and a "vele-ergometer"), a refrigerator with a supply of food products and water, and a toilet with its own forced ventilation system isolated from the rest of the station. All around the large-diameter area, there were panels on the walls giving easy access to instrumentation for controlling the station—those for life support, thermal regulation, power supply, radio communications, trajectory measurement, and command radio links. So as not to disorient crews in space, the station had a specific color scheme: the front and rear were light gray, one side was apple gray, the other wall was yellow, and the "floor" was dark gray. The exterior of the working compartment had various antennas from the station's radio complex and micrometeoroid impact panels.

The DOS contained seven specific locations for manually controlling the scientific apparatus and station systems. Station no. 1, the central command post of the station, was located in the lower part of the small-diameter portion. Equipped with two chairs, cosmonauts could control the basic on-board systems and part of the scientific equipment from here. Station no. 2, the "astropost," was also located in the small-diameter area; it was designed for manual astro-orientation and astro-navigation. Station no. 3 was in the large-diameter section and was exclusively for controlling the scientific apparatus. In addition to scientific research, Station no. 4 was for medical investigations and was located in the conical section between the small- and large-diameter sections. Station no. 5 was specifically for controlling the Orion-1 stellar telescope and was located in the transfer compartment. Station no. 6, like Station no. 2, was for astro-orientation and navigation; it was located in the small-diameter section. Finally, Station no. 7 was for controlling scientific apparatus focused on studying "atmospheric resonance" using the ERA instrument and was located opposite to Station no. 6.

The final section of the station was the aggregate compartment at the very aft end of the DOS. Not accessible by the crew, this compartment was a simple cylinder with a diameter of 4.15 meters and a length of 1.4 meters hooked to a semispherical shell attached to the station proper. The cylindrical portion was appropriated directly from the aggregate compartment of the Soyuz spaceship, and it contained the main maneuvering engine of the space station, the S5.66. The engine, almost identical to the Soyuz main engine, was developed by the same enterprise that had developed the one for Soyuz, Chief Designer Isayev's Design Bureau of Chemical Machine Building (formerly OKB-2). The S5.66 had a primary single-chamber engine with a thrust of 417 kilograms and a reserve two-chamber one with a thrust of 411 kilograms. In addition to the main engine, the station was equipped with a set of thirty-two small attitude control thrusters of ten kilograms thrust each, developed by the Scientific-Research Institute of Machine Building based at Nizhnyaya Salda. The attitude control complex consisted of two independent systems—a primary and a backup—each consisting of sixteen engines (six for yaw, six for pitch, and four for roll). The propellant tanks for these engines were installed in the aggregate compartment.

Two large solar panels, identical to the ones at the forward end of the station and derived from the Soyuz spacecraft, were installed on the aggregate compartment, lending the station a bird-like appearance. Electrical energy for the station was passed through independent systems—the SEP-1 and SEP-2—each with a potential difference of twenty-seven to twenty-eight volts. These could work simultaneously using the two pairs of solar panels, with a total surface area of twenty-eight square meters, to charge two internal nickel-cadmium batteries. SEP-2 was designed for intermittent work and was only for the scientific instrumentation.[51]

From an overall perspective, the DOS spacecraft was essentially created by combining the Almaz space station with the Soyuz spaceship. The number of systems appropriated from the latter were numerous, including the entire orientation and approach control systems, the Zarya radio-communications systems, the RTS-9 telemeasurement system, the Rubin orbital radio-

51.   I. Marinin, "Quarter Century for 'Salyut'" (English title), Novosti kosmonautiki 10 (May 6–19, 1996): 78–84; Lardier, L'Astronautique Soviétique, pp. 199–202; V. P. Glushko, ed., Kosmonautika entsiklopediya (Moscow: Sovetskaya entsiklopediya, 1985), pp. 343–44; M. P. Vasilyev. et al., eds., 'Salyut' Space Station in Orbit (Washington, DC: NASA TT F-15450, 1974), pp. 3–8. This last source is a translation of M. P. Vasilyev, et al., eds., 'Salyut' na orbite (Moscow: Mashinostroyeniye, 1973).

control system, the DRS command radio-link system, the central pilot control panel, the Igla rendezvous system, and the life support systems. The thermo-regulation system used on the DOS was an updated version of the one on Soyuz. There was widespread cooperation in the building of the DOS. Apart from TsKBEM, TsKBM's Moscow Branch, and the M. V. Khrunichev Machine Building Plant, numerous other organizations contributed to the rapid pace of progress.[52]

One of the essential components of creating the DOS-7K complex was developing a ferry version of the Soyuz spacecraft, which would ensure internal crew transfer after docking. The original Soyuz docking system was, of course, designed in such a way as to precisely prevent such internal passage. By late 1969, TsKBEM had begun redesigning the 7K-OK Soyuz into the "new" 7K-T Soyuz specifically for the DOS program. In early 1970, Department No. 231 at the design bureau issued the draft plan for the ferry vehicle under the overall leadership of Deputy Chief Designers Bushuyev and Tsybin. The 7K-T ship had an active docking unit with a rod compatible with a cone on the passive docking node on the DOS ship. Given the rapidity with which Mishin's engineers managed to design a complex docking mechanism capable of internal crew transfer, it is quite likely that they used the experience in creating a similar mechanism for the "advanced" but still-not-yet-flown 7K-S variant of the Soyuz. The Azov Machine Building Plant carried out the manufacture of the new docking mechanism, which had a 0.8-meter-diameter hatch.

Unlike the basic Soyuz, the 7K-T had a simplified life support system because it did not need to ensure autonomous flight for very long. The systems related to the Igla rendezvous system were transferred to the living compartment at the forward end of the ship, and one of the command radio links was removed completely, allowing for the elimination of the toroidal compartment around the engine unit at the rear of the original Soyuz. The 7K-T transport ship had a launch mass of 6,700 kilograms, which was about fifty kilograms in excess of its predecessors; its descent apparatus weighed 2,800 kilograms. As a whole, the ship was 6.98 meters in length. The vehicle would be capable of carrying three cosmonauts without pressure suits and return only twenty kilograms of scientific results back from the station, suggesting that the Soviets were pushing the upper limits of what they could squeeze out of the Soyuz booster-spacecraft system at the time. The new Soyuz was rated for sixty days of flight time, of which three days would be autonomous. In a clear departure from previous Soviet practices, TsKBEM elected to forego automated missions of the 7K-T Soyuz and go directly to piloted launches.[53]

If the DOS program had, at least in the initial phases, a temporary feel to it, by the end of 1970, TsKBEM had tabled several ambitious plans to extend its capabilities far beyond its modest origins. One crucial design issue was the addition of a second docking node to allow resupply visits to the station. Mishin recalled later that he had proposed the use of two docking ports on the very first DOS vehicle, but he had been overruled by Ustinov "in order to hasten our success."[54] By May 1970, engineers apparently planned for the *second* DOS to have two docking ports, allowing for visits simultaneously by two "advanced" 7K-S Soyuz vehicles. Changes crept into the plan in the subsequent months. In a preliminary conception of the five-year space plan for the Soviet Union covering 1971 to 1975, prepared in late September 1970,

52. Semenov, ed., *Raketno-Kosmicheskaya Korporatsiya*, p. 268. The other organizations included VNII Televideniya under I. A. Rosselevich (for TV systems), MNII Radiosvyazi under Yu. S. Bykov (for communications systems), KB Nauka under G. I. Voronin (for life support systems), KB Zvezda (for spacesuits), TsKB Geofizika under V. A. Khrustalev (for attitude control sensors), OKB MEI under A. F. Bogomolov (for telemetry systems), IMBP under O. G. Gazenko (for biomedical support), VNII IT under N. S. Lidorenko (for power sources), VNII EM under A. G. Iosifyan (for remote-sensing equipment), and NII KhimMash under N. M. Samsonov (for ground testing).

53. *Ibid.*, p. 187; Marinin, "Quarter Century for 'Salyut'."

54. A. Tarasov, "Missions in Dreams and Reality" (English title), *Pravda*, October 20, 1989, p. 4.

the Soviets were tentatively visualizing the launches of ten such space stations over the course of five years—that is, two per year. By this time, the *third* DOS vehicle would be the first standard model with two docking ports; the remaining vehicles would all be identical save for the internal complement of scientific equipment. The DOS number 3 would also be the first station to be serviced by the 7K-S Soyuz. Work on the second DOS had already started by the end of 1970, before the launch of the first DOS, with work advancing to the crew selection stage. Apart from the basic Almaz-based DOS model, Mishin's engineers also conceptualized much more advanced versions. In early September 1970, Ustinov visited TsKBEM to hear reports on the progress of the DOS program and, at least tentatively, approved "prospective" developments, specifically a station named the DOS-N, to be launched by an uprated N1 rocket. Later, in March 1971, Mishin reported to Ustinov on another DOS variant, the DOS-A, evidently proposed as a direct competitor to NASA's Skylab space station, which was far larger than the original DOS variants based on Almaz.

The DOS program may have been primarily a politically motivated program, but there was much debate over what kind of scientific instrumentation to have on the station. In July 1970, Mishin spoke with Academician Georgiy I. Petrov, the Director of the Institute of Space Research under the Academy of Sciences, to discuss the possibility of installing a radio telescope with a fifteen-meter-diameter parabolic antenna on a DOS. Later in the month, Mishin also sent a letter to various scientific (and probably military) organizations asking what kind of goals they would like to be solved on the station—a significant change from the haphazard nature of scientific research on piloted space missions throughout the 1960s. In early February 1971, Mishin met with Viktor A. Ambartsumyan, a Vice President of the Academy, to discuss the long-term scientific goals of the DOS. Several requirements were established:

- The Soviets needed a program for space-based astronomy research.
- Resources were needed for creating optical and radio telescopes in Earth orbit.
- Four of the future DOS vehicles should be equipped with telescopes of 1 meter in diameter for astronomy research.
- A space-based parabolic antenna should be developed with a diameter of thirty to fifty meters.
- A telescope with a mirror of three meters diameter should be created for seven to eight years of operation in orbit.

While all of these were not intrinsically related to the DOS, the space station program seems to have served as a catalyst for this new cooperation between the scientific and space communities.

Crews for the first DOS were slow to train for the first space station missions because of the inevitable acrimony over crew selection that had continuously plagued the Soviet piloted space program. On April 23, 1970, Mishin had initially proposed a set of four crews for the two planned missions to DOS-1. Not surprisingly, Air Force representative Col. General Kamanin refused to approve the choices. He believed that one man, TsKBEM civilian engineer Feoktistov, was not medically suited for flight. Feoktistov had also recently been divorced from his second wife, which would make him unsuitable for space flight. Another man, Air Force Colonel Volynov, who had earlier flown the Soyuz 5 mission, was unacceptable to Kamanin because he was Jewish. Kamanin had been instructed by Ivan D. Serbin, the chief of the Defense Industries Department in the Central Committee, not to allow Volynov to fly again. Finally, a third man, Air Force Colonel Khrunov was deemed inadequate because of his behavior after a recent hit-and-run accident in which he had failed to come to the help of the victim. After a major reshuffle of the crews, Kamanin and Mishin agreed to four new crews on May 13. The first two crews who would fly the first space station missions were: (1) Shonin, Yeliseyev, and Rukavishnikov and (2) Leonov, Kubasov, and Kolodin.

Both commanders and flight engineers were veterans of previous missions. Leonov, in particular, had finally terminated his training for piloted lunar missions in May 1970 after close to four years of work. Different members of the crews began their training at various points in 1970, but the two main crews plus a third backup crew did not begin integrated training until September 18, 1970, a few scant months prior to the expected launch of the first DOS. The fourth crew, who were not actually expected to fly, but whose commander would play an important role in the history of this first space station program, did not even begin training until January 1971.[55]

From the outset of the DOS program, it was clear that the Soviet space bureaucracy was managing this program with much more verve than its lackluster performance during the piloted lunar programs. There were regular meetings at the highest levels to assess the pace of preparations with necessary actions to compensate for potential delays. Publicly, Soviet spokespersons were claiming left and right that the future of space travel depended squarely on the development of Earth-orbital space stations. But these words would have to be backed up by actions. In late August 1970, word came down from the Central Committee that the first DOS article would have to be launched in time for the 24th Congress of the Communist Party of the Soviet Union, to be held in the spring of 1971. Later, in September, Minister Afanasyev called Mishin to ensure that the launch would be in January 1971 before the opening of the Party Congress, thus distinctly linking socialist doctrine with the Soviet expansion into space.

On September 23, 1970, Ustinov presented Mishin with a deadline for the launch of DOS-1—February 5, 1971—that is, a few weeks before the Congress. According to the plan, ground testing of the station would be complete by December 10, and the station would be taken to Tyura-Tam January 1–10, 1971. The timing with the Party Congress significantly upped the ante of the program. Afanasyev personally visited the Khrunichev Machine Building Plant on October 1, demanding that engineers complete the assembly of the first flight vehicle within forty-five days. Despite assurance from the leading officials at the plant that this would be impossible, Afanasyev did not back down. Working overtime, the workers at the plant eventually managed to fulfill the minister's demands, completing the manufacture of the first DOS by the end of November 1970. The assembled station was then transferred to TsKBEM for final ground testing.[56] Despite the hectic pace, there were delays. By mid-November, it was clear that Ustinov's schedule would not stand. On December 21, the first State Commission meeting for the DOS-7K complex took place in Moscow. The launch was postponed to March 15, 1971, still just a scant twelve months after the DOS program had been inaugurated. One issue of discussion during this period was the length of the two missions to DOS-1. Based on the resources of the station, Mishin had originally envisioned the first lasting thirty days and the second forty-five days. Several cosmonauts met with the chief designer in October 1970, proposing that the first flight be shortened to twenty to twenty-two days—a request apparently based on the abysmal condition of the Soyuz 9 crew after returning from their eighteen-day mission. The length issue was discussed at the State Commission meeting on December 21, but it was left unresolved because of opposition from Kamanin on performing a thirty-day mission.[57]

55. Marinin, "Quarter Century for 'Salyut'."

56. Ibid.; S. A. Zhiltsov, ed., Gosudarstvennyy kosmicheskiy nauchno-proizvodstvennyy tsentr imeni M. V. Khrunicheva (Moscow: RUSSLIT, 1997), p. 76; Semenov, ed., Raketno-Kosmicheskaya Korporatsiya, p. 269. TsKBEM engineers designed several models of the DOS apart from the flight model. These included a unit for working on the assembly of the payload fairing, a heat model for creating the thermo-regulation and life support systems, a unit for working on the engine system, an antenna model for checking the directions of the antennas of the radar system, and the designers' model for installing instruments.

57. Marinin, "Quarter Century for 'Salyut'."

The stage was set for the Soviet revenge on Apollo. The leap that the Soviets took in 1971, however, proved to be of a different nature. It was a story mired in the most bitter of ironies, the most dramatic of events, and certainly the most tragic of consequences.

Salyut

On February 5, 1971, just over a month prior to the scheduled launch of DOS-1, Air Force Colonel Georgiy S. Shonin, the commander of the first space station crew, did not report to training at the Cosmonaut Training Center. Kamanin personally took over the investigation and found to his surprise that this was not the first time that there had been such an absence. After further investigations, he found that Shonin had, without authorization, checked into a hospital for an unspecified "illness," which had come to light after a recent trip to the Tyura-Tam launch site. Leonov, the commander of the second crew, made a vain attempt to defend Shonin's actions, but it was too late. When Mishin discovered this lapse in training, he immediately asked Kamanin to dismiss Shonin from the mission and, in "a fit of temper," proposed an all-civilian crew to fly the first mission. In the end, Mishin backed down on his all-civilian proposal, and Kamanin removed Shonin from the primary crew. The "ill" cosmonaut was sent off to Burdenko Hospital and was found to have an unstated "reactive condition" as well as "psychological faults." Shonin, one of the original 1960 group of cosmonauts with Gagarin, never flew another space mission, although it seems that he recovered from this censure and trained again for space missions in the late 1970s.[58]

On February 12, Kamanin named revised crews for the first missions, inserting two-time veteran Shatalov as the commander of the first mission. It could not hurt that Shatalov was the *only* cosmonaut in the entire detachment who had experience in docking in space. With the hapless Shonin gone, the first two crews to the DOS became: (1) Shatalov, Yeliseyev, and Rukavishnikov and (2) Leonov, Kubasov, and Kolodin. The third and fourth crews were accordingly shuffled. In a switch that only had meaning in retrospect, the third crew—essentially a back-up crew for the two primary crews—was Dobrovolskiy, Volkov, and Patsayev.[59] None of them expected to fly. In fact, Dobrovolskiy had only begun training in January 1971.

The frantic pace of preparing the space station began to catch up with its developers, as errors and delays crept into the preparations. On March 2, 1971, a readiness review meeting of the Council of Chief Designers took place, during which significant delays were acknowledged. There had been continual postponements in vibration testing of the station flight article, while serious malfunctions had cropped up in the ground testing of the Igla docking system to be used on the Soyuz transport spacecraft. Of four Igla systems built by the Scientific-Research Institute for Precision Instruments, three had failed testing; the fourth was working only marginally. Furthermore, there were also major delays in the packing of the parachutes in the Soyuz capsule and the testing of the station's life support system. The flight version of the DOS had arrived at the Baykonur Cosmodrome in March 1971, allowing engineers to begin a forty-day working cycle to test the product. Here, engineers had to perform their tasks almost by a seat-of-the-pants approach, with makeshift equipment doing tasks that required "difficult physical work."[60] Mishin arrived at the launch site in late March, noting that many instruments had been removed from the station and that there had been mistakes during the assembly of ground systems. The delays

---

58. *Ibid.* Shonin trained for a military Transport-Supply Ship (TKS) mission in 1977–79 before officially resigning from the cosmonaut team on April 28, 1979. The all-civilian crew that Mishin proposed was Yeliseyev, Kubasov, and Rukavishnikov.

59. S. Shamsutdinov and I. Marinin, "Flights Which Never Happened" (English title), *Aviatsiya i kosmonavtika* no. 3 (March 1993): 43–44.

60. Semenov, ed., *Raketno-Kosmicheskaya Korporatsiya*, p. 269.

meant that the launch was delayed by another month, April 15 at the earliest; the Shatalov crew would be launched on April 18–20 to begin the first space station occupation.

Originally, Mishin had decided to call the station *Zarya* ("Dawn") in the open press, but at some time immediately prior to the launch, there was some discussion on the issue. Apparently, the Chinese had used the same name for one of their satellites. Instead, Mishin himself suggested at the launch site that the station be renamed *Salyut* ("Salute") as a mark of respect for the late Yuriy A. Gagarin, the tenth anniversary of whose historic space mission was also in April 1971. The original Zarya name remained inscribed on both the station and the Proton rocket's payload fairing because it was too late to change it.[61]

The three main crews training for the flight were in the process of finalizing their training program by March 1971, although there was still no final word on the length of the first mission—twenty to twenty-two days or thirty days. The overwhelming feeling was that the State Commission would vote for the latter. Mishin had evidently spoken at length with representatives of the Institute for Biomedical Problems on the issue, and he had decided that thirty days would not do significant harm to the human body. On March 16, the cosmonauts took their final exams, all of them doing splendidly, confirming their full preparedness for the mission. During a meeting on March 19, the State Commission adopted a final decision to launch the station between April 15 and 18. The cosmonauts themselves flew into Baykonur on March 20 to acclimatize themselves with prelaunch preparations at the Kosmonavt Hotel. They were witnesses to another failure during testing of the Igla rendezvous system, and one wonders whether their morale was not affected by the accumulating errors. After a brief training period, they returned to Moscow. Shatalov and Yeliseyev attended the 24th Congress of the Communist Party, which opened on March 30. Five days earlier, the Military-Industrial Commission formally approved plans for the two missions to the station.[62]

The cosmonauts, Air Force officials, and many other industry representatives returned to the launch site on April 6. Three days later, the State Commission, headed by Ministry of General Machine Building representative Maj. General Kerimov, approved the launch of the DOS ship, spacecraft no. 121, on April 19. There were no major delays or unexpected occurrences during the last days leading to the launch. In anticipation of the launch, the Soviets, in their customary manner, dropped hints of an impending spectacular. On March 14, the anonymous "Chief Designer of Spaceships"—that is, Mishin—declared in an interview with the Moscow economic daily *Sotsialisticheskaya industriya* (*Socialist Industry*), that the Soviet Union was preparing to launch another piloted spacecraft for a long-duration mission as a prelude to building a permanent Earth-orbiting space laboratory.[63] The world did not have long to wait. At 0440 hours Moscow Time on April 19, 1971, a three-stage Proton booster successfully lifted off from site 81 with its precious payload, DOS-1. The initial orbital parameters for the station, called *Salyut* in the press, were 222 by 200 kilometers at a 51.6-degree inclination. By the end of the first orbit, ground controllers discovered that the large cover on the exterior protecting the scientific apparatus compartment—that is, the OST-1 telescope—had not been jettisoned, thus jeopardizing the scientific value of any visiting expedition. Apparently, the explosive devices for the cover had failed to fire. During the second day of flight, there were also failures

61.   *Ibid.*, p. 70; V. M. Petrakov, "Soviet Orbital Stations," *Journal of the British Interplanetary Society* 47 (September 1994): 363–72. Another source suggests that the name change was because Zarya was also used as a call sign for the ground segment of the Soviet Command-Measurement Complex, thus raising the possibility of confusion between the station and Earth during communications sessions. See Dmitriy Payson, "Without the 'Secret' Stamp: 'Salyut' and Star Wars" (English title), *Rossiyskiye vesti*, November 21, 1992, p. 4.

62.   Marinin, "Quarter Century for 'Salyut'." The commission approved the plan for a thirty-day flight for the first mission and a possible forty-five-day for the second one.

63.   Theodore Shabad, "New Manned Orbital Flight Seen," *New York Times*, March 15, 1971, p. 17.

of two ventilation units used for the life support system, although this seems not to have caused any major concern on the ground.[64]

The day after the *Salyut* launch, the primary crew for the first mission was presented to the Soviet press, with the accompanying announcement that their launch would take place on April 22 at 0320 hours Moscow Time—a night launch. Shatalov, Yeliseyev, and Rukavishnikov took their places in their Soyuz spacecraft, despite some concern with respect to heavy showers during the night; the State Commission, however, agreed to proceed with the launch. The Soyuz launch vehicle was filled with propellant, and all prelaunch procedures seemed to be going according to schedule until T minus one minute. At that point, one of the masts on the launch system did not retract as planned. Officials feared that if there *was* a launch, the launch escape system would be spuriously activated and cause an explosion, as had occurred during a Soyuz launch in December 1966.[65] Mishin opted to reluctantly postpone the launch. The commission quickly decided to keep the booster on the pad fully fueled and try again the following day.

During the second launch attempt, the exact same thing occurred again; a mast from the launch structure refused to retract. Mishin was apparently aware of the reasons for this deviation from normal procedures, and he took control of the situation. Taking complete responsibility for any negative consequences, he called out for the launch to proceed. There were no problems, and the first 7K-T Soyuz spacecraft, vehicle no. 31, lifted off at 0254 hours on April 23, 1971, with its three-cosmonaut crew of forty-three-year-old Colonel Vladimir A. Shatalov (commander), thirty-six-year-old Aleksey S. Yeliseyev (flight engineer), and thirty-eight-year-old Nikolay N. Rukavishnikov (test engineer). Shatalov and Yeliseyev were the first Soviet cosmonauts to make a third spaceflight, having flown their first missions just two years before. Rukavishnikov, the only rookie on board, had extensively trained for the L1 lunar program in the late 1960s. The vehicle, named *Soyuz 10* by the press, entered a nominal orbit of 209.6 by 248.4 kilometers at a 51.6-degree inclination.

Despite a successful launch, the prognosis for the mission was dim. The lid on the scientific compartment was still lodged in its place and threatened to sabotage at least 90 percent of the scientific experiments program. Furthermore, of the eight ventilation units in the life support system, *six* had failed, raising the prospect of an internal atmosphere full of carbon dioxide and "other harmful materials."[66] During Soyuz 10's fifth and sixth orbits, there were difficulties in modifying the ship's orbit to intersect with that of *Salyut*. The first time, there was an error in the programming logic of the command, while the second time, the burn was abandoned because of insufficient time to prepare. Soyuz-10's ionic orientation system was apparently inoperational because of contamination of the optical surfaces. Shatalov eventually took control of the situation and asked for permission to manually change the orbit, which he did without any problems.

The following morning at a distance of sixteen kilometers from the station, Shatalov switched on the Igla system, which successfully brought the Soyuz to within 180–200 meters of *Salyut*. At that point, he took over manual control, successfully linking up at 0447 hours on April 24. About ten to fifteen minutes following soft-docking, Shatalov radioed to the ground that the docking indicator light was not on in the Soyuz, suggesting that hard-docking had not taken place. Ground telemetry confirmed that full docking had not occurred and that there was still a nine-centimeter gap between the two vehicles. Shatalov attempted to tighten the two ships by firing the Soyuz engines, but this did not prove successful. On their fourth orbit together, orders

---

64.   I. Marinin, "Quarter Century for 'Salyut': Part II" (English title), *Novosti kosmonavtiki* 11 (May 20–June 2, 1996): 46–51; Petrakov, "Soviet Orbital Stations."

65.   There was a launch failure on December 14, 1966, when a Soyuz (or 11A511) booster exploded at site 31 at Tyura-Tam, killing one person and injuring many. The payload was an automated Soyuz spacecraft.

66.   Marinin, "Quarter Century for 'Salyut': Part II."

were received from the ground to try and undock the Soyuz 10 spacecraft and attempt a redocking. At this point, the crew ran into "incredible difficulty" in trying to undock from the station.[67] On the fourth orbit of combined flight, Shatalov attempted to unlatch the Soyuz 10 ship from the *Salyut* station, but the spacecraft refused to dislodge. The problem was not taken lightly because such a situation could lead to the loss of the station—and perhaps the crew as well. Being unable to undock by normal means, there were two options: (1) dismantling the docking apparatus, detaching it from the Soyuz, and moving away from the station or (2) detaching the spheroid living compartment from the Soyuz spacecraft and separating, thus leaving the living compartment docked to *Salyut*. In both cases, the station would be unusable in the future because the single docking node would be occupied. The situation was compounded by the fact that there was only a limited amount of oxygen left in the Soyuz spaceship (about forty hours), within which time all of this would have to be done. Luckily for everyone, on the fifth orbit of combined flight at 1017 hours on April 24, Shatalov once again tried to undock and was successful. The two spacecraft had been docked for five and a half hours.[68]

Shatalov maintained station-keeping distance from *Salyut* as ground control debated whether to attempt a second docking with the station. After assessing the state of the on-board gyroscopes, propellant levels, and internal air, the Chief Operations and Control Group at Yevpatoriya, headed by TsKBEM Deputy Chief Designer Tregub, decided to abandon the mission and prepare for an emergency return to Earth. Before leaving the vicinity of the station, the Soyuz 10 crew flew around *Salyut* and photographed the docking node to assist engineers on the ground in determining the cause of the malfunction.[69] The crew successfully landed without incident at 0240 hours on April 25, 120 kilometers northwest of the town of Karaganda in Kazakhstan. It was the first-ever night landing in the Soviet human space program. The mission had lasted only one day, twenty-three hours, forty-six minutes, and fifty-four seconds. While the Soviet media at the time characteristically claimed that entry into the station was not even on the agenda and all the objectives of the flight had been successfully achieved, the mission had clearly been a bitter disappointment.

The investigation into the Soyuz 10 failure was completed by May 10, by which time engineers ascertained that the Soyuz docking apparatus had been damaged during the docking maneuver. There had been a breakdown in the coupling shock-absorbing claws in the active part of the docking node when the two ships had attempted hard dock. The system had been subjected to 160–200 kilograms of force during the maneuver, although the force at docking was projected to be only eighty kilograms. The coupler could withstand up to 130 kilograms. The increased force had been partly caused by the failure to stop the motion of the Soyuz after soft-docking. Engineers decided to reinforce the docking system twofold, while introducing the capability of the crew to manually control the pins of the docking system. In the meantime, Mishin proposed that despite the failure of Soyuz 10, plans should now include *two* further missions to the *Salyut* station to complete the original objectives of the program. The first would begin on June 4 and the second on July 18, 1971. Mishin also proposed to have the following crew reduced to two cosmonauts to carry bulky spacesuits that would allow an EVA by one cosmonaut to visually inspect the docking node on the station, as well as to remove the cover for the scientific experiments package. Kamanin categorically rejected this idea, arguing that the cosmonauts had not been trained for EVA and adding that the Zvezda Machine Building Plant,

67.   *Ibid.*; Nazarov, "You Cannot Paper Space With Rubles."
68.   N. Kamanin, "This Should Never Happen Again!" (English title), *Vozdushniy transport* 23 (1993): 12–13; Marinin, "Quarter Century for 'Salyut': Part II."
69.   Petrakov, "Soviet Orbital Stations."

which produced the spacesuits, would not be able to certify flight-ready suits by the launch date. In the end, the matter was dropped—a cruel irony considering the later course of events.[70]

At a meeting of the major leaders of the program on May 11, 1971, in Moscow, there was further disagreement between Mishin and Kerimov on one side and Kamanin on the other. The former proposed two missions lasting thirty days each. Kamanin opposed this idea based on his belief that on-board supplies on *Salyut* might be all used up before the end of the second expedition, thus creating a dangerous situation for any crew. In the end, officials decided that the goal of each mission would be to dock with the station and "revive" its systems; any decision on duration would be made during a particular flight. For the benefit of planning, Mishin used information from ballistics computations to tentatively plan for a twenty-five-day flight beginning on June 6, 1971.

There were a number of failures once again in the Igla system during preparations for the next mission, but the State Commission assessed the anomalies and on May 24 certified the Soyuz vehicle (with modifications to the docking system and improved autonomous capabilities) as fully ready for flight. The failure of the Soyuz 10 crew to carry out their primary mission of manning the space station meant that the third crew for the DOS, who would have been consigned to only a backup role during the program, moved up to the second spot. Thus, the crews who were named for the two newly scheduled missions to DOS-1 were: (1) Leonov, Kubasov, and Kolodin and (2) Dobrovolskiy, Volkov, and Patsayev. The latter would also serve as the backup crew to the former. Both crews arrived at Tyura-Tam late on May 28 in preparation for the launch.[71]

All the plans for the mission were thrown into complete uncertainty on June 3 when doctors from the Moscow-based Institute for Biomedical Problems detected a swelling in primary crew flight engineer Kubasov's right lung.[72] Suspecting that this was the beginning of tuberculosis, they unanimously called for his removal from the crew. According to the rules of the Ministry of General Machine Building and the Ministry of Health:

> . . . if one of the members of the crew is taken ill prior to departure to the cosmodrome, he should be replaced by the corresponding member of the other crew. Carrying out the replacement of the individual at the cosmodrome is not possible. In case of such a necessity, it is only possible to carry out the replacement of the [entire] crew.[73]

The verdict was simple but difficult to accept for the crews: the Leonov-Kubasov-Kolodin team would have to be replaced by the Dobrovolskiy-Volkov-Patsayev crew.

Yaroslav Golovanov, then a correspondent for the newspaper *Komsomolskaya pravda* who was at Tyura-Tam at the time, recalled later that "what happened at the Kosmonavt [Hotel, where the crews were staying] is hard to describe." Leonov broke down and visibly lost his temper. Kubasov, who was the center of the controversy, was simply stunned. That night, Kolodin, the third primary crew member, arrived at the hotel completely inebriated on vodka, bemoaning the fact that he may never go to space. Leonov later took the matter directly to his superiors and pleaded that the State Commission only replace the indisposed Kubasov with his backup Volkov, thus making the new crew Leonov, Volkov, and Kolodin. It seems that the

---

70. Kamanin, "This Should Never Happen Again!," no. 23; Marinin, "Quarter Century for 'Salyut': Part II"; Semenov, ed., *Raketno-Kosmicheskaya Korporatsiya*, p. 188; K. Lantratov, "20 Years From the Flight of 'Soyuz-12'" (English title), *Novosti kosmonavtiki* 20 (September 25–October 8, 1993): 39–40.

71. Kamanin, "This Should Never Happen Again!," no. 23; Marinin, "Quarter Century for 'Salyut': Part II."

72. One source suggests that the doctors found "a dark spot on the lungs." See Iosif Davydov, "How Could That Have Been?: Slandered Space" (English title), *Rossiyskaya gazeta*, June 11, 1992, p. 5.

73. Marinin, "Quarter Century for 'Salyut': Part II," p. 50. Author's emphasis.

commission was in fact leaning toward this solution despite the ministry's edict. All the cosmonauts, physicians, Cosmonaut Training Center chiefs, and Kamanin himself decided to call for *only* Kubasov's replacement. Mishin and State Commission Chairman Kerimov tentatively agreed with this recommendation until Mishin had further discussions with participants in Moscow, when he changed his mind and insisted on replacing the *entire* primary crew. The next day, June 4, two days before the launch, after the Soyuz booster had been transported to the launch pad, a final session of the core members of the State Commission was held. Again, Kamanin recommended replacing only Kubasov. This time, Mishin had the support of most of the other attendees, including Maj. General Nikolay F. Kuznetsov, the Director of the Cosmonaut Training Center. The commission finally decided to replace the entire Leonov crew and launch the Dobrovolskiy crew.[74] Later that evening, during Mishin's visit to speak to the cosmonauts, Kolodin, in a moment of outrage, "lectured [Mishin] with a lot of extraneous items, which he later much regretted."[75] According to one report, Kolodin told Mishin that "history would never forgive him" for his decision to send the backup crew.[76]

The original backup crewmembers of forty-three-year-old Lt. Colonel Georgiy T. Dobrovolskiy (commander), thirty-five-year-old civilian Vladislav N. Volkov (flight engineer), and thirty-seven-year-old civilian Viktor I. Patsayev (test engineer) were successfully launched at 0755 hours on June 6, 1971, in their 7K-T spacecraft, vehicle no. 32. The Soviet press announced the mission as *Soyuz 11*. Both Dobrovolskiy and Patsayev were making their first flights, while Volkov was making his second, having flown as part of the "troika" Soyuz mission in late 1969. Most unusually for a space mission, the crew had been formed less than four months before the launch day, having no expectations to fly on such short notice. The Soyuz 11 spaceship entered an initial orbit of 191.5 by 220.5 kilometers at a 51.64-degree inclination. After two orbital changes, the spacecraft was within seven kilometers of the *Salyut* station. The Igla system was switched on, and successful docking was accomplished at 1045 hours. Ground control at Yevpatoriya in Crimea had to wait a tense half-hour before Dobrovolskiy announced that the docking had successfully taken place. During the fourth orbit of joint operations, pressurization checks proved to be acceptable, and the crew opened the hatch to the station. Patsayev was the first one in the station; the crew immediately turned on the air regeneration system and replaced two of the six faulty ventilation units of the life support system. Unfortunately, the crew sensed a strong odor of burning in the air, which forced them to spend a tense night in their ferry craft. The next day, they returned to the station to discover the odor gone and immediately set about activating instruments on the station in support of their experiments program.[77]

As the days turned into weeks, the three men managed to carry out a remarkably full experiments program despite many attendant problems. By June 9, medical and biological experiments had begun, while experiments in other areas were started on June 11, consisting of spectrographic measurements of natural formations and water surfaces in the Soviet Union. Each day for the crew consisted of eight hours of work, two hours for meals, two hours for exercises, and two hours of personal time. By the sixth day on the station, the men had settled into a rotating routine—that is, when Dobrovolskiy was having breakfast, Volkov would be having supper and Patsayev dinner. Thus two men were always awake while the third one slept.[78]

74. *Ibid.*

75. Gordon R. Hooper, *The Soviet Cosmonaut Team, Volume 2: Cosmonaut Biographies* (Lowestoft, UK: GRH Publications, 1990), p. 131.

76. Lardier, *L'Astronautique Soviétique*, p. 190.

77. Kamanin, "This Should Never Happen Again!," no. 23; Marinin, "Quarter Century for 'Salyut': Part II."

78. Lardier, *L'Astronautique Soviétique*, p. 202; Peter Smolders, *Soviets in Space* (New York: Taplinger Publishing Co., 1973), p. 241.

During their mission, the cosmonauts performed about 140 scientific experiments, far more than on any other Soviet space mission. The medical studies included experiments involving the cardiovascular system (using the Polinom-2M), blood tests (*Amak-3*), the density of bone tissue (Plotnost), pulmonary circulation (Rezeda-5), the measurement of wrist strength, tests of visual acuity, measurements of radiation dosages, and the study of microflora. Strictly biological experiments included those on the growth of plants (using the *Oazis-1* hydroponic greenhouse starting on June 13), the study of the vestibular apparatus of tadpoles, the genetic mutations of flies, the growth of chlorella algae, and the development of grain in microgravity.

Although the crew was unable to use the OST-1 telescope because of the sealed cover, they did manage to conduct an extensive series of astrophysics-related experiments using the Orion-1 ultraviolet and the Anna-3 gamma-ray telescopes. The latter was named after its designer's daughter Anna because "like [her] daughter, gamma-ray astronomy has still a great deal to learn."[79] Cosmonauts used the Anna-3, capable of registering gamma rays with energy of up to 100 megavolts, for the first time on June 11. The Orion-1 telescope, built by the Armenian Academy of Sciences, was used on June 18 and 21 to make six and nine ultraviolet spectrograms of

The primary crew of Soyuz 11 is shown at the top of the service mast prior to entering their spacecraft. From left are Georgiy Dobrovolskiy (commander), Viktor Patsayev (test engineer), and Vladislav Volkov (flight engineer). (NASA photo)

celestial targets, such as Beta (Centaurus) and Vega (Alpha-Lyra). The study of charged particle flux was accomplished with a third telescope named the TEB. The cosmonauts also used the FEK-7 emulsion chamber to register primary cosmic rays for over a period of 1,728 hours. One inert instrument was the MMK-1, which measured micrometeoroid flux on the exterior of the station.

79. Smolders, *Soviets in Space*, p. 245. The scientist in question was A. M. Galper. The Anna-3 was developed by the Moscow Institute of Engineering and Physics under the supervision of V. G. Kirillov-Ugryumov.

Other experimental disciplines included studying the geophysical properties of Earth with various cameras, such as the AFA-M, the KFA-21, and the handheld RSS-2 spectrograph, which was developed by the Department of Atmospheric Physics at Leningrad State University. Researchers used the same spectrograph from An-2 and Il-18 aircraft at altitudes of 300 and 8,000 meters simultaneously with the cosmonauts to determine pollution and precipitation levels in the Caspian Sea. The crew also used the RSS-2 on June 14 and 15 to determine humidity of the soil in areas around the Caspian and Aral Seas. On June 11, 19, and 22, they used the spectrograph to measure the optical characteristics of Earth's atmosphere and the degree of polarization of sunlight reflected by Earth. The crew also measured the chemical composition of the atmosphere with a mass spectrometer and used the ERA instrument for studying atmospheric resonance in the ionosphere. Earth observation research included coordinated studies with a Meteor-1 satellite over several days using two hand-operated devices on the station. Strictly technological experiments were related to the station itself. The crew observed luminous particles outside the station with a photometer, and also they studied the dynamic characteristics of the station with a stellar camera.[80]

Throughout the mission, reports from the cosmonauts in *Salyut* were shown on Soviet television. Many of their exchanges on TV were humorous in nature, contrasting sharply with the morose image of the Soviet spacefarer, and it was clear that the three men were having the time of their lives. By mid-June, the three men had become household names in the Soviet Union—a new breed of folk hero for a country whose prestige had been trampled by the success of Apollo. For the first time in many years, the Soviet human space program could claim a genuine advance and victory over the United States. It would not be an overstatement to claim that much of the general population anticipated the return of the three cosmonauts in a unified way that had not been witnessed for many years.

The continuing TV reports did not, of course, tell the whole story behind the mission. There were, in fact, many problems for the crew aboard the station—problems that on occasion hindered productive work. For example, during the first two weeks of the mission, there were a number of personality clashes between the members of the crew, which were mediated by cosmonauts on the ground at Yevpatoriya who served as "capcoms."[81] Although these difficulties were resolved, a more serious emergency occurred on June 16, when flight engineer Volkov suddenly radioed to capcom Shatalov that he sensed a strong odor of smoke. Assuming the worst-case scenario of a fire in the station, cosmonauts Nikolayev and Yeliseyev on the ground ordered the crew to immediately evacuate to their Soyuz ferry craft and begin preparations for undocking. Having quickly moved into the Soyuz, the crew first began attempts to establish the cause of the emergency by switching on the backup electrical supply system on *Salyut* and turning on filters to purify the atmosphere. Following a tense period, during which instruments tested the atmosphere in the station for safety, the cosmonauts entered the *Salyut* station once again.[82]

The drama seemed to have intensified the discord brewing among the crew. Veteran cosmonaut Bykovskiy, at Yevpatoriya, recalled that during the emergency, Volkov had become extremely nervous and had tried to resolve the situation by himself, ignoring the assistance of

80. Lardier, *L'Astronautique Soviétique*, pp. 202–03; Bert Dubbelaar, *The Salyut Project* (Moscow: Progress Publishers, 1986), pp. 12–16.

81. The cosmonauts involved in ground control were V. F. Bykovskiy, V. V. Gorbatko, A. G. Nikolayev, V. A. Shatalov, and A. S. Yeliseyev.

82. Kamanin, "This Should Never Happen Again!," no. 23; I. Marinin, "Quarter Century for 'Salyut': Part III" (English title), *Novosti kosmonavtiki* 12–13 (June 3–30, 1996): 77–81.

his crewmates Dobrovolskiy and Patsayev. In an unusual move, Chief Designer Mishin communicated personally with Volkov, informing him that all operational decisions should be taken by the commander (Dobrovolskiy) and that mission-critical operations should be carried out only at his discretion. Volkov irritatedly responded that the entire crew was aggravated and that all decisions should be made collectively. In an amplification of the event, Mishin recalled in a 1989 interview that a personality clash had developed between Dobrovolskiy and Volkov, during which Volkov, the only spaceflight veteran on board, declared himself the commander of the mission, usurping Dobrovolskiy's role. There were apparently several "complicated conversations" between Mishin and Volkov after the incident.[83] In Kamanin's opinion, Volkov had acted hastily and had a disdainful attitude to those at ground control. Mishin also added that there may have been indeed a fire on the station originating from a power cable, and the crew apparently asked for permission to return to Earth immediately but were dissuaded by ground control.[84] The entire situation was diffused following extensive consultations with cosmonauts on the ground, who were able to bring the crew back to their experiments program.

The Soyuz 11 crew was scheduled to observe the third N1 lunar rocket during its launch on June 20 from Tyura-Tam using the *Svinets* instrument designed for military purposes. The launch was, however, moved to June 22 and eventually to June 27, and the crew's ground track was not over the launch site at the time of the N1 launch. Dobrovolskiy was able to skillfully use Svinets on June 24 and 25 to observe night launches of solid-propellant ballistic missiles from Tyura-Tam.[85]

The cosmonauts' medical program was not completely successful. The cosmonauts were apparently reluctant to exercise, and the problem was compounded by several failures on the station. Kamanin wrote in his diary on June 23 that:

> . . . the readaptation will be particularly difficult for Volkov: during the flight he has been more reluctant to do physical exercises than the other crew members, he has totally rejected meat food, he has often been irritated and has already been making a lot of mistakes.[86]

The running track was rarely used because of unexpected vibrations when exercising, which shook the solar panels and communications antennas. The *Chibis* vacuum suit used for shifting blood to the upper regions of the body was the source of many problems and was rarely used. The load-bearing Pingvin space suits also tore at various places during exercises, neutralizing their impact. Naturally, the lack of calisthenics was a great concern to doctors on the ground, who believed that the crew would be in extremely poor shape after a near-month-long mission. Given the problems on the mission, Mishin backed away from his insistence on a thirty-day mission, instead opting for a more conservative flight of twenty-two to twenty-four days. On June 22, the State Commission confirmed the decision to land the crew on June 30, on that day's third orbit, early in the morning.

The three cosmonauts began preparations to return to Earth on June 26. They had exceeded the world-record endurance for a single piloted spaceflight two days earlier on their eighteenth day in space. Despite increasing numbers of mistakes on the part of the crew, attributed to fatigue, the crew completed all their return procedures on time, and on the evening of June 29, they transferred to the Soyuz 11 spaceship and closed the hatch between the two space-

83. Tarasov, "Missions in Dreams and Reality."
84. *Ibid.*
85. Nikolay Kamanin, "This Should Never Happen Again!" (English title), *Vozdushniy transport* 24 (1993): 12; Marinin, "Quarter Century for 'Salyut': Part III." The cosmonauts also used the OD-4 optical "visor-range finder" to perform military observations of ground targets.
86. Kamanin, "This Should Never Happen Again!," no. 24, p. 12.

craft. The crew then moved into the descent apparatus and shut the hatch between it and the spherical living compartment. There was a major crisis at this point when the "Hatch Open" indicator light between the Soyuz living compartment and the descent apparatus failed to turn off. Fatigued and anxious, Volkov excitedly called out to ground control: "The hatch isn't pressurized, what should we do, what should we do?!!"[87] Cosmonaut Yeliseyev, who was the capcom at the time, calmed Volkov down and gave the crew detailed instructions to go through the entire hatch-closing procedure once more. Dobrovolskiy and Patsayev expertly followed the instructions, but the indicator light remained turned on. All the members of the crew grew increasingly nervous because in a few minutes that hatch would be the last barrier between the crew and open space.

After intensive discussions, the Chief Operations and Control Group on the ground recommended placing a piece of paper over the sensor that detected hatch closing, presumably in the belief that it was a sensor error. Dobrovolskiy found a piece of plaster, which he placed over the sensor, and shut the hatch once more. This time the indicator turned off, and all subsequent pressurization checks proved satisfactory. The twenty-minute crisis with the hatch had strained the nerves of the crew, but following the tests, the cosmonauts apparently calmed down and proceeded with preparations to undock from the station. At 2125 hours, 15 seconds, the Soyuz 11 spaceship undocked from *Salyut* and flew around the station, and Patsayev took a number of photographs. At around 0135 hours on June 30, Volkov reported that the "'Return' indicator light is on." Ground control replied, "Let it be on. It's correctly on. Communications are ending. Good luck!"[88] After that, communications were cut off as the Soyuz drifted out of voice contact, and they were evidently never regained. According to the preprogrammed sequence of reentry, the main Soyuz engine was to begin firing at 0135 hours, 24 seconds Moscow Time on June 30, followed by separation of the three Soyuz modules at 0147 hours, 28 seconds. Ground control was, however, unsure whether this had indeed taken place because of the loss of communications. Search-and-rescue services proceeded on the assumption that all was going according to plan on the Soyuz 11 ship, and the teams from the Soviet Air Defense Force and Air Force detected the descent apparatus of the spacecraft on time in the assigned location. The capsule landed at 0218 hours Moscow Time about 202 kilometers east of Dzhezkazgan in Kazakhstan. The mission had lasted twenty-three days, eighteen hours, twenty-one minutes, and forty-three seconds. As soon as rescue teams opened the vehicle hatch, they found the crew lifeless in their seats.

The recovery teams attempted to revive the cosmonauts after bringing them out of the capsule, but it was all in vain. In the meantime, the State Commission at Yevpatoriya received a message back from the rescue services concerning the deaths. Immediately, Afanasyev, Kerimov, Mishin, Kamanin, DOS lead designer Semenov, and others flew directly to the landing site. Other officials, including doctors, also flew in from Moscow. An on-the-spot investigation indicated that there was blood in the crew's lungs, nitrogen in their blood, and hemorrhages in their brains, which were all obvious indicators of death by depressurization. An inspection of the ship's interior showed that all the radio transmitters had been manually turned off, the shoulder straps of all the cosmonauts were unfastened, and Dobrovolskiy himself had been tangled in his straps. Everything in the Soyuz 11 descent apparatus appeared normal, everything except one of two valves in the respiratory system, which was in an open position, strongly supporting the hypothesis that there had been a rapid decompression.[89]

87.   Marinin, "Quarter Century for 'Salyut': Part III"; N. Kamanin, "This Should Never Happen Again" (English title), *Vozdushniy transport* 25 (1993): 12.

88.   O. Ye. Leonov, "Until We Meet on Earth" (English title), *Otechestvennyy arkhivu* 3 (1994): 71–80.

89.   Semenov, ed., *Raketno-Kosmicheskaya Korporatsiya*, p. 188; Marinin, "Quarter Century for 'Salyut': Part III."

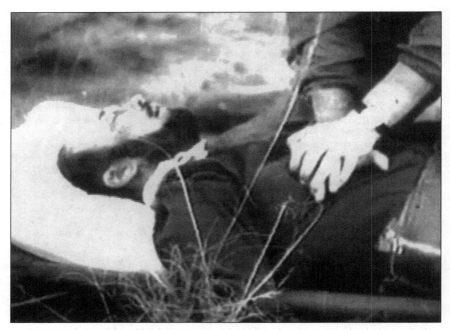

Cosmonaut Georgiy Dobrovolskiy, Soyuz 11 commander, is shown here just minutes after death as medical workers try to revive him. The beard was from approximately twenty-four days spent on the Salyut space station in June 1971. At the time, this was the longest piloted space mission in history. (copyright Rudy, Inc., via Quest)

The shock not only to the space industry but also the Soviet Union as a whole was devastating. An unprecedented wave of grief swept through the country, not unlike the collective mourning in the United States after President Kennedy's assassination in 1963. Official Soviet TV and radio changed their formats to accommodate for the tragedy, while countless condolence messages poured in from leaders all over the world. Apart from the human loss itself, the Soyuz 11 tragedy was a severe blow to the Soviet space program, coming at a time when it had been so close to reclaiming the lost glory of the Korolev years. In a cruel twist of fate, the Soviet space program was not even accorded a consolation prize in the space race. It was beset with problems far more imposing than simply political cost. If 1969 was the year of humiliation for the Soviet space program, 1971 was its nadir—an absolute low unthinkable a few years before.

The bodies of the three cosmonauts were flown back to Moscow only a few hours after landing, and the following day, on July 1, they were already lying in state in Moscow. Thousands of Soviet citizens flocked to pay their last respects. Unlike the Komarov funeral, when no NASA representative was present, veteran astronaut Brig. General Thomas P. Stafford was on hand, representing the NASA Astronaut Office. Behind the scenes, on July 3, 1971, Ustinov established a governmental State Commission to investigate the accident and recommend changes in the DOS-7K program. As one would expect, Academician Keldysh headed the commission. Chief Designer Babakin from the Lavochkin Design Bureau was the deputy chair.

The remaining members included Chief Designer Glushko and Minister Afanasyev.[90] On July 12, the commission issued a preliminary report, which gave general details of the accident to the general public for the first time:

> During the descent of the spaceship, 30 minutes before landing, pressure in the return capsule dropped rapidly, which led to the unexpected death of the cosmonauts. This has been confirmed by medical and pathological-anatomical examinations. The drop in the pressure was the result of failure of the hermetic sealing of the spaceship. . . . Technical analysis indicates that there are several possible explanations of the de-sealing. Investigation into the exact cause continues.[91]

The following day, the commission met and agreed that the most probable cause of the accident was depressurization resulting from the premature opening of the second respiratory valve in the descent apparatus. Already, two senior members of TsKBEM's staff, Bushuyev and Korzhenevskiy, were recommending that future Soyuz crews be brought down to two cosmonauts with full spacesuits.

Through the following weeks, an analysis of the *Mir* on-board memory device showed that at the moment of separation of the living compartment from the descent apparatus, at an altitude of more than 150 kilometers, the pressure in the descent apparatus dropped in the course of thirty to forty seconds to a near vacuum. The rate of the pressure drop corresponded to the respiratory system's valve opening. The conclusion was obvious: at the moment of separation of the two modules, the valve had prematurely opened. More difficult was determining exactly why it had been jarred open. Engineers carried out dozens of experiments simulating various loads on the suspect valve, but no one particular cause stood out. Only when all types of deviations from normal parameters were introduced *simultaneously* did the valve fail.[92] Based on the Keldysh commission's analysis of voice tapes and telemetry, as well as Kamanin's own diary entries, it was, however, possible to reconstruct the sequence of events that led to the tragedy.

It seems that the reentry burn was on time and completely successful. Subsequently, at the very moment that the Soyuz spacecraft separated into its three component modules, also on time, twelve explosive bolts used for separation produced an overload, displacing a ball joint from its seating.[93] This accidentally jerked open the ventilation valve, which was to have opened only *after* landing; suddenly, there was a direct passage from the crew compartment to the vacuum outside. The crew immediately noticed the drop in pressure inside the capsule; Dobrovolskiy quickly unfastened his seat belts and rushed to the frontal hatch, thinking that the problem was the faulty hatch seal from the undocking incident. The hatch was completely secure, yet the pressure continued to drop in a whistle that continued to get louder. In fact, the sound of the air whistling out of the spacecraft was coming not only from

---

90. Semenov, ed., *Raketno-Kosmicheskaya Korporatsiya*, p. 188. The remaining members of the commission were: A. I. Burnazyan (Ministry of Health), S. G. Frolov (VVS), P. D. Grushin (MKB Fakel), V. A. Kazakov (MAP), M. N. Mishuk (VVS), V. A. Shatalov (VVS), V. I. Shcheulov (TsUKOS), and A. I. Tsarev (VPK).

91. Smolders, *Soviets in Space*, p. 248. One report suggests that the early hypothesis of the doctors on the ground was that the cosmonauts died because of the effects of gravity after such a relatively long period of weightlessness. Dr. Portugalov, the chief of the Morphology Department at the Institute for Biomedical Problems of the Ministry of Health, however came to the conclusion that it had been a valve failure resulting in depressurization. The same conclusion was also reached by anatomists at the Kirov Academy. See Mozzhorin, *et al.*, eds., *Dorogi v kosmos: I*, p. 64.

92. Semenov, ed., *Raketno-Kosmicheskaya Korporatsiya*, pp. 188–89.

93. Tarasov, "Missions in Dreams and Reality."

the suspect valve, but also from on-board radio transmitters and receivers, making it difficult to isolate the true source. At this point, Volkov and Patsayev unfastened their belts and switched off all communications systems to find the source of the whistling; the sound was apparently coming from a point under Dobrovolskiy's seat—the ventilation valve. Dobrovolskiy and Patsayev attempted to manually close the valve, but the time was just too short.[94] Both fell back in their seats, with Dobrovolskiy having time to refasten his belts in a hurried move, which left them tangled.

The speed of the pressure loss in the capsule was incredibly swift. Just four seconds after the ventilation valve failure, Dobrovolskiy's breath rate shot up from sixteen (normal) to forty-eight per minute. After the beginning of pressure loss, the cosmonauts lost the capacity to work in ten to fifteen seconds and were dead in forty-eight to forty-nine seconds. They were apparently "in agony" three to five seconds after separation until about twenty to thirty seconds before death. All the pressure in the capsule dropped from a normal level of 920 millimeters to zero in a matter of 112 seconds.[95] As one Russian journalist later put it, the cosmonauts "passed away fully aware of the tragic consequences of what had happened."[96] Both Kamanin and Mishin seemed to believe that the crew could have prevented their deaths by simply blocking the suspect "hole." In an interview in 1990, Mishin added: "They could hear the hiss of escaping air. They could have put a finger over the hole and that would have done it."[97] Some believed that the crew had not been properly trained in the operation of the valve, which was to be operated only *after* Soyuz landing. The technical documentation on the valve stipulated: "If in case of a water landing, the hatch does not open due to rough seas, or rescue teams are late in coming for over an hour, the cosmonauts may open the valve."[98]

The reason why the seal failed is still unknown, although an article in *The Washington Post* in 1973 by Thomas O'Toole provided some interesting clues. O'Toole's description, based on a "classified report," was the first and only Western report to accurately describe the hatch-closing emergency prior to reentry in great detail. The author added: "[When] the exhausted cosmonauts were fighting the warning light on the hatch they apparently failed to notice that the cabin pressure had crept up to almost 20 pounds per square inch. What this did was to exaggerate any weakness in the hatch seal."[99] O'Toole's report was apparently culled from a classified CIA brief issued the day after the accident, in which the CIA detailed the undocking and reentry problems at the end of the mission.[100]

94. Mishin says that it was only Patsayev who attempted to close the valve. See G. Salakhutdinov, "Once More About Space" (English title), *Ogonek* 34 (August 18–25, 1990): 4–5.

95. Kamanin, "This Should Never Happen Again," no. 25; Marinin, "Quarter Century for 'Salyut': Part III."

96. Mikhail Rebrov, "With a One-Way Ticket" (English title), *Krasnaya zvezda*, September 26, 1996, p. 4. A somewhat different explanation was given to NASA officials in October 1973: "At approximately 723 seconds after retrofire, the 12 Soyuz pyrocartridges fired simultaneously instead of sequentially to separate the two modules. The force of the discharge caused the internal mechanism of the pressure equalization valve to release a seal that was usually discarded pyrotechnically much later to adjust the cabin pressure automatically. When the valve opened at a height of 168 kilometers, the gradual but steady loss of pressure was fatal to the crew within about 20 seconds." See Edward Clinton Ezell and Linda Neuman Ezell, *The Partnership: A History of the Apollo-Soyuz Test Project* (Washington, DC: NASA SP-4209, 1978), p. 230.

97. "The Russian Right Stuff: The Dark Side of the Moon," *NOVA* television show, #1808, WGBH-TV, Boston, February 27, 1991; Tarasov, "Missions in Dreams and Reality." The valve was apparently a millimeter across. See Payson, "Without the 'Secret' Stamp."

98. Kamanin, "This Should Never Happen Again!," no. 24.

99. Thomas O'Toole, "Soviet Union Still Trails U.S. in Space," *Washington Post*, June 17, 1973, pp. A1, A8.

100. U.S. Central Intelligence Agency, "National Intelligence Estimate 11-1-71: The Soviet Space Program," Washington, DC, July 1, 1971, pp. 30–32, as declassified in 1997 by the CIA Historical Review Program.

The members of the Keldysh commission signed the final version of the accident investigation on August 17, 1971, about a month and a half after the accident.[101] The commission collectively made some specific recommendations: increase the stability of the valve with respect to shock loads, install quick-acting (within seconds) manual chokes for valves, and use spacesuits during conditions when depressurization was possible.[102] The final point, the use of spacesuits, was evidently a much-debated issue, with individuals such as Mishin and Feoktistov arguing against it. Mishin summarized his opinion on the matter twenty years later when he wrote:

> In principle, all the recommendations of the commission were correct, but with one I do not agree to this day—the introduction of the spacesuit which Korolev had abolished from the "Voskhod" spaceship. . . . In multi-seat spaceships it is necessary to ensure collective safety, which can be better ensured by duplicating the systems that pressurize the entire Descent Apparatus. . . . The spacesuits required additional complex devices, thus increasing weights and volumes. The commission's recommendations to introduce spacesuits . . . made it necessary to reduce the crew of the spaceship to two and [to reduce] the conduct of special experiments.[103]

Unfortunately for Mishin, one of those pushing the use of spacesuits was Central Committee Secretary Ustinov. When Mishin and his Deputy Chief Designer Bushuyev met with Ustinov on August 6, Ustinov was firm on the issue: it was impossible for any more Soviet crews to fly in space without pressurized spacesuits. The Zvezda Machine Building Plant under Chief Designer Gay I. Severin was asked to accelerate its current efforts to prepare a new suit, named the *Sokol-K*, specifically for the Soyuz spacecraft. At the same time, two different departments at Mishin's design bureau began the process of redesigning the 7K-T Soyuz spacecraft to meet the recommendations of the Keldysh commission.[104] Later, the Ministry of General Machine Building handed out special reprimands for the disaster to six leading personalities, including Mishin, Bushuyev, and Tregub.[105]

As far as future missions to the *Salyut* station, they were out of the question at that point. During the flight of Soyuz 11, the State Commission had met to set the launch date of the second expedition to the station for July 20. This crew, composed of Leonov, Rukavishnikov, and Kolodin, along with their backup crews, began joint training on June 16. Training for all the crews was terminated on July 9, 1971, nine days after the Soyuz 11 tragedy. The immediate goal was to make changes to the Soyuz spacecraft and introduce mandatory spacesuits for all the crewmembers. Because the introduction of spacesuits would take additional volume and mass, TsKBEM reluctantly decided to truncate further crewmembers from three to two, eliminating the test engineer position. The *Salyut* station, meanwhile, continued to circle Earth. Ground controllers fired its main engines at least five times to prevent orbital decay over the course of three months. Finally, on October 11, its supplies already expired in August, the station was commanded to reenter Earth's atmosphere over the Pacific Ocean.

Among the many stranger-than-fiction ironies of this first space station project, clearly one of most chilling was the last-minute replacement of the Soyuz 11 primary crew. In an interview

---

101. The signatories were M. V. Keldysh (President, Academy of Sciences), L. V. Smirnov (Chairman, VPK), S. A. Afanasyev (Minister, MOM), I. D. Serbin (Chief, TsK KPSS Defense Industries Department), V. A. Kazakov (Deputy Minister, MAP), M. N. Mishuk (VVS), P. D. Grushin (General Designer, MKB Fakel), A. I. Tsarev (Deputy Chairman, VPK), and V. P. Mishin (Chief Designer, TsKBEM).

102. Semenov, ed., *Raketno-Kosmicheskaya Korporatsiya*, pp. 188–89.

103. Mozzhorin, *et al.*, eds., *Dorogi v kosmos: I*, p. 124. See also Tarasov, "Missions in Dreams and Reality."

104. Semenov, ed., *Raketno-Kosmicheskaya Korporatsiya*, pp. 188–89.

105. The reprimands were stipulated in a ministry resolution (no. 259ss) on August 19, 1971.

in June 1988, one of those replaced, Petr I. Kolodin, confided that the deaths of Dobrovolskiy, Volkov, and Patsayev still played on his conscience: "I was to fly, and Dobrovolski and his colleagues were to have remained on Earth. They were killed and I'm alive."[106] Although he was scheduled to fly a Soyuz mission in 1978, Kolodin did not, in fact, ever join the ranks of "true" cosmonauts. Leonov and Kubasov, the two remaining members of the crew, were recycled back into training for future DOS missions. Kubasov's lung problem, which had effectively saved his life—and those of Leonov and Kolodin, too—later turned out to be only an allergic reaction.[107]

Military Space

At the time of the Soyuz 11 disaster, the Soviet space program was almost fifteen years old. Forged out of the innards of the Soviet military-industrial complex, the space effort, by and large, remained hostage to the whims of military requirements and the opinions of those leaders who were responsible for building and maintaining the defense might of the Soviet Union. Typically, the triumvirate of individuals responsible for the defense industry—Ustinov, Smirnov, and Afanasyev—were ultimately accountable for dictating the direction of the space program. However, the needs of the Ministry of Defense—the primary clientele for all space products—also played a major role in the formation of long-range state policy. In March 1967, Marshal Andrey A. Grechko, a former Commander-in-Chief of Soviet Ground Forces, became the new USSR Minister of Defense. Subordinate to Grechko were all the heads of the armed services, including Marshal Nikolay I. Krylov, the Commander-in-Chief of the Strategic Missile Forces. Both were extremely influential in defining the long-term goals of the Soviet space program—the Five-Year Space Plans.[108] The fact that both were regarded as virulently against "big funding" for the piloted space effort was a major factor in the military's lack of interest in an active human space program. Air Force Col. General Kamanin, one of the few high-ranked men within the military supportive of strong piloted operations in space, lamented in June 1970:

> Grechko has still not been at the [Cosmonaut Training Center] although he promised three times to visit it. I do not know if he will keep his word this time, but his possible trip to us does not make me very happy; the minister obviously underrates the importance of the space program for the country's science, economy and defense. However we are totally dependent on Marshal Grechko and it would be foolish not to attempt to "relate" to him with space.[109]

The effects were repercussive: because all space products, whether they were Soyuz ships or space stations, were ultimately built for and operated by the Strategic Missile Forces, most of the major chief designers, such as Mishin, Glushko, Chelomey, and Yangel, had to pander to Grechko and Krylov for their blessing.

The Strategic Missile Forces remained in tight control over all operational activity in the Soviet space program. Its subordinate Central Directorate of Space Assets, headed by Lt. General Andrey G. Karas, had inherited this job from the old artillery days. The other armed

---

106. Hooper, *The Soviet Cosmonaut Team*, Volume 2, p. 132.
107. Davydov, "How Could That Have Been?"
108. There are few published details of these Five-Year Space Plans. One account—of a meeting between MOM Minister S. A. Afanasyev and USSR First Deputy Minister of Defense M. V. Zakharov in August 1969—clearly indicates the influence of the Ministry of Defense over the content of these plans. See Mozzhorin, *et al.*, eds., *Dorogi v kosmos: I*, pp. 218–19.
109. L. N. Kamanin, "Removing the Cosmetic Retouching: N. Kamanin—From His Journal Entries for 1970" (English title), *Sovetskaya kultura*, July 14, 1990, p. 15.

services—the Air Defense Forces, the Air Force, and the Navy—were naturally hostile to this monopoly, and in 1970, a detailed plan was drawn up to have this directorate subordinated directly to the Ministry of Defense, thus circumventing the stranglehold by the Strategic Missile Forces over space operations. Even Marshal Krylov initially supported the idea, but at the last moment, senior Strategic Missile Forces officers opposed the idea. Karas stalled the plan by suggesting that his directorate remain under the Strategic Missile Forces for two to three more years, to allow a more detailed look at the issue. In March 1970, the directorate was reorganized into the Chief Directorate of Space Assets (GUKOS), but it still remained an operational arm of the Strategic Missile Forces, carrying out launch, command, and control over every single Soviet spacecraft launched into orbit. The "two to three years" that Karas had proposed eventually stretched into nearly twelve years. It was only on November 10, 1981, that GUKOS was removed from Strategic Missile Forces jurisdiction.[110] The successor to GUKOS eventually became the Russian Military Space Forces—in the 1990s.

Influencing the direction of the Soviet space program was not just a matter of power but also patronage. Minister of Defense Grechko was a strong supporter of Minister of General Machine Building Afanasyev, who in turn helped prop up many of Chelomey's tenuous programs, such as the Almaz space station. On the other side, Central Committee Secretary Ustinov, a well-known anti-Chelomey partisan, was on the side of Chief Designers Yangel and Mishin. This peculiar bicameral noninstitutional factionalism helped sustain tension between the Mishin and Chelomey factions for many years. In terms of the ICBM program, the Grechko-Ustinov enmity resulted in a severely acrimonious battle—a "civil war" between Chelomey and Yangel over the development of a third generation of strategic missiles. Unable to make the decision between a Chelomey proposal and a Yangel proposal, Soviet leader Brezhnev succumbed to pressure on both sides by approving the development of two concurrent ICBMs with almost identical capabilities, thus squandering billions of rubles.[111]

The negative attitude of the military toward piloted space projects meant that a number of important programs suffered during the late 1960s and early 1970s. One program that fell under Grechko's vendetta against space was the Spiral piloted spaceplane program. By 1967, engineers at the "space branch" of the Mikoyan design bureau (MMZ Zenit) gave out subcontracts to build testbeds for Spiral. The first such testbed was an 800-kilogram, three-meter-long scale model of Spiral's Experimental Piloted Orbital Aircraft (EPOS), named BOR-1. It was designed and built by two major research institutions, the N. Ye. Zhukovskiy Central Aerohydrodynamics Institute (TsAGI) and the M. M. Gromov Flight-Research Institute. Manufacturing was carried out at Plant No. 166 at Omsk. The creation of BOR-1 was part of a larger research program in support of Spiral to investigate aerodynamics, thermal protection, the prospects of using hypersonic scramjets, and the rescue of the object after its return from space. The program would include studying atmospheric return from altitudes of 200 to ten kilometers and velocities of 7,500 down to 250 meters per second—that is, about Mach 27.5 down to Mach 0.8. The initial suborbital flights of BOR vehicles would last about three minutes; these would lead to "orbital" missions lasting fifteen to twenty minutes.

---

110. Mozzhorin, et al., eds., Dorogi v kosmos: I, pp. 221–22; I. D. Sergeyev, ed., Khronika osnovnykh sobytiy istorii raketnykh voysk strategicheskogo naznacheniya (Moscow: TsIPK, 1994), p. 17.

111. Brezhnev's final decision led to the development of Chelomey's UR-100N missile (approved by the Soviet government on August 19, 1970) and Yangel's MR UR-100 missile (approved by MOM in September 1970). For a discussion of the "civil war," see Mozzhorin, et al., eds., Dorogi v kosmos: I, pp. 149–50; Roald Z. Sagdeev, The Making of a Soviet Scientist: My Adventures in Nuclear Fusion and Space From Stalin to Star Wars (New York: John Wiley & Sons, 1993), pp. 205–06; B. Ye. Chertok, Rakety i lyudi: goryachiye dni kholodnoy voyny (Moscow: Mashinostroyeniye, 1997), 68–70; Vladimir Gubarev, "Southern Launch" (English title), Nauka i zhizn no. 10 (October 1997): 36–45.

This flown model of the BOR-2 lifting body was recently on display at a Russian exhibition. Note the damage from thermal loads during reentry on the bottom face of the vehicle. The insignia of the Gromov Flight-Research Institute is visible near the nose of the spacecraft. This vehicle was launched sometime between 1969 and 1974 on a suborbital mission. (copyright Steven Zaloga)

BOR-1 was specifically designed to separate from a conventional ballistic launch vehicle at an altitude of 100 kilometers and a velocity of 3.7 kilometers per second and then complete a gliding flight into the atmosphere. Within two years, engineers were able to develop adequate thermal shielding for the vehicle, which would potentially face angles of attack at up to forty-five degrees upon entry into the atmosphere and endure temperatures as high as 1,500–1,600 degrees Centigrade. After intensive ground trials, the first and only BOR-1 space-plane was launched on July 15, 1969, on an R-12 missile, just six days before the Apollo 11 landing. One Russian historian later summarized the outcome:

> *Test results showed that the "lifting body" was marvelously balanced even at angles of attack exceeding 60°. And although the first model was made of wood and was equipped with the gear of a size/weight mock-up, it was the model from which scientific results were obtained, before its burnup at altitudes of 60–70 kilometers.*[112]

Efforts in other fronts in the Spiral program also continued at the time. A twenty-kilometer-long landing strip was in the process of construction. Engineers had also evidently built a subsonic model of the spaceplane equipped with instrumentation transferred from the Tu-95 bomber. Unconfirmed rumors suggest that at least three drop flights were performed during this

---

112. V. Ageyev, "Unknown Pages of Space Science: In Flight—The 'BOR'" (English title), *Aviatsiya i kosmonavtika* no. 1 (January 1992): 42–43; E-mail correspondence, Igor Afanasyev to the author, December 11, 1997.

period from altitudes of 9,000 meters, which "fully confirmed the design characteristics of the Spiral airplane."[113] At the same time, MMZ Zenit, under its BOR program, emerged with plans for two new subscale lifting bodies, BOR-2 and BOR-3. Again, the purpose of the work was to carry out research on aerodynamic characteristics, heat exchange, and thermal shielding of the Spiral design at hypersonic velocities. The data gathering was limited to altitudes of ten to 100 kilometers, speeds of Mach 5 to 13.5, and angles of attack of fifteen to sixty-five degrees. Another variant, the BOR-4 model, which was designed on the basis of BOR-2, would be the basic "working horse" of the BOR program and use new heat-shielding material.[114]

TsAGI also carried out a huge amount of research on the carrier aircraft for Spiral, the so-called Hypersonic Booster-Aircraft (GSR), which would accelerate the actual spaceplane to speeds of Mach 4–6 during operational missions. Scientists studied two variants of the carrier, GSR-1 and GSR-2, both of which went through a full cycle of testing in wind tunnels at the institute. A large part of this work, performed between 1965 and 1975, was research focused on methods of testing models with air ducts over the "gondola" propulsion units during flight at hypersonic velocities.[115]

Trouble struck the Spiral program in 1969. By this time, engineers needed a formal decree of the Central Committee and the Council of Ministers to continue serious work. Unfortunately for Spiral Chief Designer Lozino-Lozinskiy, this is where Minister Grechko stepped in. Although the appropriate ministers and Communist Party leaders, in 1969, evidently signed the project order, Grechko scrawled on the document "This is a fantasy."[116] Lozino-Lozinskiy, perhaps being generous to Grechko, recalled later that "the Soviet leadership felt it would take too much time and money to bring the program all the way to completion."[117] A variety of other problems, all related to money, seems to have slowed down the project. Despite the considerable theoretical work on the GSR, the creation of flight models required a huge financial commitment, which was unavailable. By the early 1970s, scientists were also coming to the opinion that an air-launched reusable spaceplane system might not be the best route to take; a vertical missile-launched system might offer a much cheaper and efficient alternative. Research on liquid hydrogen engines for the carrier aircraft also stalled sometime in 1967 or 1968, apparently because the Soviet government was "biased" against this work, carried out by Struminskiy and Lyulka, at the Institute of Theoretical and Applied Mechanics at Novosibirsk, which was under the Academy of Sciences.[118]

Despite Grechko's prohibition on Spiral work, MMZ Zenit's space branch continued low-level work "semi-legally" on the Spiral project. The scope of the post-1969 work was, in fact, quite remarkable, and one wonders how Lozino-Lozinskiy managed to sustain it. Between 1969 and 1974, the Gromov Flight-Research Institute and TsAGI launched seven BOR-2 and BOR-3 subscale spaceplanes using the R-12 missile on suborbital and/or vertical launches to

---

113. Gleb E. Lozino-Lozinskiy and Vladimir E. Plokhikh, "Reusable Space Systems and International Cooperation," *Aerospace America* (June 1990): 37–40; G. Titov, ". . . This Is Needed for All of Us" (English title), *Aviatsiya i kosmonavtika* no. 4 (April 1993): 2–3. Unconfirmed Western sources also suggest that there may have been drop tests of the spaceplane in the late 1960s. See, for example, Peter N. James, *Soviet Conquest From Space* (New Rochelle, NY: Arlington House Publishers, 1974), p. 129. Note that these drop tests, if they did occur, were different from subsequent tests in 1976–78.

114. Afanasyev correspondence, December 11, 1997.

115. *TsAGI-Osnovnyye etapy nauchnoy deyatelnosti, 1968–1993* (Moscow: Nauka, 1996), p. 156.

116. Titov, ". . . This Is Needed for All of Us." Another source says that Grechko's inscription was "We will not engage ourselves with a fantasy." See Vyecheslav Kazmin, "The 'Quiet' Tragedy of EPOS" (English title), *Krylya rodiny* no. 1 (January 1991): 4–5.

117. Lozino-Lozinskiy and Plokhikh, "Reusable Space Systems and International Cooperation," p. 38.

118. Lardier, *L'Astronautique Soviétique*, p. 175; V. Struminskiy, "Hydrogen on Earth and in Space" (English title), undated and unsourced article provided by Christian Lardier; E-mail correspondence, Igor Afanasyev to the author, December 7, 1997.

100-kilometer altitudes, testing them at hypersonic velocities between Mach 3 and Mach 14. Unlike their BOR-1 predecessor, both BOR-2 and BOR-3 were metalloid vehicles. Their characteristics were:

Model	Length	Mass	Scale to EPOS
BOR-2	3 meters	1.2 tons	1/3
BOR-3	4 meters	1.5 tons	1/2

The BOR-2 and BOR-3 flights allowed engineers to clarify the balance and characteristics of longitudinal stability and compare the data to those from ground wind tunnels. Experimental data were obtained on the conversion of the laminar boundary layer into a turbulent layer and on the effects of altitude and flight speed on the distribution of pressure across the surface of an airframe apparatus with a complex geometric shape. In addition, algorithms for the control of the vehicles' movements were tested, and extensive research was conducted on aerodynamic heating, heat exchange, and thermal protection of various surface elements.[119] Despite the significant research in the early 1970s, the program, as a whole, lost sight of its future after Grechko's pronouncement in 1969. The ambitious plans of the mid-1960s—of having a versatile reusable small-scale spaceplane—disappeared amid the military's favoritism for automated systems.

Grechko and Krylov also influenced the course of the N1 program. Since the genesis of the program in the early 1960s, Korolev had attempted to interest the military in the rocket's capabilities, knowing that strong military interest would ensure robust funding for the effort. After Korolev's death, Mishin continued to lobby the military, proposing various forms of military complexes that could be orbited by the N1. Research on large-scale space-based armaments systems had begun as early as 1968; in April 1969, Mishin had briefed Soviet leader Brezhnev on the uses of the N1 rocket for launching powerful anti-ballistic missile complexes into space. Later, in the autumn of 1969, Mishin had also personally visited the top-secret Institute of Nuclear Physics at Novosibirsk to talk to scientists about the possibility of designing transportable particle beam accelerators that could be launched on the N1.[120]

Many such concepts from TsKBEM were studied in cooperation with various Academy of Sciences and industrial scientific institutes in 1970 and 1971. While these were not programs to which the Soviet government fully committed, they were in fact considered at very high levels. In June 1970, Mishin discussed the prospects of the Luch ("Ray") system, a space-based laser weapons system, with Afanasyev and Keldysh. By September of the same year, concrete work on Luch was planned for 1973, simultaneously with operational launches of the N1 booster. Later, in November 1970, Mishin met with Commander-in-Chief of the Soviet Air Defense Forces Marshal Pavel F. Batitskiy to brief him on Luch. All Soviet anti-ballistic missile and anti-satellite forces were under Batitskiy's command at the time. From the available evidence, Mishin faced a very difficult road in convincing military leaders of the need for the N1. As with their American counterparts, Soviet generals and marshals could find little use for very heavy-lift launch vehicles to accomplish military goals. One scientist recalled later that at the initial stages of research on space-based particle beam accelerators, there was a peer review

119. Ageyev, "Unknown Pages of Space Science"; Lardier, L'Astronautique Soviétique, p. 250; Andrey Batashev, "Steep Turns of the Spiral. A Quarter Century Did Not Suffice for Implementing the Project Created by the 'Father' of the Soviet Shuttle" (English title), Trud, June 30, 1994, p. 4; K. K. Vasilchenko, et al., eds., Letnyye issledovaniya i ispytaniya: fragmenty istorii i souremennoye sostoyaniye: nauchno-tekhnicheskiy sbornik (Moscow: Mashinostroyeniye, 1993), pp. 34, 54–55, 62; TsAGI-Osnovnyye etapy nauchnoy deyatelnosti, pp. 154; Henry Matthews, The Secret Story of the Soviet Space Shuttle (Beirut, Leb.: Henry Matthews, 1994), p. 31.

120. Sagdeev, The Making of a Soviet Scientist, pp. 123–24.

of Mishin's proposal, and that by the end of 1970, scientists had managed to terminate the project, although Mishin did give a modest contract to the Institute of Nuclear Physics to continue work on the topic.[121]

Opening Up

In light of the fundamental connection between the space and military programs of the Soviet Union, it was all the more curious when, in the early 1970s, the Soviets began to very slowly open up their space program to the general public. In an unprecedented act that would have been unthinkable just five years earlier, the Soviet censors allowed the name of Valentin P. Glushko to be published openly for the first time. In March 1971, a one-volume encyclopedia of "cosmonautics" was published, with Glushko listed as its editor. Previous editions had merely listed the editor as G. V. Petrovich, a pseudonym for the chief designer. The Moscow newspaper *Pravda*, in a postpublication article, clearly linked Glushko to Petrovich, confirming what many in the West had long suspected.[122] That it was Glushko, and not Chelomey or Yangel, whose name was declassified hints at the growing eminence and power the rocket chief designer wielded. Of the six original members of the old Council of Chief Designers, Glushko was the first one to see his name in print after the launch of Sputnik. Few biographical details were, of course, released, and it would not be until the early 1990s before even the name of his organization, the Design Bureau of Power Machine Building (KB EnergoMash), was allowed to be published.

Mishin was also in the news, albeit in an oblique manner. In 1972, a French journalist, Pierre Dumas, authored an article in the journal *La Recherche Spatiale* (*Space Research*) in which he named Academician Mishin as one of the authors of a project to send "A Manned Space Train to Mars in 1978."[123] It was the very first publication linking his name with the Soviet space program. Coincidentally or not, Mishin also wrote his first article for the Soviet media under the pseudonym "Professor M. Vasilyev" in April 1972. In this article in *Pravda*, "Vasilyev" glowingly praised the achievements of the late Korolev.[124] Ironically, at exactly the same time, a Ukrainian émigré published a remarkable analysis of the organization of the Soviet space program. Taking a cue from the French article mentioning Mishin, the author accurately named Mishin as the still-unknown "Chief Designer" of the Soviet space program.[125] Without exception, all Western analysts, including the CIA, ignored this claim, and for at least the next 15 years or so, "expert observers" in the West continued to tout the names of Yangel or Chelomey as the successor to Korolev.

Unlike Mishin, one employee of TsKBEM was allowed to speak and appear under his own name: Department Chief Boris V. Raushenbakh. In a revelation that caused a mini-sensation in the West, the fifty-five-year-old Raushenbakh was identified as a "specialist in space engineering" during the press conference following the Soyuz 10 flight in late April 1971.[126] It

---

121. *Ibid.*, p. 124.

122. "Soviet Space Chief Identified as Editor of an Encyclopedia," *New York Times*, March 19, 1971, p. 3.

123. Pierre Dumas, "Un Train Spatial Habité Vers Mars En 1978," *La Recherche Spatiale* no. 3 (May–June 1972): 26. See also Christian Lardier, "Soviet Space Designers When They Were Secrets," presented at the 47th Congress of the International Astronautical Federation, IAA-96-IAA.2.2.09.

124. Professor M. Vasilyev, "Sputnik: Start of the Space Era" (English title), *Pravda*, April 10, 1972. Further articles under the same pseudonym were published in *Izvestiya* on December 28, 1973, and *Krasnaya zvezda* on April 12, 1974. See also Lardier, "Soviet Space Designers When They Were Secrets."

125. S. Yu. Protsyuk, "Technical Chronicle: Who Runs the Program of Mastering Space in the USSR?" (English title), *Ukrainian Engineering News* 23 (March–April 1972): 60–72. This article is available as NASA Technical Translation TT-14882, dated May 1973.

126. Theodore Shabad, "Soviet Identifies 'Space Specialist'," *New York Times*, May 2, 1971.

may have been Raushenbakh's considerable talents as a scholar, an orator, a writer, a scientist, and an engineer that posited him with this opportunity. Hailing from German origins, in 1948, he had edited a Russian translation of a classic work by Hermann Oberth on space navigation. He had obtained the equivalent of a Ph.D. in 1958 and become a Corresponding Member of the USSR Academy of Sciences in July 1966. His engineering specialty was satellite orientation systems—a field that he had pioneered in the Soviet Union in the mid-1950s—but his interests were far and wide. He eventually became a doctor of theology, studying the relationship between science and religion, and he wrote several books on the mathematical analyses of perspectives in ancient and modern art.[127]

In another unprecedented move, the Soviet government allowed an American journalist to visit the Cosmonaut Training Center. In March 1972, John Noble Wilford, a reporter for *The New York Times*, took a one-day visit to Zvezdnyy gorodok (Starry Town) in support of a page-one write-up, which was published later that month. A dark bronze statue of first cosmonaut Yuriy A. Gagarin welcomed Wilford into the

Chief Designer Valentin Glushko appears here in his official portrait dating from the late 1960s. His name was officially declassified by Soviet authorities in 1971, the first major chief designer in the Soviet space program to receive this honor before his death. (files of Peter Gorin)

closed city, located about forty kilometers northeast of Moscow near the industrial town of Shchelkovo. As with many secret Soviet cities, Zvezdnyy gorodok was not identified on any public maps and was hidden from the major highway by a forest. By Wilford's estimates, the population of the town was 1,500 to 2,000. He was the first Westerner to see many of the ground trainers used by cosmonauts prior to their flights. While his hosts, cosmonauts Shatalov and Yeliseyev, spoke mostly about the future of Earth-orbital space stations, they did not shy away from the obvious question of a piloted lunar landing. When asked whether Soviet cosmonauts might land on the Moon by 1975, Yeliseyev replied, "Yes. By that time we will probably send our people to the moon."[128] Wilford himself got the impression of an active and expanding Soviet space program.

U.S. perceptions of the Soviet space program in the early 1970s differed dramatically, depending on the perspective. Having fallen prey to Soviet denials about their Moon program, most public observers tended to discount claims by a few lone analysts that the Soviets had ever tried to send cosmonauts to the Moon. The CIA, on the other hand, was clearly in a better position to assess what the Soviets were doing. Through the failures and delays of their lunar program, U.S. intelligence was keyed into the hidden arcana of the Soviet space

---

127. Yaroslav Golovanov, *Korolev: fakty i mify* (Moscow: Nauka, 1994), pp. 575–76. Among other revelations in 1971 was the identification of K. N. Rudnev as the chair of the State Commission for the first Vostok mission in 1961. See "Brezhnev Space Director Since 1963," *Space Daily*, April 15, 1971; Nicholas Daniloff, *The Kremlin and the Cosmos* (New York: Alfred A. Knopf, 1972), pp. 80–81.

128. John Noble Wilford, "Soviet Space Center: Hope Amid Expansion," *New York Times*, March 22, 1972, pp. 1, 20.

program. In a top secret National Intelligence Estimate issued in March 1970, the CIA very accurately predicted that:

> Technical problems with both the [N1] vehicle and the [Proton] booster will delay a manned lunar landing mission until 1973 at the earliest and probably beyond. Nevertheless, a lunar landing mission remains on the books as a venture to be carried out in due course.[129]

CORONA photo-reconnaissance satellites were able to discern remarkable detail of hardware. By the time of their July 1971 estimate, the CIA produced a detailed drawing of the still secret N1 and its ground infrastructure. Analysts apparently attributed a far greater ability to the N1 rocket than it actually had; according to CIA analysts, the rocket was capable of injecting as much as 125 tons into Earth orbit when its real capability was closer to ninety tons. The errors in analysis were compensated by the speed of information collection; the July 1971 estimate was issued just four days after the third N1 launch failure but contained detailed information on the accident. Listing all major liquid hydrogen upper stage programs, the CIA also added quite correctly: "All things considered . . . we think it is unlikely that development of high-energy upper stages has progressed far enough for the Soviets to begin flight-testing them on the [Proton] or the [N1] in the near future."[130]

The Soviets' increased openness and the CIA's much better intelligence collection means were both big factors in the early 1970s as the United States and the Soviet Union engaged in their first major cooperative venture in space in the backdrop of détente. Intensive discussions on a cooperative human spaceflight effort had begun as early as 1969 between then–NASA Administrator Thomas O. Paine and USSR Academy of Sciences President Mstislav V. Keldysh. Apart from the purely political value in support of détente, any potential joint mission would have functional advantages for both sides. For NASA, the year 1972 would be the end of an era in space history as the Apollo lunar landing missions began to wind down. Apollo 16 was set for April 1972, while the last mission, Apollo 17, was scheduled for December 1972. Flights in the NASA Skylab space station program were set for 1973 and 1974, followed by a hiatus in the piloted space program for at least five years before the introduction of the reusable Space Shuttle. A joint flight in the interim period would provide NASA engineers with valuable piloted spaceflight experience. For the Soviets, a joint mission would be most useful from a public relations perspective—that is, to demonstrate that its space technology was on a par with that

---

129. U.S. Central Intelligence Agency, "National Intelligence Estimate 11-1-69: The Soviet Space Program," Washington, D.C., March 26, 1970, p. 3, as declassified in 1997 by the CIA Historical Review Program.

130. CIA, "National Intelligence Estimate 11-1-71: The Soviet Space Program," pp. 10, 12, 13. The actual and suspected characteristics of the N1, called the "J-vehicle" by the CIA, are shown in the following table. The CIA data are from July 1971.

Item	Actual	CIA Estimation
Total Length	105.3 meters (m)	96.6 m
Stage I Length/Base Diameter	30.1 m/16.9 m	25.6 m/17.1 m
Stage II Length/Base Diameter	20.5 m/10.3 m	21.0 m/11.3 m
Stage III Length/Base Diameter	11.5 m/6.0 m	13.1 m/7.9 m
Stage IV Length/Base Diameter	8.0 m/6.0 m	17.4 m/6.1 m
Launch Mass	2,820 tons	4,536 tons
Stage I Thrust	4,615 tons	5,897–6,350 tons
Stage II Thrust	1,432 tons	1,588 tons
Stage III Thrust	164 tons	544 tons
Stage IV Thrust	41 tons	200 tons
Payload to Low-Earth Orbit	90 tons	125 tons

of the U.S. space program—a claim that had been difficult to support in the previous few years. By early April 1972, Vladimir A. Kotelnikov, the Deputy Chairman of Interkosmos, and George M. Low, Deputy Administrator at NASA, had agreed to a formal technical agreement on the docking of a Soyuz and an Apollo spacecraft in orbit around Earth in July 1975. A formal document, "Agreement Concerning Cooperation in the Exploration and Use of Outer Space for Peaceful Purposes," confirming this arrangement was signed by President Richard M. Nixon and Council of Ministers Chairman Aleksey N. Kosygin on May 24, 1972.[131] The American side called the project the Apollo-Soyuz Test Project, while the Soviets used the phrase Apollo-Soyuz Experimental Flight (EPAS).

The birth of EPAS coincided with major changes within TsKBEM, secretly the prime contractor firm for the joint program. For several years, Chief Designer Mishin had been proposing for a fundamental change in the hierarchical makeup of his design bureau. With the blessing of the Ministry of General Machine Building, on July 14, 1972, the TsKBEM structure was reorganized, for the first time introducing a new level of chief designers within the design bureau. Mishin would remain the Chief Designer *and* Chief of TsKBEM. Under him, there were six chief designers, each responsible for one of six projects: the N1 rocket, the L3M lunar landing complex, the DOS-7K space station, the 7K-S military Soyuz, the EPAS international project, and the RT-2PU ICBM.[132] As before, Sergey O. Okhapkin remained Mishin's First Deputy Chief Designer for all programs. Both Mishin and Okhapkin oversaw four other deputy chief designers who were in charge of specific technical areas.[133] One of Mishin's key deputies was Konstantin D. Bushuyev, whose name was also added to the growing roster of people revealed to the world. In June 1971, the Soviets named him as the director of the Soviet portion of the Apollo-Soyuz Test Project. The Americans had, for obvious reasons, no knowledge of Bushuyev's extraordinarily important role in the creation of the Soviet space program.

The Soviet public, like those abroad, continued to be fed a steady diet of propaganda concerning their space program. While the space effort may have engendered a strong degree of support in the late 1950s and early 1960s, by the early 1970s, as the country's economy ground into the "great stagnation," people were less prone to be vocally in favor of it. A story in *The Washington Post* in 1971 illustrates the point. In February 1971, a large portion of potatoes sold in Moscow had been too rotten to eat. Outraged by the dearth in quality in a staple Russian food item, one indignant grandmother declared to a crowd waiting to buy potatoes at a central farm market: "We have rockets, right? Of course, right. We have *Sputniks*, right? Of course, right. They fly beautifully in outer space. So I say to you, dear friends, Why don't we just send these rotten potatoes into outer space too." There was a small round of applause for her modest proposal. A *New York Times* correspondent added from Moscow that "Although criticism [of the space program] is kept muted by the controlled Soviet media, it is well known here that many Russians are irritated by the costly space ventures when life here is still far from satisfactory."[134]

---

131. Ezell and Ezell, *The Partnership*, pp. 182–93; Semenov, ed., *Raketno-Kosmicheskaya Korporatsiya*, p. 195.

132. Semenov, ed., *Raketno-Kosmicheskaya Korporatsiya*, pp. 160, 639. The six chief designers were B. A. Dorofeyev (N1), V. A. Borisov (L3M), Yu. P. Semenov (DOS-7K), Ye. V. Shabarov (7K-S), K. D. Bushuyev (EPAS), and I. N. Sadovskiy (RT-2P).

133. The four deputy chief designers were M. V. Melnikov ("special themes"), V. V. Simakin, A. P. Abramov (ground equipment and experimental work), and Ya. I. Tregub (testing and flight control). There were also several deputy chiefs of TsKBEM who were *not* designers: M. I. Samokhin (standard testing), A. P. Tishkin (coordination), G. M. Paukov (cadres), G. M. Yakovenko (regimes), and B. Ye. Chertok (guidance systems). The First Deputy Chief of TsKBEM was G. V. Sovkov (redesign, construction, and general problems). See *ibid.*, p. 160.

134. The first story is from the March 5, 1971, issue of *The Washington Post*. The second is from the February 28, 1971, issue of *The New York Times*. Both are referenced in *Soviet Space Programs, 1966–70: Goals and Purposes, Organization, Resources, Facilities and Hardware, Manned and Unmanned Flight Programs, Bioastronautics, Civil and Military Applications, Projections of Future Plans, Attitudes Toward International Cooperation and Space Law*, prepared for the Committee on Aeronautical and Space Sciences, U.S. Senate, 92d Cong., 1st sess. (Washington, DC: U.S. Government Printing Office, December 1971), p. 35.

While the criticisms may have been valid, the Soviet public actually knew little about the workings of the Soviet space program. In all unclassified documentation, TsKBEM was merely known as the nondescript "post office box number 651." Despite the anonymity, the town of Kaliningrad near Moscow seems to have been a major beneficiary of the massive industrial infrastructure built to support operations at TsKBEM. Dozens of high-quality households, apartment complexes, and well-stocked stores were built in the 1960s as more and more engineers from the best educational institutions all over the country joined the design bureau. At the time of the 1972 shakeup, Mishin oversaw an enterprise of 28,959 employees, most of whom were based in Kaliningrad. Because all work at TsKBEM was classified top secret, engineers were constantly shadowed by individuals from the "First Department," whose job it was to maintain tight security. As a compensatory measure, wage rates at TsKBEM were about 25–30 percent higher than those in similar institutions engaged in scientific or engineering work. Korolev's death, however, seems to have had some deleterious effects on the workforce. A former engineer who emigrated to the West in the late 1970s recalled:

> As long as Korolyov was alive, TsKBEM personnel of conscription age were not required to serve in the army. The situation changed dramatically under Mishin. Towards the end of the 1960s all deferments were canceled and men were called up in droves. In June 1968, a virtual round-up was carried out in Kaliningrad. . . . Even though several months later many of the men began returning, one of the incentives for working at TsKBEM was gone. Many began to seek jobs elsewhere. It was under these circumstances that the author left TsKBEM in 1970. . . .[135]

Losses in human potential were not limited to TsKBEM. In 1971, the Soviet space program lost three of its major leaders. On June 25, 1971, Chief Designer Aleksey M. Isayev of the Design Bureau of Chemical Machine Building in Kaliningrad passed away at the age of sixty-two after a heart attack. His organization, previously known as OKB-2, had designed almost all space-based propulsion systems in the Soviet space program, including those for the Vostok, Voskhod, Soyuz, Salyut, L1, and LOK spacecraft. One of the first engineers to travel to Germany in 1945, Isayev had later headed a group at the famous NII-88, where he had led efforts to develop rocket engines for various ballistic, cruise, surface-to-air, and anti-ballistic missiles, eventually moving into the space field. One of his major contributions was the development of the first Soviet high-energy cryogenic engine, created for an upper stage of the N1 rocket. Isayev had been offered the honor of becoming an academician of the Academy of Sciences, but he had refused on the grounds that he was an engineer, not a scientist. His name was revealed to the general public only upon his death.[136]

Less than two months later, on August 3, 1971, fifty-six-year-old Chief Designer Georgiy N. Babakin passed away. As head of the design bureau of the S. A. Lavochkin State Union Machine Building Plant since 1965, Babakin had overseen the tremendous successes of the Soviet automated lunar and interplanetary programs. In the piloted space programs, he had played prominent roles in determining policy by participating in various councils involved in the N1-L3 lunar programs. The crowning successes of Babakin's tenure were the Luna 16 soil sample return and the Lunokhod 1 lunar rover missions in late 1970, both of which were critical to

---

135. Victor Yevsikov, *Re-Entry Technology and the Soviet Space Program (Some Personal Observations)* (Falls Church, VA: Delphic Associates, 1982), pp. 1, 3, 5, 12.
136. "Aleksei Isayev, Engineer in Russian Space Efforts," *New York Times*, June 27, 1971, p. 46; "Alexei Isayev, Space Scientist Dies," *Washington Post*, June 27, 1971, p. 10; Peter Almquist, *Red Forge: Soviet Military Industry Since 1965* (New York: Columbia University Press, 1990), p. 179, footnote 7. V. N. Bogomolov succeeded Isayev as Chief Designer of KB KhimMash.

supporting the Soviet claim that they were focusing exclusively on automated lunar exploration. One of Babakin's final dreams had been to recover soil samples from the far side of the Moon. Work on such a project had, in fact, begun in 1970 during his lifetime. The plan consisted of an orbiter and a lander—the former to serve as a communications satellite between the latter and Earth. The mission was evidently scheduled for launch sometime in 1972, but after Babakin's death, the idea gradually fell to the wayside, partially because of the high level of technical complexity.[137] Academy of Sciences Corresponding Member Babakin had been working as the deputy chair of the Soyuz 11 investigation commission at the time of his death.

A third loss in 1971 was perhaps the most important from a historical perspective. One of the most influential figures in the Soviet missile and space programs, Chief Designer Mikhail K. Yangel died on October 25, 1971, at the age of sixty.[138] As the architect behind the new generation of Soviet strategic ballistic missiles, Yangel perhaps had more of an influence on the history of the Soviet Union than Korolev. Under his tutelage, KB Yuzhnoye created several high-performance ICBMs, such as the R-16, the R-36, and the R-36M, for the Strategic Missile Forces. In the space sector, his team was responsible for a variety of military satellites and satellite launch vehicles. Yangel had never had a strong interest in the piloted space program, although, from time to time, had tabled proposals such as the R-56 plan for a lunar landing or an even more ambitious Mars mission proposal in 1969. He was also closely involved in the development of the N1-L3 system, participating actively in all meetings related to the program—an interest partly stoked by his organization's help in creating the main lunar lander engine. In the last years of his life, he had been beset by serious illnesses and had had to relinquish some of his day-to-day duties. On his sixtieth birthday, October 25, 1971, there was a big reception in his honor at the offices of Minister Afanasyev. During the celebrations, Yangel complained about not feeling well and went to lie down on the sofa in an adjacent room. For a long time, there was no word from the room. After some time, attendees discovered him dead on the couch. It was his fifth heart attack.[139]

A final transitory event in the space program was not a death, but a retirement. In October 1971, sixty-year-old Col. General Nikolay P. Kamanin formally resigned as the Air Force Commander-in-Chief's Aide for Space, a post he had held since May 1966. Officially, he had been responsible for the Cosmonaut Training Center, the Air Force Biomedical Service, and the Air Force Solar Service. Throughout a ten-year period, Kamanin had not only served as the doctrinal leader of the cosmonaut corps, but also as a vocal and insistent supporter of piloted space programs. Despite speculation in the West that Kamanin was a casualty of a post-Soyuz 11 disaster shakeup, the general had, in fact, decided to retire before the end of that tragic mission.[140] His role in the Soviet space program has often been compared to that of Donald K. "Deke" Slayton at NASA—that is, as a major player in the selection and training of flight crews. But Kamanin, in many ways, exceeded that mandate by his important contributions to

137. "Georgi Babakin, Soviet Scientist," *New York Times*, August 5, 1971, p. 36; N. G. Babakin, A. N. Banketov, and V. N. Smorkalov, *G. N. Babakin: zhizn i deyatelnost* (Moscow: Adamant, 1996), pp. 73–75. Babakin was succeeded by S. S. Kryukov as Chief Designer of the design bureau of GSMZ S. A. Lavochkin.

138. "Mikhail Yangel, Soviet Space Aide: Chief Designer of Rockets for Exploration Dies," *New York Times*, October 27, 1971, p. 50.

139. Khrushchev, *Nikita Khrushchev: tom 2*, pp. 86–87; Yu. V. Biryukov, "Seventieth Birthday of Vladimir Fedorovich Utkin" (English title), *Zemlya i vselennaya* no. 3 (May–June 1994): 45–50. Yangel was succeeded by V. F. Utkin.

140. Kamanin, "This Should Never Happen Again!," no. 24. The decision to replace Kamanin with a veteran cosmonaut was adopted on June 25, 1971, five days before the return of the Soyuz 11 crew. See also "Memorable Dates" (English title), *Novosti kosmonavtiki* 12–13 (June 3–30, 1996): 76. For Kamanin's appointment to become the Air Force Commander-in-Chief's Aide for Space, see N. P. Kamanin, *Skrytiy kosmos: kniga vtoraya, 1964–1966gg* (Moscow: Infortekst IF, 1997), pp. 321, 339, 341.

the definition of state policy as well as his direct participation on flight control teams for almost all Soviet piloted space missions between 1961 and 1971. Having retired from the public eye, Kamanin did not return to it. He died on March 13, 1982, at the age of seventy-three.[141]

Perhaps in retrospect, Kamanin's greatest contributions to the history of the Soviet space program were his personal diaries. Meticulously written between 1960 and 1974, they provide an undeniably rare view into the emergence of the Soviet space effort. With an eye for analysis and reflection, Kamanin recorded much of the arcana of the decade through the lens of an active participant. Even with the declassification of archival material from the early days of the Soviet space program, his journals, which have been published piecemeal by his son in the Soviet and Russian media since 1989, add richly to a history often devoid of documentation. But like most figures of that era, Kamanin wrote with his own biases—prejudices that often leap out of his writings. A diehard Stalinist to the end, Kamanin was quick to criticize everyone but himself in the failures of the Soviet space program, repeatedly castigating Korolev, Mishin, Ustinov, Smirnov, Afanasyev, and many cosmonauts. The cosmonauts, especially, did not have an

General Nikolay Kamanin was the Air Force Commander-in-Chief's Aide for Space. His personal diaries, spanning a fourteen-year period from 1960 to 1974, have been central to understanding the intricacies of the Soviet piloted space program during the 1960s. (files of Peter Gorin)

easy relationship with him. In summing up Kamanin's relations with the cosmonauts, one famous Russian journalist, Yaroslav K. Golovanov, later accurately summed up the general's own personality:

I think that the majority of the cosmonauts did not like him. . . . Some of them confided this to me even back in the 1960s. . . . Kamanin kept a tight rein on them, demanding utter discipline and unquestioning obedience. He indulged himself in what was essentially a lack of responsibility that allowed him to demean young men far superior to himself, and he forced this style of leadership onto the whole first echelon of cosmonauts. To Kamanin it was flattering that these world famous people had to obey him, just like new recruits obey their corporal. It was even easier for him to control the people who still had to make a flight. After all, it largely depended on Kamanin when, with whom, and on which mission they flew. . . . Kamanin was feared, but not loved. Unlike his big idol Stalin, he did not succeed in being loved and feared at the same time.[142]

---

141. Glenn Fowler, "N. P. Kamanin, Soviet General," *New York Times*, March 15, 1982, p. B6.

142. Golovanov, *Korolev*, p. 665; Bart Hendrickx, "The Kamanin Diaries 1960–1963," *Journal of the British Interplanetary Society* 50 (January 1997): 33–40.

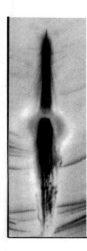

CHAPTER EIGHTEEN
ASHES TO ASHES

The early 1970s in the Soviet piloted space program was a period characterized by a notice-able lack of self-confidence. As substantial achievements began to dwindle dramatically, offi-cials and engineers began to grasp desperately for any dim possibility of success. The Soyuz 11 tragedy was obviously a severe blow, but if Ustinov, Mishin, and others believed that the spate of misfortune was over, they were wrong. In the two years following the deaths of Dobrovolskiy, Volkov, and Patsayev, the Soviet space establishment was beset by failure after failure—at the very same time that the Soviets were engaged in a bid to prove their parity with the United States in space achievements. Ironically, it was precisely during these troubled years that engineers produced, for the first time, a realistic and expansive vision of future space exploration—one that had good reason to succeed. These projects, such as the construction of giant space stations in Earth orbit and the long-term exploration of the Moon, were all, of course, dependent on the political caprices of the key influential players. In the end, as politi-cal imperatives had played a role in creating much of the early Soviet space program, they would also play a role in destroying the new vision.

The Multirole Orbital Complex

Throughout the setbacks of the DOS program, Chief Designer Mishin continued to focus efforts at his design bureau on two major long-range goals: the accomplishment of advanced lunar landing missions and the establishment of large-scale stations in Earth orbit. The former consisted of the Multirole Orbital Complex (MOK), whose central element was the Multirole Space Base-Station (MKBS)—a giant space station that had been under study since the mid-1960s. Like the long-term lunar bases that Mishin expected to establish in the 1980s, the MOK, in spirit at least, had more of a connection with the science fiction ideas from the pre-Sputnik era than the incremental developments of the 1970s. These two projects were essentially what he conceived as the first steps in the human migration into space—a vision foretold by the early-century pioneers such as Tsiolkovskiy, Oberth, and Kondratyuk. To Mishin's credit, he made sure that the MOK not only had a cogent vision but also detailed substantiation from a funding perspective.

The basic idea behind the MOK was the establishment of a large-scale complex in Earth orbit to support a variety of goals, all focused on improving life on Earth. The heart of this com-plex would be the MKBS, a giant piloted space station launched by the N1, which would be tended by a menagerie of smaller spacecraft flying to and from orbital factories. Mishin's own description from 1989 touches on the essence of the effort, which would involve:

a broad program for space exploration in circumterrestrial space within the Earth-Moon radius, including participation in solving food, energy, and ecology problems. Using a minimum number of fully equipped, standard space facilities in ground and orbital bases, the plan was to saturate local space with numerous useful vehicles.[1]

Some of the goals of the MOK sounded positively outlandish:

[Elements of the MOK] would even be able to influence the climate and lighting for cities, using a system of mirrors and solar light. It was a quite realistic project. [There would also be the] removal of harmful production facilities into space and full use of the opportunities in space—high and low temperatures, high vacuum, conditions close to weightlessness. And 90 percent of all these operations would be carried out without humans.[2]

The ongoing DOS program in the early 1970s was seen as something of a precursor to the MOK and therefore was seen less as a competitor than a complement to the new proposal. Mishin's timetables were fairly ambitious. By September 1970, he was planning to have the draft plan for the MOK ready by 1972 and to start flying station components into orbit using uprated versions of the N1 by 1974.[3] In November 1970, Mishin met with Military-Industrial Commission Chairman Smirnov to discuss the MKBS, but a decision was postponed until further evaluation by a review commission. One of the obstacles to a decision may have been a factor that had perennially slowed down many other programs: interest from the military. In May 1971, Mishin discussed the issue of a military tactical-technical requirement with Commander of the Chief Directorate of Space Assets Lt. General Andrey G. Karas. The possibility of including both passive and active military systems aboard the MKBS had been considered for many years, and some of these proposals were linked to the N1-related anti-ballistic missile systems of the day.

By mid-1971, Mishin's engineers were engaged in revisions of the technical plan for the first two stations, MKBS-1 and MKBS-2, presumably based on military, scientific, and technological limitations. Ustinov's blessing was evidence that the effort was gathering support. In August 1971, a month after the Soyuz 11 disaster, Mishin and Ustinov discussed the long-range plan for Soviet Earth-orbital stations during the 1971–80 period. The Soviet space effort would start off with Mishin's DOS, then move to Chelomey's military Almaz, and then finally migrate to the giant MKBS-1 in the mid-1970s and MKBS-2 by the end of the decade. Mishin already had plans to launch the first components of MKBS-1 on N1 boosters 10L and 11L, perhaps amid the initial lunar exploration phase of the L3 project. The last few months of 1971 were an intense period for sharpening the vision of the MOK/MKBS proposal. Discussions focused on technical aspects, such as the docking systems for heavy add-on modules for the station, and managerial aspects, such as the preparation of a formal decree in support of the program. On November 12, 1971, Mishin met with Minister of General Machine Building Afanasyev and his First Deputy Tyulin specifically to discuss the MOK/MKBS proposal. Both agreed to a new tactical-technical requirement, drawn up with the cooperation of the military. The meeting resulted in a recommendation for a Military-Industrial Commission decree on the issue and a rough timetable for the development of the complex. Mishin's engineers could expect to defend the technical plan for the MKBS at the Scientific-Technical Council of the Ministry of General Machine Building by mid-December 1972.

---

1. A. Tarasov, "Missions in Dreams and Reality" (English title), *Pravda*, October 20, 1989, p. 4.
2. *Ibid.*
3. The uprated N1 boosters would use the Blok S and Blok S_R upper stages.

On February 23, 1972, the Military-Industrial Commission issued a formal decree calling for work on a technical proposal for the creation of the MOK.[4] As a result, throughout the second half of 1972 and the first half of 1973, engineers at TsKBEM, including several leading Deputy Chief Designers, such as Anatoliy P. Abramov, Boris Ye. Chertok, Mikhail V. Melnikov, and Igor N. Sadovskiy, were involved in drawing up a detailed draft plan for the project. Many other organizations were also involved at this stage of the work.

The MOK as a whole was designed for a wide range of goals in support of science (astrophysical research and "fundamental scientific-technical research in conditions of outer space"), the national economy (the study of Earth's natural resources from space; activities related to guidance, navigation, and communication; research to study forestry, agriculture, geology, and deep sea fishing; and so on), and national defense. The MOK would consist of the following primary components:

- A circumterrestrial orbital system on the basis of the MKBS and autonomous spaceships
- A transport system on the basis of transport supply ships and, in the future, a reusable system and an orbital launch vehicle system
- A ground launch complex
- An automated control system and search-and-rescue complex

The MKBS, as the central link in the system, would serve as the primary place of residence for crews, the orbiting control center, and a base for supply and technical maintenance of the entire complex. Independently functioning apparatus unified with the MKBS would have separate goals, carrying out coordinated activities and maneuvers with their own transport systems.[5]

In designing the MOK, engineers took into account two main limitations: minimum funding and extended operation. Given these requirements, TsKBEM, in its technical plan for the MOK, addressed and adopted specific technical solutions in five major areas:

- To reduce the number of orbital elements while at the same time maximizing the scale of useful activities, engineers used the principles "one and the same goal solved by various apparatus" and "various goals solved by the same ship." In addition, planners selected a Sun-synchronous orbit with an orbital inclination of ninety-seven and a half degrees to achieve the widest range of goals. An increase in the active lifetime of the MOK to up to seven to ten years would be accomplished by making use of reserves and service repair work.
- Designers reduced the required traffic on the "Earth-to-orbit" and "orbit-to-orbit" routes by using the lowest number of consumed materials. Specifically, they used reserve propellants to maintain the complex's orbit and orientation (with electric engines), exposed film and reentry capsules for their delivery (by transferring urgent information by radio and delivering less urgent information to Earth by transport and supply ships), and special light modifications of 7K Soyuz–type ships with remote manipulator arms for intersatellite transport. Also, autonomous modules based on the MKBS would engage in regular repair work.
- Engineers reduced the cost of developing MOK systems by maximizing the use of auxiliary systems and apparatus of standard size and form that had already been developed, but with the necessary modifications. Continuity between previously created and proposed materials would be partly facilitated by the use of 7K Soyuz–type ships launched on the Soyuz booster. Apart from its direct use as a transport ship, engineers proposed automated

4.   Yu. P. Semenov, ed., *Raketno-Kosmicheskaya Korporatsiya "Energiya" imeni S. P. Koroleva* (Korolev: RKK Energiya, named after S. P. Korolev, 1996), p. 639.
5.   *Ibid.*, p. 278.

modifications in the form of "multi-goal visiting modules." In addition, they would use a new modified spacecraft module, the 19K, launched on the Proton booster, as a modified observation module, as well as heavier special modules launched by the N1. Using upper stages such as Blok S_R, the N1 would be able to launch special apparatus for the MOK to geostationary orbit.

- Engineers would make maximal use of already developed ground-based systems to support MOK operations, such as current launch complexes and the ground tracking network.
- Finally, planners expected to reduce the cost of transportation for orbital operations on the MOK by limiting operations as much as possible to a single orbital plane coinciding with the inclination of a standard Sun-synchronous orbit. TsKBEM would also develop new economical reusable transport systems, allowing for the lifting of payloads and consumables to polar orbits at inclinations of ninety-seven and a half degrees or higher.[6]

One of the main selling points of the MOK, according to its developers, was its great flexibility and adaptability in relation to its program of research—that is, the design of the complex would make it relatively easy to change and renovate the makeup of the orbital system without disrupting the basic interconnected functionalism. The creation of the MOK would unfold in two major phases: the first in an experimental orbit at a fifty-one-and-a-half-degree inclination and the standard at an inclination of ninety-seven and a half degrees at 400 by 450 kilometers.

Obviously, one of the main links in the creation of the MOK was the N1 launcher, which in its N1F configuration would be the primary launch vehicle for elements of the MKBS portion. Engineers also explored the possibility of using a partially reusable version of the N1—a rocket whose first stage, Blok A, would be powered by combined liquid and air-compressed engines firing on the liquid hydrogen–liquid oxygen (LOX) combination.

The MKBS, the main component of the MOK, looked roughly like a giant pencil in orbit and probably had design elements common to the abandoned Martian piloted spaceship proposal from 1969. At one end of the spacecraft, there was a nuclear energy unit and electric plasma engines to maintain attitude and altitude. The primary engine complex of the MKBS would use liquid-propellant rocket engines with thrusts of 300–1,000 kilograms. Attitude would be maintained by a combination of liquid-propellant (ten to forty kilograms thrust) and electric engines (100–300 grams thrust). The nuclear power unit would supply the primary power to the station, about fifty to 200 kilowatts. Solar panels, with a total surface area of 140 square meters and jutting out from various points along its main body, would provide an additional fourteen kilowatts. The nuclear energy unit was placed as far away as possible from the habitation quarters, which were on the other side of the "pencil." This opposite end would begin with a large compartment for "scientific and special equipment." Total scientific instrumentation on the MKBS would comprise about fifteen to twenty tons. Moving aft, there would be a multiple docking adapter, much like the one later used on the *Mir* space station, but far bigger. Here, at least four visiting spacecraft would dock, some of them based on the 7K Soyuz design and some of them "special modules." The docking adapter was connected to the main living and working quarters—a huge cylindrical compartment, about the size of Skylab, for crew activities. There would be six permanent crewmembers on the MKBS and up to ten for short periods. In the first two years of operation, crews would switch over about two times a year. The life support system would have a reserve of 1,100 crewperson-days with the capability to regenerate water from condensate. Ultimately, the atmosphere and water would be fully regenerated from the life support system.

---

6.   *Ibid.*, pp. 278–79.

Further aft, there was the instrument and aggregate compartments, containing a variety of instrumentation to support MKBS operations in Earth orbit. About one-third of the way down the "pencil," the station had two long arms, each twenty to thirty meters long and 180 degrees apart, both of which ended in small cylindrical compartments. Here, in these modules, each with a volume of twenty-five to thirty cubic meters, cosmonauts could spend time and enjoy the effects of artificial gravity from the spin around the station's main axis. According to preliminary calculations, an angular velocity of a half degree per second would generate up to 0.6–0.8 g's. The central node for these artificial gravity arms would also include an EVA airlock. Moving aft down the station, the cosmonauts would then find the main laboratory quarters, yet another cylindrical module, with its own adjacent multiple docking adapter with four ports. Here, the station proper would end, and three long pylons, about half the length of the station itself, would extend aftwards, ending in the nuclear reactor package on the other end. The total mass of the MKBS with four attendant visiting modules would be in the range of 220 to 250 tons, requiring assembly in orbit because the NI would be rated at eighty to eighty-eight tons of useful payload. The station would have a total length of about 100 meters and a main body diameter of about six meters. Each MKBS was expected to function about ten years in orbit.[7]

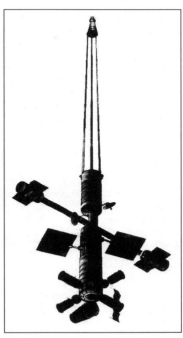

This is a drawing of the Multirole Space Base-Station (MKBS), the huge Earth-orbital complex proposed by Vasiliy Mishin in the early 1970s. (copyright Mark Wade)

Like the L3M lunar landing plan and its related Long-Duration Lunar Base, the MOK proposal was clearly a leap in ambition and capability rather than the incremental advances to which the Soviets were generally prone. While the fantastic nature of these plans would give pause to any American conception of a space program in the 1970s, the Soviets, despite losing the race to the Moon and despite the series of attendant disasters that plagued their piloted program in the early 1970s, saw these proposals as vehicles for regaining some lost glory. Thus, both at a designer level—in particular Mishin—and at a bureaucratic level—Ustinov, Smirnov, and Afanasyev—these proposals were taken very seriously and were incorporated into the long-term vision of the Soviet space program. In 1971, this vision was, however, less of a problem than the short-term one. Having just recovered three dead cosmonauts from orbit, any clarity about regaining momentum was lost amid continuing setbacks in the small space station program.

Trying to Fly

In the immediate post–Soyuz 11 disaster climate, it was clear that there would be no further missions to the first *Salyut* station.[8] One possibility was to fly the long-delayed dual

---

7.   *Ibid.*, p. 410.
8.   One unconfirmed report suggests that there may have been brief plans to fly a short Soyuz mission to the *Salyut* station in September 1971. The crew would have been A. A. Leonov, P. I. Kolodin, and A. A. Gubarev. See Mikhail Rebrov, *Kosmicheskiye katastrofy: stranichki iz sekretnogo dosye* (Moscow: Eksprint NV, 1996), pp. 72–73.

Soyuz-docking mission to test the Kontakt rendezvous radar system slated for use on the lunar landing project. Conceived sometime in 1968, the mission was repeatedly delayed because of poor results during the system's ground testing. In May 1970, the docking mission was set for August of the same year, using 7K-OK vehicles 18 and 19. The flight was then delayed to October 1970. Eventually, space program head Ustinov opted to delay the Kontakt flight in favor of the DOS space station flights in 1971, thus moving the docking flight further back to late 1971. At least four crews for the mission, including primary crews of Filipchenko with Grechko and Lazarev with Makarov, continued their training despite the increasingly gloomy prospects.[9] With the slowdown of the original L3 plan and the imminent adoption of the new L3M project, Kontakt lost much of its importance. In October 1971, Mishin officially closed down Kontakt.[10] Crews training for the mission were instead transferred to training for other projects.

With the prospect of piloted flights only within the framework of orbital stations in the near future, the focus of discussion shifted to both the DOS-7K complex and Chelomey's Almaz space station. In early August 1971, Mishin met with Ustinov to discuss long-term plans. Ustinov was clear on several points, including the urgent need to accelerate work on the *Sokol-K1* spacesuit for the Soyuz spacecraft. In addition, he made it clear that he wanted the next Soviet space station to be Mishin's DOS rather than Chelomey's Almaz. All resources should be marshaled so as to launch the next DOS before NASA's much larger Skylab space station. Based on the discussions, Mishin had a provisional schedule for work on the DOS:

Station	Launch	No. of Visits	Visiting Ship
DOS-2	First quarter of 1972	3 to 4	7K-T Soyuz
DOS-3	Fourth quarter of 1972	3 to 4	7K-T Soyuz
DOS-4	Fourth quarter of 1973	4	7K-S Soyuz

The 7K-T Soyuz variants would be equipped with the old Igla rendezvous system, while the advanced 7K-S Soyuz would have a new system, designated *Lira*. Each DOS spacecraft would have a four-month lifetime for its life support system and a six-month lifetime for all other systems. The urgency of launching the next DOS as soon as possible was underlined at a meeting in early November 1971 that was attended by all the major leaders of the Soviet space program.[11] There was a general consensus that DOS-2 should be launched so as to take some of the publicity from the Apollo 16 Moon landing planned for April 1972.

In October 1971, Col. General Kamanin retired from his post as the manager of the cosmonaut corps and was replaced by Maj. General Vladimir A. Shatalov, the forty-four-year-old veteran cosmonaut. It was a very powerful rank for a cosmonaut to hold, and his appointment order, signed earlier in June 1971, probably stemmed from Shatalov's cool disposition during his three Soyuz missions during 1969–71. One of Shatalov's first actions was to select crews for the DOS-2 space station flight. For the honor of the first visiting mission, he picked the Leonov-Kubasov team that would have flown on the ill-fated Soyuz 11 had it not been for Kubasov's

9.   The four crews, in December 1970, were A. V. Filipchenko/G. M. Grechko, V. G. Lazarev/O. G. Makarov, L. V. Vorobev/V. A. Yazdovskiy, and G. T. Dobrovolskiy/V. I. Sevastyanov.

10.   K. Lantratov, "20 Years From the Flight of 'Soyuz-12'" (English title), *Novosti kosmonavtiki* 20 (September 25–October 8, 1993): 39–41.

11.   In attendance, among others, were D. F. Ustinov (Secretary, TsK KPSS), I. D. Serbin (Chief, TsK KPSS Defense Industries Department), M. V. Keldysh (President, AN SSSR), G. A. Tyulin (First Deputy Minister, MOM), A. I. Tsarev (Deputy Chairman, VPK), B. A. Komissarov (Department Chief, VPK), B. A. Stroganov (TsK KPSS Defense Industries Department), and K. A. Kerimov (Chief, Third Chief Directorate, MOM).

brief illness.[12] Given the success of the first mission, there would be two or three additional flights to the station. For reasons that are not clear, the DOS-2 launch was significantly delayed from the first quarter to the beginning of the third quarter of 1972. The delay may have had less to do with the station itself, which was almost identical in design to the first Salyut, than problems with requalifying the 7K-T Soyuz spacecraft for flight. To test the improved life support systems with the new Sokol-K1 spacesuits, Mishin inserted a flight of an automated Soyuz into the schedule. It would be almost an entire year after the Soyuz 11 disaster that this Soyuz would be ready for launch.[13]

Soyuz 7K-T spacecraft no. 33L was launched successfully at 1453 hours Moscow Time on June 26, 1972, into an initial orbit of 195 by 342 kilometers at a 51.6-degree inclination. The spacecraft was named Kosmos-496 upon entering orbit. Little is known about the flight except that there was one orbital maneuver. After about six days in orbit, the descent apparatus separated from the rest of the vehicle and returned to Earth. The successful mission gave some much-needed confidence to the continuing preparations for the next DOS flight. Crews for the first flight flew into the Baykonur Cosmodrome in preparation for their own launch. By this time, the usual rumors were mounting in the West that a spectacular mission was imminent. On March 9, 1972, the Paris-based Agence France Presse reported that two crews were ready to fly to a new Salyut space station for missions lasting up to thirty days.[14] Shatalov added fuel to the rumors by telling the Czech press in early April that there would be additional piloted missions "probably this year."[15]

The State Commission for Soyuz, still headed by Maj. General Kerimov, approved the launch of DOS-2 for late July 1972. Subsequently, Soyuz 12 with Leonov and Kubasov would lift off during the last week of August. Another crew, Lazarev and Makarov on Soyuz 13, would fly to the station in the third week of October 1972. All these plans were not to be. The twenty-ton space station, spacecraft no. 122, was launched in the early morning, at 0620 hours, 57 seconds Moscow Time, on July 29, 1972, on top of a three-stage Proton booster. During the boost phase, at T+162 seconds, the control systems of the second stage of the launch vehicle failed, preventing orbital insertion. The mission had to be aborted. U.S. over-the-horizon sensors evidently monitored telemetry from the launch attempt, prompting subsequent news reports that one of the four second-stage engines had stopped firing during the ascent through the atmosphere.[16]

The loss of DOS-2 continued the series of strikes against the Soviet piloted space program. To take advantage of two flight-ready 7K-T Soyuz vehicles, which had been ready to deliver crews to the lost station, the State Commission in August 1972 considered launching a single Soyuz on a solo mission in Earth orbit, primarily to test the new spacesuits and redesigned

12. Four crews trained for DOS-2: A. A. Leonov/V. N. Kubasov, V. G. Lazarev/O. G. Makarov, A. A. Gubarev/G. M. Grechko, and P. I. Klimuk/V. I. Sevastyanov. See Lantratov, "20 Years From the Flight of 'Soyuz-12'." Note that originally, in September 1971, the Gubarev and Klimuk crews were slightly different. See S. Shamsutdinov and I. Marinin, "Flights Which Never Happened" (English title), Aviatsiya i kosmonavtika no. 3 (March 1993): 43–44.

13. A prototype of the Sokol-K1 suit was produced in 1971, with further revisions added between August 1971 and March 1972. See Russian Space History, Sale 6753 (New York: Sotheby's, 1996), description for Lot 147.

14. Agence France Presse, untitled press release, Paris, 1554 GMT, March 9, 1972, in English.

15. Soviet Space Programs, 1971–75: Overview, Facilities and Hardware, Manned and Unmanned Flight Programs, Bioastronautics, Civil and Military Applications, Projections of Future Plans, prepared for the Committee on Aeronautical and Space Sciences, U.S. Senate, 94th Cong., 2d sess. (Washington, DC: U.S. Government Printing Office, August 1976), p. 534.

16. V. M. Petrakov, "Soviet Orbital Stations," Journal of the British Interplanetary Society 47 (September 1994): 363–72; I. B. Afanasyev, "Unknown Spacecraft (From the History of the Soviet Space Program)" (English title), Novoye v zhizni. Nauke, tekhnike: Seriya kosmonavtika, astronomiya no. 12 (December 1991): 1–64; "Orbiting of Second Salyut Ends Soviets' Hiatus in Manned Space," Aviation Week & Space Technology, April 9, 1973, p. 21.

systems aboard the ship. Crews began training for this flight, scheduled for sometime in late August or early September 1972 on vehicle no. 34. After roughly a month of preparations, two crews—Gubarev-Grechko and Klimuk-Sevastyanov—successfully passed their final exams, but by this time, the commission began to get cold feet. Members expressed reservations for such a flight, believing it to be "inopportune," most likely because a solo Soyuz flight in Earth orbit would pale in comparison to the impending launch of Apollo 17 in December 1972. The solo flight was canceled.[17]

Besieged by failure and delays, the Soviet space station program needed some drastic help. Assistance came from neither Ustinov nor Afanasyev, but rather from the unlikely person of General Designer Chelomey and his Almaz space station program. Since the February 1970 decision to move ahead with DOS at the cost of delaying Almaz, Chelomey had doggedly and quietly pursued work on his coveted station, methodically coordinating his efforts with his primary clients, the Ministry of Defense. Although the focus of activities at the massive Khrunichev Plant during 1970–72 was on the DOS effort, representatives from Chelomey's TsKBM continued work on their own space station hulls. Engineers tested an updated version of the Almaz control system on a complex test rig. Tests of the Almaz power system included firings of the flywheel micro-liquid-propellant rocket engines at a test stand near Moscow. Various hulls were remanufactured for Almaz, including those for stress, vibration, and heat testing. A special orbital block simulator was also built at the Institute of Aviation and Space Medicine, where testers spent thirty-six days in a "flight regime," which ended on January 11, 1972. After their "mission," they reported back that "the configuration of the work and living compartments is comfortable," that "the air is good and odorless," and that they had "soon become used to the hum and vibrations caused by the instruments."[18] Crews whose missions had been sidelined because of DOS resumed their training on station components in hydrolabs and aboard Tu-104 aircraft.

In the original conceptions of the Almaz space station from the mid-1960s, Chelomey had always envisioned his station as an orbital complex rather than simply a station supplied by small ferry vehicles such as the DOS. The key to these plans was the use of a large module, about the size of the Almaz station itself, which would not only serve as a ferry craft for crews, but also add significantly to the volume and capabilities of the station once linked to the station proper. Most likely because of an overload of work, Chelomey was unable to carry out substantial work on this add-on module, called the Transport-Supply Ship (TKS). Like many of his other projects, he entrusted the work on developing the TKS to his Moscow Branch headed by First Deputy General Designer Viktor N. Bugayskiy. There, under Bugayskiy's overall supervision, engineers completed the initial technical project for the TKS (or product 11F72) in 1969.[19] While the decision to create Mishin's DOS in February 1970 may have delayed the overall Almaz program, it does not seem to have squelched Chelomey's ambitions of creating the TKS. With Minister of Defense Grechko's support, Chelomey managed to extract an official promise to commit to developing the TKS. On June 16, 1970, the Central Committee and the Council of Ministers issued a decree (no. 437-160) that officially approved the TKS program. The TKS would have the following goals:

17.   Lantratov, "20 Years From the Flight of 'Soyuz-12'."
18.   Vladimir Polyachenko, "The 'Pep' of Almaz" (English title), Krylya rodiny no. 1 (January 1992): 18–19.
19.   S. A. Zhiltsov, ed., Gosudarstvennyy kosmicheskiy nauchno-proizvodstvennyy tsentr imeni M. V. Khrunicheva (Moscow: RUSSLIT, 1997), p. 89.

- Docking of twenty-ton spaceships to each other (the TKS and the Almaz)
- Delivery and return of crews from the Almaz station
- Delivery of supplies and apparatus for carrying out functional work on the Almaz station
- Delivery of life support supplies for the crew
- Raising of station orbits
- Orientation and extended (up to ninety days) control of the flight of the entire complex
- Possibility of autonomous descent from orbit[20]

In its design, the TKS served as a direct intermediary between early Chelomey designs, such as the lunar LK-1 and LK-700 spacecraft from the 1960s, and the *Mir* modules and *Zarya* module of the International Space Station in the 1990s. The spacecraft consisted of two major components: the return apparatus and the functional cargo block (FGB). The reusable return apparatus (or product 11F74) was almost identical to the one used on the original Almaz station for returning crews to Earth. At some point in 1968, Chelomey had evidently abandoned the use of this large module on the Almaz station, opting instead to use the smaller Soyuz to return crews from the station. There were probably also technical considerations, because the hatch-in-the-heat-shield design necessitated a long and exhaustive series of tests to verify its safety before use with crews.

The functional cargo block (or product 11F77) was a large and roughly cylindrical structure connected to the base of the return apparatus. At the base of the FGB, the cylindrical shape expanded into a skirt with a maximum base diameter of 4.15 meters. The spacecraft was completed by a terminal cone fixed at the flat base of the cylindrical skirt with the apex facing aft. The main body diameter of the FGB was 2.9 meters, the same as that for the smaller section in the Almaz space station. The docking assembly of the TKS was located at the aft end of the spacecraft in the larger diameter area. After rendezvous with the Almaz station, the crew, in spacesuits, would be next to the docking assembly and observe operations through a viewport. The simplified docking procedure and expanded view would make it possible to abandon the cumbersome system of periscopes and TV cameras used on the Soyuz spacecraft. The docking assembly itself was significantly different from that used on the 7K-T Soyuz; time from the moment of docking to hatch opening would be three to four minutes, as compared to the eighteen to twenty minutes on the Soyuz-DOS combination. One of the supplementary goals of the TKS was to deliver the small recoverable capsules used on the Almaz station to return exposed film of military targets from space. Overall, the TKS would have a mass of just over twenty-one and a half tons at launch and seventeen and a half tons in orbit; it would afford as much internal space as the Almaz space station. Two Almaz-type solar arrays with an area of forty square meters would provide about three kilowatts of power. It would be both a qualitative and quantitative leap in abilities over the modest Soyuz ferry spacecraft.[21]

As a result of cumulative delays, the TKS was not expected to fly operational missions prior to the mid-1970s. In the meantime, in 1971, Chelomey had signed an agreement with Mishin to use variants of the 7K-T Soyuz spacecraft to deliver and recover crews from the Almaz space station. Work on this version of the Soyuz began the same year, and by early 1972, TsKBEM's Department No. 037 had completed the redesign of the 7K-T to support piloted missions to

---

20.   *Ibid.*, p. 88.
21.   Afanasyev, "Unknown Spacecraft"; Zhiltsov, ed., *Gosudarstvennyy kosmicheskiy*, pp. 88–89; Nina Chugunova, "Chelomey's Cosmonauts: Why There Are No Crews From NPO Mashinostroyeniya in Outer Space" (English title), *Ogonek* 4–5 (January 1993): 24–29.

Here is a model of Vladimir Chelomey's Transport-Supply Ship (TKS), which was meant to be part of the Almaz military space station complex. The conical segment at the left is the reentry capsule, apparently patterned after the U.S. Air Force's Gemini-B spacecraft. The vertically placed component at the left is the launch escape system.
(copyright Dietrich Haeseler)

the Almaz station.[22] By May 1972, four crews were in the midst of intense preparations for the first missions to Almaz.[23] Thus, by mid-1972, the Soviet Union had two full-fledged and parallel space station programs—one dedicated to primarily civilian goals, Mishin's DOS, and one for military research, Chelomey's Almaz. The path of these projects had always been inter-dependent, but in mid-1972, they forged a most unlikely alliance.

For Mishin, the DOS had always represented an unnecessary diversion from what he considered the main thematic directions of work at the design bureau: large-scale space stations such as the MOK and the lunar landing project. The DOS project had essentially been hoisted upon him at a most inconvenient juncture. That TsKBEM had managed to fulfill the original order within the given period of one year was partly because Mishin had been forced to redirect much of the resources at the design bureau to the DOS program. Mishin's primary goal was to shift the focus back to his two pet projects—the MOK and the L3M. Both had received resounding shows of support with official decisions in February and May 1972, respectively. It was time to make sure that the DOS did not hinder their implementation. At the same time, Chelomey had every reason to resent the DOS space station program—an effort that had been essentially appropriated from his own coveted Almaz project. Having seen the latter sidelined by the DOS, Chelomey was in the unlikely position of being of the same mind as Mishin on

22.   Semenov, ed., *Raketno-Kosmicheskaya Korporatsiya*, p. 190. Both the DOS and Almaz versions had the same design bureau designation—that is, 7K-T—but had different production designations: 11F615A8 (for DOS), and 11F615A9 (for Almaz).

23.   These four crews were P. R. Popovich/L. S. Demin, G. V. Sarafanov/Yu. P. Artyukhin, B. V. Volynov/ V. M. Zholobov, and V. D. Zudov/V. I. Rozhdestvenskiy.

this matter—that is, the small space station program, specifically the DOS and Almaz, needed to go back to Chelomey. With this in mind, on April 14, 1972, Mishin and Chelomey signed an agreement proposing to Minister of General Machine Building Afanasyev that after the first four DOS space stations, work on the project would be terminated. In addition, all continuing research for science and the national economy would be carried out on Almaz space stations, in addition to its own primarily military activities. Initially, the Almaz space station would be serviced by the 7K-T Soyuz, then eventually the advanced 7K-S Soyuz, and finally the TKS. One final note in the letter was to use a proposal allowing the use of Chelomey's TKS on Mishin's MOK.[24]

There was apparently much opposition *within* Mishin's design bureau against this unlikely alliance, presumably from individuals, such as Bushuyev and Feoktistov, who had wholeheartedly thrown their lot in with the DOS program. Minister Afanasyev, however, under "pressured circumstances," agreed to ratify the proposal, giving it his signature on April 21, 1972. In retrospect, this agreement was quite possibly the origin of a serious fracture within TsKBEM between the "pro-lunar program" and the "pro-DOS" factions. The hostilities that would build from this decision would prove to have cataclysmic consequences. While Mishin may have believed that an agreement to hand over the DOS to Chelomey was a pragmatic choice at the time, it is clear that he neglected to consider the personal and managerial consequences within his own organization. Worse for Mishin, while he had strong supporters for the lunar program and the MOK, his opponents were formidable, including Deputy Chief Designers Bushuyev and Chertok and the influential Department Chief Feoktistov.

The Mishin-Chelomey agreement in April 1972 meant that Almaz was less of a competitor than a complement to the DOS. New flight models of both stations were, by coincidence or not, ready to fly by early 1973. Mishin's new DOS vehicle, spacecraft no. 123, differed in many respects to its two predecessors launched in 1971 (as *Salyut*) and in 1972 (the launch failure). The original design, while adequate given the short timeframe for its creation, had some major shortcomings, limiting the effective use of the station. One of these design compromises was the configuration and location of the station's two solar panels. To have these panels face the Sun on the original *Salyut*, crews had to turn the entire station and maintain attitude continually to receive power. This resulted in high consumption levels for the on-board propellant, which was in relative short supply. The complicated solar orientation system also affected the amount of scientific experimentation possible on the station because of fluctuating power levels. The primary difference of the "new" DOS, whose development had actually begun as early as 1970, was to remove the two pairs of solar panels and instead install three self-rotating solar panels, which would turn around their own axes independently of the station. The three new panels, appropriated from Chelomey's TKS, would be installed directly in a "T" shape on the main working compartment and provide over two times more power than the earlier ones. To compensate for the additional mass from the new panels, engineers removed the number of tanks from the main engine unit. To reduce the amount of propellant required to maintain a working orbit, planners also increased the operational orbit to an altitude of 350 kilometers.[25]

There were many other changes in this "second-generation" DOS. Engineers designed a new "highly economical" orientation system named *Kaskad* and an experimental navigation system called *Delta* to replace the older ones. There was also a new thermo-regulation system

---

24. The entire letter has been reproduced in Semenov, ed., *Raketno-Kosmicheskaya Korporatsiya*, pp. 295–96. An additional point in the letter addressed the use of the Soyuz (a variant known as the 7K-M) instead of the *Salyut* space station for the Apollo-Soyuz Test Project.

25. *Ibid.*, p. 271; Afanasyev, "Unknown Spacecraft."

and an early version of a closed-cycle water supply system using the SRV-K water regeneration device. The total guaranteed lifetime of the station was increased from the ninety days for the first DOS to 180 days on the third one. The scientific complement was slightly different from the earlier model. The new one included a Roentgen telescope-spectrometer, the RT-4 Roentgen telescope mirror, and the ITS-K infrared telescope. Finally, there were some cosmetic changes, such as thicker walls, an altered frame, changes in the aggregate compartment, and the use of a unified welding installation in the main scientific apparatus compartment.[26]

There were additional changes to the 7K-T Soyuz ferry in 1972 and 1973. Anticipating that a ferry vehicle would not need to fly independently for more than two days, engineers deleted the two heavy solar panels from the spacecraft, making the ship rely completely on its modest internal chemical batteries. These batteries could be recharged once docked to a space station using power generated from either the DOS or Almaz. The mass of this second iteration of the Soyuz ferry was about 6,800 kilograms, up from the original 6,700 kilograms.

By the time that these changes were made to the DOS and Soyuz designs, Chelomey was well advanced with preparations for the launch of his own first Almaz station. On June 15, 1972, a decree of the Ministry of General Machine Building specified a schedule for immediate operations in the Almaz program. The Khrunichev Plant was to complete the assembly of the first flight model of the Almaz station and deliver it for preliminary testing by June 30, leading to delivery to the testing station at the Baykonur Cosmodrome by November of the same year.[27] If all went well, the launch would take place in late 1972 or early 1973—that is, at about the same time as Mishin's DOS-3. The concurrent and timely preparations were very much colored by activity in the United States. NASA at the time was wrapping up final preparations for the launch of its first space station, Skylab, scheduled in April 1973. If successful, it would host three crews during the year, with missions lasting twenty-eight, fifty-six, and fifty-six days, respectively. Having taken the lead in terms of space stations, with Salyut, Soviet space officials, especially Ustinov, were particularly sensitive to the possibility that Skylab would completely overshadow the achievements of Salyut. It was absolutely imperative that the Soviet Union have a space station in orbit before Skylab. Luckily for Ustinov, both Mishin and Chelomey were ready with their respective space stations at just the right time. It seems that Ustinov, as a means to upstage Skylab, wanted to fly both the DOS and Almaz in 1973. Given Ustinov's predisposition to oppose Chelomey, one would have expected the DOS to have the honor of going first, but evidently in October 1972, Soviet leader Brezhnev had the last word: Chelomey's Almaz would get the first try.[28]

The first Almaz station, vehicle no. 101-1, arrived at the Baykonur Cosmodrome in January 1973. Ground testing was completed within three months. The fact that Mishin's DOS was also undergoing ground testing at the launch site simultaneously led to problems because of stretched resources. Both stations used the same pressure chamber and fueling stations. In fact, there was a great degree of cross-pollination between the two programs, partly because TsKBEM engineers had to be involved in the Almaz effort as they were responsible for the Soyuz spacecraft.[29] Four

26. Semenov, ed., *Raketno-Kosmicheskaya Korporatsiya*, p. 272; Petrakov, "Soviet Orbital Stations."

27. Zhiltsov, ed., *Gosudarstvennyy kosmicheskiy*, pp. 78, 89. The order also specified the details of manufacture. All production would be carried out at the M. V. Khrunichev Machine Building Plant for the prime customer, which was TsKBM (for all components except the transfer compartment and the engine unit for which the customer was TsKBM's Fili Branch). In addition, the order specified that the Khrunichev Plant would complete production preparation for the manufacture of the TKS blocks in the fourth quarter of 1972.

28. The first Almaz was originally scheduled to fly in December 1972, but there were serious delays in the delivery of subsystems. MOM First Deputy Minister Tyulin finally issued a memo to Chelomey on November 23, 1972, noting that because Skylab would fly by April 30, 1973, the first Almaz had to be in orbit by March 1973.

29. Semenov, ed., *Raketno-Kosmicheskaya Korporatsiya*, p. 273; Afanasyev, "Unknown Spacecraft"; Shamsutdinov and Marinin, "Flights Which Never Happened"; V. Polyachenko and A. Tumanov, "From the History of Space Science: The Controllable 'Almaz'" (English title), *Aviatsiya i kosmonavtika* no. 8 (August 1993): 41–43.

two-person crews were on standby for two consecutive missions to the station—the first consisting of cosmonauts Popovich and Artyukhin lasting fifteen days and the second made up of Sarafanov and Demin. There were apparently serious problems with the Soyuz parachute system that threatened to disrupt the Almaz schedule. Despite these potential disruptions, Chelomey pushed ahead with the liftoff.

Launch day for the Almaz station was April 3, 1973, a little more than a month before the Skylab launch. As the clock ticked down to booster ignition, at T-15 minutes, there was a sudden alarm: propellant was apparently leaking from the Proton rocket's filling system. The danger of a terrible explosion was on everyone's mind. Chelomey fearlessly announced that he wanted to go directly to the pad. After an inspection of the situation, he returned to the blockhouse and recommended that the launch proceed. State Commission Chairman Col. General Mikhail G. Grigoryev of the Strategic Missile Forces concurred, and at exactly 1200 hours Moscow Time on April 3, 1973, Almaz lifted off into the sky, eventually entering an initial orbit of 215 by 260 kilometers at a 51.6-degree inclination.[30] A full thirteen years after proposing his first space project, Chelomey had finally launched a piloted spacecraft into orbit around Earth, the first piloted military spacecraft in space.

Chelomey might have been forgiven for believing that his beloved space station would be named *Almaz* by the Soviet press. But highly placed space officials, possibly including Ustinov, were adamantly opposed to this. Some have claimed this was because they "were dead against the presence of a second figure in the Soviet space program."[31] Others believe it was to hide the fact that Almaz was a purely military space station.[32] In any case, Chelomey, apparently humiliated, was explicitly ordered to have the name *Salyut 2* painted on the station. The shrewd general designer told his engineers to paint the offending name on the outside fairing of the station; once the fairing jettisoned in the upper reaches of the atmosphere, the station revealed Almaz clearly written on it. The Soviet press, of course, referred to it as *Salyut 2*. Launch of the first crew, on Soyuz 12, had been planned for April 13, but had to be delayed to May 8 because of continuing problems with the Soyuz parachute system. In their initial press releases on the mission of the station, the Soviets refrained from making any connection with piloted flights. At least two major orbital corrections, on April 4 and 8, resulted in a new orbit of 261 by 296 kilometers.[33]

Throughout the first few days in orbit, the Chief Operations and Control Group at Yevpatoriya, led by Yakov Ya. Sirobaba, tested the attitude control systems, life support systems, and radio communications systems, and all seemed to be working without fault. Trouble struck on the thirteenth day of flight, on April 15, on the 188th orbit of *Salyut 2*. Controllers reported that the main telemetry system had failed; according to "support" telemetry, pressure in the main hull had dropped by half, and precise measurements of the station's orbital trajectory showed that its path had deviated slightly, as if given some kind of thrust. Clearly, some type of catastrophic failure had occurred on the station, squelching the possibility that any crew would be heading in its direction any time soon. Early the next morning, the senior members of the State Commission, including Col. General Grigoryev and Space Assets Commander Lt. General Karas, met at Chelomey's offices to discuss the situation. An accident investigation commission under Karas was established. Throughout the next few days, engineers pored over ground models of Almaz to ascertain the cause of this sudden event by simulating various

30.   Polyachenko, "The 'Pep' of Almaz."
31.   Chugunova, "Chelomey's Cosmonauts."
32.   Roald Z. Sagdeev, *The Making of a Soviet Scientist: My Adventures in Nuclear Fusion and Space From Stalin to Star Wars* (New York: John Wiley & Sons, 1994), p. 207.
33.   Kenneth Gatland, *Manned Spacecraft* (New York: Macmillan, 1976), p. 234.

conditions. Specialists also flew to Yevpatoriya to look into the matter. The initial prognosis was that there might have been some ground error, but this hypothesis was eliminated when investigators ascertained that each command transmitted to *Salyut 2* had been without fault.[34] On April 18, unofficial Soviet sources in Moscow denied that piloted visits had ever been planned for *Salyut 2*. On April 28, the Soviet news agency TASS announced that *Salyut 2*, "having checked the design of improved on-board systems and carried out experiments in space, had completed its flight program," notably omitting

This drawing shows the flight variant of the Almaz space station with its unique docking node visible on the left. The viewport for the Agat-1 reconnaissance camera was located on "the underside" of the vehicle, not visible in this image. (copyright VideoCosmos Co., via Dennis Newkirk)

the word "successfully," which it normally used in such press releases.[35]

The Karas Commission arrived at the conclusion that there had been a manufacturing flaw in the main engine of the Almaz station, which, when fired, had caused punctures in the main hull.[36] One cosmonaut who trained for Almaz later recalled that there had been "an electrical fault in one of the station's devices which had eventually caused the rupture of the external hull."[37] Western reports, presumably filtered through to the open media from classified sources, suggested that the actual hull breach had been so violent that the station's solar panels and boom-mounted rendezvous radar and radio transponder had been ripped off, leaving *Salyut 2* tumbling in space. The engine, these reports suggested, could not be turned off once it was turned on.[38] Some of the station's designers begged to differ with the verdict of a malfunctioning engine, and there was apparently never any unanimity with the verdict. For example, an in-house investigation at Chelomey's design bureau concluded that the station might have been hit by residual debris from the Proton booster on April 15.[39] Perhaps the most curious claim advanced for the failure—a claim no doubt proposed to exonerate its designers of any fault—was that a meteorite had hit the station and blown a hole in its hull. Chelomey himself was said to subscribe to this opinion.[40] April was a bad month for the general designer. On April 25, one of his radar ocean reconnaissance satellites, the US-A, failed to reach orbit, depositing its nuclear isotope payload in the Pacific Ocean. U.S. Air Force planes apparently flew high above

34.	Vladimir Polyachenko, "The 'Pep' of Almaz: Part II" (English title), *Krylya rodiny* no. 4 (April 1992): 30–32; B. A. Pokrovskiy, *Kosmos nachinayetsya na zemlye* (Moscow: Patriot, 1996), pp. 411–12.

35.	Gatland, *Manned Spacecraft*, p. 234; *Soviet Space Programs: 1976–80 (With Supplementary Data Through 1983): Manned Space Programs and Space Life Sciences*, prepared for the Committee on Commerce, Science, and Transportation, U.S. Senate, 98th Congress, 2d sess. (Washington, DC: U.S. Government Printing Office, October 1984), 547–48.

36.	Polyachenko, "The 'Pep' of Almaz: Part II"; Afanasyev, "Unknown Spacecraft."

37.	Neville Kidger, "Almaz: A Diamond out of Darkness," *Spaceflight* 36 (March 1994): 86–89.

38.	Thomas O'Toole, "Soviet Union Still Trails U.S. in Space," *Washington Post*, June 17, 1973, pp. A1, A8; *Soviet Space Programs: 1976–80*, p. 548.

39.	Pokrovskiy, *Kosmos nachinayetsya na zemlye*, p. 412.

40.	Sagdeev, *The Making of a Soviet Scientist*, pp. 176, 207–08. To confuse matters further, Chief Designer Mishin has claimed that the failure was caused by a malfunction in the attitude control system of Almaz. See V. P. Mishin, "Why Didn't We Fly to the Moon?" (English title), *Znaniye: tekhnike: seriya kosmonavtika, astronomiya* no. 12 (December 1990): 3–43.

the Pacific to sample the upper atmosphere for radiation from the accident.[41] Meanwhile, *Salyut 2*, lost and tumbling in space, eventually decayed from orbit on May 28, 1973. Popovich and the remaining cosmonauts training for their long-awaited flight would have to wait longer. Chelomey did not expect to have the next Almaz station ready for flight before, at least, the end of the year.

With Almaz out of the picture, things were desperate for the 1973 version of a space public relations offensive. The Skylab launch was imminent. On February 14, NASA's Manned Space Flight Management Council met and set May 14 as the launch date for the huge space station.[42] Acutely conscious of the U.S. schedule, TsKBEM engineers accelerated the preparations for the next DOS, no doubt under severe pressure from Brezhnev and Ustinov. The station had arrived at Tyura-Tam for final preflight testing in December 1972, and by late April 1973, State Commission Chairman Kerimov set May 8 as the launch date. This would be just six days prior to the Skylab launch. Troubles during prelaunch operations, however, threatened to thwart the Soviet plans. Engineers detected a depressurization in one of the propellant valves in the Proton launch vehicle, resulting in a major fuel leak. As personnel from the Khrunichev Plant began repairs, Chief Designer Mishin, under stress and being "emotional," refused to have his station launched by this particular Proton rocket, booster no. 284-01, even if the repairs were successful. Mishin, perhaps remembering the July 1972 DOS launch failure, remained characteristically stubborn, and he refused to budge from his position despite insistent arguments from other members of the State Commission. It was only through the intervention of other senior officials from TsKBEM that Mishin conceded.[43] The delays with the propellant leak pushed the launch back to May 11. The first crew, cosmonauts Leonov and Kubasov, would lift off three days later, the same day Skylab was slated to reach orbit.

Officers of the Strategic Missile Forces successfully launched DOS-3 on May 11, 1973, at 0320 hours Moscow Time. Initial orbital parameters were 218 by 266 kilometers at a 51.6-degree inclination. The spate of troubles with the Soviet space station continued with DOS-3. Kerimov recalled many years later that "suddenly, on the very first orbit, on a segment in which our control points did not control the operation of the spacecraft, the attitude-control rockets began working irregularly. As a result, all the fuel reserves were burning up."[44] Later analysis showed that the attitude control engines had spuriously begun firing because of a failure in an ion sensor. As telemetry continued to stream into Yevpatoriya on the situation, one controller exclaimed in horror, "The tanks are almost empty!"[45] Representatives from TsKBEM were, evidently, slow to react and were unwilling to believe the telemetry. One engineer, Yevgeniy V. Bashkin, explained that such a quick consumption of propellant was impossible: it was 1,500 times faster than what was maximally possible. When subsequent telemetry confirmed rapid propellant loss, TsKBEM Deputy Chief Designer Yakov I. Tregub, the flight director from the design bureau, finally accepted the initial conclusion. Unfortunately, by this time, little would have been accomplished by turning off the orientation system because all of the station's attitude control propellant was depleted. The possibility of crewing the station was effectively eliminated. The fact that the failure was detected in the first few orbits allowed the Soviet press to disguise the mission by calling it by the next number in the Kosmos satellite series, *Kosmos-557*, instead of using the *Salyut* name.

---

41.   Thomas O'Toole, "2nd Russian Space Shot Fails," *Washington Post*, May 4, 1973, p. A1; "Soviet Space Attempt on April 25," *Space Daily*, May 8, 1973, p. 46. Note that both these articles incorrectly identified both the launch vehicle (the Proton) and the payload (Lunokhod). The actual launch vehicle was a Tsiklon-2 booster.
42.   David Baker, *The History of Manned Spaceflight* (New York: Crown Publishers, 1985), p. 463.
43.   Semenov, ed., *Raketno-Kosmicheskaya Korporatsiya*, p. 272.
44.   V. Ovcharov and L. Chernenko, "Recommended by Korolev" (English title), *Sovetskaya rossiya*, August 22, 1987, p. 2.
45.   Pokrovskiy, *Kosmos nachinayetsya na zemlye*, p. 410.

An interdepartmental commission under Vyecheslav M. Kovtunenko, a Deputy Chief Designer at KB Yuzhnoye, was established to investigate the Kosmos-557 failure and recommend compensatory measures. KGB representatives apparently participated in the deliberations, perhaps suspecting sabotage. The commission eventually found that the failure could have been averted if the flight control team had reacted faster. In the end, members prepared a plan to deorbit the station safely from orbit to preclude it from burning up over populated areas of Earth.[46] After a careful series of commands to the station, Kosmos-557's main engine was fired on May 22, 1973, to raise its orbit, but because of improper orientation, the spacecraft reentered the atmosphere and burned up over the Indian Ocean.[47] The repercussions of the accident were wider than simply the loss of a station. TsKBEM Deputy Chief Designer Tregub was dismissed from his post as the flight director of all subsequent piloted missions and fired from the design bureau. Department Chief Raushenbakh was demoted to the position of a "consultant," and he left TsKBEM soon after.There were apparently others who lost their jobs. It was the first time that such dismissals had taken place in the piloted space program, despite the earlier deaths of the Soyuz 1 and Soyuz 11 crews.[48]

In the official history of TsKBEM, the episode with the loss of DOS-3 is described as "a big blow to the program."[49] The timing of the Almaz and DOS losses in the spring of 1973 could not have been worse. NASA launched Skylab 1, the first American space station, into orbit on May 14, 1973. NASA, of course, had its own problems with Skylab. During launch, the meteoroid shield tore off, causing one of the solar panels to be ripped off and the other one to be jammed in an inert position. But the remarkable resourcefulness of NASA engineers and astronauts was demonstrated amply in late May, when three astronauts docked with the station and revived it to almost full capacity. On June 22, 1973, they returned to Earth after a twenty-eight-day flight, regaining once more the absolute endurance record in space for the United States.[50] Now the Soviet Union was lagging behind the United States in both the lunar landing *and* space station areas of piloted space exploration. Another reason for the ill-timing was the acceleration of work on the Apollo-Soyuz Experimental Program, better known in the West as the Apollo-Soyuz Test Project. Although there had been nary a word on both the *Salyut 2* and Kosmos-557 failures from the Soviet press, there was much speculation in the Western press on these two missions. Official representatives from the Soviet side were no doubt embarrassed by this attention. In October 1973, Academician Boris N. Petrov, one of the "figurehead" leaders of the Soviets, told NASA's George M. Low that "there had been no plans to send men to occupy" *Salyut 2*.[51] In another outright lie, he added that the flight of Kosmos-557 had not been related to the piloted space program.

The Light at the End of the Tunnel

The loss of both *Salyut 2* and Kosmos-557 meant there would be no Soviet space station missions during the remainder of 1973. Crews for both the Almaz and DOS programs would have to wait much longer to carry out their long-delayed space station flights. One particular crew,

46. *Ibid.*, pp. 410–11.

47. Afanasyev, "Unknown Spacecraft."

48. Semenov, ed., *Raketno-Kosmicheskaya Korporatsiya*, pp. 273, 355; Pokrovskiy, *Kosmos nachinayetsya na zemlye*, p. 411. Tregub was replaced by TsKBEM Deputy Chief Designer B. Ye. Chertok, and Raushenbakh was replaced by V. P. Legostayev.

49. Semenov, ed., *Raketno-Kosmicheskaya Korporatsiya*, p. 273.

50. Linda Neuman Ezell, *NASA Historical Data Book, Volume III: Programs and Projects 1969–1978* (Washington, D.C.: NASA Special Publication (SP)-4012, 1988), p. 104.

51. Edward Clinton Ezell and Linda Neumann Ezell, *The Partnership: A History of the Apollo-Soyuz Test Project* (Washington, DC: NASA SP-4209, 1978), p. 232.

cosmonauts Leonov and Kubasov, had perhaps the most trying experience in their arduous training program for the DOS. In June 1971, they had trained to fly the first *Salyut*, DOS-1, only to be dropped days before the launch because of Kubasov's illness. They would have flown the second mission to the station in July, had it not been for the deaths of the Soyuz 11 crew. In July 1972, they were ready to fly to the DOS-2 station when it exploded in air before ever reaching orbit. Then, for the fourth and final time, they were days from flying to DOS-3 in May 1973 before the fatal attitude control system failure destroyed that hope. After three years of training for DOS missions, on May 25, 1973, just days after the Kosmos-557 failure, Soviet officials announced that Leonov and Kubasov would be the primary crew for the Soviet side of the Apollo-Soyuz Experimental Program.[52] Naturally, there was no word on their activities of the past few years.

Having no space station to which to go meant there was the possibility of an even longer hiatus in Soviet piloted spaceflights. To take advantage of the gap, Mishin drew up a plan to thoroughly test the new 7K-T Soyuz ferry variant on an independent flight. In addition, he inserted a second solo Soyuz mission, which would carry out some of the astrophysics experiments they had been forced to abandon because of the loss of two consecutive DOS spaceships. In July 1973, crews began training for these two missions.[53]

As a prelude to these two missions, TsKBEM inserted a third solo Soyuz mission—an automated flight to verify all the new design changes on the vehicle that had been introduced in 1972–73. That Mishin did not fly such a robot flight prior to the May 1973 space station attempts indicates that those missions were under time pressure to get off before Skylab. Having lost the battle over space stations, there was no incentive not to fly a precursor mission anymore. Soyuz 7K-T spacecraft no. 35 lifted off without incident at 0900 hours Moscow Time on June 15, 1973, into an initial orbit of 209 by 268 kilometers at a 51.55-degree inclination. During its two-day, nine-minute flight, the 6,790-kilogram spacecraft, named *Kosmos-573* in the Soviet press, performed a single orbital maneuver to lower apogee before returning to Earth on June 17. Presumably, the first flight of the "solar panel-less" Soyuz variant was sufficiently successful to warrant dedicated preparations for a "return-to-flight" mission in the program.

It had been more than two years since a single Soviet cosmonaut had been in space. The honor to break this dubious record fell on the shoulders of two seasoned veterans of the cosmonaut corps, neither of whom had ever flown in space before. At age forty-five, Commander Lt. Colonel Vasiliy G. Lazarev's involvement in the space program dated back to the early 1960s, when he had taken part in the *Volga* high-altitude balloon flights to test prototype pressure suits, during which pilots parachuted from altitudes as high as thirty-two kilometers. Later, in 1964, he had been considered a prime contender to fly the historic three-cosmonaut Voskhod flight. It was only at the last minute, after insistent opposition from the late Korolev, that another candidate replaced him on the primary crew. An Air Force doctor by profession, he had "officially" joined the cosmonaut team on January 17, 1966, just days after Korolev's death.[54] Flight Engineer Oleg G. Makarov, at age forty, was an old-timer from TsKBEM

52.   K. Lantratov, "Do You Remember How All This Began? (20 Years From the Docking of 'Soyuz' and 'Apollo')" (English title), *Novosti kosmonavtiki* 15 (July 16–29, 1995): 42–52.

53.   Lantratov, "20 Years From the Flight of 'Soyuz-12'"; E-mail correspondence, Sergey Voevodin to the author, January 30, 1997. Crews for the first mission were V. G. Lazarev/O. G. Makarov, A. A. Gubarev/G. M. Grechko, and P. I. Klimuk/V. I. Sevastyanov. Crews for the second mission were L. V. Vorobev/V. A. Yazdovskiy, P. I. Klimuk/Yu. A. Ponomarev, and V. V. Kovalenok/V. I. Sevastyanov.

54.   Lazarev had actually replaced a new cosmonaut candidate V. A. Degtyarev, who resigned on the same day that Lazarev joined. See V. Semenov, I. Marinin, and S. Shamsutdinov, *Iz istorii kosmonavtiki: vypusk I: nabory v otryady kosmonavtov i astronavtov* (Moscow: AO Videokosmos, 1995), pp. 10, 12; Rex Hall, "Soviet Air Force Cosmonauts," in Michael Cassutt, ed., *Who's Who in Space: The International Space Year Edition* (New York: Macmillan, 1992), pp. 245–46.

who had worked on the development of the Vostok, Voskhod, and Soyuz spacecraft. He joined the cosmonaut team as part of the first civilian engineer intake on May 23, 1966. Later, Makarov had actively trained as one of the primary contenders for the first L1 circumlunar and L3 lunar landing missions, before finally moving to train for the *Kontakt* project in April 1970. When that effort was also canceled in September 1971, he began training for DOS space station flights.[55]

The launch of 7K-T Soyuz spacecraft no. 36 took place at 1518 hours Moscow Time on September 27, 1973. On board were Commander Lazarev and Flight Engineer Makarov. The spaceship, openly named *Soyuz 12*, entered an initial orbit of 193 by 248.6 kilometers at a 51.61-degree inclination. Within seven hours of launch, the cosmonauts fired the Soyuz main engine to alter their orbital parameters to 326 by 345 kilometers, similar to the apogees of the earlier Kosmos-496 and Kosmos-573, both automated precursors to the Soyuz station "ferry version." The crew seemed to have been simulating the first portion of a rendezvous profile with an imaginary station. Perhaps to preclude rumors of a failed mission, the Soviet press announced publicly during the first day of flight that the Soyuz 12 mission would last only two days, sufficient to test its capabilities as a crew transport ship to the DOS.[56]

Events were evidently normal during the first day of flight. Few scientific experiments were included in the program. The most prominent one announced was the use of the nine-objective LKSA multispectral camera developed by Moscow State University. Makarov took Earth resources photographs using the hand-held camera, while Lazarev simultaneously took photographs of the same targets using a standard camera. Other experimenters in airplanes took photographs of the same areas to compare distortions introduced by the atmosphere. Small biological payloads were apparently carried aboard Soyuz 12, although the Soviet press did not release any details. Contact with the ground was maintained by the ship *Akademik Sergey Korolev*, stationed in the Atlantic, and by a Molniya-1 communications satellite.[57]

One of the primary goals of the flight was to test the Sokol-K1 pressure suits. At some point during the mission, Lazarev and Makarov depressurized part of their ship to test these suits. On the second day, however, there were "serious defects" in the life support system, followed by a failure in the ship's attitude control system.[58] Soon afterwards, the cosmonauts wrapped up their activities and successfully returned to Earth wearing their new suits, landing at 1434 hours Moscow Time on September 29, after a one-day, twenty-three-hour, fifteen-minute, and thirty-two-second flight. There was a curious postscript to the flight. Both cosmonauts had candidly and bluntly written about the problems during the mission in their on-board journals. When the State Commission examined their comments, officials reportedly tried to "muffle" their complaints, calling the flight a closed subject. For a time, the cosmonauts were unsure whether their reports would affect their future careers, but soon both were assigned to another flight.[59]

The Soyuz 12 mission may not have been an unequivocal success, but the flight did serve to instill some confidence in the space program. It was the first Soviet piloted mission in more than three years that had fully achieved its objectives. The flight was followed in quick succession by two more launches of the 7K-T ship before the end of the year. The first of these was the flight of vehicle no. 34L to simulate a full two-month stay in orbit. Launched at 0820 hours Moscow

55. Semenov, Marinin, and Shamsutdinov, *Iz istorii kosmonavtiki: vypusk I*, p. 13; Voevodin correspondence, January 30, 1997.

56. Peter Smolders, *Soviets in Space* (New York: Taplinger Publishing Co., 1973), p. 250.

57. *Ibid.*; Gatland, *Manned Spacecraft*, p. 237; Christian Lardier, *L'Astronautique Soviétique* (Paris: Armand Colin, 1992), p. 191; *Soviet Space Programs: 1976–80*, p. 518.

58. Rebrov, *Kosmicheskiye katastrofy*, p. 73; Lantratov, "20 Years From the Flight of 'Soyuz-12'."

59. Rebrov, *Kosmicheskiye katastrofy*, pp. 73–74.

Time on November 30, 1973, the spacecraft was disguised under the designation *Kosmos-613*. Initial orbital parameters were 195 by 295 kilometers at a 51.6-degree inclination. Few details have been released on the flight. Over a period of six days, the spacecraft maneuvered into a "working orbit," similar to ones planned for future DOS missions, and then powered down, simulating conditions when such ferries would be docked to a space station. After an apparently successful sixty-day, nine-minute mission, Kosmos-613 returned to Earth successfully on January 29, 1974.

The final Soyuz flight in 1973 was a piloted mission, launched primarily to perform scientific experiments that had been delayed because of the repeated failures in the DOS program. The main payload on the Soyuz ship was the *Orion-2* astrophysical telescope designed by Dr. Grigor Gurzut, a Corresponding Member of the Armenian Academy of Sciences. The instrument, designed to observe stars in the ultraviolet band of the electromagnetic spectrum, was installed in place of the deleted large docking apparatus at the forward end of the spaceship. In addition, the living compartment of the vehicle was transformed from the normal living quarters into a dedicated scientific laboratory, and the spacecraft was equipped with solar panels. The mission itself was timed to coincide with Comet Kohoutek's approach to Earth in late 1973. Since July 1973, the primary crewmembers for the mission were cosmonauts Lt. Colonel Lev V. Vorobyev and Valeriy A. Yazdovskiy. The former, a forty-two-year-old Air Force pilot, had almost been victim to political intrigue in the 1960s. Having joined the cosmonaut corps on January 10, 1963, as one of a new batch of trainees who would fly to the Moon, Vorobyev immediately got into hot water when, in early 1964, he and another trainee, Eduard P. Kugno, publicly criticized the Communist Party. When asked to make a speech in front of a local Party meeting, Kugno had evidently told a senior Party official, "I will not speak to a Party of swindlers and sycophants!"[60] He was expelled from the cosmonaut team on April 16, 1964. Vorobyev survived the "purge" because he was already a member of the Communist Party. He eventually went on to train for the Almaz and Kontakt programs.

Civilian engineer Yazdovskiy, forty-three years old, played an important role in drawing up the experiments program for the Orion-2 mission. He joined TsKBEM during the Korolev era in 1957 and was a part of the teams that designed the Vostok, Voskhod, and Soyuz spacecraft. Like Vorobyev, this would be his first spaceflight, although he had served in backup capacities. Unfortunately for both, the two had an extremely difficult time getting along with each other. At one point during the training, they even refused to sit at the same table during a lunch break, preferring to sit on opposite sides of the lunch room. A month before the scheduled launch, cosmonaut overseer Lt. General Shatalov had no choice but to remove the two men from the flight and substitute the backup crew into the primary spot.[61]

The two new cosmonauts—thirty-one-year-old Major Petr I. Klimuk (commander) and thirty-one-year-old civilian Valentin V. Lebedev (flight engineer)—lifted off in 7K-T spacecraft no. 33 at 1455 hours Moscow Time on December 18, 1973. The vehicle, named *Soyuz 13* in the Soviet press, entered an initial orbit of 193.3 by 272.7 kilometers at a 51.6-degree inclination. Both cosmonauts, like the original primary crewmembers, were rookies. Klimuk, something of a child prodigy, was the first of his batch of cosmonauts, selected on October 28, 1965, to make a spaceflight. He trained for many years in the L1 and L3 lunar programs before his assignment to the current mission. Lebedev was a civilian engineer from TsKBEM who had joined the cosmonaut team on March 22, 1972, just over a year prior to the flight. It was one of the shortest times from selection to flight in the history of the Soviet space program. Both

60.   N. P. Kamanin, *Skrytiy kosmos: kniga vtoraya, 1964–1966gg* (Moscow: Infortekst IF, 1997), pp. 26, 40; Hall, "Soviet Air Force Cosmonauts," p. 244, 272; V. Molchanov, "Disgraced Cosmonaut" (English title), *Apogey* 2 (January 1993): 4. Kugno was officially dismissed on June 17, 1964.

61.   Hall, "Soviet Air Force Cosmonauts," p. 316.

men had trained extensively at the Byurakan Observatory in Armenia where the Orion-2 telescope had been built. As soon as the two cosmonauts entered orbit, it marked the first time in the history of spaceflight that men from both the United States and the Soviet Union were in space at the same time. NASA astronauts were then in the middle of their marathon Skylab 4 mission. By Soyuz 13's fifth orbit, the cosmonauts had performed a series of orbital maneuvers, depositing their ship in a 225- by 272-kilometer orbit at 51.6 degrees.[62]

During the course of their immensely successful flight, Klimuk and Lebedev performed a wide range of scientific experiments in the fields of medicine, biology, Earth resources, astronomy, and navigation. Medical experiments included one called *Levkoy-3* to investigate the circulation of blood to the brain in microgravity. The main biological experiment centered around the use of the *Oazis-2* unit used for research into protein mass in space, which the cosmonauts activated on their second day in orbit. In the experiment, the waste products of one type of bacteria served as the initial material used by other bacteria to accumulate protein mass. During the Soyuz 13 mission, this regenerative process increased the biomass by thirty-five times, an encouraging sign for those attempting to design a closed-cycle life support system. Plants used in the experiment included chlorella and duckweed.

The Earth observation experiments included use of the RSS-2 spectrograph for photographing the day and twilight horizons. The cosmonauts also used a nine-lens camera with different color filters to expose three strips simultaneously to Earth's surface. Two of the films were sensitive to visible light and the third to infrared light. Navigational exercises consisted of activities in autonomous navigation to determine the accuracy of control systems. The primary goal of the mission was the use of the Orion-2 telescope. Unlike Orion-1 on the *Salyut* station, Orion-2 was mounted completely outside the spacecraft. The telescope was mounted on a three-axis stabilized platform with a pointing accuracy of two to three seconds of arc. The pointing was performed both by moving the ship and the telescope, using thirteen electric motors. The Orion-2 telescope complex also included an instrument for studying x-ray emissions from the Sun—the crew performed such experiments on the third day during the sixty-fifth orbit concurrent with Earth-based observations. During the Soyuz 13 mission, the crew took 10,000 spectrograms of more than 3,000 stars in the constellations of Taurus, Orion, Gemini, Auriga, and Perseus. All the spectrograms, using NASA-supplied film, were in the spectral classes of 2,000–3,000 Angstrom units, which cannot be studied from Earth.[63]

The two men successfully returned to Earth after a seven-day, twenty-hour, fifty-five-minute, and thirty-five-second mission, landing at 1150 hours Moscow Time on December 26, 1973. The flight was an unqualified success—an encouraging sign that Mishin and his engineers had bounced out of the dismal dregs of the past few years. In retrospect, the Soyuz 12 and Soyuz 13 missions came at a particularly important juncture in the history of the Soviet space program. For the first time in many years, consecutive piloted missions had instilled hope instead of despair. Clearly, both of these flights had modest objectives, but for years, the Soviets had difficulty in achieving even modest goals in space. After years of doubt, it also seemed that engineers had managed to eliminate all the bugs from the troubled Soyuz spacecraft. Finally, in what no one could guessed at the time, the mission was the very final piloted mission under Mishin's command. An era was about to end.

The Saga Continues . . . Barely

The third N1 failure, on June 27, 1971, occurred three days before the deaths of Soyuz 11 cosmonauts Dobrovolskiy, Volkov, and Patsayev. One can only imagine the spirits of those

---

62. *Soviet Space Programs: 1976–80*, p. 518.
63. *Ibid.*, pp. 520–21; Gatland, *Manned Spacecraft*, pp. 238–39.

engineers who had to peruse through all the debris and telemetry of the N1 accident in the late summer of 1971. The obligatory accident commission met several times throughout July and August to determine the cause of the explosion of booster no. 6L. By October 15, Academician Keldysh had signed the final conclusion of the commission on the causes of the accident.

During the launch, all the engines worked normally for the first time after ignition, but roll stabilization of the rocket was not nominal. The roll error gradually increased and was at fourteen degrees by T+14.5 seconds—that is, the rocket had turned fourteen degrees around its main axis despite the counteraction of vernier nozzles to correct the roll. In fact, by T+7.5 seconds (at an altitude of 250 meters), the verniers had hit their mechanical stops (at forty-five degrees), unable to turn anymore. Furthermore, at T+39 seconds, the gyro instruments of the N1 terminated operation, and for the remainder of the flight, the rocket was not stabilized along its axes. At T+47.8 seconds, the booster began to break up in the area between the third stage and the L3 payload. The latter separated from the main body of the rocket and fell not far from the launch pad, while the "beheaded" rocket continued to fly. Finally, at T+50.1 seconds, when the uncontrolled roll had reached 200 degrees, the KORD system switched all the first stage engines off as a result of an emergency command from the limit switches of the gyro instruments.[64] N1 Chief Designer Boris A. Dorofeyev later described why the roll error had occurred:

The 6L vehicle lost roll control due to a design error. The designers misjudged the air pressure signature in the bottom part of the rocket in flight. They also misjudged the influence of the pyrotechnical starter exhaust tubes, which were located asymmetrically on each of the 30 engines. The shape of the rocket's bottom and two rings of closely-installed engines created two zones of air depression behind the booster. The asymmetrical location of the exhaust tubes created a high-torque rotating force on the borders of those depression zones. The six control thrusters were unable to compensate for that force. That effect did not take place on the first two launches because not all the engines worked at that time. The non-working engines of the outer rings created "air gaps," wide enough to diminish the depression zones' effect.[65]

The third failure of the N1 evidently raised the possibility of terminating the rocket program completely. Yuriy A. Mozzhorin, the influential Director of the Central Scientific-Research Institute of Machine Building (TsNIIMash), recalled that there was a meeting of the Military-Industrial Commission on the issue after the third N1 failure. He explained later:

. . . when the question of shutting the project down was being decided, I came out against it. Why? By that time, we had acquired the experience, many of the engineering objectives had already been achieved, and we had the ability to expose the weak points. . . .[66]

Despite the third failure, confidence was, in fact, growing among the rocket's leading engineers that they were close to success. The next booster, no. 7L, would be a significantly improved model, while the following one, no. 8L, was an altogether different variety with completely new multifiring engines on the first three stages, as well as highly optimized systems.

64. R. Dolgopyatov, B. Dorofeyev, and S. Kryukov, "At the Readers' Request: The N1 Project" (English title), *Aviatsiya i kosmonavtika* no. 9 (September 1992): 34–37; Igor Afanasyev, "N1: Absolutely Secret" (English title), *Krylya rodiny* no. 11 (November 1993): 4–5.

65. Boris Arkadyevich Dorofeyev, "History of the Development of the N1-L3 Moon Program," presented at the 10th International Symposium on the History of Astronautics and Aeronautics, Moscow State University, Moscow, Russia, June 20–27, 1995.

66. Col. M. Rebrov, "A Portrait in the Light of *Glasnost*: A Call After Midnight" (English title), *Krasnaya zvezda*, December 1, 1990, p. 4.

The extensive changes on booster no. 7L were crucial for achieving orbit. Many of these alterations were performed not only to improve chances for a success, but also to increase the mass of the payload itself. Designers had improved the aerodynamic characteristics of the first stage by reducing the area of the bottom of the first stage by replacing part of the lower conical skirt with a cylindrical section, thus reducing the base diameter from 16.9 to 15.8 meters. They also introduced tapered fairings to replace the rounded ones, improved the N1's thermal protection characteristics, and optimized the thermal insulation of the propellant tanks. Flight control would be performed by an on-board computer from commands issued by a gyro-stabilized platform developed by Chief Designer Pilyugin's Scientific-Research Institute for Automation and Instrument Building. To improve roll control, engineers introduced four 11D121 vernier liquid-propellant rocket engines, developed under Deputy Chief Designer Mikhail V. Melnikov, to replace the six old exhaust nozzles on the first and second stages. The rocket would also have the *Freon* passive fire extinguisher system as well as new mechanical and thermal protection for instrumentation and the on-board cable system. Finally, the telemetry measurement systems had been modified, by the Experimental Design Bureau of the Moscow Power Institute under Chief Designer Bogomolov, with the use of miniature radio-telemetry gear. The new system made it possible to receive information from approximately 700 newly mounted sensors, making a total of 13,000 sensors on the booster.[67]

With respect to the problems of the main engines of the first three stages, one of the most irksome was the burn-throughs of the internal propellant lines, especially of the LOX lines, caused by the design choice of having the engines' components very close together to reduce tubing length. The N1 State Commission, having investigated the matter, concluded on January 1, 1972, that this problem had finally been eliminated.[68] The engines on 7L also had aerodynamic shields on their exterior to protect them from high-velocity air streams. Meanwhile, Kuznetsov's new engines, capable of being refired, and with very high-performance characteristics, underwent ground testing from 1971 through 1972. Engineers completed the interdepartmental tests of the NK-33 (first-stage) and the NK-43 (second-stage) engines in September 1972.[69] Mishin's original planning from the 1970 period was to use the new engines beginning with N1 booster no. 8L, contingent on a schedule in which ground testing of the new engines would finish in time for installation on booster no. 8L. Not surprisingly, there were delays in preparing for the next N1 launch; booster 7L's launch was set for the fourth quarter of 1972, by which time Kuznetsov's new engines were ready for flight. The natural question was: what point was there in launching the N1 with old engines when the new engines were ready? Senior designers in the program recalled later that:

> . . . certain ministry heads were of the opinion that [booster no. 7L with the older engines] should be mothballed. But such a decision would have led to a further delay in the creation of the launch vehicle of at least two and a half years. And while the new engines were being manufactured and stand tests of the sections were being performed, the launch of rocket no. 7 could be used to check out the dynamics of the flight control

67. Dolgopyatov, Dorofeyev, and Kryukov, "At the Readers' Request: The N1 Project"; Afanasyev, "N1: Absolutely Secret"; Semenov, ed., *Raketno-Kosmicheskaya Korporatsiya*, p. 258; I. A. Marinin and S. Kh. Shamsutdinov, "Soviet Programs for Piloted Flight to the Moon" (English title), *Zemlya i vselennaya* no. 5 (September–October 1993): 77–85; Alexander Yasinsky, "The N-1 Rocket Programme," *Spaceflight* 35 (July 1993): 228–29.

68. N. I. Panichkin, "Some Results of N1 Development with Multi-Engine Powerplants," presented at the 10th International Symposium on the History of Astronautics and Aeronautics.

69. Dolgopyatov, Dorofeyev, and Kryukov, "At the Readers' Request: The N1 Project." Tests for the third-stage engines (NK-39) were completed by November 1973.

with the new vernier engines and the essentially new control system, as well as check out many other designs. After a number of discussions, the State Commission decided to go through with the launch [of 7L using the old engines].[70]

In August 1971, the Keldysh Commission had effectively terminated the L3 program with the recommendation that further work on the Lunar Orbital Ship (LOK) and Lunar Ship (LK) cease in favor of more capable lunar spaceships. The official decision to close down further production of L3 components was apparently issued in a September 1972 governmental decree on the N1-L3 program.[71] At the time of that order, there were several fully built models of both the LOK and the LK at the TsKBEM plant. Some of these would fly the remaining N1-L3 launches, performing automated and piloted flights to the Moon. The payload for booster no. 7L was the first flightworthy model of the LOK, vehicle no. 6A, and a mock-up of the LK lander installed underneath the L3 payload fairing. Quite possibly, the total lifting capability of booster no. 7L was not sufficient to carry *both* a functional LOK and an LK.

The flight plan for booster 7L was signed on July 18, 1972, by Mishin and his three principal Deputy Chief Designers—Okhapkin, Chertok, and Tregub. The plan detailed a complete lunar-orbital mission for the LOK from launch to landing. The N1 was to lift off from site 110L with a 89,803-kilogram L3 payload consisting of the Blok G fourth stage, the Blok D fifth stage, the LK mock-up, the LOK, and associated fairing. The nominal orbit would be 200 by 740 kilometers at a 50.7-degree inclination. If all operations were within acceptable parameters, the L3 complex would circle Earth for a period of twenty-four hours, with translunar injection taking place by firing the Blok G stage on the sixteenth or seventeenth orbit. Once the Blok G tanks were empty, the stage would cease firing and ignite the Blok D stage for a period of forty-four seconds to impart sufficient escape velocity to the payload. There were contingencies to go for translunar injection on the eighteenth or nineteenth orbits if the earlier attempt failed. In case of a complete failure to escape Earth's orbit, the LOK would simply separate from the stack, carry out a thorough testing flight in Earth orbit, and splash down in the Indian Ocean.

The LOK-LK-Blok D complex would spend just over four days in transit to the Moon, during which the Blok D would fire twice for mid-course corrections—the first at eight to ten hours after launch and the second ten to twenty-four hours prior to achieving lunar orbit. For most of this period, the stack would be in a slow roll mode of a half degree per second, accelerating during one period to two degrees per second to ensure proper thermal equilibrium in Blok D. At T+98.5 hours, the stack would enter lunar orbit. The initial and transitional lunar orbits were selected to ensure the best conditions for surface photography during the mission of booster no. 7L. The initial orbit would be near circular at 175 plus or minus seven and a half kilometers, while the later orbit would be elliptical with a perilune of forty plus or minus five kilometers. Both orbits would have inclinations to the lunar equator of 180 degrees plus or minus two degrees. Corrections to the orbit were to take place on the fifth and twenty-seventh orbits.

During the LOK's time in lunar orbit, special cameras were to take detailed photographs of the selected landing sites on the fourteenth, seventeenth, thirty-fourth, and thirty-sixth orbits. The LOK would separate from the LK mock-up and Blok D after a command from Yevpatoriya subsequent to the completion of photography on the thirty-sixth orbit. The LOK's living compartment would also detach from the rest of the vehicle on the thirty-ninth orbit, followed three orbits later by a firing of the Blok I engine to impart sufficient velocity to send the spacecraft back to the direction of Earth. Total time in lunar orbit would be about 3.7 days.

---

70. *Ibid.*, p. 36. One reliable source suggests that N1 booster no. 7L used the old engines because the new engines were not ready for flight at the time. See Afanasyev, "N1: Absolutely Secret."

71. V. Pappo-Korystin, V. Platonov, and V. Pashchenko, *Dneprovskiy raketno-kosmicheskiy tsentr* (Dnepropetrovsk: PO YuMZ/KBYu, 1994), p. 82; Semenov, ed., *Raketno-Kosmicheskaya Korporatsiya*, p. 574.

On the way back to Earth, the LOK would carry out two mid-course corrections—the first about a day after leaving lunar orbit and the second about six hours prior to approach into Earth's atmosphere. About eight minutes prior to entry into the atmosphere, the ship would separate into its remaining components, the descent apparatus and the instrument-aggregate compartment. The landing of the descent apparatus would be in the Indian Ocean after a flight from a northwesterly direction.[72]

Throughout 1972, as this mission was being prepared, there were the occasional leaks in the Western press suggesting that the Soviets had resumed their piloted lunar landing program. One of the most precise predictions came from Charles S. Sheldon II, an analyst at the Library of Congress who distinguished himself by being one of the few Western observers who continued to strongly believe that the Soviet Union was still planning piloted lunar expeditions. Without knowing the details of the L3M program, Sheldon accurately exclaimed, "When they get that big booster back in shape, the Soviets will go to the Moon." He summarized his beliefs, confirmed twenty years later by Russian disclosures, by saying, "The Soviets are simply waiting to play one-upmanship with us when we have nothing going on in manned spaceflight."[73] Rumors of the next N1 launch also filtered through. In September 1972, U.S. reconnaissance satellites evidently witnessed the N1 being taken *back* to the assembly-testing building at Tyura-Tam, thus spurring reports that no launch was imminent.[74] In fact, activity at Tyura-Tam was significantly accelerated in the waning months of 1972, primarily related to the fourth N1 launch attempt.

For the first time during an N1 launch, Chief Designer Mishin, in the hospital because of illness, was not present to direct technical operations. He assigned Deputy Chief Designer Chertok to serve as "technical director" of the State Commission.[75] Minister Afanasyev, who served as the chair of the commission, was apparently unsure of whether to risk a completely flight-ready LOK on an N1 equipped with the old engines. In a last minute appeal to N1 Chief Designer Dorofeyev on launch day, he proposed replacing the expensive LOK with a mock-up. In the final analysis, Dorofeyev convinced the Afanasyev that it would be advantageous to have a real "live" ship on the rocket.[76]

The fourth N1 lifted off at 0911 hours, 55 seconds Moscow Time on November 23, 1972. To observers, the flight seemed to be completely successful. Telemetry indicated that the engines were operating normally, and all parameters appeared normal. Passing the seventy-second mark, it was already flying longer than any of its predecessors. The six core engines shut down automatically at T+90 seconds, apparently without problems. It was only at T+104 seconds that the first sign of trouble appeared, but within the rapid seconds passing by, there was literally no chance to react. Within three seconds, a powerful explosion in the tail section of the first stage destroyed the lower portion of the spherical oxidizer tank. The booster exploded and broke up into pieces in the air. There had been just seven seconds left before first-stage shutdown and second-stage firing. This time, the difference between success and failure was measured in seconds. The emergency rescue system activated on cue and saved the LOK descent apparatus from virtual destruction.[77]

---

72.   E-mail correspondence, Vladimir Agapov by the author, September 30, 1996.

73.   Jack Hartsfield, "Russians Seen Landing Crew on Moon in 1975," *Huntsville Times*, April 2, 1972, pp. 17–18.

74.   Richard D. Lyons, "Moon Shot Delay by Soviet Likely," *New York Times*, October 1, 1972, p. 51.

75.   Tarasov, "Missions in Dreams and Reality."

76.   V. F. Gladkiy, "The Last Launch of the N1 Rocket" (English title), *Aviatsiya i kosmonavtika: vchera, segodnya, zavtra* no. 8 (August 1997): 28–31.

77.   Afanasyev, "N1: Absolutely Secret"; Semenov, ed., *Raketno-Kosmicheskaya Korporatsiya*, p. 258.

The investigation into the 7L failure, like the ones for the previous N I accidents, was long and arduous. The process, however, differed in one substantive way from the previous times: this time the investigation was bogged down in inter–design bureau rivalries and politics. At the initial hearing of the State Commission to discuss the accident, Chertok reported that preliminary data indicated that one of the engines on the periphery of the first stage had shut off spuriously before the destruction of the tail compartment. But engine Chief Designer Kuznetsov was reluctant to agree, believing that if the fault of the accident was placed on the shoulders of his design bureau, then Minister of Aviation Industries Petr V. Dementyev would shut down his entire operation—a threat that Dementyev had in fact hinted before the launch. In his defense, Kuznetsov argued that the N I had been destroyed because of design vibrations in the frame of the rocket as a result of the scheduled shutdown of the six central engines just before the explosion.

Afanasyev drew up a compromise solution in which the suspect engine had been destroyed because of the unexpected influence of oscillations in the rocket. Parties on both sides, however, refused to accept this version.[78] Kuznetsov eventually sharpened his version of the causes of the accident, suggesting that the failure had occurred *as a result of* an explosion in a pipeline leading to an engine—that is, not in the engine itself, but in the armature of the rocket. The engineers who wanted to exonerate the N I rocket gathered a formidable array of supporters to back their cause, including researchers from the Scientific-Research Institute of Thermal Processes (the former NII-1) and the Scientific-Research Institute of Measurement Technology (NII IT). Their combined investigation of sensor readings from the N I showed that a shock wave had passed through the booster's body *as a result* of the engine explosion. Kuznetsov argued back that the sensor readings were incorrect, but NII IT Director Oleg N. Shishkin persuasively showed through further investigation that all sensor readings were in fact completely reliable. Given the evidence up to this point, the State Commission accepted a provisional version that the accident had occurred because of a failure in the suspect engine and that Kuznetsov's assumption on depressurization of the oxygen pipeline before the explosion was not supported by sensor measurements.

The situation was complicated because TsKBEM Chief Designer Mishin had very good personal relations with Kuznetsov. The former was clearly put in a difficult position; most of his subordinates were opposed to Kuznetsov's argument that the blame lay in the rocket rather than the engines. On Kuznetsov's personal request, Mishin agreed to have the matter investigated by the N I Council of Chief Designers—a body that did not include representatives from the dueling ministries. The central issue at hand was the reliability of the data from N I sensors. The council's findings were also not to Kuznetsov's liking, and he apparently scoured through their report trying to unsuccessfully find any fault in their logic. According to one witness, "he simply could not believe that [the engines] had blown up at the end of their resources. . . ."[79] Minister of Aviation Industries Dementyev, Kuznetsov's somewhat unsympathetic boss, then established an independent panel of aeronautics specialists to examine Kuznetsov's claim that the failure occurred as a result of a break in a 250-millimeter line that fed LOX to engine no. 4 on the first stage. The rupture, according to Kuznetsov, had been caused by "a water hammer" from the sudden cutoff of the six central engines of the N I, which turned off on schedule between eighty and ninety seconds after launch to reduce the g loads during injection and to save propellant. Dementyev's commission came to the same conclusion: that the engine cutoff had not led to the explosion.[80]

---

78. The principal individuals on the TsKBEM side were D. I. Kozlov (Deputy Chief Designer, TsKBEM), B. A. Dorofeyev (N I Chief Designer, TsKBEM), V. V. Simakin (Chief, N I Design Complex, TsKBEM), and A. S. Kirillov (Chief, Chief Directorate, MOM). See Gladkiy, "The Last Launch of the N I Rocket."

79. *Ibid.*, p. 29.

80. *Ibid.*; Afanasyev, "N I: Absolutely Secret"; Panichkin, "Some Results of N I Development."

Despite the compelling evidence and the rising opposition against the N1 engines, Kuznetsov refused to budge. Debates and arguments continued for some time over what Kuznetsov believed was inadequate dynamic testing of the N1 on the ground for precision loads, especially as compared to the Saturn V. Newly discovered sensor tapes near the impact site of the accident promised to throw the investigation into a lurch, but the new data only confirmed that Kuznetsov was wrong. In the final analysis of the fourth N1 launch, the State Commission stuck to the evidence of the "anti-engine faction," noting that there were other opinions. In its report, the commission stated that the flight had gone normally until T+106.93 seconds. Analysis of the probable causes of the failure indicated that:

- The damage to the aft compartment of the first stage because of a failure in engine no. 4 caused the explosion.
- The hypothesis that the engine failure occurred because of internal causes [that is, the engine] did not contradict the telemetry data from engine no. 4 and from the stand tests, the findings of an inspection of the physical materials, or the physical pattern of the development of the failure of the rocket.
- The hypothesis of the depressurization of the main lines feeding propellant to the main engines and the vernier engines before the beginning of the failure [Kuznetsov's version] was not confirmed by the telemetry data.[81]

As the fingers all pointed to Kuznetsov, questions were rising all over the place on not only the old engines used on the N1 boosters so far, but also the newly improved engines his design bureau had been developing for two to three years. The issue had important long-term consequences precisely because of the tenuous connection between the old and new engines. If Kuznetsov was unable to build engines for the N1 after a ten-year research program, what guarantee was there that he would succeed with his new versions? Mishin himself recalled:

The difficulties encountered during the modification of those [liquid-propellant rocket engines], which were accompanied by repeated failures to meet delivery deadlines, generated in a certain circle of people (primarily, leaders such as D. F. Ustinov, L. V. Smirnov, S. A. Afanasyev) the opinion that N. D. Kuznetsov, given the existing attitude of the leadership of the Ministry of Aviation Industry toward the work, would not be able to bring the engines up to the specified level of reliability any time soon, and consequently, there would be neither an N1 launch vehicle nor its modified versions.[82]

Perhaps to compensate for what many believed were Kuznetsov's shortcomings, the Soviet space leadership sanctioned parallel efforts in two other design bureaus in 1973 to develop substitute engines for the N1.

One of these two was a surprise participant in the N1 program: Chief Designer Valentin P. Glushko's Design Bureau of Power Machine Building (KB EnergoMash). More than ten years after the conflict with Korolev over the N1, which permanently fractured the Soviet space program, Glushko was finally ready to swallow his pride and join forces in the N1 program. He created a special team at his design bureau to investigate various ways to increase the reliability of the N1 rocket. One of these approaches was to outfit the first and second stages of the booster with engines that already had been repeatedly tested in flight, specifically altered versions of the RD-253 engines from the Proton rocket. Research, however, showed that an N1 equipped with such engines would lose significant lifting capacity because of the use of noncryogenic

81. Dolgopyatov, Dorofeyev, and Kryukov, "At the Readers' Request: The N1 Project."
82. Mishin, "Why Didn't We Fly to the Moon?"

propellants and also would cost the rocket in terms of reliability because the N I would have to have a huge number of such engines on the first stage. A second option was to use a new and much more powerful engine. Since about 1968, Glushko had been talking of a 1,000-ton-thrust engine for a superheavy-lift launch vehicle. The idea eventually evolved by early 1970 into a 600-ton engine using kerosene-LOX, the same propellants that Glushko had opposed using for the N I in the early 1960s. With the clouded future of the Kuznetsov engines, Glushko also directed a team under Sergey P. Agafanov at his design bureau to study a 5,000-ton-thrust engine with an annular combustion chamber and a nozzle of external expansion, with a central body that could be used on the first stage of the N I. Needless to say, the prospect of developing such a massively powerful engine was not very encouraging.[83] The most realistic conception was a more modest 500-ton-thrust four-chamber engine, also using the kerosene-LOX combination.

Another organization, the Design Bureau of Chemical Automation (KB KhimAvtomatiki), the old Kosberg bureau led by Chief Designer Aleksandr D. Konopatov, also looked into substitute engines. They proposed a 250-ton-thrust motor working on LOX and kerosene, which would be developed on the basis of an old storable propellant engine developed many years ago for Chelomey's abandoned UR-700 rocket.

Despite the rising doubts about Kuznetsov's engines for the N I, Mishin's design bureau worked on two new N I boosters, 8L and 9L, "under a new technical task."[84] Both of these rockets would be equipped with the new Kuznetsov engines on its first three stages. Just in time, ground static testing of the third-stage engine, the NK-31, had finished in November 1973, thus qualifying engines for all three stages. In preparing booster no. 8L for launch, engineers took account of all the results of the prior four N I launches, painstakingly making sure that such failures would not occur again. Booster no. 8L was significantly heavier than its predecessors, partly because of new oscillation dampers installed in propellant lines to preclude the type of depressurization suspected by Kuznetsov. The new rockets were also the first equipped with filters at the inlets to the oxidizer pumps of the engines, the absence of which had caused the catastrophic July 1969 failure. Other changes included an improved fire extinguisher system and a faster acting version of the KORD engine control system. There was also talk of installing a system to separate the first and second stages in case the former was damaged; if there had been such a system at the time of the fourth failure, the malfunctioning first stage could have separated from the rest of the booster, whose upper stages would have compensated for the loss of seven seconds of first-stage firing.[85]

By early 1974, engineers had assembled booster no. 8L, allowing workers to begin installing Kuznetsov's new NK-33, NK-43, NK-31, and NK-41 engines on the rocket. The payload for the rocket was the first complete L3 complex, consisting of working versions of the LOK, the LK, and Blok D. The complex would enter lunar orbit, perform complex maneuvers, and then return to Earth without accomplishing a landing.[86] Launch was scheduled for August 1974. Subsequently, booster no. 9L would fly before the end of the year. Confidence was at a high in early 1974. As some participants later recalled:

> The people from the plants, Design Bureaus, and enterprises that had taken part in the development were preparing the rocket for flight with their former enthusiasm, because they had reason to believe that the launch would produce a positive result.[87]

83.   Afanasyev, "N I: Absolutely Secret."
84.   Tarasov, "Missions in Dreams and Reality."
85.   Afanasyev, "Unknown Spacecraft"; Afanasyev, "N I: Absolutely Secret."
86.   Afanasyev, "N I: Absolutely Secret." In another article, the same author states that the flight program of booster no. 8L would have been a complete L3 mission, including *landing*. See Afanasyev, "Unknown Spacecraft."
87.   Dolgopyatov, Dorofeyev, and Kryukov, "At the Readers' Request: The N I Project," p. 29.

By all accounts, the N1 designers strongly believed that their faith in the rocket would be vindicated after so many years—that this last flight in 1974 would be the final test launch of the giant rocket, allowing the State Commission to declare the vehicle operational. Four additional boosters—10L, 11L, 12L, and 13L—were in various states of assembly at the time, in the queue for launches in 1974 through 1976. Even the most pessimistic forecasts suggested that the N1 would be flying regular operational missions by 1976.[88]

Curtains

Early 1974 was a particularly important time for TsKBEM, precisely because it seemed, for the first time in a long time, that the unending setbacks of the previous three or four years were over. Chief Designer Mishin was presiding over six major new programs, all focused on piloted space exploration, which promised significant dividends in the late 1970s and early 1980s. Three programs involved the development of new variants of the Soyuz spacecraft, the most important of which was the 7K-S. This spacecraft had been in development since 1968, originally as a ferry vehicle to a long-abandoned military space station, but it had eventually emerged as a new generation of Soviet piloted spacecraft. In August 1972, Mishin had signed a supplement to the original draft plan for the 7K-S, which allowed engineers to proceed with the manufacture of the test and flight models. Although the spacecraft was externally almost identical to the older Soyuz, it was a completely new ship inside, with every essential system replaced by a new or modernized substitute.[89] By May 1974, engineers had already built eight models of the 7K-S, one of which was almost ready for launch, although Mishin noted later that "the work was greatly slowed down by delayed deliveries by suppliers."[90] In later years, this model was called the *Soyuz T.*

There were two other Soyuz variants in the works at the time, the first of which was the 7K-TM, built specifically for the Apollo-Soyuz Test Project. In the fall of 1972, engineers began work on this variant; Mishin signed the final draft plan on December 15, 1972.[91] The variant had common systems with the new 7K-S, but it was designed particularly with the short time-frame of the joint project in mind; the most important addition was the new androgynous docking system developed jointly by the two sides. By mid-1974, six of these ships were ready for flight. The first one, vehicle no. 71, was launched on April 3, 1974, as *Kosmos-638.* TsKBEM introduced a new variant of the emergency rescue system for the 11A511 launch vehicle. The ten-day flight was successful, although it performed an unplanned ballistic, instead of a guided, return to Earth.[92]

A third variant of the Soyuz was the 7K-TG—a spaceship designed to serve as a cargo ship to future space stations—that is, to bring propellant, food, and other supplies to crews staying on DOS ships in Earth orbit. It was a revolutionary idea for the Soviet space program and one of the most fundamental components of the USSR's ultimate goal of a permanent presence in space. Engineers began work on the tanker, later called *Progress,* in mid-1973 and issued the draft plan in February 1974.[93]

---

88. Afanasyev, "N1: Absolutely Secret"; Semenov, ed., *Raketno-Kosmicheskaya Korporatsiya,* p. 258; Sergey Leskov, "How We Didn't Get to the Moon" (English title), *Izvestiya,* August 18, 1989, p. 3. In one source, Mishin states there were *seven* total N1 boosters in various states of readiness in 1974, suggesting that the additional boosters were 10L, 11L, 12L, 13L, and 14L. See Mishin, "Why Didn't We Fly to the Moon?"

89. Semenov, ed., *Raketno-Kosmicheskaya Korporatsiya,* p. 211.

90. Yu. A. Mozzhorin, *et al.,* eds., *Dorogi v kosmos: I* (Moscow: MAI, 1992), p. 125; Tarasov, "Missions in Dreams and Reality."

91. Semenov, ed., *Raketno-Kosmicheskaya Korporatsiya,* p. 198.

92. *Ibid.,* p. 202; Tarasov, "Missions in Dreams and Reality"; Mozzhorin, *et al.,* eds., *Dorogi v kosmos: I,* p. 125; A. Yasinskiy, "Getting Away From the Fire" (English title), *Apogey* 4 (1993): 1, 3.

93. Semenov, ed., *Raketno-Kosmicheskaya Korporatsiya,* p. 334.

Mishin had also significantly advanced work within the DOS program. Station no. 124, the fourth in the series, was almost ready for launch, being in "a state of 20-day readiness for launching" in late April 1974.[94] Station no. 125, the first third-generation station with two docking ports, was already in the process of assembly at the Khrunichev Plant in Moscow.[95] These two stations were later launched as *Salyut 4* and *Salyut 6*, in 1974 and 1977, respectively. There was also significant work on the Multirole Orbital Complex (MOK) during the 1972–74 period. As per the agreement in 1972, Mishin expected to fully focus on the MOK after the flight of DOS no. 125.

The status of lunar programs remained in flux. Although the L3 program had been effectively terminated in late 1972, Mishin would fly out the remaining available hardware on several N1 launches. Boosters 8L and 9L would carry out automated missions to the Moon. If those two were successful, the first Soviet piloted landing on the Moon would be on booster 10L or 11L. The subsequent five or six boosters would carry out further piloted landings or launches of the components of the MOK. Depending on the success of the early missions, designers planned to eliminate the use of the backup LK and the Ye-8 rover to support the later piloted landings.[96] The fate of the advanced L3M program is less clear. The available evidence suggests that after the closure of the Apollo program (after December 1972), the Soviet space leadership lost interest in the Moon. As one respected Russian historian noted at the time: "Money for the N1-L3M variation was not allotted."[97] As with many other programs of the period, however, it seems that Mishin doggedly carried on work on the L3M proposal without the benefit of an official Communist Party or government decree on the matter. According to his forecast in 1973–74, a successful L3M landing on the Moon could be achieved in 1978–80 "with only a small increase in spending in 1975–1976" above what was already allocated for the N1 program.[98]

If progress on these programs were to Mishin's credit, his record as TsKBEM Chief Designer during the previous eight years was nothing to brag about. It was during his tenure that two of the worst accidents in space history occurred—the Soyuz 1 and Soyuz 11 fatalities, which killed four Soviet cosmonauts. There were also the docking failures in Soyuz 2/3, Soyuz 7/8, and Soyuz 10, the repeated failures in the L1 and DOS programs, and finally—most glaringly—the incredible catastrophes and delays in the N1 rocket project. One could argue that Mishin was possessed of nine lives to have even survived this spate of failures; any other man would have been fired long ago. Some claim that he was protected in his position because of Andrey P. Kirilenko, the powerful Politburo member, whose son-in-law, Yuriy P. Semenov, was a chief designer at Mishin's design bureau.[99] Mishin was also not the easiest man with whom to get along, continually alienating his subordinates and associates with his abrasive behavior. In addition, he apparently had an unhealthy affinity for alcohol. But Mishin's vehement critics— and there are many—forget that he did not play a personal role in each and every failure that beset the design bureau in the late 1960s and early 1970s. His deputies—particularly Bushuyev, Chertok, Okhapkin, and Tregub—were responsible for managing many of the key programs during this period. And ultimately, Mishin had the poor luck of the draw. Handed too little money, too little time, and too many demands, possibly any other manager would have had the same results.

---

94. Mozzhorin, *et al.*, eds., *Dorogi v kosmos: I*, p. 125.
95. Tarasov, "Missions in Dreams and Reality."
96. K. Lantratov, "Anniversaries: The 'Deceased' Lunar Plan" (English title), *Novosti kosmonavtiki* 14 (July 2–15, 1994): 60–61.
97. I. Afanasyev, "The 'Lunar Theme' After N1-L3" (English title), *Aviatsiya i kosmonavtika* no. 2 (February 1993): 42–44.
98. Mishin, "Why Didn't We Fly to the Moon?"
99. Sagdeev, *The Making of a Soviet Scientist*, p. 180.

During the worst series of failures, in February 1973, the Ministry of General Machine Building issued a devastatingly censorious document on the TsKBEM's activities, which was partly a direct criticism on Mishin's performance as its leader:

> . . . in the past years the effectiveness of the work at the enterprise has noticeably dropped. . . . Deficiencies exist at the enterprise in questions ensuring high quality and reliability of the apparatus created which have been repeatedly discussed in the Ministry Collegium (this has been reflected in a whole series of orders) which TsKBEM has been eliminating slowly . . . on the question of the internal organization of TsKBEM, there are yet more existing deficiencies which have negatively manifested themselves in the work of the enterprise. . . .[100]

There was also dissension growing *within* the design bureau. In 1973, three of Mishin's most powerful deputies, Bushuyev, Chertok, and Kozlov, along with former OKB-1 Deputy Chief Designer Kryukov and TsKBEM Department Chief Feoktistov, drew up and signed a letter, with the preliminary agreement of Central Committee Secretary Ustinov, to the Central Committee and the Council of Ministers pointing out the unsatisfactory work of Mishin as the Chief and Chief Designer of TsKBEM. They finished their letter with a request to dismiss Mishin from his post.[101]

The names of the signatories to the letter were not surprising. Bushuyev, Chertok, and Feoktistov had been vehement supporters of the DOS program, and all, especially Feoktistov, were increasingly lukewarm to continuing the trouble-plagued lunar program. Kozlov had had a falling-out with Mishin over the military 7K-VI program in the late 1960s and subsequently had an increasingly difficult time getting along with him. Kryukov had evidently had a spat with Mishin in 1966 soon after Korolev's death over an unknown matter, after which Mishin had demoted him from the post of deputy chief designer to department chief. Kryukov, like Bushuyev, Chertok, and Feoktistov, had also authored the important proposal in late 1969 to propose the DOS program in the first place.[102] The fracture clearly developed over the DOS program. By all accounts, Mishin believed that the N1-L3 lunar program was his life's work. As one journalist recalled it "to be his duty in Korolev's memory, as perhaps the most important accomplishment of his life."[103] His deputies, Bushuyev and Chertok, were perhaps a little more pragmatic, believing that it was time to admit failure and move on to more manageable projects—that is, the DOS program. They had also clearly felt betrayed by Mishin's 1972 agreement with Chelomey in which the former promised to transfer the small space station program to the latter after the flight of DOS-5.

The N1 versus DOS debate split the design bureau in half. Mishin did have support within TsKBEM. Okhapkin, Dorofeyev, Shabarov, and others—deputies who were responsible for the N1-L3 program—apparently stood behind the besieged chief designer. Mishin also had the support of DOS Chief Designer Semenov, no doubt because the latter owed his career to Mishin. Both Semenov and local Party Secretary Anatoliy P. Tishkin evidently came out against the letter that called for Mishin's dismissal. In the official history of the design bureau, the

---

100. Semenov, ed., *Raketno-Kosmicheskaya Korporatsiya*, p. 161.

101. *Ibid.*

102. Bushuyev, Chertok, Feoktistov, and Kryukov were four of the six men who proposed the DOS program in late 1969. See *ibid.*, p. 264. For the Kozlov-Mishin falling-out over the 7K-VI, see K. Lantratov, "Dmitriy Kozlov's 'Zvezda'" (English title), *Novosti kosmonavtiki* 6 (March 10–23, 1997): 74–80. For the Kryukov-Mishin falling-out in 1966, see Mozzhorin, *et al.*, eds., *Dorogi v kosmos: I*, p. 101.

103. Mikhail Rebrov, "The Last Argument: A Study of the Designer in Black and White" (English title), *Krasnaya zvezda*, March 25, 1995, p. 6.

authors claim that Mishin managed to neutralize the effects of the damaging letter by coming to an agreement with space program head Ustinov. The latter was on visit in early 1973 to TsKBEM to mediate this growing conflict. Mishin was not informed of this sudden visit, perhaps to allow Ustinov free reign to discuss the matter with the "anti-Mishin" contingent. Upon finding out that Ustinov was at the premises of his design bureau, Mishin rushed to meet his boss and found Ustinov inspecting the DOS-3 model. The story goes that DOS Chief Designer Semenov mentioned in passing that it would be useful to have two docking ports on a future DOS vehicle. Ustinov liked the idea. In a subsequent conversation with Mishin, Ustinov, in a conciliatory mood, offered Mishin an implicit deal: if the chief designer would agree to have two docking ports on a future DOS vehicle, then Mishin could keep his job. The official historians add: "Thus V. P. Mishin found the possibility of continuing his work and at the same time was compelled to support the idea of a new station."[104]

As with other tales of Soviet space history, it is difficult to discern the exact details of this story. The account clearly hinges on the idea that Mishin was in some way opposed to having two docking ports on a DOS spacecraft. Completely contradictory evidence comes from Mishin himself. In an interview in 1989, he clearly states that he wanted to have the *first* DOS with two docking ports but was overruled by Ustinov "in order to hasten our success."[105] Notes from Mishin's own office records of 1970–71 clearly attest to the serious considerations given to a station with two docking ports, as well as Mishin's own enthusiasm for such a station.[106] Despite these two irreconcilable accounts, one thing is clear: the 1973 letter calling for Mishin's dismissal was a key factor in the growing opposition against Mishin.

The trajectory of Mishin's career was, of course, undeniably intertwined with that of the N1-L3 program—an effort that was also under increasing attack at the time. Given the rising lack of confidence in Kuznetsov's engines, there were murmurs of discontent asking whether the program as a whole should be continued. As one historian noted:

> The creators of the N1 were being "called onto the carpet" more and more, and they had to prove their correctness each time. The rhythm of the work was disrupted owing to the confusion, and rumors were circulating in the corridors of the "firms" of the supposedly imminent "shutdown" of the N1.[107]

At one meeting on December 8, 1973, Central Committee Secretary Ustinov bluntly asked whether it was still worth it to "ride the horse" any longer. One unnamed chief designer argued that it was time to terminate the program. When it was TsNIIMash Director Mozzhorin's turn, he made a case for continuing with the N1, but abandoning the lunar landing project:

> To repeat what the Americans have done—this is to openly admit to the world our lag behind them. But as far as our N1 carrier, what will canceling it do for the situation? After all, satellites are getting heavier each year; in time, such a carrier will nevertheless be needed! To throw away the N1 at the halfway point—then the development of a new rocket of such lifting power will take a long time and vast resources . . . work on the N1 must be continued![108]

104. Semenov, ed., *Raketno-Kosmicheskaya Korporatsiya*, p. 294.
105. Tarasov, "Missions in Dreams and Reality."
106. Interview, Peter Gorin by the author, November 18, 1997.
107. Afanasyev, "The 'Lunar Theme' After N1-L3," p. 43.
108. Mikhail Rudenko, "Four Steps From the Moon" (English title), *Moskovskaya pravda*, July 19, 1994, p. 10. It is possible that this meeting may have occurred on December 8, 1972. The year is not given.

As influential as Mozzhorin may have been, among the upper echelons of the Soviet space industry, his word could not compare to more powerful players.

Perhaps aware that the fate of the N1 was on shaky ground, Mishin continued to appeal to both the Ministry of Defense and Communist Party officials that the continuing work on the booster would be invaluable for ultimately building the MOK, which would have both military and civilian mission goals. To get a firm word on the matter, Mishin, in cooperation with N1 engine Chief Designer Kuznetsov, prepared a detailed memorandum for Soviet General Secretary Brezhnev on the MOK and on the general lag of the Soviet Union in the exploration of space. They proposed and argued various measures that would allow the USSR to move ahead of the United States. Mishin was not unaware that Kuznetsov was under fire at this time for his poor contributions to the N1 program. In an attached section on the causes of the fourth N1 failure, Mishin agreed to share the blame for the accident with Kuznetsov, hoping this would put Kuznetsov in a favorable light to Brezhnev. Mishin's closest aides thus put together a report on the entire N1 program, the reasons for each failure, and the measures adopted to preclude future accidents. As far as the critical fourth failure in 1972, they noted—contrary to the official State Commission conclusion—that oscillations in the hull of the rocket caused by the switch-off of the central engines, accompanied by additional loads acting on pipelines— and the fact that the engines and their instrumentation were at the end of their resources— caused the subsequent explosion. Therefore, it was a compromise variant of the accident report. The two designers emphasized that in the succeeding launches, the level of vibrations would be decreased by throttling down the thrust levels of the central engines prior to cutoff.[109]

In late March 1974, Mishin and Kuznetsov sent their memorandum to Brezhnev with a request to accept their proposals on the MOK and the N1. Brezhnev handed the report over to Ustinov to evaluate the proposal, and Ustinov turned it over to the defense ministries to handle the matter. Parties within the Ministry of Aviation Industry were of the opinion that the two chief designers' conclusions on the N1—that is, reducing the thrust of the engines prior to engine cutoff on future N1s—were completely unfounded, because without sufficient dynamic testing, it would be almost impossible to predict the outcome of such a profile. Thus, given the chance for failure, it would be foolhardy at best to give authorization to launch further N1s based on their recommendation. Ustinov eventually invited a number of prominent chief designers to discuss the Mishin and Kuznetsov proposal. Glushko, having waited for more than a decade to air his personal vendetta against the N1, did not hold back his words. He argued that new engines or not, the N1 was doomed for failure because of the great number of engines in the first stage. Instead, Glushko proposed a new family of launch vehicles with very high-thrust engines.

In essence, Mishin made a fatal mistake by compromising his position and accepting Kuznetsov's views on the reason for the fourth accident. It was the last nail in the coffin. The pace of events in April and May 1974 was breathtaking. The maneuvering behind the scenes was done in absolute secrecy, with few people really aware of the wheeling and dealing. Perhaps as few as half a dozen people at TsKBEM were cognizant of the impending changes. One of Mishin's senior deputies, Yevgeniy V. Shabarov, an old-timer from the Korolev days, recalled later:

> . . . absolutely unexpectedly for us one day in 1974 we received an invitation, well not even an invitation, but an order to assemble all the Deputy Chiefs of the [design bureau] in the office of the Chief Designer. We gathered in complete ignorance. There we sat and waited. Suddenly the door opened, and [Minister of General Machine Building] Sergey Aleksandrovich Afanasyev entered, accompanied by Valentin Petrovich Glushko and a number of other employees from the Ministry. "Good afternoon, comrades," [Afanasyev]

---

109. Gladkiy, "The Last Launch of the N1 Rocket."

said. . . . [He] announced that, "The Politburo has taken a decision—Vasiliy Pavlovich Mishin has been relieved of his post as Chief Designer of your organization, and Valentin Petrovich Glushko has been named the General Designer. Your organization from now on will be known as the 'Energiya' Scientific-Production Association. I wish you all success." With that he left. All this happened so unexpectedly and quickly (in the course of two–three minutes) that we were stunned and did not really understand what had occurred.[110]

What happened was certainly the largest reorganization within the Soviet space industry since Korolev's death. On May 22, 1974, Mishin, at the time ill in the hospital, was officially released from his duties as TsKBEM Chief Designer. On the same day, his former design bureau, (TsKBEM), with all its affiliates, was combined with another powerful space organization (Glushko's KB EnergoMash) to form the new Energiya Scientific-Production Association (NPO Energiya). It was evidently Glushko who had personally thought of the "Energiya" name. The sixty-five-year-old Glushko was named the new Director and General Designer of this new and gargantuan empire, which included:

- The former TsKBEM, renamed the Lead Design Bureau (GKB) at Kaliningrad
- The former TsKBEM branch at Kuybyshev
- The Experimental Machine Building Plant at Kaliningrad
- KB EnergoMash at Khimki
- KB EnergoMash's Primorsk Branch
- KB EnergoMash's Kamskiy Branch
- KB EnergoMash's Privolzhsk Branch
- The EnergoMash Experimental Plant[111]

Thus, Glushko would supervise the development of almost all Soviet piloted spacecraft, launch vehicles, automated reconnaissance satellites, and high-thrust rocket engines *and* oversee their manufacture and testing. It was more power than Korolev held in his heyday. Being ill at the time, Mishin was out of the loop throughout this period. As Mishin told a journalist many years later:

To be frank with you, the decision to fire me came to me as a complete surprise. . . . [After leaving the hospital] I was invited for a talk to the Staraya Square [the residence of the Central Committee of the Communist Party], and Ustinov, the Central Committee Secretary in charge of space affairs, told me, "Leonid Ilyich [Brezhnev] asked me to convey his thanks for your work, and provide help in finding other employment."[112]

Presumably, Mishin would have been demoted to a senior position in the design bureau, but Glushko would have none of that. When Mishin left the hospital, Glushko revoked Mishin's clearance pass to enter the design bureau.[113] The new general designer wanted to make sure that Mishin never stepped into his old haunting grounds again.

The natural question is: why Glushko? How did Glushko manage to end up as head of the enterprise that was founded by one of his most famous opponents, Korolev? Glushko was

---

110. Mozzhorin, *et al.*, eds., *Dorogi v kosmos: I*, p. 183. The Politburo meeting to discuss the reorganization was held on May 14, 1974.

111. Semenov, ed., *Raketno-Kosmicheskaya Korporatsiya*, p. 288. Note that the TsKBEM branch at Kuybyshev separated from NPO Energiya on July 30, 1974, and became the independent Central Specialized Design Bureau (TsSKB).

112. Rebrov, "The Last Argument."

113. Tarasov, "Missions in Dreams and Reality."

clearly well placed and also ambitious. Since the birth of the missile program in the mid-1940s, he had always played second fiddle to Korolev. He was always the engine designer, while Korolev was the designer of the rocket or the spacecraft. His claim to become *the* chief designer had no support while Korolev was alive. But with the less-powerful Mishin, Glushko could take advantage of the former's failings, such as the repeated failures in the late 1960s and early 1970s in the piloted space program. By 1971, Glushko was clearly the most respected and influential chief designer in the business, as evidenced by the unprecedented declassification of his name. His organization had designed the engines for the first stages of almost all Soviet strategic ICBMs, including the R-7 (SS-6), the R-9A (SS-8), the R-16 (SS-7), the R-36 (SS-9), and the R-36M (SS-18). This does not include his design bureau's work on engines for a family of launch vehicles based on the R-7, as well as Yangel's Tsiklon and Kosmos series of boosters. Still, he remained only the engine designer. Presumably during the discussions in early 1974 over the N1, when he offered a replacement for the old rocket, he proposed uniting his rocket engine organization with that of Korolev's old spacecraft design bureau. From a managerial and institutional perspective, it seemed to make sense to unite these two powerful entities into one. Forces would be consolidated, and waste would be eliminated. Who was better to head the whole organization than Glushko, one of the pioneers of Soviet rocketry? On a more fundamental political level, Glushko had the support of two key individuals, Brezhnev and Ustinov. Their support was invaluable to his appointment.[114]

The N1-L3 project was the first victim of the May 1974 reorganization. The fate of the project was clearly decided at the highest levels of the Soviet Communist Party and government, but it was also a decision that stemmed from a confluence of forces that all intersected in mid-1974. Clearly one of the most important factors was the Mishin-Kuznetsov report sent to Brezhnev in March. The repercussions of this report spiraled out of control until it reached the offices of the primary client for the N1, USSR Minister of Defense Andrey A. Grechko. Given his generally negative attitude toward the N1 booster and its military uses, he was only too happy to side with those who were clamoring for some definitive action. On May 19, 1974, three days *before* Mishin's official dismissal, Grechko signed an order suspending further launches of the rocket.[115] The timing could not have been better. Glushko's first act as General Designer of NPO Energiya, signed on June 24, 1974, was to suspend all work on the N1-L3 program.[116] The suspension of work on the N1 meant that all programs associated with its development were also terminated. These included the L3M advanced lunar landing missions, the giant MOK in Earth orbit, and proposed conceptions of anti-ballistic space-based weaponry. The massive expansion of the Soviet space program, envisioned for the late 1970s by Mishin, all disappeared with a few signatures.

In the official history of NPO Energiya, the authors wrote that the decision was taken with the "tacit agreement" of Afanasyev, Keldysh, Smirnov, and Ustinov.[117] One person who may have been against this abrupt decision was Minister of General Machine Building Afanasyev, who, while not always supportive of Mishin, was a strong proponent of the N1 program. Some reliable sources claim that both Glushko's appointment and the cancellation of the N1-L3 program "was made by the Politburo behind Minister Afanasyev's back. . . . It was [Glushko's] initiative, not of his boss—Afanasyev."[118] In recent years, Mishin has been very candid about who

---

114. See Col. M. Rebrov, "Specific Impulse" (English title), *Krasnaya zvezda*, August 26, 1989, p. 4.

115. Gladkiy, "The Last Launch of the N1 Rocket."

116. Semenov, ed., *Raketno-Kosmicheskaya Korporatsiya*, p. 639.

117. *Ibid.*, p. 258.

118. Georgiy Stepanovich Vetrov, "Development of Heavy Launch Vehicles in the USSR," presented at the 10th International Symposium on the History of Astronautics and Aeronautics, Moscow State University, Moscow, Russia, June 20–27, 1995.

he believes were responsible for scuttling a program that had sucked in billions of rubles, but was so close to success:

> I think the main culprit was Dmitriy Fedorovich Ustinov. The reason for winding up the program—at least from his standpoint—was that the Americans had beaten us to it. This was a turning point in his career. Prior to this, he had not been a Politburo member, much less the Minister of Defense. He reached these positions after winding up the [N1] program. Afanasyev could not have cared less. All these failures were affecting his career. So he did not oppose winding up the N1 program.[119]

Both Ustinov and Afanasyev kept their jobs. However, Mishin was not the only one whose job came under fire. Maj. General Kerim A. Kerimov, the Chief of the Third Chief Directorate at the Ministry of General Machine Building, was apparently demoted as part of the N1 cancellation shakeup. He continued to serve as chair of the State Commission for Soyuz, but he would no longer oversee the Korolev design bureau within the ministry.[120] Others who fell under the blade included several leading engineers responsible for the N1-L3. Once Glushko came into power, he sidelined some of the senior personnel involved in the N1 project. N1 Chief Designer Dorofeyev was "forcibly dismissed," while Mishin's First Deputy Okhapkin, who had guided the program since 1962, was demoted to an innocuous position.[121] The men who inherited senior positions at NPO Energiya were, for the most part, those individuals who had little involvement in the N1-L3 effort during the past few years.[122]

The termination of the N1-L3 program was a complete surprise to most people at NPO Energiya, and it sent shock waves throughout the entire space industry. Engineers, confident beyond hope that success in the program was within reach, were simply stunned at the irony of cancellation at the cusp of victory. Especially galling was the fact that "not a single session of a scientific council, not a single conference of specialists, not a single meeting of the Council of Chief Designers" was convened prior to taking the final decision—it was all decided behind closed doors among less than half a dozen individuals. As one journalist wrote: "It was far less dangerous to transfer the responsibility onto other shoulders and to declare the N1 a mistake."[123] Perhaps the biggest victims were the engineers; without any intention of hyperbole, one observer noted:

---

119. *What Stars Are We Flying to?* (English title), Moscow Teleradiokompaniya Ostankino Television, First Program Network, Moscow, April 9, 1992, 0825 GMT.

120. Leonard Nikishin, "Inside the Moon Race," *Moscow News* 7 (April 11, 1990): 15. Kerimov's new post was First Deputy Director of TsNIIMash.

121. For Dorofeyev, see S. Kryukov, "The Brilliance and Eclipse of the Lunar Program" (English title), *Nauka i zhizn* no. 4 (April 1994): 81–85. Dorofeyev evidently did not leave NPO Energiya. By December 1977, he was Chief of Complex 10 at NPO Energiya, responsible for ground testing, which was definitely a demotion from his post in 1972–74. Okhapkin's fate is less clear. His official biography states that he was a deputy chief designer until 1976. See V. P. Glushko, ed., *Kosmonautika entsiklopediya* (Moscow: Sovetskaya entsiklopediya, 1985), p. 286. However, complete lists of deputy chief designers at NPO Energiya from 1974 to 1977 do not include his name. See Semenov, ed., *Raketno-Kosmicheskaya Korporatsiya*, pp. 288–93.

122. The new structure of NPO Energiya was approved on June 28, 1974. Glushko had two First Deputy General Designers. One of them was Yu. N. Trufanov, appointed to his post on July 16, 1974. Trufanov was an odd choice for the position, because he had come from Chelomey's TsKBM Fili Branch. The other was V. P. Radovskiy, who had served under Glushko for a long time at KB EnergoMash. Glushko had five Chief Designers under him—K. D. Bushuyev, Ya. P. Kolyako, I. S. Prudnikov, I. N. Sadovskiy, and Yu. P. Semenov—responsible for particular thematic areas in the organization. There were also seven Deputy General Designers: A. P. Abramov, B. Ye. Chertok, M. S. Khomyakov, A. A. Rzhanov, Ye. V. Shabarov, V. V. Simakin, and A. S. Yeliseyev. See Semenov, ed., *Raketno-Kosmicheskaya Korporatsiya*, pp. 288–90.

123. Leskov, "How We Didn't Get to the Moon."

> *As for such a "detail" as the honorable work of thousands of people who had devoted their best years to the N1, this was not even considered, these people did not even receive any explanation, let alone consultation . . . and many of them, I am convinced, received such a psychological blow that they have been unable to create anything of equal worth. And these were Korolev's best cadres.*[124]

Another participant remembered how Brezhnev and Ustinov compensated for their actions:

> *On the eve of those sorrowful events, many people who had taken part in the work in the lunar project . . . were presented with decorations. I admit that at that time I did not really understand why. It later became clear: we were decorated as a consolation and so that we would hold our tongues.*[125]

Unable to comprehend the rationality of such a seemingly uninformed decision, many unusual reasons filtered through the grapevine. Perhaps the most compelling one was that Soviet space officials were simply afraid that the N1 would succeed on its next launch. As one engineer working on the program recalled: "A successful launch of no. 8 . . . would require new investments that would be both considerable and immediate."[126] Military-Industrial Commission Chairman Smirnov seemed to confirm this claim, when, in 1991, he admitted that the general consensus, even among the upper leadership, was that the next launch would have been a success.[127]

When he took control of the giant Energiya organization, Glushko did not come empty-handed. He had promised Ustinov that he could do better than the N1, and in one sense, he did not disappoint. During his first days as general designer, he invited the technical leadership of the organization and presented his vision of the future of Soviet space exploration: a new family of superheavy-lift launch vehicles, ultimately leading to the establishment of large-scale permanent bases on the surface of the Moon. While most attendees viewed the lunar base idea with "great skepticism," it seems that Glushko had Ustinov's support, at least at the proposal level.[128] Why, after canceling Mishin's L3M and Long-Duration Lunar Base, Ustinov would support Glushko's "new" ideas might mystify even the most cursory observer of Soviet space history. Many within NPO Energiya were against the idea, correctly noting that the proposal was completely absurd after the N1 debacle. By October 1974, Glushko's engineers worked up a formal technical proposal for a lunar base, called *Zvezda*, which was examined by an independent expert commission of scientists and engineers headed by USSR Academy of Sciences President Keldysh. Looking at the costs, the technical complexity, and the timeframes proposed, the commission unanimously rejected Zvezda. In desperation, Glushko tried to get signatures from leading Soviet scientists on the viability of his proposal. But even Brezhnev, when told that this project would cost "only" 100 billion rubles, sobered up and declined to approve it. Zvezda died soon after.[129]

---

124. *Ibid.*

125. Vad. Pikul, "The History of Technology: How We Conceded the Moon: A Look by One of the Participants of the N1 Drama at the Reasons Behind It" (English title), *Izobretatel i ratsionalizator* no. 8 (August 1990): 20–21.

126. *Ibid.*

127. V. L. Menshikov, *Baykonur: moya vol i lyubov* (Moscow: MEGUS, 1994), p. 199.

128. Semenov, ed., *Raketno-Kosmicheskaya Korporatsiya*, p. 288. The family of new launch vehicles were informally known as Groza, Grom, and Vulkan. Their ground-to-Earth orbit payload capabilities were: RLA-120 Groza (thirty to thirty-five tons), RLA-135 Grom (100 tons), and RLA-150 Vulkan (170–250 tons). Groza and Grom, respectively, became Energiya-M and Energiya.

129. Sagdeev, *The Making of a Soviet Scientist*, pp. 182–84; Afanasyev, "The 'Lunar Theme' After N1-L3"; Afanasyev, "Unknown Spacecraft"; German Nazarov, "You Cannot Paper Space With Rubles: How to Save Billions" (English title), *Molodaya gvardiya* no. 4 (April 1990): 192–207.

If Zvezda proved to be too much for the Soviet space leadership, there was more interest in Glushko's new family of superheavy-lift launch vehicles. The military had at last found a use for such powerful boosters. Since 1972, the United States had embarked on the development of the reusable Space Shuttle. Believing the Space Shuttle to be a military threat to the Soviet Union, officials in the USSR Ministry of Defense found little interest in lunar bases or giant space stations. What they wanted was a parallel deterrent to the Shuttle. The story of exactly why the Soviets believed the Space Shuttle was such a threat has, like many others, assumed mythological proportions, with the truth probably buried forever in secret archives. The most commonly propagated story, disseminated even by the most respected historians in Russia, has an air of a folk tale:

> Leonid Smirnov, former [Military-Industrial Commission] Chairman . . . in his regular report to Brezhnev on the state of our space efforts once mentioned . . . that the Americans are intensively working on a winged space vehicle. Such a vehicle is like an aircraft; it is capable through a side maneuver of changing its orbit in such a way that it could find itself at the right moment over Moscow—possibly with a dangerous cargo. The news disturbed Leonid Ilyich [Brezhnev] very much—he contemplated it intensively, and then said, "We are not country bumpkins here. Let us make an effort and find the money."[130]

Several different organizations offered their services to develop a counterpart to the American Shuttle. In the initial stages, none of them resembled the U.S. spacecraft in the slightest. Glushko proposed a radically new design for a Soviet counterpart, the Reusable Vertical-Landing Transport Craft (MTKVP), a wingless system based on his new superheavy launcher proposal. Chelomey offered up the twenty-ton Light Space Aircraft (LKS)—an advanced reusable spaceplane concept to be launched on the Proton rocket. The MiG design bureau's old "space branch" in Dubna, after years of fruitful work on such concepts, offered up its old Spiral spaceplane. In February 1976, the chief of the space branch, Yuriy D. Blokhin, visited the Central Committee to persuade top Party leaders that the Spiral would be the most cost effective and efficient response to the American Space Shuttle, citing NASA's work on such experimental aircraft as the X-24. It was all in vain. Brezhnev, Smirnov, and particularly Keldysh were unwilling to budge on their requirement for a system identical to the NASA Space Shuttle, despite overwhelming opposition from most senior chief designers in the Soviet space program.[131] In 1993, Efraim Akim, a scientist at Keldysh's Institute of Applied Mathematics, elaborated on the precise rationale behind the "parallel response":

> When the U.S. Shuttle was announced we started investigating the logic of that approach. Very early our calculations showed that the cost figures being used by NASA were unrealistic. It would be better to use a series of expendable launch vehicles. Then, when we learned of the decision to build a shuttle launch facility at Vandenberg [Air Force Base] for military purposes we noted that the trajectories from Vandenberg allowed an overflight of the main centers of the USSR on the first orbit. So our hypothesis was

130. B. Olesyuk, "The 'Buran' Blind Alley" (English title), *Kurantiy*, December 21, 1991, p. 8. See also Yaroslav Golovanov, "Just Where Are We Flying to?" (English title), *Izvestiya*, December 12, 1991, pp. 1, 3.

131. Afanasyev, "Unknown Spacecraft"; Mikhail Rudenko, "'Star Wars'—History of the 'Death' of a Unique Spaceplane" (English title), *Trud*, August 26, 1993, p. 6; Anatoliy Kirpil and Olga Okara, "Designer of Space Planes." Vladimir Chelomey Dreamed of Creating a Space Fleet of Rocket Planes" (English title), *Nezavisimaya gazeta*, July 5, 1994, p. 6; Vyecheslav Kazmin, "The 'Quiet' Tragedy of EPOS" (English title), *Krylya rodiny* no. 1 (January 1991): 4–5. Note that by this time, the "space branch" was no longer under the MiG design bureau's jurisdiction. It had become a part of DPKO Raduga on June 19, 1972.

that the development of the shuttle was mainly for military purposes. Because of our suspicion and distrust we decided to replicate the shuttle without a full understanding of its mission. When we analyzed the trajectories from Vandenberg we saw it was possible for any military payload to reenter from orbit in three and a half minutes to the main centers of the USSR, a much shorter time than [a submarine-launched ballistic missile] could make possible (ten minutes from off the coast). You might feel that this is ridiculous but you must understand how our leadership, provided with that information, would react.[132]

Despite almost no interest from the Ministry of Defense, Keldysh managed to bulldoze the Soviet space shuttle idea was bulldozed through the Communist Party and government. On February 17, 1976, the Central Committee and the Council of Ministers issued a formal decree, which approved the creation of a reusable space system consisting of:

- A launcher stage
- An orbital aircraft
- An interorbital tug-ship
- A complex control system
- A launch-landing and assembly-work complex

The orbital aircraft would ensure delivery of up to thirty tons of payload to a 200-kilometer-altitude orbit, and it would be capable of returning twenty tons back to Earth.[133] Glushko's NPO Energiya would serve as the primary contractor for the entire system. The decree committed the Soviet Union to certainly the most expensive space project in the country's history—one that would almost bankrupt the space program. Chasing after the U.S. Space Shuttle over the following twelve years, it would work on a new launcher, the 11K25, later called *Energiya*, and a new reusable space shuttle, the 11F25, later called *Buran*.

To build the new shuttle, Glushko evidently did not want to work with organizations such as the Mikoyan or Chelomey design bureaus, which had decades of experience in developing hypersonic reusable vehicles. Instead, he subcontracted the development of the Buran shuttle to a new organization, the old Molniya Scientific-Production Association (NPO Molniya), created specifically for this task on February 24, 1976. NPO Molniya was established in Tushino near Moscow on the basis of the old Molniya Design Bureau (the former OKB-4) led by Chief Designer Matus R. Bisnovat—an entity that had hitherto zero experience in designing such spacecraft. Bisnovat's specialty had, in fact, been developing air-to-air missiles for Soviet fighters. NPO Molniya also included the Burevestnik Design Bureau (the former KB-82) led by A. V. Potopalov, which had specialized in the design of surface-to-air missiles and the manufacture of Sukhoy's advanced T-4 supersonic bomber. The third component was the Experimental Machine Building Plant (the former KB-90) led by Chief Designer Vladimir M. Myasishchev, who had been pushed out of space design work many years previously by Chelomey.[134] As a single act of concession to earlier spaceplane research, Ustinov appointed Spiral program chief

132. James Harford, *Korolev: How One Man Masterminded the Soviet Drive to Beat America to the Moon* (New York: John Wiley & Sons, 1997), p. 314.

133. Semenov, ed., *Raketno-Kosmicheskaya Korporatsiya*, p. 362.

134. Stepan Mikoyan, "'Molniya': From 'Spiral' to MAKS" (English title), *Vestnik vozdushnogo flot* 1 (1997): 60; G. P. Svishchev, ed., *Aviatsiya entsiklopediya* (Moscow: Bolshaya Rossiyskaya entsiklopediya, 1994), p. 372; E-mail correspondence, Mark Hillyer to the author, March 29, 1998. Additional manufacturing for Buran would be carried out at the Tushino Machine Building Plant (the former Plant No. 82) under Director S. G. Arutyunov. Note that both KB Molniya and KB Burevestnik were also located on the premises of the Tushino Machine Building Plant (TMZ).

Lozino-Lozinskiy to be the Director and Chief Designer of NPO Molniya, transferred from his old duties at the MiG design bureau. Despite Lozino-Lozinskiy's undeniable expertise, NPO Molniya seems to have been ill-equipped to handle such a monumental task as building a copy of the American Shuttle. One Soviet historian wrote:

> And can we manage to explain why the building of such a unique design as our first space plane was assigned to NPO Molniya and the Tushino Machine Building Plant (TMZ)? I'm not trying to insult those renowned, talented collectives, but everyone knows that NPO Molniya came about in the consolidation of two small design offices, Molniya and Burevestnik, which not only never had anything to with brainstorming about a space-plane, but also had no experience in developing ordinary airplanes from start to finish.[135]

Ignoring the decades of spaceplane research by Tsybin, Tupolev, Myasishchev, Mikoyan, and Chelomey, institutional discord and bad judgment once again set the Soviet space program on a poorly managed endeavor. Thus, Chelomey's Light Space Aircraft died an ignominious death by 1981, while Spiral puttered on until September 1978. Despite some extraordinarily successful subsonic drop tests in 1977 and 1978 from a Tu-95K bomber, the space branch at Dubna was eventually shut down. In December 1981, forty-eight senior engineers from the Spiral design bureau were ordered to pack up and join NPO Molniya to help with the creation of Buran.[136]

Fittingly, the same decree approving work on the 1K11K25 system (as the complete Energiya-Buran system was called at the time) also conclusively terminated all work on the N1-L3 program. The official reasoning was "the necessity to commence large-scale activities (involving allocation of huge sums of money) on the [Energiya-Buran] project" and more ironically "the absence of heavy payloads suitable for the lifting capacity of the launcher."[137] Amazingly, this was the same decree that approved the 11K25 superheavy-lift launch vehicle! In one of the multiple ironies of the time, Glushko elected to develop cryogenic propellant engines for the 11K25, despite having literally cracked the Soviet space program in half during the early 1960s by refusing to build engines with those propellants. Given the use of LOX-kerosene engines, there was some talk of using the Kuznetsov's new N1 engines for the job. At his own risk, Kuznetsov had continued his test certification program for the new engines, which continued as late as January 1977. His results were impressive: in running forty different NK-33 first-stage engines for test regimes of 1,200 seconds, they ran an average of 7,000–14,000 seconds without failures. One engine fired for a sum total of 20,360 seconds during repeated testing. To pass the certification process, they needed to run for only 600 seconds. In addition, he had boosted the thrust of the original NK-33 engines from 154 tons to 205–207 tons through the minor reworking of the turbopump assembly, moving the engines into a completely different class of thrust. Glushko naturally felt threatened by all this. In 1977, as his power increased to unprecedented levels, he forced a formal decision from the Council of Ministers to terminate all work done on powerful liquid-propellant rocket engines at not only Kuznetsov's design bureau, but also any place under the Ministry of Aviation Industry. Kuznetsov was also forced to hand over some of his test equipment to Glushko.[138]

---

135. Yaroslav Golovanov, "Just Where Are We Flying to?," p. 1.
136. Kazmin, "The 'Quiet' Tragedy of EPOS"; Lardier, L'Astronautique Soviétique, p. 254.
137. For the former quote, see S. Shamsutdinov, "First Flight of Buran With Tourists on Board Will Take Place on April 12, 1994" (English title), Novosti kosmonavtiki 21 (October 9–22, 1993): 40–45. For the latter quote, see Afanasyev, "N1: Absolutely Secret."
138. Afanasyev, "N1: Absolutely Secret."

The total cost of the N1-L3 program up to January 1, 1973, was 3.6 billion rubles, of which 2.4 billion rubles was specifically for the N1.[139] By rough estimates, total expenditures by the mid-1970s may have been as high as 4.0–4.5 billion rubles.[140] It is difficult to convert this figure to a dollar value, but a rough estimate, in 1960s dollars, would be about $12–13.5 billion— that is, about half of that spent on the Apollo program. But there was a human cost, too, and many, having received such a crushing blow, were reluctant to let the dream go. In a desperate gambit that ultimately met with little success, former Chief Designer Mishin lobbied hard to obtain permission to launch two of the fully prepared N1 rockets into the Pacific Ocean. In 1976, N1 Chief Designer Dorofeyev wrote letters to members of the 25th Congress of the Communist Party for the test launches. In November 1976, Mishin and Chertok sent a proposal to the Ministry of General Machine Building to convert the N1 to launch the new reusable space shuttle for the Ministry of Defense.[141]

None of it worked. Glushko was dead set against it; he was not simply satisfied with consigning the N1 program to history, but he also wanted to erase it from history. He ordered all the remaining N1 rockets—the two fully prepared for launch and five others—to be destroyed. All associated technical documentation was also destroyed, thus squelching any possibility that the rocket would make a phoenix-like reappearance in the Soviet space program. Former OKB-1 Deputy Chief Designer Sergey S. Kryukov, one of the "fathers" of the N1, later wrote: "Glushko incinerated every notion of the N1 with a hot iron."[142] Glushko also made sure that there was no indication of the program's existence in the design bureau's private museum. The project would only exist in the memories of its participants. The dream that had begun with Sergey Pavlovich Korolev in Germany in 1945 ended with a few signatures in 1976. Russian journalist Yaroslav K. Golovanov, Korolev's most well-known biographer, perhaps wrote the most eloquent of epilogues on the life and death of the N1 project:

> The unfulfilled dream of Sergey Korolev, who died on the operating table—a dream that was decimated by Valentin Glushko, that was undefended by Vasiliy Mishin, and that took years of labor by Nikolay Kuznetsov—vanished in the gulf of ministerial paperwork and the flames of failed launches that turned billions of rubles into ashes.[143]

Mishin added:

> We felt a deep sense of sadness. It was a colossal project to which we dedicated our best years. I was young at the time. And it was the work of a great many people and it vanished overnight. The Americans had won. I was made the scapegoat.[144]

---

139. Dolgopyatov, Dorofeyev, and Kryukov, "At the Readers' Request: The N1 Project." Total planned cost, including that for sixteen flight models, was 4.97 billion rubles.

140. For 4 billion rubles, see Kryukov, "The Brilliance and Eclipse of the Lunar Program." For 4.5 billion rubles, see Leskov, "How We Didn't Get to the Moon."

141. Leskov, "How We Didn't Get to the Moon"; Panichkin, "Some Results of N1 Development"; Lardier, L'Astronautique Soviétique, p. 174.

142. Kryukov, "The Brilliance and Eclipse of the Lunar Program."

143. Golovanov, "Just Where Are We Flying to?," p. 1.

144. "The Russian Right Stuff: The Dark Side of the Moon," NOVA television show, #1808, WGBH-TV, Boston, February 27, 1991.

CHAPTER NINETEEN

TOMORROW
NEVER KNOWS

In the history of the Soviet space and missile programs, three singular events stand out as defining moments: the birth of the effort in 1946, the death of Korolev in 1966, and the end of the N1-L3 program in 1974. History, of course, does not separate itself into neat little segments of time, but it would be difficult to find a moment so cataclysmic in the U.S. space program as the Soviet events of 1974. In essence, the year divided the old with the new and a lack of vision with clarity. Completely unknown to the West until the late 1980s, the changes in 1974 were effectively a watershed moment that closed the door on Korolev's determined journey, begun in 1946. What happened after 1974 warrants particular attention, not only as a matter of historical interest, but because the nature of the Soviet piloted space program changed in ways that would have been difficult to foresee at the time of NPO Energiya's formation. Having trudged through failure after failure in the late 1960s and early 1970s—the Soviet Union finally made its arrival as a formidable space superpower in the late 1980s—a full two decades after its only competitor had done the same.

The Rise and Fall of a Space Power

Glushko's ascendance to power at the top of the pyramid coincided with a dramatic shift in fortunes for the Soviet piloted space program. All the failures and catastrophes of 1971 through 1973, especially in the space station effort, seem to have exorcised the demons of the Soviet space program. In 1975, NPO Energiya performed its first fully successful space station mission on *Salyut 4*, one of the two DOS vehicles readied under Mishin. The other one, launched in September 1977 as *Salyut 6*, would finally put the Soviet space program on the slow but persistent track to success. The station's mission was one of the finest success stories in the Soviet space program. In the four years after launch, it hosted sixteen crews, four of which set absolute endurance records for time in space, significantly exceeding the eighty-four-day record set by NASA's Skylab 4 crew during 1973–74. NPO Energiya also introduced two new spacecraft: the Progress, an automated tanker and supply ship, and the Soyuz T, an advanced version of the Soyuz. Ironically, both programs had been initiated by Mishin. It was not simply a matter of setting records but of remarkable maturity in operations. Engineers perfected the very first refueling operations in space, mastered the logistics of having two ships dock to the same station, directed complex repair spacewalks outside the station, managed real-time solutions to contingencies in space, and accumulated a wealth of ground-breaking information on the effects of microgravity on the human organism. The Soviets also extracted maximum political gain from the mission of *Salyut 6* by sending "guest-cosmonauts" from other socialist countries on "friendly" visits. There were no fatalities in the program. It was a stunning return to form, prompting many Western observers to conclude that the Soviets were "ahead" in space. During the same period, the United States accomplished only one piloted flight.

The string of successes in the space station program continued with the operation of *Salyut 7* during the 1982–86 period, culminating with the launch of *Mir* ("World") in February 1986. Crews began visits to *Mir* almost immediately after its launch. In September 1989, two cosmonauts, Viktorenko and Serebrov, began a historic run of ten years of continuous crewed operations. Through the dissolution of the Union of Soviet Socialist Republics in late 1991, *Mir* remained occupied. In 1994–95, Valeriy A. Polyakov, a doctor from the Institute for Biomedical Problems, set the world's endurance record for continuous time spent in space: 438 days.

What had been a closed and secret program began to open up during the early 1990s. *Mir* played a central role in cooperative agreements with Western nations. As part of an arrangement between the United States and the Russian Federation, NASA astronauts began visiting the *Mir* space station in 1995. Seven NASA astronauts, beginning with Norman E. Thagard, spent approximately two and a half years aboard the *Mir* space station between 1995 and 1998. Their quarters were living, breathing, orbiting artifacts of the amazing history of the Soviet space program. The main *Mir* hull is almost identical to the original DOS vehicle that was designed and launched as the first *Salyut* in 1971. The same triumvirate that had built the original *Salyut* created the newer station, but these organizations exist now with different names: RKK Energiya, the Salyut Design Bureau, and the Khrunichev State Space Scientific-Production Center.[1] The primary four *Mir* modules—*Kristall, Kvant-2, Spektr,* and *Priroda*—were all based on the design of the Transport-Supply Ship's main hull, itself part of Chelomey's conception of the Almaz space station complex proposed in the late 1960s. The launch vehicle for *Mir* and its modules is the Proton—a rocket originally known as the UR-500K, proposed by Chelomey as an ICBM in 1960. The delivery vehicles for the complex are the Soyuz TM and the Progress M spacecraft, both derived from Korolev's beloved 7K-OK Soyuz spacecraft, designed in the early 1960s.

Mir, with all its historical significance, was planned for deorbiting by the end of this century. By that time, there will be a more impressive sight in Earth orbit, the International Space Station, a cooperative project involving sixteen countries. As the primary participants, the United States and the Russian Federation will provide most of the materials for this largest ever joint program in the history of space exploration.[2] The first component of the station, the Zarya Functional Cargo Block, was launched in November 1998 on a Proton booster. The station will be supplied by various modifications of the Soyuz spacecraft. *Mir* operations will probably cease once activities on the International Space Station commence. That singular event will probably mean the end of an independent Russian piloted space program—the end of the journey that Yuriy Alekseyevich Gagarin began in 1961. It will be the beginning of a new and perhaps more exciting voyage.

The *Salyut* and *Mir* space station programs were the most publicized components of the Soviet space program in the 1980s, but they were not, in fact, the most important. The lion's share of the Soviet space budget during the 1980s was taken by the Energiya-Buran effort, the most expensive program in the history of the Soviet space program. After years of delays and cost overruns, NPO Energiya finally launched the first Energiya booster in May 1987. It was the first successful Soviet rocket comparable in power and performance characteristics to NASA's long-defunct Saturn V giant. It was also the first time that the Soviets fired a high-performance LOX-liquid hydrogen rocket engine in operational conditions. What little joy there may have been in such a test was tempered by history. All of the pleas by Korolev and Mishin during the 1960s to develop such engines had fallen on deaf years, leaving Soviet rocket capabilities far behind that

---

1. The Salyut Design Bureau (KB Salyut) is actually part of the Khrunichev State Space Scientific-Production Center (GKNPTs Khrunichev).

2. Paul Mann, "U.S., Russia Draft Historic Space Pact," *Aviation Week & Space Technology,* September 6, 1993, pp. 22–23.

of the United States. It finally took Glushko's change-of-heart about cryogenic propellants before Korolev's dream became a reality. The Energiya booster was fired a second time, in November 1988, when it launched the Soviet space shuttle Buran on a highly impressive fully automated orbital flight. After decades of trying to build a spaceplane, Buran turned out to be the only such Soviet vehicle that ever made it into orbit. It was only fitting that much of the success of Buran benefited from the intensive testing of the small-scale BOR spaceplanes in the 1980s—vehicles that were left over from the ambitious Spiral project from the 1960s.

Despite early expectations of a vigorously expanding Soviet space program, the inevitable disenchantment crept in. As the Soviet economy began to implode, an increasingly free press became the forum for rising criticism of the Energiya-Buran program. By 1993, the effort was in near shambles, with ground models of the Energiya and the Buran rotting away in various plants. In May 1993, the project's Council of Chief Designers requested a final decision from the Russian government.[3] The project was formally shelved after seventeen years and 14 billion rubles. For the second time, thousands of Soviet space engineers saw their handiwork disappear into rubble. Many of those who witnessed the demise of the Energiya-Buran project were the same ones who had watched in silence at the abrupt termination of the N1-L3 program. Both projects had their own complex *raison d'être* and their own reasons for fall from grace, but both had one thing in common: they never fulfilled their original promise. The two projects together span the entire period of the piloted space program of the former Soviet Union. For those looking at waste of technology, of knowledge, of money, and ultimately of people, during the postwar Communist era, they need look no further than the N1-L3 and the Energiya-Buran programs.

The End of a Generation

Some would say that Vladimir Nikolayev Chelomey had a career worthy of a great Russian tragedy. After the cancellation of the N1-L3 program, his star seemed to rise for a brief period. In June 1974, he was elected as one of the approximately 1,500 deputies of the Supreme Soviet, the USSR's rubber-stamp parliament. While the legislature had no independent power in the country, membership usually indicated national prominence. In fact, Western observers scouring through the lists of the Supreme Soviet, upon finding Chelomey's name, believed that he was the "new head" of the Soviet space program, a "job previously held by . . . Yangel."[4] For perhaps a couple of years, he may have also resurrected his ambitious UR-700M Mars landing project. He continued work on the Almaz military space station, two of which were launched between 1974 and 1976 as *Salyut 3* and *Salyut 5*, respectively. He was evidently planning for a major expansion of activities at his design empire, planning much larger versions of Almaz stations serviced by the new Transport-Supply Ship. He even returned to one of his lifelong dreams—the development of an orbital spaceplane.[5]

---

3.   *Utro*, Moscow Ostankino Television, First Channel and Orbita Networks, Moscow, May 25, 1993, 1845 GMT; S. Shamsutdinov, "First Flight of Buran With Tourists on Board Will Take Place on April 12, 1994" (English title), *Novosti kosmonavtiki* 21 (October 9–22, 1993): 40–45.

4.   Theodore Shabad, "Russians Indicate Rocket Specialist Heads Space Effort," *New York Times*, July 14, 1974, p. 6.

5.   For the "resumption" of the UR-700M program, see Christian Lardier, *L'Astronautique Soviétique* (Paris: Armand Colin, 1992), p. 252. For the Almaz program, see Vladimir Polyachenko, "The 'Pep' of Almaz" (English title), *Krylya rodiny* no. 4 (April 1992): 30–32; Olaf Przybilski, *Almaz: Das supergeheime militärische Orbitalstationsprogramm der UdSSR* (Dresden, Ger.: Institut für Luftfahrt, 1994). For advanced Almaz projects, see I. B. Afanasyev, "Unknown Spacecraft (From the History of the Soviet Space Program)" (English title), *Novoye v zhizni. Nauke, tekhnike: Seriya kosmonavtika, astronomiya* no. 12 (December 1991): 1–64. For the orbital spaceplane, see Anatoliy Kirpil and Olga Okara, "Designer of Space Planes. Vladimir Chelomey Dreamed of Creating a Space Fleet of Rocket Planes" (English title), *Nezavisimaya gazeta*, July 5, 1994, p. 6.

All of this simply proved too good to be true. In early 1976, one of Chelomey's chief sponsors, Minister of Defense Grechko, succumbed to a heart attack. Chelomey's opponents—primarily Glushko sponsors Ustinov and Kirilenko—reacted immediately. A few weeks after Grechko's death, they bestowed Glushko with an unprecedented honor that hitherto no designer in the space sector had ever held: membership in the Central Committee of the Communist Party. Glushko was officially "elected" at the 25th Communist Party Congress in 1976.[6] As one observer noted: "From this moment onward, Glushko concentrated in his hands not only the power of an enormous space empire, but also the political power of a commissar, capable of overwhelming anyone in the space establishment."[7] Glushko's first move was to deny Chelomey any role in the space station program. By 1978, to Chelomey's great alarm, the piloted portion of the Almaz program was terminated. Chelomey had no help from the Ministry of Defense, his usual supporters. They were of the opinion that piloted orbital platforms were less efficient for overhead reconnaissance than automated satellites.[8]

The news just got worse for Chelomey. In 1976, Ustinov, as the new Minister of Defense, took it upon himself to complete the job he had set out to do more than a decade before. Ustinov:

methodically started to strangle Chelomey. He annulled all the military contracts given to Chelomey's enterprise for space flights; he canceled even those that were scheduled in unmanned mode and originally requested by the military.[9]

Perhaps the biggest blow to the Chelomey empire came on June 30, 1981, when Ustinov and Kirilenko pushed through an order that severed Chelomey's important Fili Branch from the main organization and instead made it a branch of NPO Energiya.[10] Given that Chelomey had farmed almost all the key projects to this branch, then known as the Salyut Design Bureau (KB Salyut), he lost all his space- and missile-related projects in one fell swoop. Finally, on December 19, 1981, the Central Committee and the Council of Ministers issued a decree formally terminating not only all work on the Almaz program, but also forbidding Chelomey from any further involvement in the Soviet space program. The official reason for the decision was to "concentrate forces on the creation of the 'Buran' space system."[11]

Brought to his knees by Ustinov, Chelomey quietly continued to develop naval cruise missiles for the armed forces, which was the original profile of his organization in the 1950s.

---

6. Two other chief designers in the defense industry were also elected to the Central Committee in 1976: P. D. Grushin from MKB Fakel (air defense and anti-ballistic missiles) and V. F. Utkin from KB Yuzhnoye (ICBMs, spacecraft, and launch vehicles). Grushin had been the first chief designer in the defense industry accorded this honor, with his elections in 1966 and 1971. See Julian Cooper, "The Defense Industry and Civil-Military Relations," in Timothy J. Colton and Thane Gustafson, eds., *Soldiers and the Soviet State: Civil-Military Relations From Brezhnev to Gorbachev* (Princeton, NJ: Princeton University Press, 1990), p. 168.

7. Roald Z. Sagdeev, *The Making of a Soviet Scientist: My Adventures in Nuclear Fusion and Space From Stalin to Star Wars* (New York: John Wiley & Sons, 1993), p. 209.

8. Maxim V. Tarasenko, "The U.S. and Soviet Space Systems Developments as Driven by the Cold War Competition," presented at the 45th Congress of the International Astronautical Federation, IAA-94-IAA.2.2.622, Jerusalem, Israel, October 9–14, 1994.

9. Sagdeev, *The Making of a Soviet Scientist*, pp. 209–10.

10. Yu. P. Semenov, ed., *Raketno-Kosmicheskaya Korporatsiya "Energiya" imeni S. P. Koroleva* (Korolev: RKK Energiya, named after S. P. Korolev, 1996), p. 643; Nina Chugunova, "Chelomey's Cosmonauts: Why There Are No Crews From NPO Mashinostroyeniya in Outer Space" (English title), *Ogonek* 4–5 (January 1993): 24–29.

11. G. A. Yefremov, "Anniversary: V. N. Chelomey—80 Years" (English title), *Novosti kosmonavtiki* 12–13 (June 4–July 1, 1994): 68–70; S. A. Zhiltsov, ed., *Gosudarstvennyy kosmicheskiy nauchno-proizvodstvennyy tsentr imeni M. V. Khrunicheva* (Moscow: RUSSLIT, 1997), p. 100; Chugunova, "Chelomey's Cosmonauts."

By all accounts, he never really gave up on his dreams of an ambitious space program, proposing various strategic defense programs throughout the early 1980s. He was not kind to Korolev's memory. In an interview with a journalist late in his life, Chelomey was blunt:

> Well, what can I tell you about Korolev? Korolev was a man with a limited education. But he commanded a remarkable technical intuition and was enormously talented as an organizer. Yes. But he couldn't perform even a simple mathematical operation with integrals. He took the circumlunar [program] away from me and then he didn't do it himself. You call that talent?[12]

Ejected from the Soviet space program, Chelomey's will and his reach for success never diminished to his last days. In early December 1984, still in lively health, he was at his dacha, getting ready to go somewhere in his Mercedes. Leaving the car running, he walked out to open the garage door, but the car, still running, moved by itself and pinned his legs against the gate. He was admitted to the hospital with a simple fracture. While in the hospital, he learned that his nemesis Ustinov had suffered a massive heart attack, was paralyzed, and could not speak. Chelomey could be forgiven for believing that his fortunes were about to improve. On the third day in the hospital, on the early morning of December 8, he was speaking to his wife on the telephone when the conversation suddenly stopped. She desperately called the hospital staff, who rushed to his room to find him dead. Doctors suspected a sudden fatal stroke apparently caused by the broken leg. He was seventy years old at the time of his death. Legend has it that Ustinov was brought a piece of paper with a handwritten message stating "Chelomey just died." Ustinov read it and closed his eyes in satisfaction. The first name on the list of signatories of Chelomey's obituary was that of Ustinov.[13]

Today, Chelomey's former organization is called the Scientific-Production Association for Machine Building (NPO Mashinostroyeniya) and is still located at its old grounds at Reutov outside Moscow. Having relinquished hold of its Moscow Branch in 1981, it has little connection to the Russian space program. Its current General Designer, Gerbert A. Yefremov, who succeeded Chelomey, continues to focus mostly on naval cruise missiles. Its only major space-related project is a continuation of the Almaz program—a robotic remote-sensing platform for Earth resources surveying. Three such spacecraft were launched—in 1986, 1987, and 1991—but despite Yefremov's best efforts, funding for a fourth is on a shoestring budget. By September 1994, the organization was in a severe financial crisis, planning to lay off thousands of employees.[14] While the organization may have been in dire straits, Chelomey's legacy, in some ways, remains much more visible than even that of Korolev. Given that Chelomey had his Fili Branch produce most of his space work, the thriving nature of that branch has maintained Chelomey's long shadow across the current Russian space program. The Proton rocket, the *Mir* space station (derived as it was from the original Almaz design), and the *Mir* modules (such as Spektr and Priroda) all attest to a vision that has remained intact despite the best intentions of Ustinov or Glushko. If Chelomey were alive today, he might have some comfort in knowing that the first

12. Yaroslav Golovanov, *Korolev: fakty i mify* (Moscow: Nauka, 1994), p. 724.
13. *Ibid.*, p. 729; Valeriy Rodikov, "Academician Chelomey and His Times" (English title), in V. Shcherbakov, ed., *Zagadki zvezdnykh ostrovov: kniga pyataya* (Moscow: Molodaya gvardiya, 1989), pp. 35–36; "Vladimir N. Chelomei, Soviet Rocket Scientist," *New York Times*, December 15, 1984, p. 28; Christian Lardier, "Soviet Space Designers When They Were Secrets," presented at the 47th Congress of the International Astronautical Federation, IAA-96-IAA.2.2.09, Beijing, China, October 7–11, 1996.
14. I. Cherniy, "NPO Mashinostroyeniya Reduces Staff" (English title), *Novosti kosmonavtiki* 20 (September 24–October 7, 1994): 49–50; Gerbert Aleksandrovich Yefremov, "NPO Mashinostroyeniya Is Moving Into the High-Technology Market" (English title), *Vooruzheniye, politika, konversiya* 3(3) (1995): 31–37.

element of the International Space Station, the Zarya Functional Cargo Block, is based on the design of the service module of the Transport-Supply Ship—a program that he pushed into approval in 1970. Zarya was designed, built, and delivered to NASA by the Khrunichev State Space Scientific-Production Center, a conglomerate of Chelomey's former Fili Branch (now called the Salyut Design Bureau) and the Khrunichev Machine Building Plant, established by governmental order on June 7, 1993.[15]

Chelomey's nemesis Ustinov had a meteoric career. With the exception of Korolev, Ustinov may have been the single most important individual in the emergence of the Soviet space program. At the same time, he is probably also the most overlooked. Scarcely mentioned in the Western historiography of the Soviet space program, Ustinov was at the center of the vortex of events of the Soviet space effort from 1946 to 1984, close to a forty-year span of time. Even in Russia, there have been no biographies of the man, nor is their evidence to suggest that he left personal memoirs. Of course, Ustinov's importance was not limited to the space program. He directly oversaw the tremendous growth and arrival of the Soviet Union as a formidable military player in world politics. Between 1965 and 1976, Ustinov was the Secretary of the Central Committee for Space and Defense, but he did not achieve his lifelong dream of entering the ranks of the Politburo until April 1976 with his appointment as the first Soviet civilian to serve as the Minister of Defense. His tenure at the post was a time fraught with unprecedented tensions with the United States, particularly over the Soviet invasion of Afghanistan and the Americans' stationing of Pershing missiles in Western Europe. He was reportedly in ill health in the early 1980s and was fast becoming a victim of Communist Party politics.[16] Ustinov died on December 20, 1984, after a two-month illness at the age of seventy-six.[17] In one of the bitterest of all ironies, his death came just twelve days after Chelomey had passed away. In an indication of new times, his death was first announced by Mikhail S. Gorbachev, considered at the time a fast-rising personality in the upper of echelons of power. Not surprisingly, Westerners writing about his life's achievements at the time almost completely missed his contribution to the creation and sustenance of the Soviet space program.[18]

Among the other heavy hitters of the Soviet space program, Leonid V. Smirnov, the former Chairman of the Military-Industrial Commission, served in that position from March 1963 to December 1985, managing the development and creation of several new generations of Soviet weaponry. During the reshuffling after Ustinov's death, Smirnov retired—the last of the powerful defense industry juggernauts who built up the military might of the USSR. Although still alive at the time of this writing, the eighty-three-year-old Smirnov has remained completely out of the public eye. His personal reminiscences would no doubt be a priceless asset to understanding Soviet motives during the Cold War.

Sergey A. Afanasyev, the Minister of General Machine Building—that is, the "space and missile" ministry,—served in that capacity from March 1965. After the death in 1976 of his chief sponsor, Minister of Defense Grechko, Afanasyev's star dropped rapidly. In April 1983, Ustinov finally had him fired. He was given the far less important job of Minister of Heavy and Transport Machine Building, which was a sector outside the defense industry. With his ambition of one day entering the ranks of the Politburo crushed, Afanasyev trudged through his new dreary job, before finally being forced to retire prematurely in July 1987.[19] The "Big Hammer,"

15.   Zhiltsov, ed., *Gosudarstvennyy kosmicheskiy*, pp. 126–27.
16.   Sagdeev, *The Making of a Soviet Scientist*, pp. 258–59.
17.   Serge Schmemann, "Defense Minister of Soviet Union Is Dead at Age 76," *New York Times*, December 22, 1984, pp. 1, 6.
18.   See, for example, Eric Page, "Ustinov Had Key Roles in Military and Politics," *New York Times*, December 22, 1984, p. 6.
19.   Sagdeev, *The Making of a Soviet Scientist*, p. 200.

as he was nicknamed by many in the space industry, has retained contacts with the Russian space industry as a "Chief Scientific Consultant to the General Designer" of RKK Energiya. In one of his very rare published memoirs of the space era, he had only favorable words to say of Ustinov, despite the obvious clashes between the two men.[20] At the time of this writing, he was eighty-one years old.

The Soviet space program had originated as an arm of the artillery sector of the Soviet armed forces, and as such, there were a number of important artillery officers who played prominent roles in guiding the entire effort. There was probably no one officer more important than Lt. General Georgiy A. Tyulin, whose involvement in the Soviet missile program began in 1944, when he was a young lieutenant charged with assessing German rocket technology. In March 1965, he was appointed Afanasyev's First Deputy in the Ministry of General Machine Building. During the 1960s, he served as the chair of various State Commissions, including those for the later Vostok missions, the Voskhod program, the L1 circumlunar project, and various lunar and interplanetary probes. He remained at his ministerial post until 1976, when, rumor has it, Afanasyev fired him for being part of the "Ustinov camp."[21] Forced into retirement, the quiet and reticent Tyulin returned to teach theoretical mechanics at the M. V. Lomonosov Moscow State University.[22] In 1987, he began writing publicly about his deep well of experience in the missile and space programs—articles that have been remarkably valuable in filling in the gaps of this secret history. After a long illness, he died in April 1990 at the age of seventy-five.[23]

Tyulin was certainly better known than Vasiliy M. Ryabikov, who chaired the State Commission for Sputnik. One of the most mysterious figures in the early Soviet space program, Ryabikov was instrumental in the process of approving the first Sputnik launch. His early career was under Ustinov's shadow, but for a brief period in the 1950s, he emerged as one of the power players in the defense industry, only to disappear into relative oblivion. Almost nothing is known about his personal history. After his "ejection" from the defense industry, he served as the First Deputy Chairman of the State Planning Organ (better known as "Gosplan") until his death on July 19, 1974, at the age of sixty-seven. Even in recent years, Russian historians have generally shied away from any in-depth analysis of Ryabikov's role in the genesis of Sputnik. It is a curious omission for a man who may have facilitated the inauguration of the space era in 1957.

Of the two other major artillery officers from the space era, one remains alive. Lt. General Kerim A. Kerimov was demoted out of his ministry position in 1974, but remained the chair of the State Commission for Soyuz until 1991, a position he had assumed in 1966. He oversaw the launch of every single Soyuz spacecraft to the *Salyut* and *Mir* space stations during that period.[24] At the time of his retirement, he was officially the First Deputy Director of the Central Scientific-Research Institute for Machine Building (TsNIIMash), the leading research and development institution in the Soviet space industry.[25] At this writing, he was eighty-two years old. Lt. General Yuriy A. Mozzhorin, the powerful Director of TsNIIMash, remained in that post until December 1990, completing almost thirty years of service as one of the primary policymakers in the Soviet space program. He continued to be active in chronicling the history of the Soviet missile and space programs and served as editor of the series of memoirs titled *Dorogi v*

20. See Yu. A. Mozzhorin, *et al.*, eds., *Dorogi v kosmos: I* (Moscow: MAI, 1992), pp. 34–48.

21. Sagdeev, *The Making of a Soviet Scientist*, p. 201.

22. Col. M. Rebrov, "Where the Cranes Fly" (English title), *Krasnaya zvezda*, September 19, 1987, pp. 3–4.

23. "G. A. Tyulin," *Krasnaya zvezda*, April 25, 1990, p. 4.

24. The only exceptions were some of the Soyuz T missions and all the Soyuz missions to the military Almaz space station.

25. *Russian Space History, Sale 6516* (New York: Sotheby's, 1993), description for Lot 58.

kosmos (*Roads to Space*), the first volume of which was published in 1992. He died on May 15, 1998, at the age of seventy-seven.

The Designers

All six members of the original Council of Chief Designers are deceased. Korolev, of course, was the first to go in January 1966. Academician Nikolay A. Pilyugin, Korolev's closest friend on the council, died on August 2, 1982, at the age of seventy-four, after a long bout with diabetes.[26] His obituary was signed by Brezhnev, Andropov, Gorbachev, and Chernenko, all heads of the Soviet state at various points. Chief Designer Mikhail S. Ryazanskiy died after a long battle with cancer of the prostate gland on August 7, 1987. Academician Viktor I. Kuznetsov passed away four years later on March 22, 1991. The last member (aside from Glushko), Academician Vladimir P. Barmin, lived to the age of eighty-four, heading the organization he had founded until his death on July 17, 1993.[27] In one of his last interviews, Barmin waxed philosophical about the constraints of the Communist era:

> . . . I have been working as a Chief Designer for more than fifty years, and have been "open" to the press only in recent years. My articles in the newspaper Pravda used to be under a pseudonym, Professor Vladimirov. . . ."[28]

Although he was not a member of the original council, General Designer Academician Nikolay D. Kuznetsov, responsible for creating the N1's rocket engines, played a major role in the rise and fall of the huge project. Despite Glushko's order to have all N1-related materials destroyed, Kuznetsov, at his own risk, preserved ninety-four engines of the first, second, third, and fourth stages at the storage facilities of the Trud Scientific-Production Association. All were completely ready for operational use. In addition, he also hid away fifty to sixty experimental units, ready for future developmental work. Kuznetsov's gamble paid off when in the early 1990s, major U.S. aerospace companies expressed interest in using the engines for the next generation of expendable U.S. launch vehicles. In late 1993, the Aerojet Propulsion Division imported a flight-ready version of the NK-33, believing the design to be of "very modern technology compared with what the U.S. has in LOX/kerosene engines."[29] In 1995, Kuznetsov's organization went head-to-head with Glushko's firm bidding for their respective engines on new versions of the Atlas or Delta rockets. Although Glushko's engines won that bid, the N1 engines may still see the light of day.[30] In 1996–97, Kistler Aerospace Corporation of Kirkland, Washington, signed an agreement with Kuznetsov's former organization to use the N1's NK-33 and NK-43 engines on the company's K-1 reusable launch vehicle. In what could be a fitting legacy of the N1 rocket, the first K-1 vehicle is expected to use the *very same* engine units that were meant for use on the canceled 8L launch of the N1 in 1974.[31] Sadly, Kuznetsov himself will not be witness to their use. At the age of eighty-four, he died on July 31, 1995.[32]

---

26.   "Academician N. A. Pilyugin," *Pravda*, August 3, 1982, p. 3; "Nikolai A. Pilyugin, 74, Dies; Was Key Soviet Space Figure," *New York Times*, August 3, 1982, p. A19.

27.   "Academician V. P. Barmin," *Krasnaya zvezda*, July 22, 1993, p. 4.

28.   V. Smirnov, "Topical Interview: Space Starts With the People on the Ground" (English title), *Aviatsiya i kosmonavtika* no. 10 (October 1992): 2–3.

29.   Michael A. Dornheim, "Aerojet Imports Trud NK-33 Engine," *Aviation Week & Space Technology*, October 25, 1993, p. 29.

30.   Dennis Newkirk, *Russian Space Review 1996* (Roselle, IL: Dennis Newkirk, 1997).

31.   V. S. Anisimov, T. C. Lacefield, and J. Andrews, "Evolution of the NK-33 and NK-43 Reusable LOX/Kerosene Engines," presented at the 33rd AIAA/ASME/SAE/ASEE Joint Propulsion Conference & Exhibit, Seattle, Washington, July 6–9, 1997.

32.   "Academician N. D. Kuznetsov," *Krasnaya zvezda*, August 3, 1995.

The legacy of the NI also survives in the high-performance LOX–liquid hydrogen engines that were developed and tested in the early 1970s but were never used in flight. Most notable was Chief Designer Lyulka's 11D57 engine. The engine's production stopped in 1975 after 105 were built. During the testing period, the engine had accumulated more than 53,000 seconds of full-engine run time. In late 1993, Aerojet expressed interest in using the engine for its single-stage-to-orbit program.[33] Chief Designer Isayev's 11D56, another LOX–liquid hydrogen engine developed for the NI, became the center of controversy in 1993, when the sale of the engine to the Indian Space Research Organization was blocked by the U.S. government, which had concerns over their potential application in military missile systems.[34] After further negotiations, the Russian Federation delivered the first such engine to India in September 1998.

Mikhail Tikhonravov was one of the most important engineers behind the emergence of the Soviet space program. He designed the first Soviet liquid-propellant rocket (in the 1930s), performed research to optimize early ICBM designs (in the 1940s and 1950s), and was the leading engineer behind the genesis of Sputnik. In later years, he was also instrumental in the design of the Vostok spaceship. (files of Peter Gorin)

Although their names have not been prominent in Western histories of the Soviet space program, a number of men from the old Korolev design bureau played very critical roles in the road that led to Sputnik. Certainly from an engineering standpoint, there was no other individual more important in the genesis of Sputnik than Mikhail K. Tikhonravov. Overshadowed by the much more famous Korolev, Tikhonravov's role in the early space program was quite likely as important as that of his boss. With his landmark 1954 report on artificial satellites, he set off a process that ended with the launch of Sputnik in 1957. After Sputnik, Tikhonravov led the teams that designed the first piloted spacecraft, the first automated lunar probes, and the first Soviet reconnaissance satellites. He also contributed to policy by co-authoring important long-range plans for Korolev's design bureau. He continued work under Korolev, vigorously supporting piloted space exploration against those who believed in robotic exploration. He seems to have retired from the design bureau after Korolev's death and returned to teaching and writing. He died on March 4, 1974, at the age of seventy-four, after a spectacular career that had begun with his design of the first Soviet liquid-propellant rocket, the "09" in 1933. As with many other important individuals in Soviet space history, his life and his remarkable contributions remain drowned out by the flood of writings on Korolev. As a mark of respect to his memory, in February 1995, the Russian Military Space Forces renamed their leading space research institute, TsNII-50, after Tikhonravov.[35]

---

33.   "Aerojet, Lyulka Push D-57 for SSTO Validation," *Aviation Week & Space Technology*, October 11, 1993, pp. 50–51.

34.   Jeffrey L. Lenorovitz, "Space Systems/Loral Books Proton Launches," *Aviation Week & Space Technology*, September 20, 1993, pp. 90–91.

35.   Valeriy Baberdin, "The Once Secret Space NII Will Now Bear the Name of Tikhonravov" (English title), *Krasnaya zvezda*, January 18, 1995, p. 6. TsNII-50 separated from the original military NII-4 on April 3, 1972, to focus exclusively on military space research as opposed to ballistic missile research.

With respect to the Korolev "high guard"—his key deputies—most have passed away. The *de facto* head of all piloted space programs at OKB-1 throughout the 1960s and 1970s, Konstantin D. Bushuyev, lived to serve as the director of the Soviet side of the Apollo-Soyuz Test Project in 1975. Although his true position, a Chief Designer at NPO Energiya, was kept tightly under wraps, he told his U.S. counterparts on one occasion that "he had started working with Korolev right after World War II. . . ."[36] Officially, during the entire joint project, he was forced to pretend that he was actually an employee of the Institute of Space Research under the USSR Academy of Sciences. This charade played out right up to his sudden death of a heart attack on October 26, 1978, at the age of sixty-four. He was apparently suffering from a toothache and was headed to the hospital when he suddenly dropped dead in a corridor. Unsure of how to facilitate the funeral of a figure in the Soviet space program whose identity was known *before* his death, Soviet officials chose the most ludicrous path. As one observer noted later: "After his death, instead of having a decent funeral at the former Korolev Design Bureau, where he had spent most of his active working life, the final sad ceremony was moved to the [Institute of Space Research], simply as a cover. . . ."[37]

As for the two "fathers" of the N1, Sergey O. Okhapkin died in March 1980 at age seventy.[38] Given a different course of events in 1974, Okhapkin might very well have succeeded Mishin as head of the organization, because he had served as Mishin's First Deputy since 1966. The other N1 designer, Sergey S. Kryukov, remains alive today, and he occasionally writes in the Russian media on the topic. He had one of the more interesting careers of any of Korolev's protégés. A few years after Korolev's death, on March 30, 1970, Kryukov left TsKBEM and joined the Lavochkin Design Bureau as the famous Babakin's First Deputy, thus turning his back on the N1 and piloted spacecraft to focus on robotic probes. After Babakin's death, on August 26, 1971, Kryukov took over the design bureau and guided the organization through a mixed bag of lunar and interplanetary missions. Having become the victim of political maneuvering over a proposed Martian sample return project, Kryukov returned to his original place of work, then NPO Energiya. On November 17, 1977, he was appointed the First Deputy General Designer under Glushko. After overseeing the immensely successful *Salyut 6* space station missions, he retired in January 1982.[39] Still a "scientific consultant" to Energiya, Kryukov, at the time of this writing, is eighty-one years old.

Of all of Korolev's deputies, perhaps the most well known is Boris Ye. Chertok. His career started with Mishin and Bushuyev at the famous Bolkhovitinov Design Bureau in the late 1930s. Chertok remained a powerful figure at Energiya through the 1980s, but he never rose to the top of the organization. Although he retired in 1991 from his official duties as Deputy General Designer, he continues to maintain his offices at the giant organization as a "Chief Scientific Consultant." Still full of verve and energy at the age of eighty-seven, Chertok recently admitted that "in the N1-L3 project we . . . made serious mistakes."[40] He is one of the few men who, having lived through those historic times, has put pen to paper, and he is in the midst of publishing a multiple-volume set of priceless reminiscences. Incredibly detailed and remarkably devoid of partiality, these memoirs, titled *Raketi i lyudi* (*Rockets and Men*), cover everything from Chertok's early forays into Germany in search of A-4 missiles in 1945 to the

36. Edward Clinton Ezell and Linda Neumann Ezell, *The Partnership: A History of the Apollo-Soyuz Test Project* (Washington, DC: NASA Special Publication (SP)-4209, 1978), p. 288.
37. Sagdeev, *The Making of a Soviet Scientist,* p. 174.
38. Golovanov, *Korolev,* p. 480.
39. Yu. Markov, *Kurs na Mars* (Moscow: Mashinostroyeniye, 1989), pp. 25–26; Semenov, ed., *Raketno-Kosmicheskaya Korporatsiya,* p. 641. Note that the first source gives a slightly different date, December 1, 1977, as Kryukov's dismissal and subsequent appointment at NPO Energiya.
40. Peter Smolders, "I Meet the Man Who Brought the V-2 to Russia," *Spaceflight* 37 (July 1995): 218–20.

demise of the Energiya-Buran system during the early 1990s. Much more accessible than many other old-timers, Chertok continues to travel all over the world, including the United States, to speak of his life. He also has one foot in the future. His current project is a modest system of communications satellites in low-Earth orbit to serve the general populace.[41]

From the scientific community, there was probably no one individual who wielded as much influence as Academician Mstislav V. Keldysh, President of the USSR Academy of Sciences from 1961 and one of the most brilliant mathematicians of his generation. Unlike Korolev, Keldysh's personal contributions span the gamut from purely technical to purely managerial. During the 1950s, Keldysh personally participated and directed top-secret studies on the optimal design characteristics of multistage rockets, the question of returning a satellite from Earth orbit, the theory of passive gravitational stabilization of satellites, the calculation of various satellite orbits, and the mathematical analyses of optimal trajectories for flight to the Moon, Mars, and Venus. This research was performed at two institutions, both of which Keldysh headed simultaneously: the Department of Applied Mathematics of the V. A. Steklov Mathematics Institute under the Academy of Sciences and NII-1 under the Ministry of Aviation Industries. At the latter institute, Keldysh also initiated work on high-performance ramjet engines and nuclear rocket engines.

From 1961, after he was appointed to head the Academy of Sciences, Keldysh's most important contributions were as the chair of the Interdepartmental Scientific-Technical Council for Space Research. With Keldysh as chair, various permutations of this council served as "expert commissions" for several dozen different military and space programs.[42] In 1975, Keldysh stepped down as President of the Academy of Sciences because of ill health. A man with a calm disposition who rarely, if ever, lost his temper, Keldysh's favorite form of relaxation was collecting prints of the Impressionists. He died on June 24, 1978, at the age of sixty-seven, sitting at the wheel of his car in the garage of his country home.[43] Keldysh's ashes were interned in the Kremlin Wall, an honor reserved for only the most revered Soviet citizens of this century. All fourteen members of the Politburo signed his obituary. Throughout his extraordinary life, there were probably few sectors of the Soviet military-industrial complex Keldysh did not influence with his scientific contributions or advisory activities.

Glushko

Academician Valentin Petrovich Glushko effectively headed the Soviet space program from 1974 for a fifteen-year period, and during that time, some would argue, there was almost a cult of personality surrounding his name. Glushko, having a hand in the editorial supervision of all books related to space exploration, made sure that his role and contributions to the development of Soviet space technology were placed in a favorable light. If in a 1957 speech at Korolev's fiftieth birthday, Glushko could say "Korolev occupies first place after Tsiolkovskiy" in the development of Soviet rocketry, he did not hesitate in later years to insert his name before Korolev in all histories of the Soviet space program.[44] But with so much power, Glushko was still unable to carry out one of his most coveted dreams—piloted landing expeditions to the

41.   Ibid.
42.   N. Chentsov, "World Famous, But Secret in Every Way" (English title), Nauka i zhizn no. 2 (February 1991): 102–07; V. S. Avduyevskiy and M. Ya. Marov, "Mstislav Vsevolodovich Keldysh and Space Research" (English title), Zemlya i uselennaya no. 3 (May–June 1991): 46–52.
43.   "M. V. Keldysh Dies; Mathematician Led Academy in Soviet," New York Times, June 27, 1978.
44.   For an interesting analysis of Glushko and his role in rewriting history, see German Nazarov, "You Cannot Paper Space With Rubles: How to Save Billions" (English title), Molodaya gvardiya no. 4 (April 1990): 192–207. See also Sagdeev, The Making of a Soviet Scientist, p. 182. For an edited version of his 1957 speech, see A. Yu. Ishlinskiy, ed., Akademik S. P. Korolev: ucheniy, inzhener, chelovek (Moscow: Nauka, 1986), pp. 191–95.

Moon and Mars. At various points throughout the 1980s, he continued to bring this idea up to the Soviet leadership, but each time it was rejected. Knee-deep in the Energiya-Buran program, the Soviet military had little interest in funding another repeat of the N1-L3 debacle.[45]

Despite Glushko's remarkable rise to prominence as the reigning emperor of the Soviet space program, he was still a man trapped within his times. Few photographs of him were published, and apart from the cursory details of his professional history, outsiders had no clue about his personal life. Recently, there has been a tendency to paint Glushko as some kind of evil player of the Soviet space program, the man who single-handedly destroyed the N1 program—first when he broke off relations with Korolev in the 1960s and second when he canceled the program in the 1970s. But this revisionism comes perhaps more from haphazard retroactive assessments than any in-depth analysis. While Korolev has been humanized by countless biographies, Glushko still remains an enigma—a man whose only motive, it seems, was to sabotage Korolev's dreams. Is it possible to bring Glushko down to the level of a human, flawed perhaps, but at the end of the day having the same ideals of space exploration as those of Korolev? He was apparently well versed in the finer arts, such as music, painting, and literature, was good at drawing, spoke fluently in five languages, and regularly kept up with non-scientific foreign journals. His deputies remember him as a man who had an "eye for style and flair for detail . . . he would always be elegantly dressed."[46] He was married several times. Apart from his clearly notable contributions to the space program, Glushko also spent years completing a forty-volume series for the Academy of Sciences on the topics related to rocket propulsion theory. Overall, he published more than 250 scientific works.

In 1989, Mikhail F. Rebrov, a Soviet military journalist acquainted with Glushko, wrote a very candid account of the general designer's life:

> He was never weak nor banal—traits that frequently accompany material and professional success. As he himself said, his life was a long, difficult search which essentially consisted of attempting to reach the desired level of simplicity upon mastering incredibly complex designs. He apparently gave himself over fully to his life's main work, and was ready to sacrifice for it. But that was only the way things seemed. Where Korolev could at some point, after judiciously evaluating his capabilities, reserve the main strategic problem for himself, and turn some problems over to his students, Glushko did not let anything out of his hands.[47]

Referring to the final years of his life, Rebrov wrote:

> Nothing it seems could quench his thirst for activity, his frenzied passion to go down in history by completing what would come to be called "the first in the world." He was compared to the director of an enormous orchestra who was enchanted by the dream of playing something in such a way that would make the world talk about "the new Russian triumph in space." And, to a certain extent, he succeeded in this. . . .[48]

---

45.   Glushko also tried to generate public interest in his Mars plans by writing in newspapers. See, for example, V. Glushko, Yu. Semenov, and L. Gorshkov, "Fantasy on the Drawing Board: The 'Road to Mars" (English title), *Pravda*, May 24, 1988, p. 3.

46.   Boris Katorgin and Leonid Sternin, "Pushing Back the Missile Technology Frontiers," *Aerospace Journal* no. 5 (September–October 1997): 88–90.

47.   Col. M. Rebrov, "Specific Impulse" (English title), *Krasnaya zvezda*, August 26, 1989, p. 4.

48.   *Ibid.*

The Korolev-Glushko fallout has been discussed much in recent years, but most accounts attribute this fracture to personal vendetta more than professional opinions. It seems more likely, however, that both men were acting perfectly within the bounds of their experience over the N1 propellant issue, with Glushko supporting storables and Korolev cryogenics. Both men had solid reasons for their choices—rationales that had almost nothing to do with personality conflicts or outright hatred. The two had, after all, worked together through the Purges, through prison, and through the Stalinist era and maintained their cooperation. Recent accusations notwithstanding, there is no evidence to suggest that Glushko's testimony led to Korolev's imprisonment in the 1930s. In fact, both men acted with remarkable honor, given the exigencies of the day. Perhaps the tragedy of Glushko's life, if there is one, is that his ultimate ambition of being known as the most important person in the tapestry of the Soviet space program will never come true. He will always be behind Korolev, and he probably knew this fact very well. As early as 1968, a couple of years after Korolev's death, when a journalist asked Glushko about Korolev, Glushko replied, "But why are you always going on about Korolev! Korolev! And what was Korolev? He was just a thin metallic pipe. Inside it I placed my engines, Pilyugin—his instruments. Barmin put it on the launch pad and it flew. . . ."[49]

By the late 1980s, Glushko was seriously ill and partially paralyzed, to the point that a special stamp was made for him because he could not even sign his own name. While he was able to attend the first launch of his life's dream, Energiya, he was too ill to be at the Baykonur Cosmodrome for the launch of Buran. He continued working from his bed, asking for reports from his deputies on every little detail. In August 1988, knowing his days were numbered, he told one of his deputies that he wanted his ashes to be placed in an urn and kept in a safe place so that one day it may be taken to the surface of Venus.[50] Just fifty-six days after the first and last flight of the Buran space shuttle, on January 10, 1989, Glushko passed away in Moscow at the age of eighty.[51] Even Gorbachev paid his respects. Thus ended the journey that had begun in 1923, when a fifteen-year-old boy had written to Tsiolkovskiy about rockets traveling in space.

Mishin

If Glushko is conventionally known as the man who sabotaged the N1 program, then the popular retrospective evaluations of the contributions of Academician Vasiliy Pavlovich Mishin have been even less generous. One can almost randomly pick up any article on the history of the Soviet space program, and there will probably be a disparaging remark about Mishin. The hapless Mishin, after all, presided over the most ignominious period in Soviet space history. What better way to explain away all those failures than to attribute it to a short-tempered, impulsive, and unskilled manager who had a drinking problem? In all likelihood, there is probably much truth in the negative assessment of Mishin's role as a chief designer. He made some exceptionally poor decisions and pursued causes that collectively had seriously regressive repercussions on the effort as a whole. But like any figure in a complex history, his contributions were not one dimensional. In fact, quite possibly, his role has been demeaned unfairly.

After he was fired in May 1974, Mishin declined to take up Brezhnev's offer to help find a job, and he returned to full-time teaching at the prestigious Moscow Aviation Institute, his alma mater. He had originally founded the Cosmonautics Faculty at that institute in 1959 and taught

---

49. Golovanov, Korolev, p. 707.

50. Semenov, ed., Raketno-Kosmicheskaya Korporatsiya, p. 434; Katorgin and Sternin, "Pushing Back the Missile Technology Frontiers." Note that in the latter source, the authors state that Glushko wanted his ashes taken to the Moon or Mars.

51. "Academician V. P. Glushko," Pravda, January 13, 1989, p. 8; "Valentin Glushko, 80, Rocket Pioneer for Soviet Program," New York Times, January 13, 1989, p. B5.

on a part-time basis for fifteen years. After his dismissal, he went back only once to his old design bureau, as part of a project to document Korolev's scientific heritage. While he was no longer involved in the mainstream Soviet space program, Mishin continued to pursue an academic career focused on space technology. As part of his teaching, he directed a design project that led to the creation of the *Radio-1* and *Radio-2* amateur communications satellites.[52] His name was, of course, never mentioned in any histories of the period. Glushko apparently wished to remove Mishin from history. Although he was allowed to publish under his own name, Mishin wrote only technical books or contributed to historical works without being able to admit his own personal role in any space or missile project.[53]

His relatively obscure existence was interrupted dramatically in 1985 when the KGB abruptly summoned him for questioning about his relationship to a journalist named Suslov who had interviewed him. The KGB agents told him that Suslov was under arrest on charges of passing Soviet secrets to the West; they believed that Mishin was his accessory and threatened to strip him of membership in the Communist Party and put him on trial. The KGB finally dropped the case when they could not find evidence to implicate Mishin. He had simply been one of the people Suslov interviewed.[54]

Given that Mishin was not allowed to talk about his role in the history of the Soviet missile and space program, few people in the West were even aware of his significant role. His name was first mentioned by a Soviet émigré in 1982 and later by a French journalist in 1985, but most Western analysts remained unconvinced, believing that it was Yangel or perhaps Chelomey who had succeeded Korolev in 1966.[55] As the new era of *glasnost* ("openness") dawned on the Soviet Union during the mid-1980s, it slowly opened up the cellar doors of long-forgotten tales. A nation began to rewrite its history. In 1986, journalist Yaroslav K. Golovanov was allowed, by special clearance of the Central Committee, to write on the original group of cosmonauts. In a six-part article in the official Soviet newspaper *Izvestiya*, Golovanov revealed the events behind Gagarin's historic mission.[56] Among the more tantalizing revelations were the names of all twenty men who had been selected as cosmonauts in 1960. Until then, Soviet censors had allowed the publication of only the twelve who had flown into space.

Unflown cosmonauts were not the only ones who benefited from this free exchange of information. In late 1987, as part of celebrations for the thirtieth anniversary of the launch of the first Sputnik, the Soviet astronomy and space journal *Zemlya i vselennaya* (*Earth and Universe*) published a short article by Mishin in which he openly revealed that he had succeeded Korolev.[57] *Glasnost* may have meant openness, but Glushko made sure that there was no talk of the piloted lunar program, for to admit such a history was to admit that not only did the USSR race the United States to the Moon, but that it had lost. It finally took Glushko's death in January 1989 to change the climate. In July 1989, a relatively obscure newspaper named *Poisk* (*Search*) published a short article consisting of a few diary entries from the personal journal of Air Force General Nikolay P. Kamanin. There was no ambiguity in his writing: the Soviet Union had had a piloted lunar

52. Jacques Villain, ed., *Baïkonour: la porte des étoiles* (Paris: Armand Colin, 1994), p. 136.

53. Two of his books from the 1970s and 1980s are: V. P. Mishin, *Vvedeniye v mashinnoye proyektirovaniye letatelnykh apparatov* (Moscow: Mashinostroyeniye, 1978), and V. P. Mishin, *Osnovy proyektirovaniya letatelnykh apparatov (Transportnyye sistemy)* (Moscow: Mashinostroyeniye, 1985).

54. Mikhail Rebrov, "The Last Argument: A Study of the Designer in Black and White" (English title), *Krasnaya zvezda*, March 25, 1995, p. 6.

55. For the émigré, see Victor Yevsikov, *Re-Entry Technology and the Soviet Space Program (Some Personal Observations)* (Falls Church, VA: Delphic Associates, 1982). For the French journalist, see C. Wachtel, "The Chief Designers of the Soviet Space Program," *Journal of the British Interplanetary Society* 38 (December 1985): 561–63.

56. The articles were titled "Cosmonaut No. 1" and were published between April 2 and 6, 1986.

57. "Our Interviews" (English title), *Zemlya i vselennaya* no. 6 (November–December 1987): 2–5.

program, timed to compete with Apollo.[58] The following month, veteran cosmonaut Valeriy F. Bykovskiy, one of those who had trained for the project, admitted the same thing in his just-published biography.[59] These two publications burst the floodgates. Within weeks, there was a major in-depth article in *Izvestiya* on the N1 program, and by October 1989, Mishin gave a long interview on the project, covering his role as one of the pioneers of the Soviet space program.[60]

In this and subsequent interviews, Mishin did not hide his bitterness at having been wiped from the history of the Soviet space program. He took aim at Glushko, Ustinov, and all the other individuals who had silenced him for fifteen years. He also did not have kind words for the current Soviet space program:

Very little has been done about what we thought about and dreamed about 20 years ago, even 30 years ago with Korolev. It is simply vexing that so few useful and efficient space vehicles are in Earth orbit . . . we have become addicted to the long, monotonous long-duration manned missions in the tight Salyut-Mir which repeat each other. It is very wasteful.[61]

When asked how there could be a vigorous forward-looking space program, he replied:

Space exploration has been hampered by monopoly and secrecy, and by nepotism and politically dealing in the allocation of buildings and subsidies. We need broad, open competition in projects for a unified technical task. And discussion of tasks, ideas, and proposals, and independent expert evaluations, and open selection of the winners. Only after this, in full view of everyone, should there be implementation of projects in which the whole of society is convinced of their need and soundness.[62]

He might as well have been talking of an alien world as compared to the Soviet system. Mishin also added his own two cents to the emerging debate over whether if Korolev had lived, the Soviets might not have had more success in their ventures into space:

. . . in the main thing, in the desire to create a well-considered strategy for space exploration, we were, I hope, fellow thinkers. No, I probably did not possess the kind of will and sharp tongue that distinguished Korolev. I am prepared to admit that. But in our space situation, the replacement of one character for another and the replacement of leading personalities did not play any decisive role.[63]

In the initial flurry of publicity concerning the Soviet Union's aborted piloted lunar program, Mishin wrote extensively and granted many interviews, but in recent years, he has remained out of the public eye, except to occasionally author pieces paying tribute to his mentor, Sergey P. Korolev. The latter clearly had a very high regard for Mishin, having picked the thirty-year-old Mishin to serve as principal deputy in 1946. Korolev told a journalist once:

I found this to be true and recurring regardless of circumstances: every now and again, all of a sudden a man would come from out of nowhere, from the great unknown, a

58.   Lev Kamanin, "From the Earth to the Moon and Back" (English title), *Poisk* 12 (July 1989): 7–8.
59.   Grigoriy Reznichenko, *Kosmonaut-5* (Moscow: Politicheskoy literatury, 1989), p. 98.
60.   Sergey Leskov, "How We Didn't Get to the Moon" (English title), *Izvestiya*, August 18, 1989, p. 3; A. Tarasov, "Missions in Dreams and Reality" (English title), *Pravda*, October 20, 1989, p. 4.
61.   Tarasov, "Missions in Dreams and Reality."
62.   *Ibid.*
63.   *Ibid.*

*man that would be remarkable precisely for the qualities you sought: he is gifted, coura-
geous, honest. . . . He would introduce himself, extend his hand in a trustworthy manner, modestly speak of the work he has done, and a miracle would happen, the
unknown is no longer the unknown. And then you would say to yourself: "This is he,
precisely the man I need. "[64]*

He was evidently speaking of Mishin.

Perhaps wanting his story to be told, Mishin put his personal diaries—thirty-one volumes
covering the period from 1960 to 1974— up for sale at a special auction of Soviet and Russian
space artifacts at Sotheby's in late 1993.[65] The Perot Foundation purchased the diaries for
$190,000. Unfortunately at the time of this writing, the institution has yet to make these price-
less writings available to scholars. Portions have been exhibited as part of the National Air and
Space Museum's "Space Race" exhibit opened in 1997. Mishin himself continues to teach at
the Moscow Aviation Institute, having just turned eighty-two. He wrote perhaps the best epi-
taph to his own contribution to the Soviet piloted lunar program in a monograph he authored
in 1990 on the N1-L3 program:

*I do not want the readers to think that I am trying to relieve myself as Chief Designer of
responsibility for certain errors committed (including by me personally) during work on
the lunar program. Only he who does nothing makes no mistakes. We, the successors
of S. P. Korolev, did everything we could, but our efforts proved to be inadequate.[66]*

64.   Rebrov, "The Last Argument."
65.   *Russian Space History, Sale 6516* (New York: Sotheby's, 1993), description for Lot 29; John Noble
Wilford, "Soviet Space Papers Going on Sale," *New York Times*, December 5, 1993, p. 36.
66.   V. P. Mishin, "Why Didn't We Fly to the Moon?" (English title), *Znaniye: tekhnike: seriya kosmonautika, astronomiya* no. 12 (December, 1990): 3–43.

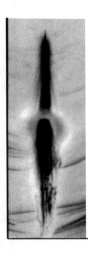

CHAPTER TWENTY
CODA

For a brief period in the late 1950s and early 1960s, one could reasonably argue that the Soviet Union was the leading spacefaring nation in the world. In the light of the ultimate demise of the Soviet empire, however, thinking of the USSR as launching humanity's first steps into the cosmos seems a strange abstraction—the memory feels oddly empty, almost irrelevant perhaps. When we do remember, we tend to divorce Sputnik from its origin·as a uniquely Soviet artifact—an eighty-four-kilogram sphere that was designed and built by men who lived through a war in which their country lost more than 25 million people, experienced the terror of Stalinist times, and defined themselves as Communists. Instead, we focus overwhelmingly on the *impact* of Sputnik rather than the construction of the artifact itself. I do not mean to suggest that meanings are unimportant. But in privileging only Sputnik's impact, we have told only half the story. Certainly, this is partly because the Soviet space program was given birth—and given flight—behind closed doors. Peeking through the now opened curtains, what I have tried to present here is an account of the missing half of that tale. This is only one version of the story, and certainly not the only one. But in sifting through the evidence and constructing the narrative, three broad themes have served as guidelines.

The first theme has to do with the institutional framework and the interplay among different factions within the Soviet space program and its antecedent missile project. Four primary constituencies were fundamental to establishing a Soviet ballistic missile program in the 1940s and 1950s. They were the engineers, the artillery officers, the defense industrialists, and the Communist Party. Each faction had its own agenda, but for a period of about fifteen years following the end of the war, their motivations intersected at crucial points to give rise to the world's first space program. The engineers were driven by their somewhat idealistic dreams of exploring space—dreams that had taken flight when they were young rocketry enthusiasts in the 1930s. The artillery officers were in need of a new generation of strategic weaponry to transform their backdated service into a powerful deterrent force. The defense industrialists had the not inconsiderable task of expediting the development of a strong military. And the Party leaders—in particular Stalin and Khrushchev—were driven by the political exigencies of the Cold War to direct the three other factions to elevate the Soviet Union from a nation afraid to defend itself to one that could threaten with offense.

The collusion of these four groups was necessary for the development of the world's first intercontinental ballistic missile, the R-7. This missile, of course, was simply a military weapon—a tool for mass destruction. In the hands of one visionary—Sergey Pavlovich Korolev—it became something entirely different. In the unlikely marriage of military imperative and idealistic ambition, the R-7 missile fired the first salvo in the space era—not by exploding a nuclear warhead, but by sending a small ball of metal around Earth on October 4, 1957. A country that had been dismissed as a nation of farmers and factory workers had suddenly arrived on the world's stage with an achievement that was too impressive to ignore.

There was more to come. Within four years, using the same rocket that had launched Sputnik, the Soviet Union, now armed with a new tint to the old socialist doctrine of harnessing technologies in the interest of the state, reached the apotheosis of its dizzying trajectory into the heavens. Historian Walter A. McDougall, writing in the introduction of his seminal work. . . the Heavens and the Earth, compared the event to the migration of the fish Eusthenopteron from the seas to the land:

> In A.D. 1961 Homo sapiens, in turn, left the realm of solids and gases and lived, for 108 minutes, in outer space. Life again escaped, or by definition extended, the biosphere. The earth's crust and canopy of air became another platform to a new universe as infinite as soil and sky must have seemed to Eusthenopteron.[1]

Only the vicissitudes of history will decide whether the flight of Yuriy Alekseyevich Gagarin in the spaceship Vostok will be remembered with such sweeping comparisons in the centuries hence. Even as the decade of the 1960s passed through tumult and chaos, humankind's first trip into space began to recede into the background. By the time that the first humans landed on the surface of the Moon in 1969, Gagarin's flight had been eclipsed in the popular conception of space exploration by the spectral images of two American men who left their footprints on another planetary body. The technology, the men, the pictures, and even the parades seemed so much more compelling to a new generation. In the historiography of space exploration, Gagarin's excursion assumed more importance for how it affected the American decision to aim for the Moon than for its own place in the history of human evolution.

But Gagarin's flight, both from a historical and a technological perspective, warrants more scrutiny. This is not only because it was achieved by a nation that was not expected to do so, but, simply and ultimately, because it was, as McDougall pointed out, an event that, like perhaps Apollo 11, transcended nations, languages, cultures, and continents and, for 108 minutes, represented the planet Earth: for the first time since the origin of life on this planet, one life form had managed to escape it. At the same time, we should not minimize more earthly considerations. Gagarin's flight did not, after all, happen in isolation from the political, economic, and social dimensions of the Cold War. And ironically, as this book has shown, the same forces that allowed the Soviet Union to send the first human into space—the need to arm themselves with powerful new weapons—deprived the country of further national triumphs in the space race.

Considering the post–Gagarin era leads to the second major theme of this work: the Soviet effort to beat the United States in landing the first person on the Moon. After an unprecedented catalogue of firsts in the late 1950s and early 1960s, the Soviets failed dismally in this quest. The road to failure began almost as soon as Gagarin had floated down in his parachute. After 1961, the Soviet space program, jostling for a role in the new Soviet military technocracy, began to stumble and slide in trying to attain a stable position of growth. Different factions were all vying for the same piece of pie. The artillery officers, now subsumed under the Strategic Missile Forces, increasingly declined to fund the primarily civilian endeavors of the human space program. The military, it seems, was more interested in missiles than the Moon. Grand visions of space exploration, as the one Korolev proposed in 1960, died under their own weight as the military siphoned off funding from important space projects in favor of developing a new generation of strategic weapons systems. Because their primary job was to design intercontinental

---

1.   Walter McDougall, . . . the Heavens and the Earth: A Political History of the Space Age (New York: Basic Books, 1985), p. 3.

ballistic missiles, the main space design organizations had to sacrifice ambitious space plans on the altar of strategic necessity. The Cold War, having given birth to the Soviet space program, would seriously threaten its very existence. In this climate, important avenues of research, such as the development of high-energy cryogenic rocket engine technology, never reached beyond the exploratory stage.

Despite the visible firebrand rhetoric of Nikita S. Khrushchev in the early 1960s, his support of a coherent long-range civilian space program was lukewarm at best. In 1961, when U.S. President John F. Kennedy laid down the challenge of reaching the surface of the Moon prior to the end of the decade, the Soviet space leadership hardly took notice. Organizational chaos emerged as a flurry of competitors began to dilute the hard-earned gains of the space program. The engineer faction, so united at the time of Sputnik, began to fragment in the face of limited funding. Between 1961 and 1964, Korolev ran his program on a shoestring budget, as proposal after proposal ended up in ministry file cabinets, never to be seen again. In desperation, he mounted two hastily prepared spectaculars in the mid-1960s—missions that had no value other than to please the Party leaders: the launch of the first multicosmonaut crew in 1964 and the accomplishment of the first spacewalk in 1965. The diversion cost the Soviets dearly. It was only in 1964 that Khrushchev sanctioned a piloted lunar landing program, three years after Kennedy's own challenge. The commitment itself was never followed up as the military continued to withhold funding, prompting the engineers to omit crucial phases in the ground testing of their lunar rocket. The shortcuts inexorably led to the series of crushing failures just as the United States was landing its first citizens on the surface of the Moon.

With the loss of the Moon race in 1969, the Soviets diverted much of their resources in the following years to space station programs. Korolev's successor, Vasiliy Mishin, has argued in recent years that despite the success of Apollo, there was no cause to abandon the massive N1-L3 lunar program simply because the Americans had arrived at the Moon first. As sound as this logic seems to be, engineers in the 1970s had to deal with a political climate that was vehemently hostile to expensive civilian programs such as lunar missions. Soviet leaders saw little need for such projects, because their success would raise inevitable questions about the original failure to beat Apollo.

The third and final theme of this work addresses the manner in which the Soviets handled technological innovation in the space program. The evidence both confirms and counters our *a priori* conceptions of Soviet technology as one characterized by evolutionary rather than revolutionary changes. In the space program, the Soviets used a combination of both; the decision to forego the former in favor of the latter often had as much to do with accident as with political expediency. For example, having built the first piloted spaceship Vostok by 1960, the Soviets tried hard to extend its capabilities by introducing relatively minor changes that cumulatively added to moderate gains. They abandoned Vostok as a viable piloted spaceship only in 1966 when they absolutely had to—that is, when the Soyuz spacecraft, which represented a qualitative leap in design and performance, was virtually ready. The decision to fly one over the other in 1966 had as much to with the impossibility of fulfilling contemporary objectives in space with the Vostok as with the fact that flying the Vostok (or its surrogate Voskhod) in 1966 would have demonstrated a visible and obvious lag to U.S. space technology.

In the thirty-year period spanning from 1945 to the mid-1970s, I looked at two cases of technological leaps: the R-7 ICBM and the N1 Moon rocket. Both projects required immense coordination in scale and scope across a Byzantine state-controlled landscape of research, development, and production. The success of the R-7 resulted not only from the high degree of financial commitment afforded by the state, but also because of the use of unorthodox management institutions such as the Council of Chief Designers. For example, in 1952, when Korolev decided to skip an intermediate stage in missile development and move directly to the ambitious ICBM project, the council proved to be a key and influential forum through which he could substantiate the proposal and ultimately convince the Soviet government of its feasibility. The council also served as an unusual managerial mechanism that allowed chief designers to intervene at key points in the development of the ICBM. Ultimately, the Council of Chief Designers managed to circumvent the internal self-generated inertia of Soviet industry, which discouraged major technological leaps and favored short-term gains.

In his important study on the origins of the Soviet atomic bomb, *Stalin and the Bomb*, David Holloway concluded that "[a]fter Stalin's death, nuclear scientists . . . enjoyed unprecedented authority among the political leaders."[2] The evidence from the space program suggests that the privilege of authority granted to the nuclear scientists was eventually expanded to include the engineers who played influential roles in the rise of the Soviet ballistic missile program. Although Western observers have long thought that it was Sputnik that changed the fortunes of these engineers such as Korolev and Glushko, their relationship to the political leadership changed more than a year before Sputnik with the successful test of the first Soviet strategic missile, the R-5M. The landmark test dramatically escalated the space engineers' leverage with both the Communist Party and the government and eventually led to the formation of a loose coalition of designers who would wield considerable power and influence. Although after 1960 they rarely, if ever, acted as a united front, the missile and space designers represented a formidable constituency that profoundly affected the direction of space policy beginning the late 1950s.

Because the Soviet political leadership lacked a clear understanding of the new technologies of the missile and space program, they needed the engineers to actively participate in policy formulation. The chief designers obliged willingly by forming lobbies, and, in the process, they acquired sufficient power to oppose important mandates from the top. There should be no confusion as to how the designers attained their powerful positions—it was not space, but rather their contribution to missile development that empowered them. Because the space program was largely a byproduct of missile production, the privileges almost by default were extended from the latter to the former. Both Nikita Khrushchev and Sergey Korolev played key roles in this process: Khrushchev because he allowed the rise of a constituency, and Korolev because he strongly pushed for it. But as the powerful chief designers vied for limited resources, they began to abuse the patronage system through various contacts within the Central Committee. "The favor of not even Khrushchev, Brezhnev, or Ustinov, but of a totally forgotten Central Committee agent," one Russian journalist wrote, "could determine the prospects for the development of the highly complex [space] sector for years."[3] The chaos was one of the key factors in the failure of the N1 program.

Korolev did not live long enough to witness the ultimate decline of the juggernaut he helped create. He did, however, leave behind an unmatched legacy—one that continues to be debated more than thirty years after his death. Most historians, both in Russia and in the West, have not argued with the holy grail of the history of the Soviet space program: that Sergey Pavlovich Korolev was its founder and central motivator. Given what is known about the vortex of events surrounding the launch of both Sputnik and Gagarin, it would be hard to dispute that claim. But at the same time, there has been an eagerness to attribute to Korolev roles that he clearly did not play, at least in his later life. It is particularly noteworthy that Sergey Korolev, the person who was most responsible for Sputnik was neither a scientist nor an engineer, but rather a manager with a vision. Boris V. Raushenbakh, one of Korolev's close associates from the 1960s, later wrote:

---

2. David Holloway, *Stalin and the Bomb: The Soviet Union and Atomic Energy, 1939–1956* (New Haven, CT: Yale University Press, 1994), pp. 366–67.

3. S. Leskov, "'Salyut-7' Is Falling. No One Knows Where and When" (English title), *Izvestiya*, January 18, 1991, p. 8.

Sometimes one hears it said that Korolyov was an excellent engineer and scientist. It is difficult to agree with this if the terms "engineer" and "scientist" are accorded their usual meaning. Korolyov himself did not devise any especially interesting solution to a complicated structural problem, as is the case with brilliant engineers. He was also not a scientist in the usual sense of the word: his name is not linked with any original scientific theory or with any prolonged and extensive study of a complicated phenomenon. However, these statements are not to be construed as a deprecation of the role which he played in the birth of space travel. There are many outstanding scientists and engineers, and Korolyov is a unique individual. His uniqueness, moreover, is linked to the fact that he was to introduce a new era into human history: the space age.[4]

The rise of the Soviet space program was one of the most significant processes in the history of science and technology in this century, not only because it opened what Raushenbakh calls "the space age," but also because it had profound sociopolitical consequences all over the world. Within the context of the Cold War itself, the Soviet space effort was a benchmark—a milestone that turned history from one path to another. For the first decade after Sputnik, the space age was indistinguishable in many ways from the space *race*. As the breadth of retrospect grows longer and longer, the import of the latter—that is, the space race—will no doubt recede far into the background, as perhaps it should. But for a short period in this century, the *race* provided the impulse for humankind to depart from this planet and reach the Moon. The Soviets, of course, lost this race, although they managed to throw shreds of doubt onto the victor's parade by denying that they had even signed up for the event. This deception existed for more than two decades, and when the truth was finally revealed, few took notice.

In 1999, during the thirtieth anniversary of the landing of humans on the surface of the Moon, the memory of Apollo spurred a brief but important resurgence of the sense of wonder and fascination that humans attach to space exploration. But lost amid the reevaluations and archaeology digs through Apollo, perhaps the greatest technological adventure of humankind was the other side of the coin—the story of those who had given reason to embark on Apollo in the first place. Buried under history was Korolev's "last love," the N1 program. That the N1 program was consigned to the status of a footnote is not so unusual: history has a way of privileging successes over failures. Our understanding of this dichotomy, between success and failure, is intrinsically tied to a second one—that between inevitability and contingency. On the one hand, we tend to see an inevitability in history's trajectory to the present—for example, that given the set of prevailing circumstances, the N1 program had to fail and that Apollo had to succeed. On the other hand, the story is compelling precisely because the outcome was not inevitable—that is, it was contingent on thousands of circumstantial factors. The tension between contingency and inevitability has contributed much to the enduring myths we now associate with the race to the Moon. The story of the Soviet space program has long been part of that myth. We have tended to see Soviet successes in space (such as Sputnik) as contingent and Soviet failures (such as the N1) as inevitable. This myth, it seems, is far too simplistic and takes away from the genuinely worthy accomplishments of Sputnik and Gagarin. The myth served its purpose during the Cold War, but now with the collapse of the Soviet Union and the opening of the Russian archives, it is finally time to put it away.

---

4.   Boris V. Rauschenbach, *Hermann Oberth: The Father of Space Flight* (Clarence, NY: West-Art, 1994), p. 172.

BIBLIOGRAPHIC ESSAY

Writing a history of the Soviet space program poses some interesting historiographical challenges in terms of source selection. As much as possible, I have tried to rely on Russian language sources. With very little exception, Western literature on the history of the Soviet space program has been a hodgepodge of speculation and sensationalism. Problems abound within Russian-language literature. Almost everything published prior to about 1988 was filtered through the Soviet censorship apparatus; details were sparse, and accounts often filled with inaccuracies. A major problem in the post-1988 literature is the dearth of primary sources. All archival sources, both at the governmental and Communist Party levels as well as within specific design bureaus, remain off limits to Western researchers.

Almost all of the Russian-language books and journals I have listed below are available at the Library of Congress in Washington, D.C. Others are available at libraries with large Russian-language collections, such as the University of Pennsylvania, the University of Pittsburgh, the NASA Headquarters Library, the NASA History Division archives, and the University of Massachusetts–Amherst. Many articles from the Russian media have also been translated into English by the Joint Publications Research Service (JPRS) under the JPRS-USP (Central Eurasia: space), JPRS-UAC (aviation and cosmonautics), and the JPRS-UMA (Soviet/Russian military affairs) titles. The JPRS apparently discontinued the first two series by 1995. Space articles are now continued under the JPRS-UST (science and technology) series. I would encourage all researchers of Soviet space history to begin with the JPRS issues, especially those covering 1988–95. All JPRS issues are available at the Library of Congress in both paper and microfiche forms. Most large university libraries also carry the entire series on microfiche.

I have used a combination of eight different types of materials to piece together this narrative:

1. Primary documents that have been published as collected works by Russian historians with access to archives
2. Official histories from Soviet-era space organizations
3. Biographies of major participants of the Soviet space program
4. Oral histories and memoirs from veteran participants of the Soviet space program
5. Articles and books by historians of the Soviet space program
6. English-language sources
7. Declassified documents
8. Interviews and correspondence

Primary Russian Documents

Falling into the first category, four works were invaluable as the backbone of this current work. The most important of these was *Tvorcheskoye naslediye Akademika Sergeya Pavloviicha Koroleva: izbrannyye trudy i dokumenty* (*The Creative Legacy of Academician Sergey Pavlovich Korolev: Selected Works and Documents*) (Moscow: Nauka, 1980), collectively edited under the leadership of Academy of Sciences President Mstislav Keldysh. This particular book is a collection of many of Korolev's technical works spanning 1930 to 1965. What the book suffers in terms of Soviet-era censorship is more than compensated by the remarkable breadth of materials. A less than stellar English translation of this book is available at the NASA History Division as prepared by the Translation Division of the Foreign Technology Division at Wright-Patterson

Air Force Base in Ohio. The translation reference number is FTD-ID(RS)T-0504-81; it was issued on September 3, 1981. The main compiler of the Russian-language work was Georgiy S. Vetrov, a historian at RKK Energiya who died in October 1997. At the time of his death, he had completed a second complementary volume of similar documents titled *Korolev i yego delo: svet i teni v istorii kosmonavtiki (Korolev and His Works: Light and Shadows in the History of Cosmonautics)*, which was published by the Nauka publishers in Moscow in mid-1998. This volume contains many documents on secret programs that could never have been published during the Soviet era. Unfortunately, I was only able to make minimal use of Vetrov's new work because my own manuscript was already completed at the time of its publication.

There are two other book-length works that are collections of primary documents. They are V. S. Avduyevskiy and T. M. Eneyev's *M. V. Keldysh: izbrannyye trudy: raketnaya tekhnika i kosmonavtika (M. V. Keldysh: Selected Works: Missile Technology and Cosmonautics)* (Moscow: Nauka, 1988), and B. V. Raushenbakh's *Materialy po istorii kosmicheskogo korabl "vostok" (Materials on the History of the "Vostok" Space Ship)* (Moscow: Nauka, 1991). I would recommend the latter especially for those interested in the development of the Vostok spacecraft. This slim volume also contains the completely unexpurgated version of Korolev and Tikhonravov's landmark 1954 letter to the Soviet government.

Several works from these three books have been translated into English. Some of them, including the complete 1954 report, can be readily seen at the NASA History Office Web site at *http://www.hq.nasa.gov/office/pao/History/sputnik/ussr.html.*

Soviet-era military organizations have published their own histories. The two most useful texts of the Strategic Missile Forces are *Raketnyye voyska strategicheskogo naznacheniya (Missile Forces of Strategic Designation)* (Moscow: RVSN, 1992) and *Khronika osnovnykh sobytiy istorii raketnykh voysk strategicheskogo naznacheniya (Chronicle of the Primary Events in the History of the Missile Forces of Strategic Designation)* (Moscow: TsIPK, 1994). The latter is a particularly important book because it includes the complete text of the famous May 1946 decree on the formation of the Soviet missile program. The book also contains unedited reproductions of many relevant documents from the famous R-16 disaster in 1960, which killed more than 100 people. The original decree on the formation of the Strategic Missile Forces is also included. There is also Ye. B. Volkov's *Mezhkontinentalnyye ballisticheskiye rakety SSSR (RF) i SShA (Intercontinental Ballistic Missiles of the USSR (RF) and the USA)* (Moscow: RVSN, 1996), which is an official history of the missile programs of the Strategic Missile Forces, handy for a technical overview. Finally, a recent history of the defunct Military Space Forces that contains much previously classified information is worth seeking out for understanding Soviet military space policy during the Cold War. See V. V. Favorskiy and I. V. Meshcheryakov, eds., *Voyenno-kosmicheskiye sily (voyenno-istoricheskiy trud): kniga I: kosmonavtika i vooruzhennyye sily (Military-Space Forces [Military-Historical Work]: Volume I: Cosmonautics and the Armed Forces)* (Moscow: Sankt-Peterburgskoy tipografii no. 1 VO Nauka, 1997).

There is also a remarkable work available on the evolution of the Soviet military-industrial complex, based exclusively on primary archival documentation. I would highly recommend the following book for any scholar attempting to gain insight into the interactions among the Soviet military, industry, and state during the Cold War. See N. S. Simonov, *Voyenno-promyshlennyy kompleks SSSR v 1920–1950-ye gody: tempy ekonomicheskogo rosta, struktura, organizatsiya proizvodstva i upravleniye (Military-Industrial Complex of the USSR from 1920–1950s: Rate of Economic Growth, Structure, Organization of Production and Administration)* (Moscow: ROSSPEN, 1996).

I would add some more sources into this first category. One Soviet-era journal has published a remarkable set of original and unedited documents from Gagarin's flight in 1961. These included the complete downlink during the launch phase of the mission and Gagarin's own report to the State Commission following landing. They can be found in V. Belyanov, L. Moshkov, Yu.

Murin, N. Sobolev, A. Stepanov, and B. Stroganov, "Yuriy Gagarin's Star Voyage: Documents from the First Flight of a Human into Space" (English title), *Izvestiya TsK KPSS* 5 (1991): 101–29. In addition, the journal *Voyenno-istoricheskiy zhurnal* (*Military-History Journal*) has published complete texts of many of the documents related to Korolev's arrest and incarceration in the late 1930s. These can be found in its October and November 1989 issues.

All of Georgiy Vetrov's works should also be included in this first category. His book *S. P. Korolev i kosmonavtika: pervye shagi* (*S. P. Korolev and Cosmonautics: First Steps*) (Moscow: Nauka, 1994) is quite possibly the best scholarly work on Korolev's pre-1945 work on rocketry. Vetrov's April and May 1994 articles in *Nauka i zhizn* (*Science and Life*) on the N1 rocket include reproductions of several original design bureau and governmental documents from the 1960s on the development of this booster. Before his death, Vetrov prepared a number of manuscripts that contain original documentation or interpretations of primary sources. These include *S. P. Korolev: Nauchnaya biografiya* (*S. P. Korolev: A Scientific Biography*), co-authored with his wife K. A. Krasnova. He also prepared a book called *Otkrytiye kosmosa* (*Opening Space*), which is a history of the early space program. Excerpts from this book have been published in issues 16 and 23 of the journal *Novosti kosmonavtiki* (*News of Cosmonautics*) in 1997 as well as the October 1997 issue of *Nauka i zhizn*. Another book, not completely finished, was *Taynyye tropy kosmonavtiki* (*Hidden Ways of Cosmonautics*), which is a nontechnical account of the relationships among Korolev, Glushko, and Ustinov. A final book apparently also completed is *Sekrety ostrova Gorodomlya* (*Secrets of Gorodomlya Island*), about the German rocket scientists in the Soviet Union following World War II. Most of these books were to have been published in 1997–98, but financial problems at the Nauka publishers have delayed their issuing. Vetrov's death seems to have delayed plans for publishing even further.

Official Organization Histories

In the second category, at least three Soviet-era space organizations have published detailed institutional and technical histories. I would highly recommend *Raketno-Kosmicheskaya Korporatsiya "Energiya" imeni S. P. Koroleva* (*The "Energiya" Rocket-Space Corporation Named After S. P. Korolev*) (Korolev: RKK Energiya, named after S. P. Korolev, 1996), which is a massive work covering the entire history of the Korolev design bureau. The book reproduces many of original documents from the space program; the entire narrative is based completely on the internal archives of the organization. Less useful is *Gosudarstvennyy kosmicheskiy nauchno-proizvodstvennyy tsentr imeni M. V. Khrunicheva* (*State Space Scientific-Production Center Named After M. V. Khrunichev*) (Moscow: RUSSLIT, 1997), a somewhat Soviet-era style history of the Khrunichev Machine Building Plant. Chief Designer Yangel's Yuzhnoye Plant has also published a detailed chronology of its participation in the missile and space programs. This is V. Pappo-Korystin, V. Platonov, and V. Pashchenko's *Dneprovskiy raketno-kosmicheskiy tsentr* (*Dneprov Rocket-Space Center*) (Dnepropetrovsk: PO YuMZ/KBYu, 1994). This work has an incredibly detailed chronology of the life of the organization and is packed with previous classified information relevant to the evolution of the Soviet space program.

Participant Biographies

Without a doubt, the most essential biography of any player in the Soviet space program is Yaroslav Golovanov's *Korolev: fakty i mify* (*Korolev: Facts and Myths*) (Moscow: Nauka, 1994). This 800-page work, sixteen years in the making, is not only an indispensable historical work, but also a magnificent piece of literature, unrivaled in its scope and lyrical quality. Another recommended biography of Korolev is Aleksandr Romanov's *Korolev* (Moscow: Molodaya gvardiya, 1996), which has been updated several times since its original publication in 1976. Romanov's work has a different tenor to Golovanov's biography in that it is slightly more anecdotal and lacks critical analysis.

Unfortunately, there have not been any in-depth treatments of other Soviet chief designers or officials in the post-1988 era. Researchers can search out N. G. Babakin, A. N. Banketov, and V. N. Smorkalov's *G. N. Babakin: zhizn i deyatelnost* (*G. N. Babakin: Life and Activities*) (Moscow: Adamant, 1996), which is a fairly good post-Soviet account of Babakin's life. There is also V. K. Kupriyanov and V. V. Chernyshev's *I vechernyy start . . .: rasskaz o glavnom kon-struktorye raketnykh dvigateley Alekseye Mikhaylovichye Isayevye* (*Evening Launch . . .: Accounts on the Chief Designer of Rocket Engines Aleksey Mikhaylovich Isayev*) (Moscow: Moskovskiy rabochiy, 1988), which suffers a little from Soviet-era censorship. One book, A. P. Romanov and V. S. Gubarev's *Konstruktory* (*Designers*) (Moscow: Izdatelstvo politicheskoy lit-eratury, 1989), contains substantial biographies of Glushko and Yangel in addition to Korolev.

Although strictly not a biography, another book, A. Yu. Ishlinskiy's *Akademik S. P. Korolev: ucheniy, inzhener, chelovek* (*Academician S. P. Korolev: Scholar, Engineer, Person*) (Moscow: Nauka, 1986), is a very useful gathering of recollections by dozens of men and women who knew Korolev. I highly recommend it to anyone interested in Korolev's life. A complete English translation of this is available at the NASA History Division prepared by the Translation Division of the Foreign Technology Division at Wright-Patterson Air Force Base in Ohio. The translation reference number is FTD-ID(RS)T-1140-87; it was issued on April 29, 1988. Comparable in spir-it, but vastly more informative is a work on Glushko, *Odnazhdy i navsegda . . .: dokumenty i lyudi o sozdatelye raketnykh dvigateley i kosmicheskikh sistem akademikye Valentinye Petrovichye Glushko* (*Once and Forever . . .: Documents and People on the Creation of Rocket Engines and Space Systems of Academician Valentin Petrovich Glushko*) (Moscow: Mashinostroyeniye, 1998), edited by V. F. Rakhmanin and L. Ye. Sterpin. This particular book on Glushko illuminates many episodes from the Soviet space program from a completely dif-ferent perspective—that is, the story from "the other side," as it were. Less helpful is Dmitriy Khrapovitskiy's *Generalnyy Konstruktor Akademik V. N. Chelomey* (*General Designer Academician V. N. Chelomey*) (Moscow: Vozdushniy transport, 1990). There is also B. V. Raushenbakh's *Iz istorii Sovetskoy kosmonavtiki: sbornik pamyati Akademika S. P. Koroleva* (*From the History of Soviet Cosmonautics: A Collection of Memories of Academician S. P. Korolev*) (Moscow: Nauka, 1983), which has an extremely detailed chronology of Korolev's entire life, including dates for many of his missile and spacecraft studies.

Oral Histories and Memoirs

The fourth category is memoirs. The most thorough and impartial memoirs authored by any participant in the Soviet space program have been those by Korolev's deputy Boris Chertok. By 1998, he had published three thick volumes: *Rakety i lyudi* (*Rockets and Men*) (Moscow: Mashinostroyeniye, 1994), which addresses roughly the period 1945 to 1957; *Rakety i lyudi: Fili Podlipki Tyuratam* (*Rockets and Men: Fili Podlipki Tyura-Tam*) (Moscow: Mashinostroyeniye, 1996), which contains events from 1957 to 1961; and *Rakety i lyudi: goryachiye dni kholodnoy voyny* (*Rockets and Men: Hot Days of the Cold War*) (Moscow: Mashinostroyeniye, 1997), which covers 1961 to 1968. These three volumes collectively should be the starting point for any scholar interested in the history of the Soviet space program. Chertok is an amazingly astute observer with a stunning memory for detail. These are invaluably rich contributions to this his-tory. A fourth volume on the lunar program is evidently on the way.

Equally essential are the diaries of General Nikolay Kamanin. Since 1989, his son Lev Kamanin has published excerpts from his diaries piece by piece in various newspapers. His journals from 1960 to 1966 have been collected into two very handy volumes, *Skrytiy kosmos: kniga pervaya, 1960–1963gg* (*Hidden Space: Volume One, 1960–1963*) (Moscow: Infortekst IF, 1995) and *Skrytiy kosmos: kniga vtoraya, 1964–1966gg* (*Hidden Space: Volume Two, 1964–1966*) (Moscow: Infortekst IF, 1997). Further additions to the series are expected in the near future. In the meantime, those interested in diary entries for 1966 to 1974 can search out issues of the Russian newspaper *Vozdushniy transport* (*Air Transport*), which has published extensive entries in issues 12 to 15, 23 to 25, and 43 to 50 in 1993 and in issues 9 to 19 in 1994. Almost all of these newspaper issues have been translated into English and are available at the NASA History Division as NASA TT-21658 dated December 1994. Researchers should note that the translations have been compiled in some cases without regard to chronological order.

Other memoirs relevant to Soviet space history include Sergey Khrushchev's two-volume *Nikita Khrushchev: krizisy i rakety: vzglyad iznutri* (*Nikita Khrushchev: Crises and Missiles: View From the Inside*) (Moscow: Novosti, 1994). A slightly different English version of these two volumes is to be published in 2000 under the title *The Creation of a Superpower* (*A View From the Inside*). One designer of the Soviet lunar lander has published a book on its development, *Vospominaniya o lunnom korablye* (*Recollections on the Lunar Ship*) (Moscow: Kultura, 1992).

An invaluable addition to the literature on Soviet space history are the *Dorogi v kosmos* (*Roads to Space*) series prepared by the Scientific-Research Center for Space Documentation in Moscow. These volumes include reminiscences from some of the most important players in the 1950s and 1960s—most notably some politicians, who have been notoriously absent in writing their memoirs. The contributors to this series include Minister Afanasyev, Military-Industrial Commission Deputy Pashkov, Chief Designers Barmin and Mishin, N1 designer Kryukov, Vostok designer Ivanovskiy, artillery officers Mozzhorin, Nesterenko, and Tyulin, and physician Yazdovskiy. Three volumes have been published so far: *Dorogi v kosmos: I* (*Roads to Space: I*) (Moscow: MAI, 1992), *Dorogi v kosmos: II* (*Roads to Space: II*) (Moscow: MAI, 1992), and *Nachalo kosmicheskoy ery: vospominaniya veteranov raketno-kosmicheskoy tekhniki i kosmonavtiki: vypusk vtoroy* (*The Beginning of the Space Era: Memoirs of Veterans of Rocket-Space Technology and Cosmonautics: Volume Two*) (Moscow: RNITsKD, 1994). A large selection from these three volumes has been translated into English and published as one book under the title *Roads to Space* (New York: McGraw-Hill, 1995). Unfortunately, I would not recommend the translation; it is filled with egregious errors and distorts many of the original passages and quotes from the Russian edition. The NASA History Division has translated two chapters from the first volume of the Russian edition. These can be found in NASA TT-21770 dated 1995.

For those interested in the development of the Soviet ground communications network, I would recommend *Kosmos nachinayetsya na zemlye* (*Space Begins From the Earth*) (Moscow: Patriot, 1996), which is written by B. A. Pokrovskiy, one of the major players in the network's creation. There have been many memoirs published on the creation of the Baykonur Cosmodrome. Perhaps the best one is the Council of Veterans of the Baykonur Cosmodrome's *Nezabyvayemyy Baykonur* (*Unforgettable Baykonur*) (Moscow: Interregional Council of Veterans of the Baykonur Cosmodrome, 1998), which among other things contains a blow-by-blow detailed chronology of the launch range from 1957 to 1961. I would also recommend the same council's *Proryv v kosmos: ocherki ob ispitatelyakh spetsialistakh i stroitelyakh kosmodroma Baykonur* (*Breakthrough Into Space: Essays on Test Specialists and Builders of the Baykonur Cosmodrome*) (Moscow: TOO Veles, 1994).

Some participants have published isolated articles in the Soviet and Russian media. Former NII-88 Director Yuriy Mozzhorin has co-authored an excellent two-part series of articles with

A. Yeremenko on the origins of the Soviet missile and space program. These can be found in the July and August 1991 issues of *Aviatsiya i kosmonavtika* (*Aviation and Cosmonautics*). Translations of these can be found in JPRS-UAC-92-002 dated February 3, 1992, and JPRS-UAC-92-003 dated February 13, 1992. An amplification of these articles by Biryukov and Yeremenko was published in *Novosti kosmonavtiki* in issue 10 from 1996. Artillery officer Aleksandr Maksimov has authored an illuminating series of articles on the first launches from Baykonur. These can be found in the September–October 1990, November–December 1990, January–February 1991, and March–April 1991 issues of *Zemlya i vselennaya* (*Earth and Universe*). Before his death, artillery officer Georgiy Tyulin authored a wonderful series of memoirs from his experiences covering the early years of the space program. These were published in the newspaper *Krasnaya zvezda* (*Red Star*) on April 2, 3, and 5, 1988, May 18, 1988, and April 1, 1989. The April 1988 issues have been translated in JPRS-USP-89-001 issued on January 18, 1989. The April 1989 article can be found in JPRS-UMA-89-013 issued on May 26, 1989.

N1 designers Dolgopyatov, Dorofeyev, and Kryukov published an article on the giant rocket in the September 1992 issue of *Aviatsiya i kosmonavtika*. N1 designer Kryukov has also written on the rocket in the April 1994 issue of *Nauka i zhizn*. Chief Designer Mishin wrote a long article on the same project in the December 1990 issue of *Znaniye: tekhnike: seriya kosmonavtika, astronomiya* (*Knowledge: Technology: Cosmonautics, Astronomy Series*). This is a very important piece because it is Mishin's only in-depth commentary on the Soviet piloted lunar program, the central thematic goal of his design bureau during the late 1960s. There is a complete translation of this in JPRS-USP-91-006 dated November 12, 1991. Vladimir Polyachenko, a senior designer of Chelomey's Almaz program, has published a two-part article on Almaz in the January and April 1992 issues of *Krylia rodiny* (*Wings of the Motherland*). These are available in English translation at the NASA History Division as NASA TT-21769 dated 1995.

Historian Articles and Books

The fifth category includes articles by Russian and Soviet journalists on the history of the Soviet space and missile programs. Many of these researchers have access to both primary documents and major participants in the effort. Certainly one of the most useful works by a Soviet researcher is Igor Afanasyev's "Unknown Spacecraft (From the History of the Soviet Space Program)," which was published in the December 1991 issue of *Znaniye: tekhnike; seriya kosmonavtika, astronomiya*. This was the very first declassification of a plethora of Soviet piloted space projects that never reached fruition or were considered secret for more than thirty years. This work has been translated into English in JPRS-USP-92-003 dated May 27, 1992. Afanasyev has also authored an excellent series of articles on the history of the N1 rocket in the journal *Krylia rodiny* in the September, October, and November 1993 issues. Translations are available in JPRS-USP-94-002-L dated July 7, 1994. Viktor Kazmin's ground-breaking articles on the Spiral program were published in the same journal in November and December 1990 and in January 1991. A translation of this is in JPRS-USP-91-007 dated November 22, 1991. A useful article on Chelomey was in issues 4–5 of *Ogonek* (*Light*) in January 1994. An English translation of this is available at the NASA History Division as NASA TT-21771 dated 1995.

Several journals and newspapers were indispensable for research on this book. First and foremost was *Novosti kosmonavtiki*, which is a monthly (formerly biweekly) publication produced from Moscow. Many unprecedented revelations about previously hidden aspects of Soviet space history have come forth through this magazine, probably the best publication in the world devoted to space exploration. For the most part, authors tend to focus on technical rather than political or institutional aspects. The editors can be reached at *i-cosmos@ mtu-net.ru*. An irregularly published journal that is very useful for historians is *Iz istorii aviatsii i kosmonavtiki* (*From the History of Aviation and Cosmonautics*). I particularly recommend its issue number 42 from 1980, which contains a series of informative articles on the works of the pioneer Mikhail Tikhonravov.

The Russian military newspaper *Krasnaya zvezda* often has had revealing articles on space history by its history correspondent, the late Mikhail Rebrov. Rebrov authored a wonderful six-part series on the original members of the Council of Chief Designers, which was published on October 22, 1988 (Barmin), January 7, 1989 (Kuznetsov), February 25, 1989 (Pilyugin), March 11, 1989 (Ryazanskiy), July 1, 1989 (Korolev), and August 26, 1989 (Glushko). A seventh article on April 8, 1989, was on the council itself. In the following years, Rebrov wrote dozens of more articles on various aspects of Soviet space history in the same newspaper. Many of these have been collected into one work, *Kosmicheskiye katastrofy: stranichki iz sekretnogo dosye* (*Space Catastrophes: Pages From the Secret Dossier*) (Moscow: Eksprint NV, 1996).

English-Language Sources

For those without knowledge of the Russian language, studying Soviet space history presents significant obstacles. Most of the English-language works are dated because they were published during the Soviet era. Fortunately, many of them are still worth perusing as excellent starting points for an introduction to the Soviet space program. I would highly recommend Nicholas Daniloff's *The Kremlin and the Cosmos* (New York: Alfred A. Knopf, 1972), which is a well-researched book that still stands up incredibly well, almost thirty years after its publication. F. J. Krieger's *Behind The Sputniks: A Survey of Soviet Space Science* (Washington, DC: Public Affairs Press, 1958) is an excellent collection of translations of pre-1958 articles on space exploration from the Soviet media. Certainly, the most famous book on the Soviet space program is James E. Oberg's *Red Star in Orbit* (New York: Random House, 1981), a still-readable account of what we knew about the Soviet space effort in the early 1980s. For those interested in more technical matters, Phillip Clark's *The Soviet Manned Space Program: An Illustrated History of the Men, the Missions, and the Spacecraft* (New York: Orion, 1988) is an incomparable treatise on all Soviet piloted space missions. Equally useful is Dennis Newkirk's *Almanac of Soviet Manned Space Flight* (Houston: Gulf Publishing Co., 1990), which is essentially a strict chronology culled from hundreds of sources. A good starting point for those interested in Soviet lunar and planetary exploration is Andrew Wilson's *Solar System Log* (London: Jane's Publishing Co., 1987).

One of the few post-1989 works on Soviet space history is James Harford's *Korolev: How One Man Masterminded the Soviet Drive to Beat America to the Moon* (New York: John Wiley & Sons, 1997). Although strictly a biography, Harford masterfully weaves a larger history from dozens of priceless interviews with many participants of the Soviet program from the 1950s and 1960s. I would highly recommend T. A. Heppenheimer's *Countdown: A History of Space Flight* (New York: John Wiley & Sons, 1997), a superbly written history of the early space era with considerable attention to Soviet achievements. A book exclusively on the Soviet piloted lunar program is Nicholas L. Johnson's *The Soviet Reach for the Moon* (Cosmos Books, 1995). It is now out of print.

Two books peripherally related to the Soviet space program that were very useful for my own work were David Holloway's *Stalin and the Bomb: The Soviet Union and Atomic Energy: 1939–1956* (New Haven: Yale University Press, 1994) and Steven J. Zaloga's *Target America: The Soviet Union and the Strategic Arms Race, 1945–1964* (Novato, CA: Presidio, 1993). Both benefit greatly from the fact that the authors were able to extensively use recently declassified information from Russian sources. Combined together, these two works are probably the best existing studies in English on the development of Soviet strategic weapons in the immediate postwar era.

The Congressional Research Service at the Library of Congress published a series of excellent summaries of the Soviet space program during the Cold War titled *Soviet Space Programs*. They covered the years 1962, 1962–65, 1966–70, 1971–75, 1976–80, and 1981–87. Packed with vast amounts of information, all of these books are now out of print but can be found at any major university library. I highly recommend these volumes to any serious scholar of the Soviet space program. Unlike many other works on Soviet space history, these books are particularly useful for analyses of space law, institutions, resource burdens, political motives, and international cooperation in the Soviet space program. Soviet-U.S. international cooperation in space is also the subject of Dodd L. Harvey and Linda C. Ciccoritti's excellent *U.S.-Soviet Cooperation in Space* (Miami: Center for Advanced International Studies, University of Miami, 1974).

The political motives of the early Soviet space program are the subject of two seminal works. These are Walter McDougall's *. . . the Heavens and the Earth: A Political History of the Space Age* (New York: Basic Books, 1985) and William H. Schauer's *The Politics of Space: A Comparison of the Soviet and American Space Programs* (New York: Holmes & Meier, 1976). Although both have dated somewhat in terms of their interpretations of the Soviet space program, I would particularly recommend McDougall's work as an excellent starting place to understand the Soviet government's views toward the role of technology in society. For a more recent scholarly view from a political science perspective, I would recommend William P. Barry's excellent *The Missile Design Bureaux and Soviet Piloted Space Policy, 1953–1974*, which is a doctoral dissertation at the University of Oxford from 1995.

The American Astronautical Society (AAS) publishes a series titled *History of Rocketry and Astronautics* as part of the AAS History Series. Many of these volumes contain very informative articles by direct participants of the Soviet space program. The AAS can be reached at AAS Publications, P.O. Box 28130, San Diego, CA 92198.

One important English-language source for Soviet space history are papers presented at the annual congresses of the International Astronautical Federation. These can be obtained at the International Astronautical Federation, 3-5 Rue Mario-Nikis, 75015, France. There are usually several papers every year that address important aspects of Soviet space history.

Uncovering the institutional machinations of the Soviet defense industry, and thus their space program, has been a difficult process, but some Soviet-era works have been useful as starting points. These included Michael McGwire, Ken Booth, and John McDonnell's *Soviet Naval Policy: Objectives and Constraints* (Halifax, NS: Centre for Foreign Policy Studies, 1975), David Holloway's *The Soviet Union and the Arms Race* (New Haven, CT: Yale University Press, 1984), Jiri Valenta and William C. Potter's *Soviet Decisionmaking for National Security* (London: George Allen & Unwin, 1984), David Lane's *Elites and Political Power in the USSR* (Edward Elgar, 1988), Timothy J. Colton and Thane Gustafson's *Soldiers and the Soviet State: Civil-Military Relations From Brezhnev to Gorbachev* (Princeton, NJ: Princeton University Press, 1990), and Peter Almquist's *Red Forge: Soviet Military Industry Since 1965* (New York: Columbia University Press, 1990). Arthur J. Alexander's still-remarkable work "Decision-Making in Soviet Weapons Procurement" in the Winter 1978/1979 issue of *Adelphi Papers* is a gold mine of information on the operations of the Soviet defense industry. A useful summary of information on the organization of the Soviet space program is contained in the August and September 1994 issues of the magazine *Spaceflight*. There is also an excellent Web site maintained by the Federation of American Scientists that contains detailed historical information on dozens of Soviet-era design bureaus and institutes specializing in space and missile technology. See *http://www.fas.org/spp/civil/russia/index.html*.

Despite its age, particularly useful in excavating the shifts in the Kremlin power structure during the Khrushchev era was Michael Tatu's *Power in the Kremlin: From Khrushchev's Decline to Collective Leadership* (London: Collins, 1969). An indispensable reference of information on the Soviet government and Communist Party was Edward L. Crowley, Andrew I. Lebed, and

Dr. Heinrich E. Schulz's *Party and Government Officials of the Soviet Union 1917–1967* (Metuchen, NJ: The Scarecrow Press, 1968), which contains lists of all senior Party and government officials in the Soviet era up to 1968. A post-Soviet English-language book highly recommended for those interested in the Cold War in general is Vladislav Zubok and Constantine Pleshakov's *Inside the Kremlin's Cold War: From Stalin to Khrushchev* (Cambridge, MA: Harvard University Press, 1996), which is based on recently declassified archival material.

A vast amount of technical information on the history of the Soviet space program has been published in English since 1989. *Quest: The Journal of Spaceflight History* should be a starting point for anyone with a cursory interest in the topic. Although not strictly focused on history, *Spaceflight*, a magazine of the British Interplanetary Society, has published many interesting articles on the history of the Soviet space program. For those interested in technical aspects, I would recommend articles by Timothy Varfolomeyev in the August 1995, February 1996, June 1996, January 1998, March 1998, and May 1998 issues on the development of Soviet launch vehicles. The *Journal of the British Interplanetary Society* also publishes an annual Soviet astronautics issue. For example, readers can search out an article by Mikhail Tikhonravov in the May 1994 issue of the magazine on the creation of Sputnik. An ongoing series in the same journal on military space topics, edited by Dwayne Day, has also included several important articles on Soviet space history.

I would highly recommend two books written in neither English nor Russian. The first is Christian Lardier's *L'Astronautique Soviétique* (Paris: Armand Colin, 1992). For those interested in the technical arcana of the Soviet space program, this is the best book *ever* written on the subject. It uses much information declassified by the Soviets following 1988 and is incomparable in its breadth and ambition to any other book published on the subject in either English or Russian. Although published during the Soviet era, I would also highly recommend Peter Stache's *Soujetischer Raketen* (Berlin: Militarverlad der DDR, 1987), which is in Polish. Fortunately, a complete English translation is available at the NASA History Division as prepared by the Translation Division of the Foreign Technology Division at Wright-Patterson Air Force Base. The translation reference number is FTD-ID(RS)T-0619-88; it is dated November 29, 1988.

Declassified Documents

A vast number of the CIA's National Intelligence Estimates (NIE) on the Soviet space and missile programs have now been declassified. For the space program in particular, these include NIEs issued on:

* December 5, 1962 (NIE 11-1-62)
* January 27, 1965 (NIE 11-1-65)
* March 2, 1967 (NIE 11-1-67)
* April 4, 1968 (NIE 11-1-68)
* June 19, 1969 (NIE 11-1-69)
* March 26, 1970 (NIE 11-1-70)
* July 1, 1971 (NIE 11-1-71)
* December 20, 1973 (NIE 11-1-73)
* July 19, 1983 (NIE 11-1-83)
* July 19, 1983 (NIE 11-1-83JX)
* December 1985 (NIE 11-1-85J)

All of these were titled *The Soviet Space Program* or (from 1973) *Soviet Space Programs*.

For the Soviet missile program in particular, most of the NIEs have also been declassified. Until 1962, assessments of the Soviet space program were included with the missile reports. I would recommend the following:

- October 5, 1954 (NIE 11-6-54)
- August 19, 1958 (NIE 11-5-58)
- November 3, 1959 (NIE 11-5-59)
- May 3, 1960 (NIE 11-5-60)
- April 25, 1961 (NIE 11-5-61)

These were titled *Soviet Capabilities and Probable Programs in the Guided Missile Field* (in 1954) and then *Soviet Capabilities in Guided Missiles and Space Vehicles*. All of these NIEs are invaluable for confirming or debunking unsubstantiated claims from the Russian media on various aspects of the Soviet space program. At the same time, I would caution researchers to use them with care, because it is clear that in certain areas, such as the institutional backdrop of the Soviet program, the CIA knew very little until well into the late 1960s.

One particularly useful CIA document is the agency's Office of Scientific Intelligence's *Scientific Research Institute and Experimental Factory 88 for Guided Missile Development, Moskva/Kaliningrad*. This report is numbered OSI-C-RA/60-2 and was issued on March 4, 1960. It addresses U.S. knowledge of the famous NII-88 institute in the late 1950s. Another useful report is the CIA Directorate of Science and Technology's *Scientific and Technical Intelligence Report: The Major Soviet Missile Design Bureaus*. This report was issued in June 1973. The study is notable because it illustrates not only what the CIA knew but also what it guessed completely wrong. For the defense industry in general, I would recommend the CIA Directorate of Intelligence's *The Soviet Weapons Industry: An Overview*, numbered DI 86-10016 and dated September 1986. A useful report on Soviet science is the NIE 11-6-59 titled "Soviet Science and Technology," issued on July 21, 1959. Several articles in the CIA journal *Studies in Intelligence* on the Soviet space program have also been declassified as part of the CIA's Historical Review Program. All of the declassified CIA documents are readily available to any researcher at the National Archives at 8601 Adelphi Road, College Park, MD 20740-6001. The phone number is (301) 713-6645. The National Archives can also be reached by e-mail at *cer@nara.gov*.

Interviews and Correspondence

The final category is interviews and correspondence. These are listed in chapter references.

TABLES

TABLE I

Table IA
Orbital Launch Attempts in the Soviet Piloted Space Program, 1960–74

Mission Name	Launch Date	Launch Time (Moscow Time)	Production Index of Spacecraft	Spacecraft Type	Spacecraft Serial No.	Launcher	Launcher Serial No.	Launch Site at Tyura-Tam
Vostok Program								
Korabl-Sputnik	May 15, 1960	0300:05.6	11F61	1KP	—	8K72	L1-11	1
—	July 28, 1960	1231	11F61	1K	1	8K72	L1-10	1
Korabl-Sputnik 2	Aug. 19, 1960	1144:06.8	11F61	1K	2	8K72	L1-12	1
Korabl-Sputnik 3	Dec. 1, 1960	1030:04.3	11F61	1K	5	8K72	L1-13	1
—	Dec. 22, 1960	1045:19	11F61	1K	6	8K72K	L1-13A	1
Korabl-Sputnik 4	Mar. 9, 1961	0928:59.6	11F63	3KA	1	8K72K	E103-14	1
Korabl-Sputnik 5	Mar. 25, 1961	0854:00.4	11F63	3KA	2	8K72K	E103-15	1
Vostok	Apr. 12, 1961	0906:59.7	11F63	3KA	3	8K72K	E103-16	1
Vostok 2	Aug. 6, 1961	0900	11F63	3KA	4	8K72K	E103-17	1
Vostok 3	Aug. 11, 1962	1130	11F63	3KA	5	8K72K	—	1
Vostok 4	Aug. 12, 1962	1102:33	11F63	3KA	6	8K72K	—	1
Vostok 5	Jun. 14, 1963	1158:58	11F63	3KA	7	8K72K	—	1
Vostok 6	Jun. 16, 1963	1229:52	11F63	3KA	8	8K72K	—	1
Voskhod Program								
Kosmos-47	Oct. 6, 1964	1000	11F63	3KV	2	11A57	—	1
Voskhod	Oct. 12, 1964	1030:01	11F63	3KV	3	11A57	—	1
Kosmos-57	Feb. 22, 1965	1030	11F63	3KD	1	11A57	—	1
Kosmos-59	Mar. 7, 1965	—	11F69	4K	4	11A57	—	—
Voskhod 2	Mar. 18, 1965	1000:00	11F63	3KD	4	11A57	—	1
Kosmos-110	Feb. 22, 1966	2309:36	11F63	3KV	5	11A57	—	1
Soyuz Program								
Kosmos-133	Nov. 28, 1966	1400	11F615	7K-OK(A)	2	11A511	UI50002	31
—	Dec. 14, 1966	1400	11F615	7K-OK(P)	1	11A511	UI50001	31
Kosmos-140	Feb. 7, 1967	0620	11F615	7K-OK(P)	3	11A511	—	1
Soyuz 1	Apr. 23, 1967	0035:00	11F615	7K-OK(A)	4	11A511	—	1
Kosmos-186	Oct. 27, 1967	1230	11F615	7K-OK(A)	6	11A511	—	31
Kosmos-188	Oct. 30, 1967	1112	11F615	7K-OK(P)	5	11A511	—	1

Mission Name	Launch Date	Launch Time (Moscow Time)	Production Index of Spacecraft	Spacecraft Type	Spacecraft Serial No.	Launcher	Launcher Serial No.	Launch Site at Tyura-Tam
Kosmos-212	Apr. 14, 1968	1300	11F615	7K-OK(A)	8	11A511	-	-
Kosmos-213	Apr. 15, 1968	1234	11F615	7K-OK(P)	7	11A511	-	-
Kosmos-238	Aug. 28, 1968	1259	11F615	7K-OK(P)	9	11A511	-	-
Soyuz 2	Oct. 25, 1968	1200	11F615	7K-OK(P)	11	11A511	-	-
Soyuz 3	Oct. 26, 1968	1134	11F615	7K-OK(A)	10	11A511	-	31
Soyuz 4	Jan. 14, 1969	1029	11F615	7K-OK(A)	12	11A511	-	31
Soyuz 5	Jan. 15, 1969	1005	11F615	7K-OK(P)	13	11A511	-	-
Soyuz 6	Oct. 11, 1969	1410:00	11F615	7K-OK(D)	14	11A511	-	31
Soyuz 7	Oct. 12, 1969	1344:47	11F615	7K-OK(P)	15	11A511	-	-
Soyuz 8	Oct. 13, 1969	1319:09	11F615	7K-OK(A)	16	11A511	-	31
Soyuz 9	June 1, 1970	2300:00	11F615	7K-OK(D)	17	11A511	-	-
Soyuz 10	Apr. 23, 1971	0254	11F615A8	7K-T	31	11A511	25	-
Soyuz 11	June 6, 1971	0755	11F615A8	7K-T	32	11A511	-	-
Kosmos-496	June 26, 1972	1753	11F615A8	7K-T	33L	11A511	-	-
Kosmos-573	June 15, 1973	0900	11F615A8	7K-T	35	11A511	-	-
Soyuz 12	Sept. 27, 1973	1518	11F615A8	7K-T	36	11A511	-	-
Kosmos-613	Nov. 30, 1973	0820	11F615A8	7K-T	34L	11A511	-	-
Soyuz 13	Dec. 18, 1973	1455	11F615A8	7K-T	33A	11A511	-	-
Kosmos-638	Apr. 3, 1974	1031	11F615A12	7K-TM	71	11A511U	-	-
Kosmos-656	May 27, 1974	1026	11F615A9	7K-TA	61	11A511	-	-
Soyuz 14	July 3, 1974	2151	11F615A9	7K-TA	62	11A511	-	-
Kosmos-670	Aug. 6, 1974	0301	11F732	7K-S	1L	11A511U	-	-
Kosmos-672	Aug. 12, 1974	0924	11F615A12	7K-TM	72	11A511U	-	-
Soyuz 15	Aug. 26, 1974	2257:54	11F615A9	7K-TA	63	11A511	-	-
Soyuz 16	Dec. 2, 1974	1240:00	11F615A12	7K-TM	73	11A511U	-	-
Zond Program								
Kosmos-146	Mar. 10, 1967	1430:33	11F91	7K-L1P	2P	8K82K/11S824	227-01	8L
Kosmos-154	Apr. 8, 1967	1200:33	11F91	7K-L1P	3P	8K82K/11S824	228-01	8L
-	Sept. 28, 1967	0111:54	11F91	7K-L1	4L	8K82K/11S824	229-01/12L	8L
-	Nov. 22, 1967	2207:59	11F91	7K-L1	5L	8K82K/11S824	230-01/13L	8P
Zond 4	Mar. 2, 1968	2129:23	11F91	7K-L1	6L	8K82K/11S824	231-01	8L
-	Apr. 23, 1968	0201:27	11F91	7K-L1	7L	8K82K/11S824	232-01/15L	8P

TABLE I

Mission Name	Launch Date	Launch Time (Moscow Time)	Production Index of Spacecraft	Spacecraft Type	Spacecraft Serial No.	Launcher	Launcher Serial No.	Launch Site at Tyura-Tam
–	July 14, 1968	–	11F91	7K-L1	8	8K82K/11S824	233-01	81L
Zond 5	Sept. 15, 1968	0042:11	11F91	7K-L1	9L	8K82K/11S824	234-01/17	81L
Zond 6	Nov. 10, 1968	2211:31	11F91	7K-L1	12L	8K82K/11S824	235-01/19	81L
–	Jan. 20, 1969	0714:36	11F91	7K-L1	13L	8K82K/11S824	237-01	81L
Zond 7	Aug. 8, 1969	0248:06	11F91	7K-L1	11	8K82K/11S824	243-01	81L
Zond 8	Oct. 20, 1970	2255:39	11F91	7K-L1	14	8K82K/11S824	250-01	81L
N1-L3 Program								
–	Feb. 21, 1969	1218:07	11F92	7K-L1S	3S	11A52	15003	110P
–	July 3, 1969	2318:32	11F92	7K-L1S	5L	11A52	15005	110P
–	June 28, 1971	0215:08	11F94/11F93 mock-ups	–	–	11A52	15006	110L
–	Nov. 23, 1972	0911:55	11F93	7K-LOK	6A	11A52	15007	110L
7K-L1E Program								
–	Nov. 28, 1969	1200:00	–	7K-L1E	1	8K82K/11S824	245-01/25L	81L
Kosmos-382	Dec. 2, 1970	2000:00	–	7K-L1E	2K	8K82K/11S824	252-01/26	81L
T2K Program								
Kosmos-379	Nov. 24, 1970	1400	11F94	T2K	1	11A511L	–	–
Kosmos-398	Feb. 26, 1971	1514	11F94	T2K	2	11A511L	–	–
Kosmos-434	Aug. 12, 1971	1250	11F94	T2K	3	11A511L	–	–
DOS/Salyut Program								
Salyut	Apr. 19, 1971	0440:00	11F715	17K	121	8K82K	254-01	81P
–	July 29, 1972	0620:57	11F715	17K	122	8K82K	260-01	81L
Kosmos-557	May 11, 1973	0320:00	11F715	17K	123	8K82K	284-01	81L
Salyut 4	Dec. 26, 1974	0715	11F715	17K	124	8K82K	284-02	81P
Almaz/Salyut Program								
Salyut 2	Apr. 3, 1973	1200:00	11F71	–	101-1	8K82K	283-01	81L
Salyut 3	June 25, 1974	0138:00	11F71	–	101-2	8K82K	283-02	81L

Table IB
Mission Parameters for Selected Orbital Flights

Name	Mass (kg)	Orbits	Initial Orbital Parameters (km)*	Orbital Inclination (°)	Duration (days:hrs:min:sec)**	Crew
Vostok						
Korabl Sputnik	4540	–	312 X 369	65	843:09:36	–
Korabl Sputnik 2	4600	27	306 X 339	64.95	01:02:23:36	–
Korabl Sputnik 3	4563	24	180 X 249	64.97	01:01:45:36	–
Korabl Sputnik 4	4700	1.92	183.5 X 248.8	64.93	00:01:47	–
Korabl Sputnik 5	4695	1.92	178 X 247	65.9	00:01:47	–
Vostok	4725	1	181 X 327	64.95	00:01:48	Gagarin
Vostok 2	4731	17	183 X 244	64.93	01:01:18	Titov
Vostok 3	4722	64	180.7 X 234.6	64.98	03:22:22	Nikolayev
Vostok 4	4728	48	179.8 X 236.7	64.95	02:22:56:27	Popovich
Vostok 5	4720	81	174.7 X 222.1	64.96	04:23:07:02	Bykovskiy
Vostok 6	4713	48	180.9 X 231.1	64.95	02:22:50:08	Tereshkova
Voskhod						
Kosmos-47	c. 5320	–	177 X 413	64.77	01:00:18	–
Voskhod	5320	16	177.5 X 408	–	01:00:17:03	Komarov, Feoktistov, Yegorov
Kosmos-57	c. 5680	–	175 X 512	64.77	03:00:02	–
Kosmos-59	c. 4730	–	209 X 339	65	8 days	–
Voskhod 2	5682	18	137 X 498	65	01:02:02:17	Belyayev, Leonov
Kosmos-110	c. 5700	–	187 X 904	51.9	21:18:00	–
Soyuz						
Kosmos-133	6386	34	181 X 232	51.9	01:21:21	–
Kosmos-140	c. 6450	33	170 X 241	51.7	01:23:32	–
Soyuz 1	6450	18	203.1 X 221.1	51.72	01:02:47:52	Komarov
Kosmos-186	c. 6530	–	209 X 235	51.7	03:22:49	–
Kosmos-188	c. 6530	–	200 X 276	51.68	03:00:58	–
Kosmos-212	6569	–	210 X 239	51.7	04:22:50	–
Kosmos-213	6458	–	205 X 291	51.4	05:00:37	–
Kosmos-238	c. 6520	–	199 X 219	51.7	03:23:04	–
Soyuz 2	6550	–	185 X 224	51.7	02:22:51	–

TABLE I

Name	Mass (kg)	Orbits	Initial Orbital Parameters (km)*	Orbital Inclination (°)	Duration (days:hrs:min:sec)**	Crew
Soyuz 3	6575	64	183.5 X 222.2	51.69	03:22:50:45	Beregovoy
Soyuz 4	6625	48	173 X 225.3	51.72	02:23:20:47	Shatalov
Soyuz 5	6585	49	198.7 X 230.2	51.69	03:00:54:15	Volynov, Yeliseyev, Khrunov
Soyuz 6	6577	80	186.2 X 222.8	51.68	04:22:42:47	Shonin, Kubasov
Soyuz 7	6570	80	207.4 X 225.9	51.68	04:22:40:23	Filipchenko, Volkov, Gorbatko
Soyuz 8	6646	80	204.5 X 223.7	51.68	04:22:50:49	Shatalov, Nikolayev
Soyuz 9	6590	286	208 X 220.6	51.7	17:16:59	Nikolayev, Sevastyanov
Soyuz 10	6575	32	209.6 X 248.4	51.6	01:23:45:54	Shatalov, Yeliseyev, Rukavishnikov
Soyuz 11	6790	384	191.5 X 220.5	51.64	23:18:21:43	Dobrovolskiy, Volkov, Patsayev
Kosmos-496	c. 6570	–	195 X 342	51.6	05:23:01	–
Kosmos-573	c. 6570	–	209 X 268	51.55	02:01:12	–
Soyuz 12	6720	31	193 X 248.6	51.61	01:23:15:32	Lazarev, Makarov
Kosmos-613	c. 6570	–	195 X 295	51.6	60:00:09	–
Soyuz 13	6560	127	193.3 X 272.7	51.6	07:20:55:35	Klimuk, Lebedev
Kosmos-638	c. 6570	–	195 X 325	51.8	09:21:34	–
Kosmos-656	c. 6570	–	194 X 354	51.6	00:23:46	–
Soyuz 14	c. 6570	252	195.9 X 242.7	51.62	15:17:30:28	Popovich, Artyukhin
Kosmos-670	c. 6700	–	217 X 307	51.6	02:23:58	–
Kosmos-672	c. 6570	–	198 X 239	51.8	05:22:38	–
Soyuz 15	6760	32	193.4 X 235.2	51.62	02:00:12:11	Sarafanov, Demin
Soyuz 16	6800	96	191.7 X 314.8	51.79	05:22:23:35	Filipchenko, Rukavishnikov
Zond						
Kosmos-146	c. 5375	–	190 X 310	51.5	08:17:01	–
Kosmos-154	c. 5375	–	186 X 232	51.6	02:05:23	–
Zond 4	c. 5375	–	–	–	07:00:20	–
Zond 5	c. 5375	–	–	–	06:28:26	–
Zond 6	c. 5375	–	–	–	06:18:48	–
Zond 7	5979	–	–	–	06:18:25	–
Zond 8	c. 5375	–	–	–	06:18:00	–
T2K						
Kosmos-379	c. 7495	–	198 X 253	51.6	4.686 days	–
Kosmos-398	c. 7255	–	196 X 276	51.63	Dec. 10, 1995 (decayed)	–

Name	Mass (kg)	Orbits	Initial Orbital Parameters (km)*	Orbital Inclination (°)	Duration (days:hrs:min:sec)**	Crew
Kosmos-434	c. 7000	–	197 X 285	51.6	3,653 days	–
7K-L1E						
Kosmos-382	c. 10,380	–	320 X 5040	51.5	in orbit	–
DOS/Salyut						
Salyut	c. 18,500	–	200 X 222	51.6	175 days	–
Kosmos-557	c. 19,400	–	218 X 266	51.6	11:02:39	–
Salyut 4	c. 18,900	–	219 X 270	51.6	769:19:12	–
Almaz/Salyut						
Salyut 2	c. 18,500	–	215 X 260	51.6	55:02:39	–
Salyut 3	c. 18,500	–	219 X 270	51.6	213 days	–

* The orbital parameters announced by Soviet/Russian sources (shown above) differ from those tracked by Western sources because the Soviet Union used a different model of Earth.

** For durations with only three figures, such as 00:01:48 for the first Vostok, the seconds were not included.

TABLE I

Table IC

Nonorbital Launch Attempts Supporting the Soviet Piloted Space Program, 1951–74

Launch Date	Launch Time (Moscow Time)	Altitude (km)	Launch Vehicle	Site	Comments
July 22, 1951	1254	101	R-1V	Kapustin Yar	Dogs Dezik and Tsygan, recovered
July 29, 1951	–	100	R-1B	Kapustin Yar	Dogs Dezik and Lisa, both killed
Aug. 15, 1951	–	100	R-1B	Kapustin Yar	Smelaya and unnamed dog recovered, spectral composition studies
Aug. 19, 1951	–	100	R-1V	Kapustin Yar	Two dogs successfully recovered
Aug. 28, 1951	–	100	R-1B	Kapustin Yar	Failure, two dogs killed
Sept. 3, 1951	–	100	R-1B	Kapustin Yar	ZIB and unnamed dog recovered
July 2, 1954	–	110	R-1D	Kapustin Yar	Two dogs, one successfully recovered
July 7, 1954	–	110	R-1D	Kapustin Yar	Two dogs, one successfully recovered
July 26, 1954	–	110	R-1D	Kapustin Yar	Dogs Lisa and Ryzhik, one recovered
Jan. 25, 1955	–	100	R-1Ye	Kapustin Yar	Two dogs, spurious payload separation at T+22 seconds
Feb. 5, 1955	–	100	R-1Ye	Kapustin Yar	Two dogs, neither recovered
Nov. 4, 1955	–	100	R-1Ye	Kapustin Yar	Two dogs, recovered
May 14, 1956	–	100	R-1Ye	Kapustin Yar	Two dogs, recovered, first with SOI spectrograph
May 31, 1956	–	100	R-1Ye	Kapustin Yar	Two dogs, recovered
June 7, 1956	–	100	R-1Ye	Kapustin Yar	Dogs Ryzha and Damka, both recovered
May 16, 1957	0515	212	R-2A	Kapustin Yar	
May 24, 1957	–	212	R-2A	Kapustin Yar	
Aug. 25, 1957	0627	212	R-2A	Kapustin Yar	
Aug. 31, 1957	–	212	R-2A	Kapustin Yar	
Sept. 9, 1957	–	473	R-5A	Kapustin Yar	
Feb. 21, 1958	1140	212	R-2A	Kapustin Yar	
Aug. 2, 1958	–	212	R-2A	Kapustin Yar	
Aug. 13, 1958	0806	451	R-5A	Kapustin Yar	Dogs Palma and Pushuk, radiation research
Aug. 27, 1958	–	473	R-5A	Kapustin Yar	Two dogs
Sept. 19, 1958	1554	473	R-5A	Kapustin Yar	Two dogs
Oct. 31, 1958	–	195	R-2A	Kapustin Yar	
June 22, 1959	–	212	R-2A	Kapustir. Yar	Dogs Otvazhnaya and Snezhinka
July 2, 1959	0740	212	R-2A	Kapustin Yar	Does Otvazhnaya and unnamed, solar ultraviolet study
July 10, 1959	–	212	R-2A	Kapustin Yar	
July 14, 1959	–				

Launch Date	Launch Time (Moscow Time)	Altitude (km)	Launch Vehicle	Site	Comments
June 15, 1960	0643	212	R-2A	Kapustin Yar	Does Otvazhnaya and unnamed, solar ultraviolet study
June 24, 1960	–	212	R-2A	Kapustin Yar	Two dogs
Sept. 16, 1960	–	210	R-2A?	Kapustin Yar	Two dogs
Sept. 22, 1960	–	210	R-2A?	Kapustin Yar	Two dogs
Dec. 27, 1961	–	–	R-12	Vladimirovka	MP-1 spaceplane testbed, recovered
Mar. 21, 1963	1440	–	R-12	Kapustin Yar	M-12 spaceplane testbed, destroyed
Sept. 26, 1964	0650	–	R-5V	Kapustin Yar	VAO payload with Soyuz descent apparatus, shroud breakup
July 15, 1969	–	–	R-12	Plesetsk site 84	BOR-1 Spiral/EPOS subscale model
Aug. 18, 1970	0645	–	8K82K no. 246-01	Tyura-Tam site 81L	82EV, Proton-K test

Note: There were a total of seven BOR-2 and BOR-3 suborbital launches between 1969 and 1974. No dates are known.

Selected Sources for Tables IA, IB, and IC

1. E-mail correspondence, Vladimir Agapov to the author, September 27, 1996.
2. E-mail correspondence, Vladimir Agapov to the author, September 30, 1996.
3. The piloted mission parameters are from V. P. Glushko, ed., Kosmonautika entsiklopediya (Moscow: Sovetskaya entsiklopediya, 1985), and V. P. Glushko, Razvitiye rake-tostroyeniya i kosmonavtiki v sssr: izdaniye utoroye, dopolnennoye (Moscow: Mashinostroyeniye, 1981).
4. The automated mission durations are from Sergey A. Voevodin, VSA074, electronic newsletter, June 5, 1997.

Table II

TABLE II

Cosmonaut Selection Groups, 1960–74

Name	Birth	Resign.	Death	Space Missions
March–June 1960 (by order of the GK VVS)—VVS Group 1[1]				
1. SL Ivan Nikolayevich Anikeyev	02-12-33	04-17-63	08-20-92	
2. M Pavel Ivanovich Belyayev	06-26-25	01-10-70	01-10-70	Voskhod 2 (1965)
3. SL Valentin Vasilyevich Bondarenko	02-16-37	03-23-61	02-23-61	
4. SL Valery Fedorovich Bykovskiy	08-02-34	01-26-82		Vostok 5 (1963), Soyuz 22 (1976), Soyuz 31 (1978)
5. SL Valentin Ignatyevich Filatyev	01-21-30	04-17-63	09-15-90	
6. SL Yuriy Alekseyevich Gagarin	03-09-34	03-27-68	03-27-68	Vostok (1961)
7. SL Viktor Vasilyevich Gorbatko	12-03-34	08-28-82		Soyuz 7 (1969), Soyuz 24 (1977), Soyuz 37 (1980)
8. C Anatoliy Yakovlevich Kartashov	08-25-32	04-07-62		
9. SL Yevgeniy Vasilyevich Khrunov	09-10-33	12-25-80	05-19-00	Soyuz 5 (1969)
10. CE Vladimir Mikhaylovich Komarov	03-16-27	04-24-67	04-24-67	Voskhod (1964), Soyuz 1 (1967)
11. L Aleksey Arkhipovich Leonov	05-30-34	01-26-82		Voskhod 2 (1965), Soyuz 19 (1975)
12. SL Grigoriy Grigoryevich Nelyubov	03-21-34	05-04-63	02-18-66	
13. SL Andrian Grigoryevich Nikolayev	09-25-29	01-26-82		Vostok 3 (1963), Soyuz 9 (1970)
14. C Pavel Romanovich Popovich	10-05-30	01-26-82		Vostok 4 (1960), Soyuz 14 (1974)
15. SL Mars Zakirovich Rafikov	09-30-33	03-24-62		
16. SL Georgiy Stepanovich Shonin	08-03-35	04-28-79	04-07-97	Soyuz 6 (1969)
17. SL German Stepanovich Titov	09-11-35	06-17-70		Vostok 2 (1961)
18. SL Valentin Stepanovich Varlamov	08-15-34	03-06-61	10-02-80	
19. SL Boris Valentinovich Volynov	12-18-34	03-17-90		Soyuz 5 (1969), Soyuz 21 (1976)
20. SL Dmitriy Alekseyevich Zaykin	04-29-32	10-25-69		
March–April 1962 (GK VVS)—TsPK Women[2]				
1. P Tatyana Dmitriyevna Kuznetsova	07-14-41	10-01-69		
2. P Valentina Leonodovna Ponomareva	09-18-33	10-01-69		
3. P Irina Baenovna Solovyeva	09-06-37	10-01-69		
4. P Valentina Vladimirovna Tereshkova	03-06-37	04-30-97		Vostok 6 (1963)
5. P Zhanna Dmitriyevna Yerkina	05-06-39	10-01-69		
January 10, 1963 (GK VVS)—VVS Group 2				
Pilots				
1. M Georgiy Timofeyevich Dobrovolskiy	06-01-28	06-30-71	06-30-71	Soyuz 11 (1971)

Name	Birth	Resign.	Death	Space Missions
2. M Anatoliy Vasilyevich Filipchenko	02-26-28	01-26-82		Soyuz 7 (1969), Soyuz 16 (1974)
3. M Aleksey Aleksandrovich Gubarev	03-29-31	09-01-81		Soyuz 17 (1975), Soyuz 28 (1978)
4. M Anatoliy Petrovich Kuklin	01-03-32	09-15-75		
5. LC Vladimir Aleksandrovich Shatalov	12-08-27	06-25-71		Soyuz 4 (1969), Soyuz 8 (1969), Soyuz 10 (1971)
6. M Lev Vasilyevich Vorobyev	02-24-31	06-28-74		
Engineers				
7. ME Yuriy Petrovich Artyukhin	07-22-30	01-26-82	08-04-98	Soyuz 14 (1974)
8. SLE Eduard Ivanovich Buynovskiy	02-26-36	12-11-64		
9. LCE Lev Stepanovich Demin	01-11-26	01-26-82	12-18-98	Soyuz 15 (1974)
10. SLE Vladislav Ivanovich Gulyayev	05-31-38	03-06-68	04-19-90	
11. CE Petr Ivanovich Kolodin	09-23-30	04-20-83		
12. SLE Eduard Pavlovich Kugno	06-27-35	04-16-64		
13. CE Aleksandr Nikolayevich Matinchenko	09-04-27	01-19-72	06-18-99	
14. C Anatoliy Fedorovich Voronov	06-11-30	05-25-79	10-31-93	
15. SLE Vitaliy Mikhaylovich Zholobov	06-18-37	07-01-81		Soyuz 21 (1976)

January 25, 1964 (GK VVS)—VVS Group 2 Supplementary

1. LC Georgiy Timofeyevich Beregovoy	04-15-21	02-25-82	06-30-95	Soyuz 3 (1968)

May–June 1964 (MK)—Voskhod[3]

IAT AN SSSR

1. Georgiy Petrovich Katys[4]	08-31-26	10-xx-64		

IMBP

2. Boris Ivanovich Polyakov	Unknown	07-02-64		

GosNII AiKM

3. LC Vasiliy Grigoryevich Lazarev[5]	02-23-28	10-xx-64	12-31-90	
4. Boris Borisovich Yegorov	11-26-37	10-xx-64	09-12-94	Voskhod (1964)

OKB-1

5. Konstantin Petrovich Feoktistov[6]	02-07-26	10-xx-64		Voskhod (1964)

OKB-156

6. Vladimir Nikolayevich Benderov	08-24-24	07-02-64	06-03-73	

TsPK

7. Aleksey Vasilyevich Sorokin	03-30-32	10-xx-64	01-23-76	

TABLE II

Name	Birth	Resign.	Death	Space Missions
October 28, 1965 (GK VVS)—VVS Group 3				
Pilots				
1. L Leonid Denisovich Kizim	08-05-41	06-13-87		Soyuz T-3 (1980), Soyuz T-10(1984), Soyuz T-15(1986)
2. L Petr Ilich Klimuk	07-10-42	03-03-82		Soyuz 13 (1973), Soyuz 18 (1975), Soyuz 30 (1978)
3. L Aleksandr Yakovlevich Kramarenko	11-08-42	04-30-69		
4. L Aleksandr Yakovlevich Petrushenko	01-01-42	06-15-73	11-11-92	
5. L Gennadiy Vasilyevich Sarafanov	01-01-42	07-07-86		Soyuz 15 (1974)
6. L Ansar Ilgamovich Sharafutdinov	06-26-39	01-05-68		
7. L Vasiliy Dmitriyevich Shcheglov	07-09-40	10-18-72	07-16-73	
8. L Aleksandr Aleksandrovich Skvortsov	06-08-42	01-05-68		
9. L Valeriy Abramovich Voloshin	04-24-42	04-09-69		
10. L Oleg Anatolyevich Yakovlev	12-31-40	05-22-73	05-02-90	
11. L Vyecheslav Dmitriyevich Zudov	01-08-42	05-14-87		Soyuz 23 (1976)
Engineers				
12. ME Boris Nikolayevich Belousev	07-24-30	01-05-68	06-17-98	
13. M Vladimir Aleksandrovich Degtyarev	04-09-32	01-17-66		
14. L Anatoliy Pavlovich Fedorov	04-14-41	05-28-74		
15. SLE Yuriy Nikolayevich Glazkov	10-02-39	01-26-82		Soyuz 24 (1977)
16. L Vitaliy Andreyevich Grishchenko	04-26-42	02-05-68	05-04-92	
17. SLE Yevgeniy Nikolayevich Khludeyev	09-10-40	10-11-88	09-19-95	
18. CE Gennadiy Mikhaylovich Kolesnikov	10-07-36	12-16-67		
19. CE Mikhail Ivanovich Lisun	09-05-35	09-19-89		
20. SG Vlad. Yevenyevich Preobrazhenskiy	02-03-39	11-18-80	10-25-93	Soyuz 23 (1976)
21. SLE Valeriy Ilich Rozhdestvenskiy	02-13-39	06-24-86		
22. CE Eduard Nikolayevich Stepanov	04-17-37	10-31-92		
January 17, 1966 (GK VVS)—VVS Group 3 Supplementary				
1. LC Vasily Grigoriyevich Lazarev[7]	02-23-28	11-27-85	12-31-90	Soyuz 12 (1973), Soyuz 18-1 (1975)
May 23, 1966—TsKBEM[8]				
1. CO Sergey Nikolayevich Anokhin	03-19-10	05-27-68	04-15-86	
2. Vladimir Yefgrafovich Bugrov	01-18-33	07-12-67		
3. Gennadiy Aleksandrovich Dolgopolov	11-14-35	05-03-67		
4. Georgiy Mikhaylovich Grechko[9]	05-25-31	06-05-86		Soyuz 17 (1975), Soyuz 26 (1977), Soyuz T-14 (1985)
5. Valeriy Nikolayevich Kubasov	01-07-35	11-03-93		Soyuz 6 (1969), Soyuz 19 (1975), Soyuz 36 (1980)

Name	Birth	Resign.	Death	Space Missions
6. Oleg Grigoryevich Makarov	01-06-33	04-04-86		Soyuz 12 (1973). Soyuz 18-1 (1975). Soyuz 27 (1978). Soyuz T-3 (1980)
7. Vladislav Nikolayevich Volkov	11-23-35	06-30-71	06-30-71	Soyuz 7 (1969). Soyuz 11 (1971)
8. Aleksey Stanislovich Yeliseyev	07-13-34	01-10-86		Soyuz 5 (1969). Soyuz 8 (1969). Soyuz 10 (1971)

January–February 1967—TsKBEM[10]

Name	Birth	Resign.	Death	Space Missions
1. Nikolay Nikolayevich Rukavishnikov	09-18-32	07-07-87		Soyuz 10 (1971). Soyuz 16 (1974). Soyuz 33 (1979)
2. Vitaliy Ivanovich Sevastyanov	07-08-35	12-30-93		Soyuz 9 (1970). Soyuz 18 (1975)

April–May 1967 (GK VVS)—VVS Group 4[11]

Pilots

Name	Birth	Resign.	Death	Space Missions
1. SL Valeriy Mikhaylovich Beloborodov	10-26-39	08-29-69		
2. C Sergey Nikolayevich Gaydukov	10-31-36	12-04-78		
3. SLE Vladimir Vasilyevich Kovalenok	03-03-42	06-23-84		Soyuz 25 (1977). Soyuz 29 (1978). Soyuz T-4 (1981)
4. SL Vladimir Sergeyevich Kozelskiy	01-12-42	04-20-83		
5. SL Vladimir Afanasevich Lyakhov	07-20-41	08-19-94		Soyuz 32 (1979). Soyuz T-9 (1983). Soyuz T-6 (1988)
6. SL Yuriy Vasilyevich Malyshev	08-27-41	07-02-88	11-08-99	Soyuz T-2 (1980). Soyuz T-11 (1984)
7. SL Viktor Mikhaylovich Pisarev	08-15-41	05-21-68		
8. C Mikhail Vladimirovich Sologub	11-06-36	09-20-68	08-04-96	

Engineers

Name	Birth	Resign.	Death	Space Missions
9. ME Vladimir Borisovich Alekseyev	08-19-33	04-20-83		
10. ME Mikhail Nikolayevich Burdayev	08-27-32	04-20-83		
11. SL Vladimir Timofeyevich Isakov	04-04-40	04-20-83		
12. ME Nikolay Stepanovich Porvatkin	04-15-32	04-20-83		

May 22, 1967 (AN SSSR)

IZMIRAN AN SSSR

Name	Birth	Resign.	Death	Space Missions
1. Mars Nurgaliyevich Fatkullin	05-14-39	xx-xx-70		
2. Rudolf Alekseyevich Gulyayev	11-14-34	08-xx-68		
3. Ordinard Panteleymonovich Kolomiytsev	01-29-39	xx-xx-68		

IPM AN SSSR

Name	Birth	Resign.	Death	Space Missions
4. Valentin Gavriilovich Yershov	06-21-28	08-xx-74	02-15-98	

May 27, 1968 (MOM)—TsKBEM Group 1[12]

Name	Birth	Resign.	Death	Space Missions
1. Vladimir Grigoryevich Fartushniy[13]	02-03-38	xx-xx-71		

TABLE II

Name	Birth	Resign.	Death	Space Missions
2. Konstantin Petrovich Feoktistov[14]	02-07-26	10-28-87		Voskhod (1964)
3. Georgiy Mikhaylovich Grechko[15]	05-25-31	06-05-86		Soyuz 17 (1975), Soyuz 26 (1977), Soyuz T-14 (1985)
4. Valeriy Nikolayevich Kubasov	01-07-35	11-03-93		Soyuz 6 (1969), Soyuz 19 (1975), Soyuz 36 (1980)
5. Oleg Grigoryevich Makarov	01-06-33	04-04-86		Soyuz-12 (1973), Soyuz-18-1 (1975), Soyuz-27 (1978), Soyuz T-3 (1980)
6. Viktor Ivanovich Patsayev	06-19-33	06-30-71	06-30-71	Soyuz 11 (1971)
7. Nikolay Nikolayevich Rukavishnikov	09-18-32	07-07-87		Soyuz 10 (1971), Soyuz 16 (1974), Soyuz 33 (1979)
8. Vitaliy Ivanovich Sevastyanov	07-08-35	12-30-93		Soyuz 9 (1970), Soyuz 18 (1975)
9. Vladislav Nikolayevich Volkov	11-23-35	06-30-71	06-30-71	Soyuz 7 (1969), Soyuz 11 (1971)
10. Valeriy Aleksandrovich Yazdovskiy	07-08-30	07-01-82		
11. Aleksey Stanislavovich Yeliseyev	07-13-34	01-10-86		Soyuz 5 (1969), Soyuz 8 (1969), Soyuz 10 (1971)

April 27, 1970 (GK VVS)—VVS Group 5

Pilots

Name	Birth	Resign.	Death	Space Missions
1. C Anatoliy Nikolayevich Berezovoy	04-11-42	10-21-92		Soyuz T-5 (1982)
2. SL Anatoliy Ivanovich Dedkov	07-27-44	04-20-83		
3. C Vladimir Aleksandrovich Dzhanibekov	05-13-42	06-24-86		Soyuz 27 (1978), Soyuz 39 (1981), Soyuz T-6 (1982), Soyuz T-12 (1984), Soyuz T-13 (1985)
4. SL Yuriy Fedorovich Isaulov	08-31-43	01-29-82		
5. L Vladimir Ivanovich Kozlov	10-02-45	05-28-73		
6. L Leonid Ivanovich Popov	08-31-45	06-13-87		Soyuz 35 (1980), Soyuz 40 (1981), Soyuz T-7 (1982)
7. SL Yuriy Viktorovich Romanenko	08-01-44	10-11-88		Soyuz 26 (1977), Soyuz 38 (1980), Soyuz TM-2 (1987)

Engineers

Name	Birth	Resign.	Death	Space Missions
8. L Nikolay Nikolayevich Fefelov	05-20-45	11-09-95		
9. Valeriy Vasilievich Illarionov	06-02-39	10-30-92	03-10-99	

March 22, 1972 (GMVK)

IMBP Group 1

Name	Birth	Resign.	Death	Space Missions
1. Georgiy Vladimirovich Machinskiy	10-11-37	06-04-74		
2. Valeriy Vladimirovich Polyakov	04-27-42	06-01-95		Soyuz TM-6 (1988), Soyuz TM-18 (1994)
3. Lev Nikolayevich Smirennyy	10-25-32	11-xx-86		

TsKBEM Group 2

Name	Birth	Resign.	Death	Space Missions
4. Boris Dmitriyevich Andreyev	10-06-40	09-05-83		
5. Valentin Vitalyevich Lebedev[16]	04-14-42	11-04-89		Soyuz 13 (1973), Soyuz T-5 (1982)
6. Yuriy Anatolyevich Ponomarev	03-24-32	04-11-83		

Name	Birth	Resign.	Death	Space Missions
TsKBM Group 1				
7. Valeriy Grigoryevich Makrushin	xx-xx-40	04-08-87		
March 27, 1973 (GMVK)				
TsKBEM Group 3				
1. Vladimir Viktorovich Aksenov	02-01-35	10-17-88		Soyuz 22 (1976). Soyuz T-2 (1980)
2. Aleksandr Seregeyevich Ivanchenkov	09-28-40	11-03-93		Soyuz 29 (1978). Soyuz T-6 (1982)
3. Valeriy Viktorovich Ryumin	08-16-39	10-28-87		Soyuz 25 (1977). Soyuz 32 (1979). Soyuz 35 (1980). STS-91 (1998)
4. Gennadiy Mikhaylovich Strekalov	10-28-40	01-17-95		Soyuz T-3 (1980). Soyuz T-8 (1983). Soyuz T-11 (1984). Soyuz TM-10 (1990). Soyuz TM-21 (1995)
TsKBM Group 2				
5. Dmitriy Andreyevich Yuyukov	02-26-41	04-08-87		

TABLE 11

Notes

1. Anikeyev, Bykovskiy, Gagarin, Gorbatko, Komarov, Leonov, Nelyubov, Nikolayev, Popovich, Shonin, Titov, and Volynov joined the TsPK cosmonaut detachment by an order of the GK of VVS dated March 7, 1960. Khrunov joined on March 9, 1960. Zaykin and Filatyev joined on March 25, 1960. Belyayev, Bondarenko, and Rafikov joined on April 28, 1960, and Kartashov joined on June 17, 1960.

2. Kuznetsova, Solovyeva, and Tereshkova joined the TsPK cosmonaut detachment by an order of the GK of VVS dated March 12, 1962. Yerkina and Ponomareva joined on April 3, 1962.

3. MK selected Lazarev, Polyakov, Sorokin, and Yegorov on May 26, 1964. On May 29, they and Katys were ordered to begin training at TsPK. On June 1, 1964, the five of them began training. Feoktistov was selected by MK on June 11, 1964.

4. Katys was selected later as a candidate cosmonaut in April 1965.

5. Lazarev was selected later as a pilot/physician on January 17, 1966.

6. Feoktistov was reselected as a test-cosmonaut of MOM on May 27, 1968. Despite the rank of cosmonaut 3rd class, Feoktistov was not transferred to the flight-test department. He was officially named instructor-test-cosmonaut on January 1, 1977, and remained so until October 28, 1987.

7. Lazarev had already been selected as a candidate cosmonaut on June 1, 1964.

8. Grechko, Kubasov, Makarov, Volkov, and Yeliseyev were reselected on May 27, 1968, as test-cosmonauts of MOM's TsKBEM.

9. Grechko was selected later as a test-cosmonaut from AN SSSR on July 6, 1986.

10. Rukavishnikov was named test-engineer on February 1, 1967, while Sevastyanov was named the same on January 31, 1967. Both were reselected later as test-cosmonauts of MOM's TsKBEM on May 27, 1968.

11. Alekseyev, Burdayev and Porvatkin joined by order of the GK of VVS on April 12, 1967, and the remaining men joined on May 7, 1967.

12. Of the eleven, all except Fartushniy, Feoktistov, and Yazdovskiy had been selected as part of the OKB-1 selections in May 1966 and January-February 1967.

13. Fartushniy was originally a researcher at the Paton Institute of Welding at Kiev, but he transferred to TsKBEM upon selection.

14. Feoktistov was selected as a Voskhod candidate cosmonaut from OKB-1 on June 12, 1964. He was named a cosmonaut 3rd class on May 25, 1966, but did not officially join the test-flight department. He was officially named an NPO Energiya instructor-test-cosmonaut on January 1, 1977, and remained so until October 28, 1987.

15. Grechko was selected later as a test-cosmonaut from AN SSSR on July 6, 1986.

16. Lebedev was later reselected as an NPO Energiya test-cosmonaut on November 1, 1989.

Sources

1. I. Marinin, "Russian Cosmonaut-Scholars" (English title), Novosti kosmonavtiki 3 (January 28–February 11, 1996): 49–54.

2. I. Marinin, "The First Civilian Cosmonauts" (English title), Novosti kosmonavtiki 12–13 (June 3–30, 1996): 81–87.

3. V. Semenov, I. Marinin and S. Shamsutdinov, Iz Istorii kosmonavtiki: vypusk I: nabory v otryady kosmonavtov i astronavtov (Moscow: AO Videokosmos, 1995).

4. S. Yegupov and I. Karpenko, "Detachment of Air Force's Cosmonauts" (English title), Aviatsiya i kosmonavtika no. 5 (May 1990): 46–47.

5. "Air Force Cosmonaut Detachment" (English title), Aviatsiya i kosmonavtika no. 6 (June 1990): 46–47.

Abbreviations

AN SSSR	USSR Academy of Sciences
C	Captain
CO	Colonel
CE	Captain-Engineer
GK VVS	Commander-in-Chief of the Air Force
GMVK	State Interdepartmental Commission for the Selection of Candidate Cosmonauts
GosNII AiKM	State Scientific-Research Institute for Aviation and Space Medicine
IAT	Institute of Automation and Telemechanics
IMBP	Institute of Medico-Biological Problems
IPM	Institute of Applied Mathematics
IZMIRAN	Institute of Terrestrial Magnetism and Radio Waves
L	Lieutenant
LC	Lieutenant-Colonel
LCE	Lieutenant-Colonel-Engineer
M	Major
ME	Major-Engineer
MK	Mandate Commission
MOM	Ministry of Machine Building
P	Private
SG	Sergeant
SL	Senior Lieutenant
SLE	Senior Lieutenant-Engineer
TsKBEM	Central Design Bureau of Experimental Machine Building
TsKBM	Central Design Bureau of Machine Building
TsPK	Cosmonaut Training Center
VVS	Air Force

TABLE III

Table III
Administrative Organizations in the Soviet Missile and Space Programs, 1945-91

Policy

Special Committee for Reactive Technology of the USSR Council of Ministers
(established on May 13, 1946)

History: This committee was established by official decree no. 1017-419ss of the USSR Council of Ministers, dated May 13, 1946, to oversee the development of all long-range ballistic, cruise, and air defense missiles. The committee was dissolved in 1949. By 1957, policy aspects of the missile and space programs were moved to the Central Committee of the USSR Council of Ministers.

Designations	Date From	Date to
• Special Committee for Reactive Technology of the USSR Council of Ministers	May 1946	June 1947
• Special Committee No. 2 of the USSR Council of Ministers	June 1947	October 1949

Chairmen		
• G. M. Malenkov	May 1946	March 1947
• N. A. Bulganin	March 1947	October 1949

Secretary for Defense Industries and Space of the Secretariat of the Central Committee of the Communist Party (established in June 1957)

History: The position was established in June 1957 by Nikita S. Khrushchev as the locus of power in the Soviet Union shifted from the USSR Council of Ministers to the Central Committee of the Communist Party. The holder of the post was the most powerful leader in the USSR in determining Soviet space policy during the 1957–91 period.

Secretaries	Date From	Date to
• L. I. Brezhnev	July 1957	July 1960
• F. R. Kozlov	July 1960	June 1963
• L. I. Brezhnev	June 1963	March 1965
• D. F. Ustinov	March 1965	October 1976
• Y. P. Ryabov	October 1976	April 1979
• A. P. Kirilenko	April 1979	August 1983
• G. V. Romanov	August 1983	July 1985
• L. N. Zaykov	July 1985	February 1988
• O. S. Baklanov	February 1988	August 1991

Central Committee Defense Industries Department

History: The origins of this department are obscure, but it clearly assumed a greater role beginning in 1958, when I. D. Serbin became its chief. Its role was to serve as doctrinal overseer of the defense industrial and space sectors. The department reported directly to the Secretary of the Central Committee for Defense Industries and Space. The department was abolished in June 1990.

Designations	Date From	Date to
• Defense Industries Department of the Central Committee	Unknown	September 1988
• Defense Department of the Central Committee	September 1988	June 1990

Chiefs		
• I. D. Serbin	February 1958	February 1981
• I. F. Dmitriyev	February 1981	August 1985
• O. S. Belyakov	August 1985	June 1990

First Deputy Chief
- I. F. Dmitriyev   1965   February 1981
- N. M. Luzhin   1988   June 1990

Sector Head for Space
- B. A. Stroganov   1960s   Unknown

Instructor for Space
- V. A. Popov   1960s   Unknown

Implementation

Military-Industrial Commission (VPK) of the Presidium of the USSR Council of Ministers (established on April 14, 1955)

History: VPK traces its ancestry back to the Third Chief Directorate (TGU) of the USSR Council of Ministers, which was established on February 3, 1951, to manage the development of all Soviet missile weapons (cruise, ballistic, air defense, and naval). On July 1, 1953, the TGU was combined with the First Chief Directorate of the USSR Council of Ministers to form the new Ministry of Medium Machine Building (MSM). The TGU, now known as GlavSpetsMash (Chief Directorate of Special Machine Building), became a subordinate department to MSM. On April 14, 1955, GlavSpetsMash was separated from MSM. A portion, including all subordinate design bureaus and subdivisions, was moved to the Ministry of Defense Industries. Simultaneously, the remainder (that is, the old structure of the TGU) was used as the basis for the new Special Committee for Armaments for the Army and VMF (the Navy) and subordinated directly to the USSR Council of Ministers. From then on, this Special Committee supervised all tactical and strategic missile programs in the Soviet Union. In December 1957, this Special Committee was renamed the Commission of the Presidium of the USSR Council of Ministers for Military-Industrial Issues-or more familiarly, the Military-Industrial Commission (VPK). Its supervisory duties were expanded from missiles to the entire Soviet defense industry.

Designations	Date From	Date to
• Special Committee for Armaments for the Army and VMF of the USSR Council of Ministers	April 1955	December 1957
• Commission of the Presidium of the USSR Council of Ministers for Military-Industrial Issues	December 1957	1986
• State Commission of the USSR Council of Ministers for Military-Industrial Issues	1986	August 1991

Chairmen

• V. M. Ryabikov	April 1955	December 1957
• D. F. Ustinov	December 1957	March 1963
• L. V. Smirnov	March 1963	December 1985
• Yu. D. Maslyukov	December 1985	February 1988
• I. S. Belousey	February 1988	January 1991
• Yu. D. Maslyukov	January 1991	August 1991

First Deputy Chairmen

• G. A. Titov	April 1955	December 1957
• S. I. Vetoshkin	December 1957	1964
• N. S. Stroyev	1977	1987
• V. L. Koblov	1987	August 1991

Deputy Chairmen	Date From	Date to
• A. K. Repin	April 1955	Unknown
• A. N. Shchukin	April 1955	1969
• G. N. Pashkov	December 1957	1970
• G. A. Titov	December 1957	1974
• N. S. Stroyev	1966	1977
• L. I. Gorshkov	1966	1970s
• S. A. Arzhakov	Unknown	Unknown

TABLE III

• B. A. Komissarov	1970s	Unknown
• A. I. Voznesenskiy	Unknown	Unknown
• L. B. Vasilyev	March 1988	Unknown

Ministry of Armaments (MV) (established on January 11, 1939)

History: This ministry was originally established in January 1939, having being split off from the People's Commissariat of Defense Industry. Through its various incarnations, it managed the development of the Soviet ballistic missile and space programs from 1946 to 1965 via its subordinate Seventh Chief Directorate. In March 1965, the Seventh Chief Directorate was removed from the ministry and became the basis for the new Ministry of General Machine Building. Since that time, the ministry had little involvement in the ballistic missile and space programs.

Designations / **Date From** / **Date to**

Designations	Date From	Date to
• People's Commissariat of Armaments (NKA)	January 1939	March 1946
• Ministry of Armaments (MV)	March 1946	March 1953
• Ministry of Defense Industry (MOP)	March 1953	December 1957
• State Committee for Defense Technology (GKOT)	December 1957	March 1965

People's Commissars/Ministers/Chairmen

• B. L. Vannikov	January 1939	June 1941
• D. F. Ustinov	June 1941	December 1957
• A. V. Domrachev	December 1957	March 1958
• K. N. Rudnev	March 1958	June 1961
• L. V. Smirnov	June 1961	March 1963
• S. A. Zverev	March 1963	March 1965

First Deputies

• V. M. Ryabikov	1940	February 1951
• A. V. Domrachev	1951	1957
• S. I. Vetoshkin	1955	December 1957
• S. A. Zverev	December 1959	March 1963
• G. A. Tyulin	June 1961	March 1965

Deputies

• V. N. Novikov	1941	1948
• K. M. Gerasimov	1949	1951
• I. G. Zubovich	October 1949	March 1953
• A. V. Domrachev	1951	1951
• S. A. Zverev	March 1952	March 1953
• K. N. Rudnev	May 1952	March 1958
• P. N. Goremykin	August 1953	April 1955
• K. M. Gerasimov	1954	1957
• V. N. Novikov	1954	April 1955
• S. A. Zverev	March 1954	December 1959
• L. A. Grishin	March 1958	October 1960
• G. N. Kozhevnikov	Late 1950s	Unknown
• V. M. Larionov	Late 1950s	Unknown
• S. N. Makhonin	Late 1950s	Unknown
• L. V. Smirnov	February 1961	June 1961

Chiefs of the Chief Directorates

• N. E. Nosovskiy (First GU)	1940	1947
• K. M. Gerasimov	1941	1949
• K. M. Gerasimov	1951	1954
• L. A. Grishin	October 1952	March 1958
• S. A. Zverev (Second GU)	March 1952	March 1952
• S. A. Zverev (Eighth GU)	March 1952	March 1954
• V. N. Novikov (Fifth GU)	1953	1953

• P. V. Finogenov (Sixth GU)	1963	March 1965
• S. I. Vetoshkin (Seventh GU)	1939	October 1949
• I. G. Zubovich (Seventh GU)	October 1949	August 1951
• L. V. Smirnov (Seventh GU)	August 1951	June 1952
• M. S. Ryazanskiy (Seventh GU)	June 1952	1954
• V. A. Kolychev (Seventh GU)	December 1955	Unknown
• A. S. Tomilin (Seventh GU)	Late 1950s	Unknown
• B. A. Komissarov (Seventh GU)	Early 1960s	Unknown

Ministry of Aviation Industry (MAP) (established on January 11, 1939)

History: This ministry was originally established in January 1939, having being split off from the People's Commissariat of Defense Industry. As more and more aviation organizations began participating in the missile and space sector beginning the late 1950s, the ministry took a greater role in such efforts. Note that from March to August 1953, it was part of the Ministry of Defense Industries. Many of the space and missile organizations were transferred from the Ministry of Aviation Industry to the new Ministry of General Machine Building upon the latter's formation in March 1965.

Designations

	Date From	Date to
• People's Commissariat of Aviation Industry (NKAP)	January 1939	March 1946
• Ministry of Aviation Industry (MAP)	March 1946	December 1957
• State Committee for Aviation Technology (GKAT)	December 1957	March 1965
• Ministry of Aviation Industry (MAP)	March 1965	January 1992

People's Commissars/Ministers/Chairmen

• M. M. Kaganovich	January 1939	January 1940
• A. I. Shakhurin	January 1940	March 1946
• M. V. Khrunichev	March 1946	March 1953
• P. V. Dementyev	August 1953	May 1977
• V. A. Kazakov	June 1977	February 1981
• I. S. Silayev	February 1981	November 1985
• A. S. Systov	November 1985	November 1991

First Deputies

• P. V. Dementyev	1941	1946
• V. P. Balandin	August 1953	1957
• S. M. Leshchenko	1957	1964
• V. A. Kozlov	1964	1965
• S. I. Kadyshev	1965	1974
• V. A. Kazakov	1974	June 1977
• I. S. Silayev	1977	December 1980
• A. S. Systov	February 1981	November 1985

Ministry of General Machine Building (MOM) (established on March 2, 1965)

History: This ministry was established on the basis of the Seventh Chief Directorate of the State Committee for Defense Technology (GKOT), which oversaw all ballistic missile and space programs. MOM managed the development of almost all Soviet ballistic missiles and spacecraft from 1965 to 1991. It was officially abolished in November 1991.

Ministers

	Date From	Date to
• S. A. Afanasyev	March 1965	April 1983
• O. D. Baklanov	April 1983	February 1988
• V. K. Doguzhiyev	April 1988	July 1989
• O. N. Shishkin	July 1989	August 1991
• R. R. Kiryushin	August 1991	November 1991

First Deputy Ministers

	Date From	Date to
• G. A. Tyulin	March 1965	1976

TABLE III

• B. V. Balmont	1976	February 1981
• O. D. Baklanov	February 1981	April 1983
• V. N. Konovalov	April 1983	1987
• V. K. Doguzhiyev	1987	March 1988
• O. N. Shishkin	February 1988	July 1989
• R. R. Kiryushin	July 1989	August 1991

Deputy Ministers

• V. Ya. Litvinov	March 1965	1973
• G. M. Tabakov (engines)	March 1965	1981
• Ye. V. Mazur (construction)	March 1965	1982
• G. R. Udarov (launch complexes)	March 1965	1979
• N. D. Khokhlov (quality)	March 1965	1983
• L. I. Gusev (guidance)	March 1965	1965
• M. A. Brezhnev (guidance systems)	1965	Unknown
• B. V. Balmont	1973	1976
• V. V. Lobanov (finances)	January 1974	1980s
• O. D. Baklanov	1976	1981
• V. N. Konovalov (naval)	Unknown	April 1983
• V. N. Soshin (construction)	1982	Late 1980s
• V. K. Doguzhiyev	1983	1987
• Ye. A. Zhelonov	1984	Unknown
• A. S. Matrenin (quality)	1984	November 1991
• O. N. Shishkin (space)	Unknown	February 1988
• G. F. Grigorenko	1980s	November 1991
• Yu. N. Koptev	Unknown	November 1991
• A. Ye. Shestakov	Unknown	November 1991
• R. R. Kiryushin	Unknown	1989
• V. Ye. Sokolov	Unknown	1990s
• S. S. Vanin (complexes)	Unknown	1980s
• V. N. Ivanov	Mid-1980s	Unknown

Chiefs of Chief Directorates

• K. A. Kerimov (Third GU)	March 1965	June 1974
• V. D. Vachnadze (Third GU)	June 1974	June 1977
• A. K. Vanitskiy	1974	1976
• B. V. Balmont (Sixth GU)	1965	1972
• B. V. Balmont (Eighth GU)	1972	1973
• A. S. Matrenin (Seventh GU)	1969	1984
• A. S. Kirillov	June 1969	November 1977
• Yu. N. Koptev (Third GU)	Mid-1980s	Unknown
• V. A. Andreyev (First GU)	January 1989	Unknown
• A. I. Dunayev (Thirteenth GU)	June 1985	November 1991
• I. N. Gabelko	Unknown	November 1991

Ministry of Medium Machine Building (MSM) (established on July 1, 1953)

History: This ministry was responsible for the manufacture of all Soviet nuclear warheads from 1953 to 1991. Its lineage goes back to August 20, 1945, with the formation of the First Chief Directorate (PGU) of the USSR Council of Ministers. On March 16, 1953, the PGU absorbed the Second Chief Directorate of the Council of Ministers. On July 1, 1953, the PGU and the Third Chief Directorate combined to form the Ministry of Medium Machine Building (MSM). MSM oversaw all missile programs through its subordinate GlavSpetsMash between July 1953 and April 1955.

Designations	**Date From**	**Date to**
• First Chief Directorate	August 1945	June 1953
• Ministry of Medium Machine Building (MSM)	July 1953	March 1963
• State Committee for Medium Machine Building (GKSM)	March 1963	March 1965
• Ministry of Medium Machine Building (MSM)	March 1965	June 1989

Chiefs/Ministers/Chairmen

• B. L. Vannikov	August 1945	June 1953
• V. A. Malyshev	July 1953	February 1955
• A. P. Zavenyagin	February 1955	December 1956
• M. G. Pervukhin	April 1957	July 1957
• Ya. P. Slavskiy	July 1957	November 1986
• L. D. Ryabev	November 1986	June 1989

State Commissions in the Early Space and Missile Programs

Product	Chairmen	Dates
A-4	N. D. Yakovlev	1947
R-1	S. I. Vetoshkin	1948–50
R-2	G. I. Ioffe	1950–51
R-5	P. A. Degtyarev	1953–55
R-11	A. I. Nesterenko	1953–55
R-5M	P. A. Degtyarev	1954–56
R-7	V. M. Ryabikov	August 1956–57
	K. N. Rudnev	1957–59
R-7A	M. I. Nedelin, A. G. Mrykin,	
	A. I. Nesterenko, K. V. Gerchik	1959–60
Scientific vertical launches	A. A. Blagonravov	1951–61
Sputnik	V. M. Ryabikov	1957–58
Luna	K. N. Rudnev	1958–60
	M. I. Nedelin	1960
	G. A. Tyulin	1963–76

Product	Chairmen	Dates
Vostok	M. I. Nedelin	1960
	K. N. Rudnev	1960–61
	L. V. Smirnov	1961–63
	G. A. Tyulin	1963
Voskhod	G. A. Tyulin	August 1964–66
Soyuz, DOS/*Salyut, Mir*	K. A. Kerimov	October 1966–91
	V. L. Ivanov	1991–96
UR-500K-L1	G. A. Tyulin	December 1966–70
N1-L3	S. A. Afanasyev	1967–72
T2K	A. A. Maksimov	1970–71
Almaz/*Salyut*	M. G. Grigoryev	1973–77
MP-1	A. G. Zakharov	1961

Clients

Ministry of Defense (established in postwar form on February 25, 1946)

History: The Ministry of Defense was the primary client of the Soviet missile and space programs. Its subordinate Strategic Missile Forces managed all missile and space operations during 1959–81. The Deputy Minister of Defense for Armaments was responsible for weapons (and spacecraft) procurement. Note that between 1960 and 1970, N. N. Alekseyev was the Chairman of the Scientific-Technical Committee (NTK) of the General Staff of the Ministry of Defense, essentially performing the same duties as the Deputy Minister of Defense for Armaments, a post that did not exist during that period.

Designations	Date From	Date to
• Ministry of Armed Forces (MVS)	February 1946	February 1950
• Ministry of War	February 1950	March 1953
• Ministry of Defense (MO)	March 1953	January 1992

TABLE III

Ministers

• I. V. Stalin	February 1946	March 1947
• N. A. Bulganin	March 1947	March 1949
• A. M. Vasilyevskiy	March 1949	March 1953
• N. A. Bulganin	March 1953	February 1955
• G. K. Zhukov	February 1955	October 1957
• R. Ya. Malinovskiy	October 1957	March 1967
• A. A. Grechko	April 1967	April 1976
• D. F. Ustinov	April 1976	December 1984
• S. L. Sokolov	December 1984	May 1987
• D. T. Yazov	May 1987	September 1991
• Ye. I. Shaposhnikov	September 1991	December 1991

Deputy Ministers of Defense for Armaments

• M. I. Nedelin	March 1955	December 1959
• N. N. Alekseyev	1960	1978
• V. M. Shabanov	1978	Unknown

Missile Forces of Strategic Designation (RVSN) (established on December 17, 1959)

History: RVSN managed all Soviet missile and space operations during 1959–81.

Commanders

	Date From	Date to
• M. I. Nedelin	December 1959	October 1960
• K. S. Moskalenko	October 1960	April 1962
• S. S. Biryuzov	April 1962	March 1963
• N. I. Krylov	March 1963	February 1972
• V. F. Tolubko	February 1972	July 1985
• Yu. P. Maksimov	July 1985	August 1992

First Deputy Commanders

• V. F. Tolubko	March 1960	1968
• M. G. Grigoryev	1968	December 1981
• Yu. A. Yashin	December 1981	1989
• A. P. Volkov	1989	1994

Chiefs of the Scientific-Technical Committee (NTK)

• V. P. Morozov	June 1962	1967
• A. A. Vasilyev	1967	1969
• A. S. Kalashnikov	1969	1974
• S. A. Sergeyev	1974	1979
• V. M. Ryumkin	1979	1989
• V. G. Popov	1989	

Chief Directorate of Reactive Armaments (GURVO) (established on May 13, 1946)

History: From 1946, GURVO, in its various incarnations, oversaw the procurement of ballistic missiles into armaments of the Strategic Missile Forces. Between 1960 and 1970, GURVO's subordinate TsUKOS was the primary client of the Soviet space program.

Designations

	Date From	Date to
• 4th Directorate of the Chief Artillery Directorate	May 1946	April 1953
• Directorate of the Deputy Commander of Artillery (UZKA)	April 1953	March 1955
• Directorate of the Commander of Reactive Armaments (UNRV)	March 1955	December 1959
• Chief Directorate of Reactive Armaments (GURVO)	December 1959	1993

Commanders

	Date From	Date to
• A. I. Sokolov	May 1946	August 1954
• A. I. Semenov	August 1954	August 1964

• A. A. Vasilyev	August 1964	1967
• N. N. Smirnitskiy	1967	December 1975
• Yu. A. Pichugin	December 1975	1984
• A. A. Ryazhskikh	1984	1993

Chief Directorate of Space Assets (GUKOS) (established in October 1964)

History: In October 1964, the Third Directorate of the Chief Directorate of Reactive Armaments (GURVO) of the Strategic Missile Forces (RVSN) was reorganized into TsUKOS. In March 1970, TsUKOS was combined with the Center for Leading the Development and Production of Space Assets (itself established in March 1963 within GURVO) to form the new GUKOS and subordinated to RVSN. On November 10, 1981, GUKOS was separated from RVSN and subordinated directly to the Ministry of Defense. GUKOS was the primary client for the Soviet space program, responsible for all operational aspects, including tracking and launch activities. It had jurisdiction over NIIP-5 (Tyura-Tam), military units at Mirnyy (Plesetsk), the Command-Measurement Complex (KIK), the A. F. Mozhayskiy Military Academy, TsNII-50, 28 Arsenal (Karian-Stroganov), and military representatives to research and development organizations.

Designations	Date From	Date to
• Third Directorate of the Chief Directorate of Reactive Armaments	September 1960	October 1964
• Central Directorate of Space Assets (TsUKOS)	October 1964	March 1970
• Chief Directorate of Space Assets (GUKOS)	March 1970	November 1986
• Directorate of the Chief of Space Assets (UNKS)	November 1986	August 1992

Commanders		
• K. A. Kerimov	September 1960	March 1965
• A. G. Karas	March 1965	January 1979
• A. A. Maksimov	January 1979	1989
• V. L. Ivanov	1989	October 1996

First Deputy Commanders		
• A. A. Maksimov	Unknown	January 1979
• G. S. Titov	July 1979	October 1991
• V. L. Ivanov	1984	1989

Command-Measurement Complex Center (TsKIK)
(established by order dated September 3, 1956)

History: The Command-Measurement Complex (KIK) was the ground communications network for the Soviet space program. It was established on the basis of the Range Measurement Complex network of tracking stations established for early R-7 ICBM launches. In 1956-57, the Range Measurement Complex was reconfigured into the KIK to support the launch of the Object D satellite (launched as Sputnik 3). The KIK, including its main center, the Command-Measurement Complex Center (TsKIK), was subordinate to NII-4 until March 7, 1962, when it was subordinated directly to the Strategic Missile Forces (RVSN). The TsKIK began operations on July 12, 1957. In January 1982, the TsKIK was reorganized into the Chief Scientific-Research Testing Center for Space Assets of the Ministry of Defense (GNIITs KS MO). The center operated tracking for all Soviet-era space operations via its various Scientific-Measurement Points (NIP) spread across the Soviet Union.

Commanders of Military Unit No. 32103/TsKIK	Date From	Date to
• A. A. Vitruk	July 1957	1959
• A. G. Karas	1959	March 1965
• I. I. Spitsa	March 1965	January 1973
• I. D. Statsenko	January 1973	January 1976
• N. F. Shlykov	January 1976	1989
• V. N. Ivanov	1989	1989

Scientific-Measurement Points (NIP)	Location
• IP-1	Tyura-Tam
• IP-2	Makat
• IP-3	Sary-Shagan

TABLE III

- NIP-4   Yeniseyesk
- IP-5   Iskhup
- NIP-6   Yelizovo (near Petropavlovsk-Kamchatka)
- IP-7   Klyuchi
- NIP-9   Krasnoye selo (near Leningrad)
- NIP-10   Simferopol
- IP-11   Sartychaly
- NIP-12   Kolpashevo
- NIP-13   Ulan-Ude
- NIP-14   Shchelkovo (near Moscow)
- NIP-15   TsDRS, Galenkiy (near Ussuriysk)
- NIP-16   TsDKS, Yevpatoriya
- NIP-17   Yakutsk
- NIP-18   Vorkuta
- NIP-19   Dunyevtsy, Khmelnitskaya oblast
- NIP-20   Solnechnyi (near Komsomolsk-na-Amur)
- NIP-22   Yevpatoriya?
- IP-41Ye   Simeiz
- IP-42Ye   Moscow

Air Force (VVS)

History: The Deputy Chief of Combat Preparations of the Air Force was directly responsible for the selection and training of cosmonauts and the selection of crews for all piloted space missions. By an order dated April 10, 1962, the holder of these duties was made the General Staff Deputy Chief for Space. On March 29, 1966, the holder of these duties was made the Commander-in-Chief's Aide for Space. The Aide for Space officially supervised the Cosmonaut Training Center, the Air Force Biomedical Service, and the Solar Service.

Commanders-in-Chief	Date From	Date to
- K. A. Vershinin	1946	1949
- P. F. Zhigarev	1949	1957
- K. A. Vershinin	1957	March 1969
- P. S. Kutakhov	March 1969	December 1984
- A. N. Yefimov	December 1984	1990
- Ye. A. Shaposhnikov	1990	August 1991

First Deputy Commanders		
- S. I. Rudenko	1958	July 1968
- P. S. Kutakhov	July 1968	March 1969
- A. N. Yefimov	March 1969	December 1984

Deputy Chiefs of Combat Preparations		
- N. P. Kamanin	1958	October 1971
- V. A. Shatalov	October 1971	June 1986

Science Sector

USSR Academy of Sciences (AN SSSR)

History: The Russian Academy of Sciences was established on January 28, 1724. In 1925, it was renamed the USSR Academy of Sciences.

Presidents	Date From	Date to
- S. I. Vavilov	July 1945	January 1951
- A. N. Nesmeyanov	January 1951	May 1961
- M. V. Keldysh	May 1961	November 1975
- A. P. Aleksandrov	November 1975	1986
- V. A. Kotelnikov	1986	October 1986
- G. I. Marchuk	October 1986	1991

Launch Sites

Kapustin Yar/State Central Test Range No. 4 (GTsP-4)
(established by order dated May 13, 1946)

History: The specific location of the range was confirmed by an order dated July 27, 1947.

Commanders	Date From	Date to
• V. I. Voznyuk	August 1946	April 1973
• Yu. A. Pichugin	April 1973	1975
• P. .G. Degtyarenko	1975	September 1981
• N. Ya. Lopatin	September 1981	1983
• N. V. Mazyarkin	1983	1991

Tyura-Tam/Scientific Research and Testing Range No. 5 (NIIP-5)
(established on June 2, 1955)

History: On January 29, 1958, the town of Zarya was renamed Leninsk. In December 1995, Leninsk was renamed Baikonur (also spelled Baykonur).

Commanders	Date From	Date to
• A. I. Nesterenko	June 1955	July 1958
• K. V. Gerchik	July 1958	April 1961
• A. G. Zakharov	May 1961	March 1965
• A. A. Kurushin	March 1965	1973
• V. I. Fadeyev	1973	1978
• Yu. N. Sergunin	1978	1983
• Yu. A. Zhukov	1983	1989
• A. L. Kryzhko	1989	1991

Mirnyy/Scientific-Research and Testing Range No. 53 (NIIP-53)
(established on January 11, 1957)

History: On August 30, 1963, this became a space launch center. In 1982, one portion of NIIP-53 became GTsIPKS-1278.

Commanders	Date From	Date to
• M. G. Grigoryev	January 1957	1962
• S. F. Shtanko	1962	1963
• G. Ye. Alpaidze	1963	August 1975
• Yu. A. Yashin	August 1975	1979
• V. L. Ivanov	1979	1984
• G. A. Kolesnikov	1984	1985
• I. I. Oleynik	1985	1991

TABLE III

Selected Sources

1. Peter Almquist, *Red Forge: Soviet Military Industry Since 1965* (New York: Columbia University Press, 1990).
2. Central Intelligence Agency, *Directory of Soviet Officials, Vol. I: Personnel in the Communist Party, Government, and Mass Organizations of the U.S.S.R. and R.S.F.S.R.* (Washington, DC: Department of State, 1960).
3. B. Ye. Chertok, *Rakety i lyudi* (Moscow: Mashinostroyeniye, 1994).
4. B. Ye. Chertok, *Rakety i lyudi: Fili Podlipki Tyuratam* (Moscow: Mashinostroyeniye, 1996).
5. B. Ye. Chertok, *Rakety i lyudi: goryachiye dni kholodnoy voyny* (Moscow: Mashinostroyeniye, 1997).
6. Julian Cooper, "The Elite of the Defence Industry Complex," in David Lane, ed., *Elites and Political Power* (London: Edward Elgar, 1988), pp. 167-87.
7. Edward L. Crowley, Andrew I. Lebed, and Dr. Heinrich E. Schulz, eds., *Party and Government Officials of the Soviet Union 1917–1967* (Metuchen, NJ: The Scarecrow Press, 1968).
8. Grigoriy Kisunko, *Sekretnaya zona* (Moscow: Sovremennik, 1996).
9. Christian Lardier, *L'Astronautique Soviétique* (Paris: Armand Colin, 1992).
10. Yu. P. Maksimov, ed., *Raketnyye voyska strategicheskogo naznacheniya* (Moscow: RVSN, 1992).
11. I. D. Sergeyev, ed., *Khronika osnovnikh sobitiy istorii raketnikh voysk strategicheskogo naznacheniya* (Moscow: TsPIK, 1994).
12. Michel Tatu, *Power in the Kremlin: From Khrushchev's Decline to Collective Leadership* (London: Collins, 1970).

TABLE IV

Table IV

Major Contractor Organizations in the Soviet Space Program, 1945–74

Spacecraft and Launch Vehicles

Designer Bureau Name Established Location	Designations	Chief/General Designers	History
Babakin **OKB-301** July 1, 1937 Khimki	• OKB-301 (1945–60) • GSMZ Lavochkin (1960–62) • OKB-52 Branch No. 3 (1962–65) • GSMZ Lavochkin (1965–74) • NPO Lavochkin (1974–)	A. A. Dubrovnin (1937–39) S. A. Lavochkin (1939–60) M. M. Pashinin (1960–62) A. I. Eidis (1962–65) G. N. Babakin (1965–71) S. S. Kryukov (1971–77) V. M. Kovtunenko (1977–95) V. A. Serebrennikov (1995–97) S. P. Kulikov (1997–)	This was established at Plant No. 301. During the war, it was transferred to other locations but returned to Khimki in October 1945. OKB-301 was Branch No. 3 of OKB-52 from December 18, 1962, to March 2, 1965. At various times, this entity specialized in aircraft (1940s), missiles (1950s), and robotic spacecraft (from 1965). In 1974, OKB and the plant merged to form NPO Lavochkin.
Chelomey **OKB-52** August 8, 1955 Reutov	• OKB-52 (1955–66) • TsKBM (1966–83) • NPO Mashinostroyenia (1983–)	V. N. Chelomey (1955–84) G. A. Yefremov (1984–)	Special Design Group No. 10 (SKG-10) was established at Plant No. 500 at Tushino on June 9, 1954. On August 8, 1955, the group became OKB-52, moving in 1956 to the Reutov Machine Building Plant to focus on naval cruise missiles. It was subordinate to GSNII-642 from November 6, 1957, to March 8, 1958, but their roles reversed when GSNII-642 dissolved and turned into a plant. During 1958–59, OKB-52 began work on spacecraft and ballistic missiles.
Iosifyan **NII-627** September 1941 Moscow	• NII-627 (1944–53) • VNII EM (1953–92) • NPP VNII ElektroMekhaniki (1992–)	A. G. Iosifyan (1941–74) N. N. Shermetyevskiy (1974–91) Yu. N. Trifonov (1993–)	This was established at Plant No. 627 in Moscow in 1941 and became NII in 1944. It originally developed power generators for ballistic missiles in the 1940s and 1950s before moving into spacecraft in the 1960s after inheriting the Meteor program from OKB-586.
Korolev **OKB-1** August 26, 1946 Kaliningrad	• NII-88 SKB Dept. No. 3 (1946–50) • NII-88 OKB-1 (1950–56) • OKB-1 (1956–66)	S. P. Korolev (1946–66) V. P. Mishin (1966–74) V. P. Glushko (1974–89) Yu. P. Semenov (1989–)	This was established as Dept. No. 3 of NII-88 at Plant n88 on August 26, 1946, to develop long-range ballistic missiles. On April 24, 1950, this department was restructured into OKB-1, still subordinate to NII-88. On August 14, 1956, OKB-1 became independent of NII-88. On May

Designer Bureau Name Established Location	Designations	Chief/General Designers	History
OKB-1 cont.			22, 1974, OKB-1 (then called TsKBEM) merged with KB EnergoMash to form the new NPO Energiya. KB EnergoMash separated from NPO Energiya on January 19, 1990. From 1945 until the early 1960s, the primary thematic thrust was the development of long-range ballistic missiles. From the early 1960s on, the organization focused primarily on piloted and robotic spacecraft.
Kozlov TsSKB July 23, 1959 Kuybyshev	• OKB-1 SKO No. 25 (1959–60) • OKB-1 Branch No. 3 (1960–66) • OKB-1 (1956–66) • TsKBEM Kuybyshev Branch (1966–74) • TsSKB (1974–96) • GNPRKTs TsSKB-Progress (1996–)	D. I. Kozlov (1959–)	This was established on July 23, 1959, as Special Design Department No. 25 (SKO-25) at Plant No. 1 to supervise the manufacture of the R-7 and derivative launch vehicles, becoming OKB-1 Branch No. 3 on July 17, 1960. It inherited all work on robotic reconnaissance satellites and R-7-based launch vehicles in 1964 from OKB-1, although it remained subordinate to its parent entity until July 30, 1974, when it became independent as TsSKB. It combined with the production facility Progress Plant (formerly Plant No. 1) on April 12, 1996, to form GNPRKTs TsSKB-Progress.
Mikoyan OKB-155 December 8, 1939 Moscow	• OKO (1939–42) • OKB-155 (1942–66) • MMZ Zenit (1966–71) • MMZ Mikoyan (1971–78) • Mikoyan ANPK MiG	A. I. Mikoyan (1939–71) R. A. Belyakov (1971–)	This was established as Experimental Design Section (OKO) in December 1939. It was evacuated to Kuybyshev in October 1941 but returned to Moscow in March 1942 at Plant No. 480. On March 16, 1942, it was renamed Plant No. 155. It worked on the Spiral spaceplane in 1965–78.
Myasishchev OKB-23 March 24, 1951 Fili	• OKB-23 (1951–60) • OKB-52/TsKBM Branch No. 1 (1960–81) • NPO Energiya KB Salyut (1981–88) • NPO EM KB Salyut • KB Salyut • GKNPTs Khrunichev KB Salyut (1993–)	V. M. Myasishkiy (1951–60) V. N. Bugayskiy (1960–73) D. A. Polukhin (1973–93) A. S. Moyseyev (1993–94) A. K. Nedavyoda (1994–)	This was established at Fili in Moscow at Plant No. 23 (established as a factory in April 1916) to develop long-range bombers and cruise missiles. On October 3, 1960, it became a branch of OKB-52 and began developing spacecraft, ICBMs, and space launch vehicles. The aviation database went to OKB-51 and OKB-156. This organization remained an OKB-52 branch until June 30, 1981, when it became a branch of NPO Energiya. On June 22, 1988, it separated from NPO Energiya and formed NPO EM. This NPO eventually dissolved, and KB Salyut became independent for a short while before joining with the M. V. Khrunichev Machine Building Plant to form GKNPTs Khrunichev on June 7, 1993.

TABLE IV

Designer Bureau Name Established Location	Designations	Chief/General Designers	History
Raspletin **KB-1** September 8, 1947 Moscow	• SB-1 (1947–50) • KB-1 • MKB Strela • TsKB Almaz (1967–71) • NPO Almaz (1971–95) • TsKB Almaz (1995–)	P. N. Kuksenko (1947–53) S. L. Beriya (1947–53) S. M. Vladimirskiy (1953) A. A. Raspletin (1953–67) B. V. Bunkin (1967–)	This was formed in 1947 to work on the Kometa system. In August 1950, it was reorganized into KB-1 to work on the Berkut Moscow defense system. In April 1955, three subdivisions were created within KB-1, one of which, SKB-31, was a primary subdivision of KB-1 and headed by Raspletin since its founding. From 1960 on, when Raspletin became KB-1 Director, SKB-31 was headed by B. V. Bunkin. It worked on robotic military reconnaissance satellites via its subordinate OKB-41.
Reshetnev **OKB-10** June 4, 1959 Krasnoyarsk	• OKB-1 Branch (1959–60) • OKB-1 Branch No. 2 (1960–61) • OKB-10 (1961–66) • KB PM (1966–77) • NPO Prikladnoy mekhaniki (1977–96) • Reshetnev NPO Prikladnoy mekhaniki (1996–)	M. F. Reshetnev (1961–96) A. G. Kozlov (1996–)	This was formed at Plant No. 1001 to supervise ICBM production for OKB-1, but it inherited a number of communications satellite projects from OKB-586 and OKB-1 in 1962–67 and began indigenous space space projects. On December 18, 1961, it separated from OKB-1 and became an independent entity (Prikladnoy mekhaniki means Applied Mechanics).
Savin **OKB-41** April 1955 Moscow	• KB-1 SKB-41 (1955–62) • KB-1 OKB-41 • KB-1 OKB Kometa • TsNII Kometa (1973–??) • TsNPO Kometa	A. A. Kolosov A. I. Savin (1962–)	This was established in 1955 as SKB-41, a subdivision of KB-1, to focus on rocket armaments for aircraft. In 1973, OKB-41 (by then named OKB Kometa) detached from KB-1 (by then MKB Strela) and became an independent organization, TsNII Kometa. It worked on robotic military EORSATs, RORSATs, and ASATs.
Tsybin **OKB-256** May 23, 1955 Dubna	• OKB-256 (1955–59)	P. V. Tsybin (1955–59)	This was established at Plant No. 256 in Dubna. When OKB-256 dissolved on October 1, 1959, the database on spaceplane research was transferred to OKB-155. OKB-256 was absorbed by OKB-23, while the plant went to OKB-1 in 1960.
Tyurin **TsKB-7** November 21, 1949 Leningrad	• TsKB-7 (1949–67) • Frunze KB Arsenal (1967–)	N. P. Antonov (1949–52) P. A. Tyurin (1953–81) S. P. Parnyakov (1981–83) Yu. F. Valov (1983–95) B. I. Poletayev (1995–) V. F. Kalabin (1969–83) V. G. Volkov	This was established in 1949 at Plant No. 7 at Leningrad and subordinated to MATsKB for designing naval anti-ship artillery armaments. In 1959–60, it began work on solid-propellant ballistic missiles. In 1969, it began work on space themes after being assigned the production of US-P spacecraft from KB-1. In the late 1980s, it was part of PO Arsenal, remaining so until the early 1990s.

Designer Bureau Name Established Location	Designations	Chief/General Designers	History
Yangel **OKB-586** May 9, 1951 Dnepropetrovsk	• SKB-586 (1951–54) • OKB-586 (1954–66) • KB Yuzhnoye (1966–86) • NPO Yuzhnoye (1986–)	V. S. Budnik (1951–54) M. K. Yangel (1954–71) V. F. Utkin (1971–90) S. N. Konyukhov (1990–)	This was established at Plant No. 586, through the transfer of personnel from NII-88 OKB-1, as Serial Design Bureau No. 586 to supervise the production of OKB-1 missiles. On April 10, 1954, another group of engineers from NII-88 was transferred to the plant, and OKB-586 was formally established.
Zaslavskiy **TsNII-108** July 4, 1943	• VSNII Radiolokatsii (1943–46) • TsNII-108 (1946–66) • TsNIRTI (1966–) • NPO imeni P. S. Pleshakova	M. Ye. Zaslavskiy (1960s) A. I. Berg (1946–53) N. Yemokhonov (1953–59) P. S. Pleshakov (1959–64) Yu. N. Mazhorov (1964–70s) A. Shulunov (1987–)	This is the overall systems integrator for all Soviet electronic intelligence satellite systems. It also developed electronic intelligence packages for all Soviet civil and military satellites. S. I. Baburin and V. L. Grechki were Directors of the Kaluga Branch (space) of TsNII-108, which eventually became NPO Palma.

Rocket and Ramjet Engines

Designer Bureau Name Established Location	Designations	Chief/General Designers	History
Bondaryuk **OKB-670** October 1950 Moscow	• OKB-670 • MKB Krasnaya zvezda • NPO Krasnaya zvezda (1972–)	M. M. Bondaryuk (1947–69) V. I. Serbin N. I. Mikhiyevich G. M. Gryaznov	This was established in 1940 as EKB-3 of NII GVF. It became part of OKB-293 before becoming part of the new NII-1 in 1944. Bondaryuk was named Chief Designer of NII-1 OKB-3 on August 30, 1947. OKB-3 separated in 1950 to become the independent OKB-670. It worked on ramjet engines for a variety of missiles. Activities and personnel related to ramjets were transferred to TMKB Soyuz at Turayevo in December 1972. The remaining part became NPO Krasnaya zvezda to work on nuclear power reactors inherited from OKB-300. Ramjet personnel were reorganized as the Plamya Branch of NII TP (later NPVO Plamya) in 1978.
Glushko **OKB-456** September 29, 1946 Khimki	• OKB-456 (1946–67) • KB EnergoMash (1967–74) • NPO Energiya KB EnergoMash (1974–90) • Glushko NPO EnergoMash (1990–)	V. P. Glushko (1946–89) V. P. Radovskiy (1989–91) B. I. Katorgin (1991–)	This was established at Plant No. 16 at Kazan in July 1944 as OKB-SD. In 1946, the group moved to Khimki near Moscow at the premises of Plant No. 456 (established on April 16, 1942) to become OKB-456. It worked on rocket engines for ICBMs and space launchers. KB EnergoMash merged with TsKBEM on May 22, 1974, to create NPO Energiya. On January 19, 1990, the two organizations separated, and KB EnergoMash became the independent NPO EnergoMash.

TABLE IV

Designer Bureau Name Established Location	Designations	Chief/General Designers	History
Isayev **OKB-2** June 23, 1944 Kaliningrad	• OKB-293 Dept. (1943–44) • NII-1 Dept. (1944–48) • NII-88 SKB Dept No. 9 (1948–52) • NII-88 OKB-2 (1952–59) • KB KhimMash (1967–71) • Isayev KB KhimMash (1971–)	A. M. Isayev (1947–71) V. N. Bogomolov (1971–85) N. I. Leontyev (1985–)	This was established as part of KB-D in OKB-293 on February 4, 1943. On May 29, 1944, OKB-293 merged with NII RA to form the new NII-1. On June 23, 1944, Isayev was named chief of the department at NII-1. He was then named Chief Designer on August 30, 1947. The department was transferred to NII-88 on July 1, 1948, as the new SKB Department No. 9, which in March 1952 became OKB-2 of NII-88. OKB-2 and NII-88's OKB-3 combined in December 1958. On January 16, 1959, OKB-2 became independent. It worked on engines for SAMs, submarine-launched ballistic missiles, spacecraft, and space launchers (KhimMash means Chemical Machine Building).
Izotov **OKB-117** August 1935 Leningrad	• OKB-117 • Leningrad OKB Klimov (1963–75) • Leningrad NPO Klimov (1975–92) • AOOT Klimov (1992–)	V. Ya. Klimov (1946–62) S. P. Izotov (1960–83) V. G. Stepanov (1983–88) A. A. Sarkisov	This was established in 1935 at Plant No. 26 at Rybinsk, which later evacuated to Ufye. There, in 1941, it merged with the Krasny Oktyabr Plant (established originally as Russkiy Reno Plant in 1914 in St. Petersburg), which also evacuated to Ufye in August 1941. In 1946, Klimov became head of the Leningrad OKB and simultaneously headed OKB-45 in Moscow in 1947–56. This organization worked on engines for the upper stage of Chelomey's ICBMs and the LK-700 lunar lander.
Kartukov **Plant No. 81** 1946 Moscow	• Plant No. 81 KB-2 • KB Iskra • MKB Iskra	I. I. Kartukov (1946–60s) B. A. Raysberg Yu. K. Kulikov (1980s)	This was established at Plant No. 81, which later became the Iskra Plant. It worked on solid-propellant accelerators for SAMs, naval missiles, and space launcher escape systems.
Kosberg **OKB-154** October 13, 1941 Voronezh	• OKB-296 (1941–46) • OKB-154 (1946–57) • GSOKB-154 (1957–66) • KB KhimAvtomatiki (1966–)	S. A. Kosberg (1941–65) A. D. Konopatov (1965–93) V. S. Rachuk (1993–)	This was established as OKB-296 at Berdsk as a result of the evacuation of Plant No. 296 from Kharkov and part of OKB of Plant No. 33 from Moscow. In late 1945, it was transferred to Voronezh to Plant No. 265, becoming OKB-154 on May 30, 1946. On August 20, 1957, it was Reorganized into the State Union Experimental Design Bureau No. 154 (GSOKB-154). It worked on engines for SAMs, SLBMs, ICBMs, and space launch vehicles (KhimAvtomatiki means Chemical Automation).

Designer Bureau Name Established Location	Designations	Chief/General Designers	History
Kuznetsov **OKB-276** April 17, 1946 Kuybyshev	• OKB-276 (1953–67) • KB Trud (1967–81) • Kuybyshev NPO Trud (1981–91) • Samara GNPP Trud (1991–94) • Samara NTK NK Dvigatel (1994–96) • AOOT Samara NTK Kuznetsov (1996–)	N. D. Kuznetsov (1949–94) Ye. A. Gritsenko (1994–)	This was established at Plant No. 2 at Kuybyshev in 1946. Plant No. 2 was re-formed into Plant No. 276 in June 1953. It became NPO in 1982 after merging with the Kuybyshev Motor Plant. Its main work was jet engines for aircraft. It moved into rocket engines for ICBMs and space launch vehicles in the late 1950s.
Lyulka **OKB-165** March 30, 1946 Moscow	• OKB-165 (1946–67) • KB Saturn (1967–82) • NPO Saturn (1982–84) • Lyulka NPO Saturn • Lyulka OAO Saturn	A. M. Lyulka (1946–84) V. M. Chepkin (1984–)	This was originally a department at NII-1 and detached in 1946 to Plant No. 165 (later MZ Saturn) in Moscow to form the independent OKB-165. In 1982, MKB Granit and MZ Saturn combined to create NPO Saturn. MKB Granit had been established in 1945 as OKB-45 at Plant No. 45 in Moscow. In 1963, OKB-45 was renamed OKB-45-165 and then MKB Granit in 1966. OKB-165's main work was jet engines, but it developed the liquid hydrogen engine for the lunar program in the 1960s.
Stechkin **OKB Fakel** 1959 Kaliningrad	• OKB Fakel (1971–)	B. S. Stechkin (1955–69) A. S. Moyegulov (1980s) M. I. Shalamov V. V. Suslennikov (1980s) A. S. Bober (1989–)	This was established in 1959 as a laboratory under the USSR Academy of Sciences. It was reorganized in 1971 as OKB Fakel. It worked on attitude control thrusters for spacecraft. It may have been related to NII-88's OKB-3 headed in 1952–58 by D. D. Sevruk.
Stepanov **TMKB Soyuz** August 1, 1964 Turayevo	• TMKB Soyuz • NPO Soyuz	V. G. Stepanov (1964–83) D. D. Gilevich (1983–91) G. V. Komissarov (1991–)	This was established in 1964 as a branch of OKB-300 for the development of attitude control engines for spacecraft. TMKB Soyuz inherited all work on ramjet engines from OKB-670 in December 1972. Eventually, the space-related activities of TMKB Soyuz moved to a location in Moscow, while ramjet research continued at Turayevo until 1978.
Tumanskiy **OKB-300** February 18, 1943 Tushino	• OKB-300 (1943–66) • MMZ Soyuz (1966–81) • MNPO Soyuz • Tushino MKB Soyuz	A. A. Mikulin (1943–55) S. K. Tumanskiy (1955–73) O. N. Favorskiy (1973–87) V. K. Kobchenko (1987–)	This organization's original profile was jet engines for aircraft. In the 1960s, it developed low-thrust rocket engines for robotic spacecraft and small nuclear reactor power sources. Its former profile moved to TMKB Soyuz, while the latter profile moved to MKB Krasnaya zvezda, both in 1972.

TABLE IV

Designer Bureau Name Established Location	Designations	Chief/General Designers	History
Launch Complexes			
Barmin **GSKB SpetsMash** June 30, 1941 Moscow	• Kompressor Plant SKB (1941–46) • GSKB SpetsMash (1946–67) • KB OM (1967–93) • Barmin KB Obshchego mashinostroyeniya (1993–)	V. P. Barmin (1941–93) I. V. Barmin (1993–)	This was established at the Kompressor Plant in Moscow to produce Katyusha launchers during World War II. In May 1946, it began work on launch complexes for ballistic missiles and SAMs. It later did the same for space launch vehicles. It also worked on robotic soil scoopers and long-term lunar bases (SpetsMash means Special Machine Building).
Solovyev **TsKB TransMash** August 10, 1948 Moscow	• GSKB (1948–67) • KB TransMash (1967–)	G. A. Yakovlevich (1948–51) V. P. Petrov (1951–63) V. N. Solovyev (1963–91) G. P. Biryukov (1992–)	This organization originally produced ground equipment for missiles. In 1963, it became the leading developer of launch complexes for ICBMs and space launch vehicles (TransMash means Transport Machine Building).
Communications, Control, and Guidance Systems			
Abramov **OKB-12**	• NISO • OKB-12	G. A. Levin (1940s) N. I. Petrov (1940s) A. S. Abramov (1960s)	This organization developed propellant loading control systems for the NI, as well as nuclear reactor control systems.
Bogomolov **OKB MEI** 1947 Moscow	• MEI ONIP (1947–58) • OKB Moskovskogo energeticheskogo instituta (1958–)	V. A. Kotelnikov (1947–54) A. F. Bogomolov (1954–88)	This was established as the Experimental Scientific-Research Profile (ONIP) sector of MEI (Moscow Power Institute) in 1947. In September 1958, this sector became OKB. It developed telemetry systems for ICBMs, space launchers, and spacecraft.
Bykov **NII-695** 1927 Moscow	• NII-695 (1950–65) • MNII RadioSvyazi • AOOT MNII RadioSvyazi	B. M. Konoplev (1955–59) Yu. S. Bykov (1959–70) A. P. Bilenko I. D. Bogachev N. N. Nesvit N. A. Pochtar N. Kh. Golshteyn M. R. Kaplanov	This entity became involved in space and missile programs in 1957. It developed communications systems for piloted and automated spacecraft (RadioSvyazi means Radio Communications).

Designer Bureau Name Established Location	Designations	Chief/General Designers	History
NII-695 cont.		M. S. Nemirovskiy (1960s) V. I. Mescheryakov (1980s) Yu. N. Matveyev (1990s)	
Gubenko **SKB-567** 1950s		Ye. S. Gubenko (1950–59) A. V. Belousev (1959–??)	This was originally Department No. 12 of NII-885, but it separated from the parent organization in the 1950s. It developed telemetry systems for ICBMs and robotic spacecraft.
Kuznetsov **NII-944** September 1955 Moscow	• NII-944 • NII PM • Kuznetsov NII Prikladnoy mekhaniki	V. I. Kuznetsov (1955–89) I. N. Sapozhnikov (1969–)	This was established on the basis of NII-10's SKB, which separated in 1955. Kuznetsov was Chief Designer of NII-10 in 1946–55. NII-944 developed gyroscopes for ICBMs, space launchers, and spacecraft (Prikladnoy mekhaniki means Applied Mechanics).
Mnatsakanyan **NII-648** 1953 Moscow	• NII-648 • NII Tochnykh priborov • NPO Tochnykh priborov	P. Z. Stas (1953–55) N. I. Belov (1955–61) A. S. Mnatsakanyan (1961–77) O. N. Shishkin (1977–81) A. V. Churkin (1981–86) V. A. Gorkovoy (1986–94) A. V. Shishanov (1994–)	This entity was established as Department No. 4 at NII-885 before becoming independent in 1953. It developed systems for robotic and piloted spacecraft. It later developed rendezvous radars for piloted spacecraft (Tochnykh priborov means Precision Instruments).
Pilyugin **NII AP** April 1963 Moscow	• NII-885 Dept. No. 3/ Complex No. 1 (1948–63) • NII AP (1963–78) • NPO Avtomatiki i priborostroyeniya (1978–) • NPTs AP	N. A. Pilyugin (1948–82) V. L. Lapygin (1982–97)	This organization was established in 1948 as Department No. 3 at NII-885. By the early 1960s, it was incorporated into Complex No. 1, which also included other departments and Pilot Plant No. 1. Complex No. 1 separated from NII-885 in April 1963 to become the independent NII AP. It developed inertial guidance systems for missiles and spacecraft (AP stands for Automation and Instrument Building).
Ryazanskiy **NII-885** February 20, 1938	• NII-885 (1938–46) • NII ST (1946) • NII-885 (1946–63) • NII P (1963–78) • NPO RadioPribor (1978–92) • RNII Kosmicheskogo priborostroyeniya (1992–)	M. S. Ryazanskiy (1946–51) N. A. Pilyugin (1948–63) B. M. Konoplev (1950–55) M. S. Ryazanskiy (1955–87) N. Ye. Ivanov (1960s) M. I. Borisenko (1966–74) L. I. Gusev (1988–)	In May 1946, this entity was transferred to the Ministry of Electrical Industries (MEP) while changing its name to the Scientific-Research Institute of Special Technology (NII ST). Plant No. 1 (located at the site originally built for Plant No. 192) of the Ministry of Armed Forces was handed over to NII ST on June 8, 1946. In April 1963, Complex No. 1 separated from NII-885. The remainder became the Scientific-Research Institute for Instrument Building (NII P). It

TABLE IV

Designer Bureau Name Established Location	Designations	Chief/General Designers	History
NII-885 cont.		Ye. N. Galin	developed radio guidance systems for missiles, launch vehicles, and spacecraft, as well as remote-sensing, communications, and navigation spacecraft payloads.
Utkin **NII IT** July 22, 1966 Kaliningrad	• NII IT (1966–78) • NPO Izmeritelnoy tekhniki (1978–)	I. I. Utkin (1953–70) O. A. Sulimov (1992–)	This was established on December 17, 1953, as Department No. 20 for measurement technology at NII-88. In October 1960, this department was reorganized into Complex No. 5 for measuring systems at NII-88. On July 22, 1966, Complex No. 5 separated from its parent institute and became independent. In July 1978, NII IT, its Ukraine Branch, and the Izmeritel Plant combined to become NPO IT. It developed data recorders for missiles and spacecraft (Ismeritelnoy tekhniki means Measurement Technology).
Biomedicine			
Gazenko **IMBP** October 28, 1963 Moscow	• IMBP (1963–94) • GNTs RF-Institut mediko-biologicheskikh problem (1994–)	A. V. Lebedinskiy (1963–65) V. V. Parin (1965–69) O. G. Gazenko (1969–88) A. I. Grigoryev (1988–)	This entity was formed by the merger of subdivisions from the Air Force Institute for Aviation and Space Medicine and then transferred to the Ministry of Health. It specialized in biomedicine research and cosmonaut training (IMBP stands for Institute for Biomedical Problems).
Samoylov **SKTB Biofiz Pribor** June 3, 1955 Leningrad		A. V. Samoylov (1950s) G. S. Mayorov (1992–)	This organization developed feeding systems for biological payloads in spacecraft.
Severin **Plant No. 918** 1952 Tomilino	• Plant No. 918 • KB Zvezda • AOOT NPP Zvezda	S. M. Alekseyev (1952–73) G. I. Severin (1964–)	This entity developed ejection seats, spacesuits, and airlocks for piloted spacecraft. It also developed Soviet EVA maneuvering units.

Designer Bureau Name Established Location	Designations	Chief/General Designers	History
Volynkin GosNII AiKM 1935 Moscow	• NII AM (1935–60) • GosNII AiKM (1960–)	A. V. Pokrovskiy (1950s) A. G. Kuznetsov (1959–60) Yu. M. Volynkin (1960–69) N. M. Rudniy (1969–74) S. A. Gozulov (1974–84) S. A. Bugrov (1974–88) V. A. Ponomarenko (1988–92)	This entity was established subordinate to the Soviet Air Force. It was was responsible for early cosmonaut selection and biomedicine research. It duties for the most part were taken over by IMBP in the 1960s (AiKM stands for Aviation and Space Medicine).
Voronin OKB-124 1950s Moscow	• OKB-124 • KB Nauka • NPO Nauka • AOOT NPO Nauka	G. I. Voronin (1939–85) I. V. Tishin (1985–) G. F. Khomutov (1990s)	Plant No. 124 was established during the 1930s. The design branch of the plant became OKB-124 in the 1950s. It developed life support systems for all piloted spacecraft.

Testing and Training

Designer Bureau Name Established Location	Designations	Chief/General Designers	History
Karpov TsPK January 11, 1960 Zelenyy	• TsPK (1960–68) • Gagarin TsPK (1968–95) • Gagarin RGNII Tsentr podgotovki kosmonavtov (1995–)	Ye. A. Karpov (1960–63) M. I. Odintsov (1963) N. F. Kuznetsov (1963–72) G. T. Beregovoy (1972–86) V. A. Shatalov (1986–91) P. I. Klimuk (1991–)	From 1960 to 1962, this center was subordinate to the Institute of Aviation and Space Medicine (GosNII AiKM) of the Soviet Air Force. By a decision dated April 10, 1962, this entity was directly subordinated to the Air Force General Staff. It has always been responsible for cosmonaut training.
Lyzhkov NII KhSM February 24, 1960 Zagorsk	• NII Khimicheskikh i stroitelnykh mashin	M. V. Sukhopalko (1960–?) V. S. Lyzhkov (1960s) G. I. Matysyak (1988–)	This was established as the Scientific-Testing Range (NIP) for testing ground equipment. On March 6, 1966, it was reorganized as NII KhSM and served as the site for testing piloted lunar landers on a simulated lunar landscape (KhSM stands for Chemical and Building Equipment).
Stroyev LII March 8, 1941 Zhukovskiy	• LII • Gromov LII • Gromov GNTs Letno-issledovatelskiy institut	M. M. Gromov (1941) A. V. Chesalov (1941–42) V. S. Molokov (1942–43) A. V. Chesalov (1943–47) I. F. Petrov (1947–51) A. A. Kobzarev (1951–54)	The decree for establishing the Institute for Flight Research was issued on June 13, 1940. On March 8, 1941, the People's Commissariat of Aviation Industry adopted a decree on the creation of LII (Flight-Research Center) from a number of subdivisions of TsAGI. This entity was originally established as an aircraft and systems test center. It engaged in research on parachutes, simulators, flight training, and

TABLE IV

Designer Bureau Name Established Location	Designations	Chief/General Designers	History
LII cont.		N. S. Stroyev (1954–66) V. V. Utkin (1966–81) A. D. Mironov (1981–85) K. K. Vasilchenko (1985–) F. D. Zolotarev V. M. Bakayev	aerodynamics. It was responsible for building the Spiral spaceplane testbeds during the 1960s–80s.
Tabakov NII-229 July 7, 1948 Zagorsk	• NII-88 Branch No. 2 (1948–56) • NII-229 (1956–67) • NII KhimMash (1967–)	G. M. Tabakov (1948–49) V. S. Shachtin A. I. Bykhovskiy G. M. Tabakov (1957–63) V. A. Pukhov (1960s) N. M. Samsonov (1970s) Yu. A. Korneyev (1980s) A. A. Makarov (1988–)	This entity was established in 1948 as Branch No. 2 of NII-88. On August 14, 1956, it separated into an independent organization as NII-229. It served as a rocket engine test facility (KhimMash means Chemical Machine Building).

Research and Development

Designer Bureau Name Established Location	Designations	Chief/General Designers	History
GIPKh May 13, 1946 Leningrad	• GIPKh • NPO GIPKh	V. S. Shpak (1952–77) B. V. Gidaspov (1977–89) G. F. Tereshchenko (1989–)	This organization developed and synthesized new propellants for Soviet missile and space programs (GIPKh stands for State Institute of Applied Chemistry).
GOI December 15, 1918 Leningrad	• GOI • Vavilov GOI • Vavilov VNTs Gosudarstvennyy opticheskiy institut (1991–)	B. A. Yermakov L. B. Glebov (1990s) V. I. Puchkov (1990s) M. M. Miroshnikov (1990s) G. T. Petrouskiy (1994–)	This entity designed optical instruments for robotic and piloted spacecraft, both civilian and military (GOI stands for State Optical Institute).
IES 1934 Kiev-5	• IES • Paton Institut ElektroSvarky	B. Ye. Paton (1953–)	This organization developed welding technology for the missile and space programs (IES stands for Institute of Electrical Welding).
IKI July 14, 1965 Moscow	• Institut kosmicheskikh issledovaniy	G. I. Petrov (1965–73) R. Z. Sagdeyev (1973–88) A. A. Galeyev (1988–)	This entity was established on the basis of a department at the Institute of Applied Mathematics dedicated to mission planning and data processing of scientific information. It was responsible for scientific payloads (IKI stands for Institute of Space Research).

Designer Bureau Name Established Location	Designations	Chief/General Designers	History
IPM 1953 Moscow	• OPM MIAN (1953–66) • IPM (1966–78) • Keldysh Institut prikladnoy mekhaniki (1978–)	M. V. Keldysh (1953–78) A. N. Tikhonov S. P. Kurdyumov (1990s)	This entity was established as the Department of Applied Mathematics Of the Mathematics Institute Named After V. A. Steklov of the USSR Academy of Sciences (OPM MIAN). In 1966, it became independent. It was responsible for mission modeling and ballistics computations (IPM means Institute of Applied Mathematics).
LOMO February 4, 1914 Leningrad	• RAOOMP (1914–) • LOMO (until 1993) • AOOT Leningradsoye optiko-mekhanicheskoye obedineniye (1993–)	M. P. Panfilov (1970–94) I. I. Klebanov (1994–97) A. S. Kobitskiy (1997–)	This organization developed optical systems for robotic and piloted spacecraft (LOMO stands for Leningrad Optical-Mechanical Association).
NII-1 September 21, 1933 Moscow	• RNII (1933–36) • NII-3 (1936–42) • GIRT (1942–44) • NII RA (1944) • NII-1 (1944–48) • TsIAM Branch No. 1 (1948–52) • NII-1 (1952–65) • NII Teplovyy protsessy (1965–95) • Keldysh issledovatelskiy tsentr (1995–)	I. T. Kleymenov (1933–37) B. N. Slonimer (1937–39) A. G. Kostikov (1939–44) V. I. Polikouskiy (1944) P. I. Fedorov (1944–45) Ya. L. Bibikov (1945–46) M. V. Keldysh (1946–55) A. S. Koroteyev (1988–)	This entity was established in 1933 with the merger of GIRD and GDL. On July 15, 1942, it was renamed the State Institute of Reactive Technology (GIRT) with subordinates Plant No. 55 and Plant No. 462. On February 18, 1944, it was renamed NII RA, and on May 29, 1944, it absorbed OKB-293 and became NII-1. In June 1946, OKB-293 separated. Between 1948 and March 10, 1952, NII-1 was a branch of TsIAM. It was responsible for research on high-speed flight, advanced rocket engines, nuclear rocket engines, and aerodynamic modeling. A branch (established in 1958) of NII-1 separated in 1981 to become NII Mashinostroyeniya, which developed micro-rocket engines for spacecraft (Teplovyy protsessy means Thermal Processes).
NII-4 June 21, 1946 Bolshevo	• NII-4 (1946–72) • TsNII-4 (1972–)	A. I. Nesterenko (1946–50) G. A. Tyulin (1950–51) P. P. Chechulin (1951–55) A. I. Sokolov (1955–70) Ye. B. Volkov (1970–82) L. I. Volkov (1982–93) V. Z. Dvorkin (1993–)	This was established as a result of a Council f Ministers decree on May 13, 1946. The Ministry of Armed Forces order for formation was dated May 24, 1946. The entity was responsible for research on military applications of ballistic missiles and spacecraft. On April 3, 1972, the space branch of NII-4 (established on March 11, 1968) separated to become the independent TsNII-50.
NII-88 May 16, 1946 Kaliningrad	• NII-88 (1946–67) • TsNIIMash (1967–)	L. R. Gonor (1946–50) K. N. Rudnev (1950–52) M. K. Yangel (1952–53)	This entity was established on the premises of Plant No. 88 at Kaliningrad. The plant was established in 1866 as Plant No. 8 at St. Petersburg. The organizational structure of NII-88 was fortified by a

TABLE IV

Designer Bureau Name Established Location	Designations	Chief/General Designers	History
NII-88 cont.		A. S. Spiridinov (1953–59) G. A. Tyulin (1959–61) Yu. A. Mozzhorin (1961–90) V. F. Utkin (1990–2000)	decision date August 26, 1946. It was responsible for basic research and development on various space profiles, as well as for the long-range planning of the Soviet space program (TsNIIMash stands for Central Scientific-Research Institute of Machine Building).
NITI-40 May 28, 1938 Moscow	• TsSKB-40 (1938–40) • GSPI-40 (1940–46) • NITI-40 (1946–67) • NII TekhnoMash (1967–90) • NPO TekhnoMash (1990–92) • GP NPO TekhnoMash (1992–)	A. V. Kolupayev (1960s, 1980s) O. Ye. Ostrovskiy (1960s) V. A. Isachenko (1990s) V. V. Bulaukin (1991–)	GSPI-40 was established in 1940 on the basis of TsSKB-40. GSPI-3, and part of GSPI-7 in NKOP. In August 1946, it was renamed NITI-40. It was responsible for research and development on manufacturing processes and tool manufacture for the Soviet missile and space industry.
TsAGI December 1, 1918 Zhukovskiy	• TsAGI • Zhukovskiy TsAGI • Zhukovskiy GNTs RF-Tstentralnyy aerogidrodinamicheskiy institut	S. N. Shishkin (1941–50) A. I. Makareuskiy (1950–60) V. M. Myasishchev (1960–67) G. P. Svishchev (1967–89) G. I. Zagaynov (1989–93) V. Ya. Neyland (1993–) V. G. Dmitriyev (1990s)	This was responsible for basic and applied research on high-speed flight, including the spaceplane and lifting body programs (TsAGI stands for Central Aerohydrodynamics Institute).
TsIAM December 3, 1930 Moscow	• IAM (1930–32) • TsIAM (1932–33) • Baranov Tsentralnoye institut aviatsionnogo motorostroyeniya (1933–)	G. P. Svishchev (1954–67) S. M. Shlyakhtenko (1967–82) D. A. Ogorodnikov (1982–)	This entity was responsible for research and testing of air-breathing propulsion systems for spaceplanes and lifting bodies (TsIAM stands for Central Institute of Aviation Motor Building).
TsNII-50 April 3, 1972 Moscow	• TsNII-50 (1972–95) • Tikhonravov TsNII-50 (1995–)	G. P. Melnikov (1972–83) I. V. Meshcheryakov (1983–88) E. V. Alekseyev (1988–93) V. A. Menshikov (1993–)	This entity was established the space branch of NII-4 on March 11, 1968, before separating from the parent institute on April 3, 1972. Starting in 1982, TsNII-50 was subordinated to UNKS. It was also known as military unit no. 73790. It became a branch of GKNPTs Khrunichev in November 1997. It was responsible for planning military applications of spacecraft.

Designer Bureau Name Established Location	Designations	Chief/General Designers	History
VIAM June 28, 1932 Moscow	• VIAM (1932–92) • Vserossiyskiy NII aviatsionnykh materialov (1992–)	*R. T. Tumanov (1938–50)* *V. V. Boytsov (1950–63)* *R. T. Tumanov (1963–76)* *P. Ye. Shalin (1977–)*	This organization was formed on the basis of the aviation material testing department at TsAGI. It was responsible for research and development, testing, and certification of all metallic and nonmetallic materials used on spacecraft and hypersonic vehicles (VIAM stands for All-Russian Institute of Aviation Materials).
Other Systems			
Darevskiy **SOKB LII** August 21, 1967 Zhukovskiy-2	• SOKB LII (1967–83) • NII Aviatsionnogo oborudovaniya (1983–97) • SOKB KT NII AO (1997–)	*S. G. Darevskiy (1965–75)* *S. A. Borodin (1975–)*	This was established as a subdivision of LII. It separated from LII in 1971. In 1982, it merged with another branch at LII to become NII AO. It developed simulators, cockpit consoles, and avionics for piloted spacecraft (AO stands for Aviation Equipment).
Goltsman **OKB-686** Moscow	• OKB-686 • GOKB Prozhektor • PO Prozhektor	*A. M. Goltsman* *V. A. Okunev*	This entity developed power source components for missiles and spacecraft.
Kemurdzhian **VNII-100** 1947 Leningrad	• VNII-100 • VNII TransMash	*A. A. Kemurdzhian (1960s)* *E. K. Potemkin (1990s)*	This organization developed mobile robots for robotic and piloted spacecraft, including Lunokhod (TransMash means Transport Machine Building).
Khrustalev **TsKB-589** Moscow	• TsKB-589 • TsKB Geofizika • NPO Geofizika	*V. A. Khrustalev (1950s–60s)* *N. G. Vinogradov (1960s)* *V. S. Kuzmin*	This plant was established in 1852. In 1957–58, it became involved in the space program. It developed attitude control sensors for robotic and spacecraft.
Lidorenko **VNII IT** Moscow	• VNII IT • GNPP Kvant (1987–)	*N. S. Lidorenko* *Yu. V. Skokov* *A. B. Slutskiy*	This entity developed battery and solar power sources for missiles and spacecraft. In 1987, it merged with the Foton Experimental Plant to become GNPP Kvant (IT stands for Current Sources).
Rosselevich **NII-380** 1946 Leningrad	• NII-380 • VNII Televideniya • NII Televideniya	*I. A. Rosselevich* *M. A. Grudzinskiy (1983–)*	This organization developed TV systems for robotic and piloted spacecraft.

TABLE IV

Designer Bureau Name Established Location	Designations	Chief/General Designers	History
Tkachev **NIEI PDS** 1946 Moscow	• NIEI PDS (1946–66) • NII AU (1966–80s) • NII Parashutostroyeniya	F. D. Tkachev (until 1968) N. A. Lobanov (1968–77) O. V. Rysev (1978–)	This entity developed parachute systems for robotic and piloted spacecraft.

Offensive Space Systems

Kisunko **OKB-30** 1962	• KB-1 SKB-30 (1955–61) • OKB-30 (1961–70) • TsNPO Vympel (1970–92) • MAK Vympel (1992–)	G. V. Kisunko (1955–75) A. G. Basistov (1960s-90s) V. G. Repin (1970–) A. A. Kuzmin (1990s) A. V. Menshikov (1990s) M. A. Arkharov (1990s)	This entity was established as SKB-30, a subdivision of KB-1, in April 1955 to focus on anti-ballistic missiles. It separated from KB-1 in August 1961 to become the independent OKB-30. In 1970, it combined with several other organizations (Radiotechnical Institute, NII DAR, KB RadioPribor, and Dnepropetrovsk Radio Plant) to form TsNPO Vympel. It worked on anti-ballistic missile systems and the Soviet "Star Wars" program.
Nudelman **OKB-16**	• OKB-16 • KB TochMash	A. E. Nudelman (1943–87)	This entity was established to focus on anti-aircraft guns. In the late 1950s, it shifted to anti-tank missiles and later to lasers. It designed space-based cannons for Chelomey and Kozlov (TochMash means Precision Machine Building).
Ustinov **TsKB Luch** 1969 Moscow	• TsKB Luch • TsKB Astrofizika • NPO Astrofizika	N. D. Ustinov (1969–88) B. Chemodanov (1988–92) N. D. Belkin (1992–) V. Orlov N. Sheburkin	This entity was established to work on ground and space-based lasers. NPO Astrofizika eventually included KB Rubin, KB Ametist, KB Granat, and Nov Electromechanical Plant. This organization was also linked to KB Raduga, which focused on the creation of laser weaponry. Orlov and Sheburkin were at KB Granat, which separated in April 1992, followed by Nov Plant and KB Ametist.

Note: Design entities often had a Director in addition to a Chief Designer or General Designer. For the sake of brevity, only the Chief/General Designers are listed. In the cases of some institutes, the Directors are listed (in italic type).

Selected Sources

1. *Raketno-kosmicheskaya otrasl rossii, 1996–97: katalog predpriyatiy, uchrezhdeniy i organizatsii* (Moscow: RKA, 1996).
2. Stephane Chenard, *Euroconsult's Space Directory of Russia* (Paris: Euroconsult, 1993).
3. V. V. Favorskiy and I. V. Meshcheryakov, eds., *Voyenno-kosmicheskiye sily (voyenno-istoricheskiy trud): kniga I: kosmonautika vooruzhennye sily* (Moscow: Sankt-Peterburgskoy tipografii no. I VO Nauka, 1997).
4. Grigoriy Kisunko, *Sekretnaya zona* (Moscow: Sovremennik, 1996).
5. Yu. A. Mozzhorin, et al., eds., *Nachalo kosmicheskoy ery: vospominaniya veteranov raketno-kosmicheskoy tekhniki i kosmonavtiki: vypusk utoroy* (Moscow: RNITsKD, 1994).
6. Yu. P. Semenov, ed., *Raketno-Kosmicheskaya Korporatsiya "Energiya" imeni S. P. Koroleva* (Korolev: RKK Energiya, named after S. P. Korolev, 1996).
7. U.S.-Russia Business Development Committee Defense Conversion Subcommittee, *Russian Defense Business Directory* (Washington, DC: U.S. Department of Commerce, 1993).
8. Various correspondences with Mark S. Hillyer.

TABLE V

Table V
Space Launch Vehicle Designations

U.S. Dept. of Defense	Sheldon	Soviet Public	Missile Derived From	Production Index	First Orbital Attempt	Payload
SL-1	A	Sputnik	R-7	8K71PS	Oct. 4, 1957	Sputnik
SL-2	A	Sputnik	R-7	8A91	April 27, 1958	*Sputnik*
SL-3	A-1	Luna	R-7	8K72	Sept. 23, 1958	*Luna*
SL-3	A-1	Vostok	R-7	8K72K	Dec. 22, 1960	*Korabl-Sputnik*
SL-3	A-1	Vostok	R-7A	8A92	July 28, 1962	Kosmos-7
SL-3	A-1	Vostok-M	R-7A	8A92M	March 17, 1966	Kosmos-112
SL-4	A-2	Voskhod	R-7A	11A57	Nov. 16, 1963	Kosmos-22
SL-4	A-2	Soyuz	R-7A	11A511	Nov. 28, 1966	Kosmos-133
SL-4	A-2	Soyuz-L	R-7A	11A511L	Nov. 24, 1970	Kosmos-379
SL-4	A-2	Soyuz-M	R-7A	11A511M	Dec. 27, 1971	Kosmos-470
SL-4	A-2	Soyuz-U	R-7A	11A511U	May 18, 1973	Kosmos-559
SL-4	A-2	Soyuz-U2	R-7A	11A511U2	Dec. 28, 1982	Kosmos-1426
SL-5	A-1m	–	R-7A	11A510	Dec. 22, 1965	Kosmos-102
SL-6	A-2e	Molniya	R-7A	8K78	Oct. 10, 1960	*Mars*
SL-6	A-2e	Molniya-M	R-7A	8K78M	Feb. 19, 1964	*Venera*
SL-7	B-1	Kosmos-2	R-12	63S1	Oct. 27, 1961	*Kosmos*
SL-7	B-1	Kosmos-2	R-12	63S1M/11K63	Oct. 19, 1965	Kosmos-93
SL-8	C-1	Kosmos-1	R-14	65S3	Aug. 18, 1964	K-38/39/40
SL-8	C-1	Kosmos-3	R-14	11K65	Nov. 16, 1966	*Kosmos*
SL-8	C-1	Kosmos-3M	R-14	11K65M	May 15, 1967	Kosmos-158
SL-8	C-1	–	R-14	K65MR	June 3, 1982	Kosmos-1374
SL-9	D	Proton	UR-500	8K82	July 16, 1965	Proton-1
SL-10	A-m	–	R-7A	11A59	Nov. 1, 1963	Polet-1
SL-11	F-1r	–	R-36-O	8K69	Dec. 16, 1965	*OGCh*
SL-11	F-1m	Tsiklon-2A	R-36	11K67	Oct. 27, 1967	Kosmos-185
SL-11	F-1m	Tsiklon-2	R-36	11K69	Aug. 6, 1969	Kosmos-291
SL-12	D-1e	Proton-K	UR-500K	8K82K	March 10, 1967	Kosmos-146
SL-13	D-1	Proton-K	UR-500K	8K82K	Nov. 16, 1968	Proton-4
SL-14	F-2	Tsiklon-3	R-36	11K68	June 24, 1977	Kosmos-921
SL-15	G-1e	N1	–	11A52	Feb. 21, 1969	*L3S*
SL-16	J-1	Zenit-2	–	11K77	June 21, 1985	–
SL-17	K-1	Energiya	–	14A02	May 15, 1987	*Polyus*
SL-17	K-1	Energiya	–	11K25	Nov. 15, 1988	Buran

U.S. Dept. of Defense	Sheldon	Soviet Public	Missile Derived From	Production Index	First Orbital Attempt	Payload
SL-18	L-1	Start-1	RT-2PM	–	March 25, 1993	Start-1-1
SL-19	–	Rokot	UR-100N	–	Dec. 26, 1994	RS-15
–	–	Start	RT-2PM	–	March 28, 1995	Gurwin-1, UNAMSAT, EKA
–	–	Start-1.2	RT-2PM	–	March 4, 1997	Zeya
–	–	Shtil-1N	R-29RM	–	July 7, 1998	Tubsat-N, Tubsat-N1

Abandoned Projects

Name	Production Index	Dates	Design Bureau	Description
A-300	–	Late 1950s	Chelomey	For Raketoplan
UR-200K	8K81K	1960–64	Chelomey	For IS and US
UR-200A	8K83	1960–64	Chelomey	Orbital bomb
UR-700	–	1964–69	Chelomey	Moon rocket
UR-700M	–	1969–71	Chelomey	Mars rocket
UR-530	11K99	Late 1970s	Chelomey	For heavy Almaz
RLA-150	–	1974–76	Glushko	Heavy lift
Vulkan	–	1974–76	Glushko	Heavy lift
–	8A93	1957–60	Korolev	For recon. satellite
–	8K73	1958–59	Korolev	Lunar rocket
YaKhR-2	–	Early 1960s	Korolev	Nuclear rocket
–	–	Early 1960s	Korolev	Nuclear rocket
KhR-3	–	Early 1960s	Korolev	Heavy-lift rocket
MR	8K711	1960–62	Korolev	For Sever
–	11A55	1962–63	Korolev	For Soyuz-A
–	11A56	1962–63	Korolev	For Soyuz-B/V
N1	11A51	1960–62	Korolev	Early N1
N2	11A52	1960–62	Korolev	Early N1
N11	11A53	1962–65	Korolev	N1 variant
N111	11A52	1962–65	Korolev	N1 variant
GR-1	8K513	1962–65	Korolev	GR-1 variant
–	11A514	1964–65	Kozlov	For Soyuz-R
M-1	–	1959–60	Myasishchev	–
Tsiklon-1	64S2	Early 1960s	Yangel	Based on R-16
SK-100	–	Early 1960s	Yangel	Heavy rocket
R-56	–	1962–64	Yangel	Moon rocket

TABLE V

Notes

1. The "SL" column refers to U.S. Department of Defense designations for "Satellite Launcher." The system is roughly chronological from SL-8 to SL-17 in order of introduction of the launcher. The first mention of the "SL" system in a declassified CIA document dates from March 2, 1967. This was in U.S. Central Intelligence Agency, "National Intelligence Estimate 11-1-67: The Soviet Space Program," Washington, DC, March 2, 1967, as declassified December 11, 1992, by the CIA Historical Review Program.
2. The one major discrepancy in the "SL" system is with SL-5 and SL-10. For almost three decades, Western analysts have equated SL-5 with the Polet launches in 1963–64 and SL-10 with two isolated Kosmos launches in 1965–66. When CIA NIE 11-1-67 was declassified in December 1992 (see first note above), it turned out that in truth it was exactly the opposite-that is, SL-5 launched Kosmos-102 and Kosmos-105 in 1965–66, while SL-10 launched the two Polet satellites in 1963–64.
3. The "Sheldon" column denotes the system devised by Charles S. Sheldon II, who was the Chief of the Science Policy Research Division at the Library of Congress. It was first described in Charles S. Sheldon II, "The Soviet Space Program: A Growing Enterprise," *TRW Space Log* 8(4) (Winter 1968–69): 3–23.
4. If the payload is listed in italics, it indicates that the payload failed to attain Earth orbit.

Selected Sources

1. B. Ye. Chertok, *Rakety i lyudi: goryachiye dni kholodnoy voyny* (Moscow: Mashinostroyeniye, 1997).
2. S. N. Konyukhov and V. A. Pashchenko, "History of Space Launch Vehicles Development," paper presented at the 46th Congress of the International Astronautical Federation, IAA-95-IAA 2.2.09, Oslo, Norway, October 2-6, 1995.
3. V. Pappo-Korystin, V. Platonov and V. Pashchenko, *Dneprovskiy raketno-kosmicheskiy tsentr* (Dnepropetrovsk: PO YuMZ/KBYu, 1994).
4. Yu. P. Semenov, ed., *Raketno-Kosmicheskaya Korporatsiya "Energiya" imeni S. P. Koroleva* (Korolev: RKK Energiya, named after S. P. Korolev, 1996).
5. U.S. Central Intelligence Agency, "National Intelligence Estimate 11-1-71: The Soviet Space Program," Washington, DC, July 1, 1971, as declassified in 1997 by the CIA Historical Review Program.

TABLE VI

Table VI
Details of Launch Vehicles Used in the Soviet Piloted Space Program, 1957–74

Launcher	Served	OKB	Length (meters)	Base Diameter (meters)	Total Launch Thrust (tons)	Total Launch Mass (tons)	Capacity* (tons)	Stages and Engines	Engine Contractors
8K71PS Sputnik	1957	Korolev	29.17	10.3	398	267.3	c. 2 to LEO	1. 4 x 8D74PS 2. 1 x 8D75PS	1. Glushko 2. Glushko
8A91 Sputnik	1958	Korolev	30.21	10.3	388	269.4	c. 2 to LEO	1. 4 x 8D76 2. 1 x 8D77	1. Glushko 2. Glushko
8K72 Luna	1958–60	Korolev	33.5	10.3	407.5	279.1	c. 5 to LEO	1. 4 x 8D74 2. 1 x 8D75 3. 1 x 8D714	1. Glushko 2. Glushko 3. Kosberg
8K72 Vostok	1960–62	Korolev	38.36	10.3	407.5	283.5	4.55 to LEO	1. 4 x 8D74 2. 1 x 8D75 3. 1 x 8D714	1. Glushko 2. Glushko 3. Kosberg
8K72K Vostok	1960–63	Korolev	38.36	10.3	409.8	287	4.73 to LEO	1. 4 x 8D74 2. 1 x 8D75 3. 1 x 8D719	1. Glushko 2. Glushko 3. Kosberg
11A57 Voskhod	1963–76	Korolev	43.5–45.22	10.303	413.3	—	c. 7.5 to LEO	1. 4 x RD-107 2. 1 x RD-108 3. 1 x RD-0108	1. Glushko 2. Glushko 3. Kosberg
8K82 Proton	1965–66	Chelomey	43.5	7.4	894	550	12.2 to LEO	1. 6 x RD-253 2. 3 x RD-0208 1 x RD-0209	1. Glushko 2. Kosberg
11A511 Soyuz	1966–75	Kozlov	49.913	10.303	413.3	310	c. 7.5 to LEO	1. 4 x RD-107 2. 1 x RD-108 3. 1 x RD-0110	1. Glushko 2. Glushko 3. Kosberg

Launcher	Served	OKB	Length (meters)	Base Diameter (meters)	Total Launch Thrust (tons)	Total Launch Mass (tons)	Capacity* (tons)	Stages and Engines	Engine Contractors
8K82K Proton-K	1967–	Chelomey	57.1–61.1	7.4	894	690	4.8 to GTO, 5.7 to Moon	1. 6 x RD-253 2. 3 x RD-0210 1 x RD-0211 3. 1 x RD-0212 4. 1 x 11D58	1. Glushko 2. Kosberg 3. Kosberg 4. Korolev
8K82K Proton-K	1968–	Chelomey	57.8	7.4	894.	—	20–21 to LEO	1. 6 x RD-253 2. 3 x RD-0210 1 x RD-0211 3. 1 x RD-0212	1. Glushko 2. Kosberg 3. Kosberg
11A52 N1	1969–72	Korolev	105.286	16.875	4,620	2,750	91.5 to 220-km orbit	1. 30 x NK-15 2. 8 x NK-15V 3. 4 x NK-21 4. 1 x NK-19 5. 1 x 11D58	1. Kuznetsov 2. Kuznetsov 3. Kuznetsov 3. Kuznetsov 3. Korolev
11A511L Soyuz-L	1970–71	Kozlov	45	10.3	413.3	310	c. 7.5 to LEO	1. 4 x RD-107 2. 1 x RD-108 3. 1 x RD-0110	1. Glushko 2. Glushko 3. Kosberg
11A511U Soyuz-U	1973–	Kozlov	49.3	10.303	413.3	310	6.8 to 220-km orbit	1. 4 x RD-107 2. 1 x RD-108 3. 1 x RD-0110	1. Glushko 2. Glushko 3. Kosberg

* LEO = low-Earth orbit; GTO = geostationary transfer orbit; km = kilometer

TABLE VI

Selected Sources

1. Igor Afanasyev, "N-1: Absolutely Secret" (English title), *Krylia rodiny* no. 9 (September 1993): 13–16.
2. M. V. Keldysh, ed., *Tvorcheskoe nasledie Akademika Sergeya Pavlovicha Koroleva: izbrannie trudi i dokumenty* (Moscow: Nauka, 1980).
3. V. P. Mishin, "Why Didn't We Fly to the Moon?" (English title), *Znaniye: tekhnike: seriya kosmonautika, astronomiya* no. 12 (December 1990): 3–43.
4. S. Sergeyev, "Tsiklon," *Aviatsiya i kosmonautika* no. 2 (March–April 1994): 38–41.
5. S. P. Umanskiy, "Russian Space Launch Vehicles" (English title), *Zemlya i uselennaya* no. 2 (March–April 1994): 97–105.
6. T. Varfolomeyev, "Soviet Rocketry that Conquered Space: Part 1: From First ICBM to Sputnik Launcher," *Spaceflight* 37 (1995): 260–63.
7. T. Varfolomeyev, "Soviet Rocketry that Conquered Space: Part 2: Space Rockets for Lunar Probes," *Spaceflight* 38 (1996): 49–52.
8. T. Varfolomeyev, "Soviet Rocketry that Conquered Space: Part 3: Lunar Launchings for Impact and Photography," *Spaceflight* 38 (1996): 206–08.
9. Phillip Clark, ed., *Jane's Space Directory: 1997–98* (Coulsdon, UK: Jane's Information Group Ltd., 1997).

Table VII

TABLE VII

Designations for Piloted Space Vehicles

Public Name	KB Name	Production Index	System	First Launch	OKB	Comments
Almaz Complex						
Salyut		11F71	Almaz OPS	Salyut 2	Chelomey	Military system
		11F711			Chelomey and Kozlov	Almaz OPS plus Soyuz 7K-VI
		11F712			Chelomey	Almaz OPS plus TKS
Kosmos		11F72	TKS	Kosmos-929	Chelomey	Transport ship
		11F74	TKS VA	Kosmos-881/882	Chelomey	TKS return ship
		11F75	Almaz OPS	On Salyut 2	Chelomey	Main part of OPS
		11F76	SpK/KSI	On Salyut 2	Chelomey	OPS capsule
		11F77	TKS FGB	On Kosmos-929	Bugayskiy	TKS main module
		11F667	Almaz-K/Mech-K		Chelomey	OPS with two ports
Luna and Lunokhod						
Luna	Ye-1			(Sept. 23, 1958)	Korolev	Lunar impact
Luna	Ye-1A			(June 18, 1959)	Korolev	Lunar impact
Luna	Ye-2/Ye-2A			AMS	Korolev	Far side photography
Luna	Ye-3				Korolev	Lunar explosion
Luna	Ye-2F/Ye-3			(April 15, 1960)	Korolev	Far side photography
Luna	Ye-5				Korolev	Far side photography
Luna	Ye-6			(January 4, 1963)	Korolev	Lunar lander
Luna	Ye-6M			Luna 9	Korolev	Lunar lander
Luna	Ye-6S			Kosmos-111	Korolev	Lunar orbiter
Luna	Ye-6LF			Luna 11	Korolev	Lunar orbiter
Luna	Ye-6LS			Kosmos-159	Babakin	Communications test
Luna	Ye-7				Babakin	Lunar orbiter
Luna	Ye-8			Luna 17	Babakin	Lunar rover
Luna	Ye-8-5			Luna 15	Babakin	Lunar lander
Luna	Ye-8LS			Luna 19	Babakin	Lunar orbiter
Luna	Ye-8-5M			Luna 23	Babakin	Lunar lander
Lunokhod	8YeL			Lunokhod (1)	Babakin	Lunar rover

Public Name	KB Name	Production Index	System	First Launch	OKB	Comments
Piloted Lunar Program						
	1L		Soyuz-A		Korolev	Circumlunar ship
	7K		Soyuz-B		Korolev	Circumlunar ship
	9K		Soyuz-V		Makeyev	Propulsion ship
	11K				Reshetnev	Tanker
Zond	7K-L1P	11F91		Kosmos-146	Korolev	Circumlunar ship
Zond	7K-L1	11F91		Zond 4	Korolev	Circumlunar ship
	7K-L1S	11F92		(Feb. 21, 1969)	Korolev	Zond on N1
	7K-LOK	11F93		LOK	Korolev	Orbiter
		11F94		LK	Korolev	Lander in L3
				(Nov. 28, 1969)		
Kosmos	7K-L1E				Korolev	Blok D test
	7K-OK-T				Korolev	Ferry for EOR L1
	7K-PLK				Korolev	Lunar orbiter
	T1K				Korolev	LOK test
Kosmos	T2K			Kosmos-379	Korolev	LK test
Progress						
Progress	7K-TG	11F615A15	Progress	Progress(-1)	Korolev	Basic
Progress M	7K-TGM	11F615A55	Progress M	Progress-M(-1)	Korolev	Modified
Progress M2	7K-TGM	11F615A75	Progress M2		Korolev	For ISS
Progress M2	7K-TGM	11F615A77	Progress M2		Korolev	For ISS
Soyuz						
	5K		Sever		Korolev	Early Soyuz
	5KM		Sever		Korolev	Early ASAT Soyuz
	7K-PPK				Kozlov	ASAT Soyuz
	7K-R	11F71	Soyuz-P		Kozlov	Recon. Soyuz
	7K-TK	11F72	Soyuz-R		Kozlov	Almaz ferry
	7K-VI	11F73			Kozlov	Military Soyuz
	7K-S	11F732	Zvezda	Kosmos-670	Korolev	OIS ferry

TABLE VII

Public Name	KB Name	Production Index	System	First Launch	OKB	Comments
	7K-S-I	11F733			Korolev	Short duration
	7K-S-II	11F734			Korolev	Long duration
	7K-G	11F735			Korolev	OIS tanker
Soyuz	7K-OK	11F615		Kosmos-133	Korolev	Basic Soyuz
Soyuz	7K-T	11F615A8		Soyuz 10	Korolev	DOS ferry
Soyuz	7K-T	11F615A9		Kosmos-656	Korolev	Almaz ferry
Soyuz	7K-TM	11F615A12		Kosmos-638	Korolev	EPAS Soyuz
Soyuz-T	7K-ST	11F732		Kosmos-1001	Korolev	DOS ferry
Soyuz-TM	7K-STM	11F732		Soyuz TM(-1)	Korolev	Mir ferry
Soyuz-TM	7K-STM	11F732		Soyuz TM-16	Korolev	APAS-89 version
		14F70	Zarya		Korolev	Mir-2 crew ferry

Space Shuttles and Spaceplanes

Public Name	KB Name	Production Index	System	First Launch	OKB	Comments
Buran	OK	11F35		Buran	Korolev and Lozino-Lozinskiy	
Kosmos	BOR-4			(Dec. 5, 1980)	Lozino-Lozinskiy	
Kosmos	BOR-5			(June 5, 1984)	Lozino-Lozinskiy	
	LO	14F33			Korolev	Buran payload
	NPG	17F32			Korolev	Buran payload

Space Stations and Modules

Public Name	KB Name	Production Index	System	First Launch	OKB	Comments
Salyut	17K	11F715	DOS Zarya	Salyut (1)	Korolev	Station
Mir	17KS no. 12701			Mir	Korolev	Station core
SM	17KSM no. 12801				Korolev	ISS module
	27KS				Korolev	Mir complex
Kvant	37KE	11F37	TsM-E	Kvant	Korolev	Astrophysics
	377KE		TKM-E	Kvant	Korolev	Kvant plus tug
	37KD				Korolev	Augment module
	37KT				Korolev	Tech. module

Public Name	KB Name	Production Index	System	First Launch	OKB	Comments
	37KP				Korolev	Remote sensing
	37KG				Korolev	Cargo module
	37KB		BDP		Korolev	For Buran
	37KBI				Korolev	Research module
	37KBE				Korolev	Power module
	37KBT				Korolev	Biotechnology
Kvant	77KSD	11F77D	TsM-D	Kvant-2	Korolev	EVA module
Priroda	77KSI	11F77I	TsM-I	Priroda	Korolev	Earth observ. module
Spektr	77KSO	11F77O	TsM-O	Spektr	Korolev	Remote sensing
Kristall	77KST	11F77T	TsM-T	Kristall	Korolev	Tech. module
SO	316GK			SO	Korolev	Docking module
FGB	77KSM no. 17501			FGB	Korolev	Core of ISS
	OIK	11F71	Soyuz-R		Kozlov	Military station
	OB-VI	11F730	Soyuz-VI		Korolev	Military station
	19K	11F731	MOK		Korolev	OIS base station
					Korolev	Large station
Sputnik						
Sputnik	PS-1			Sputnik (1)	Korolev	First satellite
Sputnik	PS-2			Sputnik 2 (April 27, 1958)	Korolev	Biosatellite
Sputnik	D-1				Korolev	Scientific
	OD-1				Korolev	Recon. satellite
	OD-2				Korolev	Biosatellite
Vostok and Voskhod						
Korabl-Sputnik	IKP	11F61		Korabl-Sputnik	Korolev	Precursor
Korabl-Sputnik	IK	11F61		Korabl-Sputnik 2	Korolev	Manned precursor
Vostok	3KA	11F63		Korabl-Sputnik 4	Korolev	Manned precursor

TABLE VII

Public Name	KB Name	Production Index	System	First Launch	OKB	Comments
Voskhod	3KV	11F63		Kosmos-47	Korolev	Three-person Voskhod
Voskhod	3KD	11F63		Kosmos-57	Korolev	EVA Voskhod
	7K		Vostok-7		Korolev	Uprated Vostok

Sources

1. Various issues of *Novosti kosmonavtiki*.
2. Sergey Voevodin, *VSA071*, newsletter over Internet, April 30, 1997.

Table VIII

Table VIII

Automated Launches Related to the Soviet Piloted Space Program, 1957–76

Name	Launch Date	Launch Time (Moscow Time)	Spacecraft	S/C No.	Launcher and Serial No.	Comments
Sputnik						
Sputnik	Oct. 4, 1957	2228:34	PS-1	—	8K71PS M1-1PS	Successful launch
Sputnik 2	Nov. 3, 1957	0530:42	PS-2	—	8K71PS M1-2PS	Successful launch
	April 4, 1958	1201	D-1	1	8A91 B1-2	Launcher breakup at T+96.5 seconds
Sputnik 3	May 15, 1958	1000:35	D-1	2	8A91 B1-1	Successful launch
Luna						
	Sept. 23, 1958		Ye-1	1	8K72 B1-3	First-stage failure, no lunar impact
	Oct. 12, 1958		Ye-1	2	8K72 B1-4	First-stage failure, no lunar impact
	Dec. 4, 1958		Ye-1	3	8K72 B1-5	Second-stage failure, no lunar impact
Space Rocket	Jan. 2, 1959	1941:21	Ye-1	4	8K72 B1-6	Missed Moon, no lunar impact
	June 18, 1959	1108	Ye-1A	5	8K72 I1-7	Second-stage failure, no lunar impact
Second Space Rocket	Sept. 12, 1959	0939:42	Ye-1A	7	8K72 I1-7B	Successful lunar impact
Automatic Interplanetary Station	Oct. 4, 1959	0343:40	Ye-2A	1	8K72 I1-8	Successful far side photography
	April 15, 1960	1806:44	Ye-3/Ye-2F	1	8K72 I1-9	No escape velocity attained, no photography
	April 16, 1960	1907:43	Ye-3/Ye-2F	2	8K72 LI-9A	First- and second-stage failures, far side Photography
	Jan. 4, 1963	1149	Ye-6	2	8K78/Ye-6 T103-09	Blok L transformer failure, TLI failure
	Feb. 3, 1963	1229	Ye-6	3	8K78/Ye-6 G103-10	Blok L guidance failure, no Earth orbit
Luna 4	April 2, 1963	1116	Ye-6	4	8K78/Ye-6 G103-11	Yupiter astro-navigation system failure
	Mar. 21, 1964	1108	Ye-6	6	8K78M/Ye-6 T15000-20	Third-stage valve failure, no Earth orbit
	April 20, 1964	1230	Ye-6	5	8K78M/Ye-6 T15000-21	Blok I and Blok L failure, no Earth orbit
Kosmos-60	Mar. 12, 1965		Ye-6	9	8K78/Ye-6 R103-25	Blok L guidance failure, TLI failure
	Apr. 10, 1965		Ye-6	8	8K78M R103-26	Blok I failure, no Earth orbit
Luna 5	May 9, 1965	1049:37	Ye-6	10	8K78M UI03-30	Incorrect command to engine, landing failure

Name	Launch Date	Launch Time (Moscow Time)	Spacecraft	S/C No.	Launcher and Serial No.	Comments
Luna 6	June 8, 1965	1040	Ye-6	7	8K78 UI03-31	Spacecraft guidance failure, landing failure
Luna 7	Oct. 4, 1965	1056:40	Ye-6	11	8K78 UI03-27	Loss of s/c orientation, landing failure
Luna 8	Dec. 3, 1965	1346:14	Ye-6	12	8K78 UI03-28	Soft-landing failure
Luna 9	Jan. 31, 1966	1441:37	Ye-6M	202/13	8K78M UI03-32	First successful landing first s/c built by Lavochkin (202 serial no.)
Kosmos-111	Mar. 1, 1966	1403:49	Ye-6S	204	8K78M NI03-41	TLI failure, no lunar orbit
Luna 10	Mar. 31, 1966	1347	Ye-6S	206	8K78M NI03-42	First successful lunar orbit
Luna 11	Aug. 24, 1966	1103	Ye-6LF	101	8K78M NI03-43	Successful lunar orbit
Luna 12	Oct. 22, 1966	1142	Ye-6LF	102	8K78M NI03-44	Successful lunar orbit
Luna 13	Dec. 21, 1966	1317	Ye-6M	205/14	8K78M NI03-45	Successful lunar soft-landing
Kosmos-159	May 17, 1967		Ye-6LS	111	8K78M Ya716-56	Insufficient velocity, no lunar orbit
	Feb. 7, 1968		Ye-6LS	112	8K78M Ya716-57	No Earth orbit, launcher failure
Luna 14	April 7, 1968	1344	Ye-6LS	113	8K78M Ya716-58	Successful lunar orbit
	Feb. 19, 1969	0948:15	Ye-8	201	8K82K 239-01 + Blok D	No Earth orbit, launcher failure
	June 14, 1969	0700:47	Ye-8-5	402	8K82K 238-01 + Blok D	Blok D failure
Luna 15	July 13, 1969	0554:42	Ye-8-5	401	8K82K 242-01 + Blok D	Sample returner impacted on Moon
Kosmos-300	Sept. 23, 1969	1707:36	Ye-8-5	403	8K82K 244-01 + Blok D	Scooper failed to leave Earth orbit
Kosmos-305	Oct. 22, 1969	1709:59	Ye-8-5	404	8K82K 241-01 + Blok D	Scooper failed to leave Earth orbit
	Feb. 6, 1970	0716:06	Ye-8-5	405	8K82K 247-01 + Blok D	First-stage failure
Luna 16	Sept. 12, 1970	1625:53	Ye-8-5	406	8K82K 248-01 + Blok D	First successful sample return
Luna 17	Nov. 10, 1970	1744:01	Ye-8	203	8K82K 251-01 + Blok D	Released Lunokhod rover
Luna 18	Sept. 2, 1971	1640:40	Ye-8-5	407	8K82K 256-01 + Blok D	Failed sample returner
Luna 19	Sept. 28, 1971	1300:22	Ye-8LS	202	8K82K 257-01 + Blok D	Successful advanced lunar orbiter
Luna 20	Feb. 14, 1972	0627:59	Ye-8-5	408	8K82K 258-01 + Blok D	Second successful sample return
Luna 21	Jan. 8, 1973	0955:38	Ye-8	204	8K82K 283-01 + Blok D	Released Lunokhod 2 rover
Luna 22	May 29, 1974	1156:51	Ye-8LS	206	8K82K 282-01 + Blok D	Successful advanced lunar orbiter
Luna 23	Oct. 28, 1974	1730:32	Ye-8-5M	410	8K82K 285-01 + Blok D	Advanced sample returner crashed on Moon
	Oct. 16, 1975	0704:56	Ye-8-5M	412	8K82K 287-01 + Blok D	No Earth orbit, launch failure
Luna 24	Aug. 9, 1976	1804:12	Ye-8-5M	413	8K82K 288-02 + Blok DM	Third successful sample return

Zond

Name	Launch Date	Launch Time (Moscow Time)	Spacecraft	S/C No.	Launcher and Serial No.	Comments
Kosmos-21	Nov. 11, 1963	0924	3MV-1A	2	8K78 GI03-18	Blok L incorrectly oriented, Mars test

TABLE VIII

Name	Launch Date	Launch Time (Moscow Time)	Spacecraft	S/C No.	Launcher and Serial No.	Comments
	Feb. 19, 1964	0848	3MV-1A		8K78 T15000-19	Communications loss with Blok L. Venus test flight, model variously reported as 3MV-1A and 3MV-4
Kosmos-27	Mar. 27, 1964	0625	3MV-1	5	8K78 T15000-22	Blok L power supply failure. Venus lander
Zond 1	April 2, 1964		3MV-1	4	8K78 T15000-23	Lost pressure, no Venus landing
Zond 2	Nov. 30, 1964		3MV-4A	2	8K78	Unopened solar panel, Mars photo failure
Zond 3	July 18, 1965		3MV-4A	3	8K78	Successful lunar far side photography

Sources

1. B. Ye. Chertok. *Rakety i lyudi: goryachiye dni kholodnoy voyny* (Moscow: Mashinostroyeniye, 1997).
2. K. Lantratov, "Jubilees: 25 Years for Lunokhod-1" (English title), *Novosti kosmonautiki* 23 (November 5–18, 1995): 79–83.
3. K. Lantratov, "Jubilees: 25 Years for Lunokhod-1" (English title), *Novosti kosmonautiki* 24 (November 19–December 2, 1995): 70–79.
4. Christian Lardier, *L'Astronautique Soviétique* (Paris: Armand Colin, 1992).
5. Jonathan McDowell, April 26, 1994 and July 31, 1994, personal correspondence to the author.
6. Timothy Varfolomeyev, "Soviet Rocketry that Conquered Space: Part 1: From the First ICBM to Sputnik Launcher." *Spaceflight* 37 (1995): 260–63.
7. Timothy Varfolomeyev, "Soviet Rocketry that Conquered Space: Part 2: Space Rockets for Lunar Probes." *Spaceflight* 38 (1996): 49–52.
8. Timothy Varfolomeyev, "Soviet Rocketry that Conquered Space: Part 5: The First Planetary Probe Attempts." *Spaceflight* 40 (1998): 85–88.

TABLE IX

Table IX

Governmental Decrees in the Soviet Missile and Space Programs, 1945–76

Date	Decree No.*	Title of Decree and Description	Issuing Body**
1945			
January 18	7350	On tasking OKB-51 to develop a copy of the Fi-103 missile	NKAP
May 26	8803	"On Measures to Reconstruct Industry in Connection with Maintaining Production of Armaments"—postwar rearmament plan	GKO
August 20	9887/ss/op	"On a Special Committee Under the GKO's Jurisdiction"—creation of Special Committee and its subordinate First Chief Directorate of GKO (later of Council of Ministers) for the development of atomic bomb	GKO
November 30		On organization of Special Design Bureau for rocket technology at Plant No. 88	NKA
1946			
May 13	1017-419ss	"Questions of Reactive Armaments"—formation of Special Committee for Reactive Technology (later Special Committee No. 2) of Council of Ministers for the coordination of work on missiles	SM
June 21		On establishment of NII-4	–
July 3	424	On redirecting Plant No. 456 at Khimki for the production of rocket engines	MAP
July 14		On establishment of BON of Fourth Directorate of GAU in Ministry of Armed Forces	–
August 9	83-K	On appointment of S. P. Korolev as Chief Designer of R-1	MV
August 26		On establishment of structure of NII-88	–
August 30		On appointment of S. P. Korolev as Chief of Department No. 3 of NII-88 SKB	–
September 29	1167	On establishment of OKB-456 at Plant No. 456 at Khimki	SM

Date	Decree No.*	Title of Decree and Description	Issuing Body**
1947			
July 26	2643-818ss	On testing of two series of A-4 rockets in 1947	SM
July 27		On establishment of GTsP-4 at Kapustin Yar	MVS
1948			
April 14		On work on the R-1 and R-2 missiles	SM
June 11	2018-791	On construction of a test site to ground-test rocket engines	SM
July 7	256	On establishment of NII Branch No. 2 to ground-test rocket engines	MV
1949			
December 27	5744-2162	On creation of Second Chief Directorate of Council of Ministers to manage uranium mining	SM
December 30		On work on geophysical variants of the R-1 missile	SM
1950			
April 26		On creation of OKB-1 within NII-88	MV
August 13	3456-1446	On termination of the work of German scientists in the Soviet missile program	SM
November 28		On adoption of the R-1 into armaments	SM
December 4		On themes N1, N2, and N3 in the ballistic missile program	SM
December 7	947-712	On transfer of OKB-456 from the Ministry of Aviation Industry to the Ministry of Armaments	SM
1951			
February 3		On creation of the Third Chief Directorate of Council of Ministers to manage missile programs	SM

TABLE IX

Date	Decree No.*	Title of Decree and Description	Issuing Body**
May 9		"On the Transfer of the Dnepropetrovsk Automobile Plant From the Ministry of Automobile and Tractor Industry to the Ministry of Armaments"—transfer of Dnepropetrovsk Plant No. 586 into the Ministry of Armaments for missile production	SM
May 24		On formation of OKB-23 at Fili	SM
June 1		On starting of series production of the R-1 at Dnepropetrovsk Plant No. 586	MV
November 27	4972-2096	On adoption of the R-2 into armaments	SM
November 30		On starting of series production of the R-2 at Dnepropetrovsk Plant No. 586	MV
1953			
February 13		On approval of work on themes T1 and T2, on approval of work on the R-5, R-11, and EKR missiles, and on transferring draft plan work for the R-12 from NII-88 to SKB-586	SM
February 18		On formation of AN SSSR Commission for Research on the Upper Layers of the Atmosphere	AN SSSR
February 19	533-271	On dissolution of OKB-51 and its transfer to OKB-155	SM
March 16		On uniting the First and Second Chief Directorates of Council of Ministers	SM
March 27		On renaming the Ministry of Armaments to the Ministry of Defense Industry	SM
April 18		On formation of UZKA of GAU	MVS
July 1		On unification of the First and Third Chief Directorates of Council of Ministers into the new Ministry of Medium Machine Building	SM
July 29		On transferring all missile work on guided missiles to the Ministry of Medium Machine Building	SM
September 23	2498-1031	On order of scientific-design work on new weapons systems	SM

937

Date	Decree No.*	Title of Decree and Description	Issuing Body**
1954			
January 26		On approval of work on the R-11FM	SM
March 17		On selection of launch area for the R-7, 40 Buran, and 350 Burya	SM
April 10		On approval of work on the R-5M missile and on establishment of OKB-586 at Dnepropetrovsk Plant No. 586	SM
May 20	956-4088s	On approval of work on the R-7 ICBM, R-5R, and M5RD missiles	SM
May 20	957-409	On transfer of intercontinental cruise missile work to the Ministry of Aviation Industry	SM
June 9		On establishment of Special Design Group 10 at Plant No. 500 at Tushino under V. N. Chelomey	MAP
June 28		"On NIP Plan for Special Product"—course of work on the R-7 ICBM	SM
July 9		On appointment of M. K. Yangel as Chief Designer of OKB-586	–
August 26		On approval of work on the R-11M missile	SM
November 20		On approval of the R-7 draft plan	SM
1955			
February 3		On creation of the submerged submarine-launched ballistic missile	SM
February 12	292-181	"On New Range for the USSR Ministry of Defense"—establishment of NIIP-5 at Tyura-Tam	SM
April 14		On creation of Special Committee for Armaments of the Army and VMF (later VPK) of Council of Ministers to manage missile programs	SM
July 13		On adoption of the R-11 into armaments	SM

TABLE IX

Date	Decree No.*	Title of Decree and Description	Issuing Body**
August 8		On establishment of OKB-52 at Reutov	TsK and SM
August 13		"On the Creation and Preparation of the R-12 (8K63) Missile"—start of work on the R-12	SM
1956			
January 30	149-88s	On creation of the Object D artificial satellite	SM
March 20		On means to ensure testing of the R-7	TsK and SM
June 20		On production of the R-5A and R-2A scientific missiles at Plant No. 586	MOP
June 21		On adoption of the R-5M into armaments	–
July 11		On approval of work on the R-11A missile for the IGY	–
August 14	310	On separation of OKB-1 from NII-88	MOP
August 31		On creation of the State Commission for the R-7	SM
September 3	1241-632	On creation of the Command-Measurement Complex	TsK and SM
September 30		On approval of the draft plan for Object D	VPK
December 17		"On the Creation of the Intercontinental Ballistic Missile R-16 (8K64) with Start of LKI in June 1961"—start of work on the R-16 ICBM	SM
1957			
January 11	61-39ss	On creation of launch complex Angara at NIIP-53	SM
January 11		On approval of flight-testing program for the R-7 ICBM	SM
February 15	171-83ss	"On Measures to Carry Out During the International Geophysical Year"—launch of simple satellites in mid-1957	SM

Date	Decree No.*	Title of Decree and Description	Issuing Body**
May 10		"On Future Improvements in the Organizational Control of Industry and Construction"—abolishing the branch ministry system	VS SSR
August 20		On transfer of two R-12 missiles to China (or August 6?)	MV
November		On termination of work on the 40 Buran intercontinental cruise missile	SM
November 6		On subordinating OKB-52 to GSNII-642	MAP
December 14		On creation of the Military-Industrial Commission on the basis of the Special Committee of the Council of Ministers and on change of the defense industry system from ministries to State Committees	TsK and SM
1958			
January 2		On adding of space work to Plant No. 1 (later Progress Plant) at Kuybyshev	–
January 29	293-140	On renaming town of Zarya to Leninsk at site 10 at Tyura-Tam	Kazakh VS
March 8		On subordinating GSNII-642 to OKB-52	MAP
March 20		On work on automated lunar probes and three-stage launch vehicles for them	–
March 20		On creation of Branch No. 1 of OKB-456	–
March 24		"On the Creation of the Burya Winged Missile"—course of work on 40 Burya	GKOT
April 1		On adoption of the R-11M into armaments	–
April 3		On creation of OKB-1 branch at Progress Plant	–
April 11		On creation of Branch No. 1 of OKB-456	GKOT
June 30		"On the Creation of Rockets With Engines on the Basis of Nuclear Energy Applications"—work on a draft plan for rockets with nuclear engines	TsK and SM

TABLE IX

Date	Decree No.*	Title of Decree and Description	Issuing Body**
July 2	726-346	On start of work on the R-14 missile, on creation of the R-7A ICBM, and on creation of the winged-ballistic rocket at OKB-52	TsK and SM
August 28		On expansion of work on the R-16 ICBM	—
September 2		On launch of automated lunar probes	—
November		On course of work on the piloted spaceship	SGK
December 6	1550-659	On reorganization of the missile and space industry	SM
1959			
January 5	22-10ss	On biomedical preparations for human spaceflight	TsK and SM
January 16		On separation of OKB-2 from NII-88	—
February 20		On adoption of the R-11FM into armaments	—
March 4		On adoption of the R-12 into armaments	—
March 14		"On Work on the R-7 Product and Flight-Design Testing of the R-7A Product"—testing of the R-7 and R-7A ICBMs	TsK and SM
April 1		On production of the R-7A missile at Plant No. 1001 at Krasnoyarsk-26 and establishment of a branch	TsK and SM
May 13		On start of dedicated work on the R-9 and R-16 ICBMs	TsK and SM
May 13		On creation of the Computer Center of NII-88	—
May 22	569-264	On work on a reconnaissance satellite and piloted spaceship	TsK and SM
June 4	191	On creation of OKB-1 Branch No. 2 at Krasnoyarsk-26	GKOT
July 3		On attaching of TsNII-58 for solid-propellant work to OKB-1	GKOT

Date	Decree No.*	Title of Decree and Description	Issuing Body**
July 15		On formation of the Design Department of OKB-1 at the Progress Plant in Kuybyshev	–
November 15		On tasking NII-4 to create a sea-based communications network	TsK and SM
December 10		"On the Development of Research on Cosmic Space"—future of Soviet space program (piloted, interplanetary, spaceplane)	TsK and SM
December 10	1386-618	"On the Creation of AMS for Landing on the Moon, and Flights to Venus and Mars"—approving automated lunar and interplanetary spacecraft	TsK and SM
December 17	254	"On the Establishment of the Post of Commander-in-Chief of Missile Forces in the Armed Forces of the SSSR"—creation of the Strategic Missile Forces	TsK
December 17	1384-615	"On the Establishment of the Post of Commander-in-Chief of Missile Forces in the Armed Forces of the SSSR"—creation of the Strategic Missile Forces	SM
1960			
January 11		On creation of the Cosmonaut Training Center	VVS
January 20		On adoption of the R-7 ICBM into armaments	–
February 5	138-48	On termination of work on the La-350 Burya at OKB-301	–
February 22	236-89	On creation of the State Commission for the R-16 ICBM	SM
May 9		On approval of work on the Elektron scientific satellite	TsK and SM
June 4	587-238	"On the Realization of the Plan to Master Cosmic Space in 1960 and the First Half of 1961"—creation of a four-stage launcher for interplanetary missions and schedule for the Korabl-Sputniks	TsK and SM
June 14		On creation of shaft units (silos) for the R-12, R-14, R-16, and R-9 missiles	GKOT
June 23	715-295	On approval of preliminary work on the Raketoplan, UR-200 ICBM, and IS anti-satellite system	–
June 23	715-296	"On the Creation of Powerful Booster Rockets, Satellites, Spacecraft and the Mastery of Space in 1960-67"—future of Soviet space program	TsK and SM

TABLE IX

Date	Decree No.*	Title of Decree and Description	Issuing Body**
August 3		On training of cosmonauts only at the Cosmonaut Training Center	SM
August 3		On naming June 2, 1955, as the birthday of NIIP-5	MO
August 8		"On the Creation of the Rocket-Carrier 63S1 Based on the R-12 Missile, and the Development and Launch of 10 Small ISZ"—start of work on launcher and satellites at OKB-586	TsK and SM
August 30	866-361	"On the Status of Cosmonauts"—medical requirements for cosmonauts	–
September 12		On adoption of the R-7A into armaments	–
October 3	1057-434	On transfer of OKB-23 as Branch No. 1 of OKB-52 and on course of work on IS anti-satellite system	–
October 11		On plan to launch the first human in space on Vostok	TsK and SM
1961			
During the year		On formalizing powers of the Council of Chief Designers	–
March 16	420-174I	On approval of work on the US satellite and UR-200 launch vehicle/ICBM	TsK and SM
April 3		On approval for launch of Vostok	TsK Presidium
April 24		On adoption of the R-14 missile into armaments	–
April 27		"On the Organization of Military Duty of the R-16U (Unified Variant)"—putting the R-16U missile into operation	GKOT
May 13		"On the Revision of Plans for Space Objects to Directions for Accomplishing Goals of Defense Designations"—heavy boosters, course of work on Elektron, and suspension of work on the Kosmoplan and Raketoplan with continuation of new Raketoplan work	TsK and SM
July 3		On naming OKB-23 plant after M. V. Khrunichev	SM

Date	Decree No.*	Title of Decree and Description	Issuing Body**
August 1		On course of work on the UR-200 missile and launcher	TsK and SM
October 20		On adoption of the R-16 into armaments	–
October 30		On plans for the military use of space during the period 1961–65	TsK and SM
October 30		On approval of work on the Molniya-1 communications satellite and Meteor-1 weather satellite	TsK and SM
October 31		"On the Creation of the Space Carrier 6S53"—start of work on a launch vehicle based on the R-14 and Meteor, Strela, and Pchela satellites	TsK and SM
December 18		On establishment of the independent OKB-10 at Krasnoyarsk-26 on the basis of OKB-1's Branch No. 2	GKOT
December 30	10/19	On selection of sixty new cosmonauts, including five women	TsK and SM
1962			
March 23		On Luna spacecraft for soft-landing on the Moon	TsK and SM
April 13		On restriction of work on the N1	–
April 16		"On Important Development of Intercontinental Ballistic and Global Missiles and Carriers-Rockets for Space Objects"—work on the N1, R-36, R-36-O, and R-56	–
April 16		"On the Development of the 'Soyuz' Complex for Piloted Flight to the Moon"—approving the Soyuz program for circumlunar flight	TsK and SM
April 29		On start of work on the UR-500 missile and carrier-rocket	SM
June 3		On course of work on the US reconnaissance satellite system launched on the UR-200	–
June 27		On formation of the Scientific-Technical Council of the Strategic Missile Forces	MO
September 24	1021-436	On start of work on the N1 and GR-1	TsK and SM

TABLE IX

Date	Decree No.*	Title of Decree and Description	Issuing Body**
October 13	640/06	On start of work on the GR-1	GKOT
1963			
January 2	15-5	On creation of the space and missile cosmodrome at NIIP-53 at Mirnyy/Plesetsk	-
February 8	24	"On the Manufacture of 'Vostok' Objects"—manufacture of Vostok spacecraft in 1963	VPK
March 21		On approval of work on the Soyuz complex	MNTS-KI
March 21		On preparation of proposals on launches of Vostok spacecraft	TsK and SM
April 13		On plans for Vostok launches in 1963	TsK and SM
June 15		On adoption into armaments of the R-16 surface variant	-
July 15		On adoption into armaments of the R-12U, R-14U, and R-16U shaft versions	-
August 30		On formation of NIIP-53 at Angara	MO
September 16	999-347	On creation of the space and missile cosmodrome at NIIP-53 at Mirnyy/Plesetsk	-
October		"Program for Space Investigations With Small ISZ, Launched on the Kosmos RN"—course of work on small satellites at OKB-586	-
November 13		On approval of the schedule of work for the N1 launch complexes	VSNKh
December 3		On approval of work on the Soyuz 7K-9K-11K circumlunar complex	TsK and SM
December 24		On ensuring the manufacture of ground equipment for the N1	-
1964			
January 9		On adoption of the R-12U and R-14U shaft versions into armaments	-
March 10	0045	On adopting the Zenit-2 satellite launched on the 8A92 into armaments	MO

Date	Decree No.*	Title of Decree and Description	Issuing Body**
March 13	59	On approval of work to convert Vostok to Voskhod and use it for three-person space missions	VPK
April		On appointing KB TM as main space launch pad design organization	–
April 13		On approval of work on four Voskhods and five EVA-equipped Vykhods	TsK and SM
May 22		On termination of work on the Kosmoplan and Raketoplan at OKB-52 and approval for the LK-1	TsK and SM
June 14		On approval of work on Voskhod and Vykhod	–
June 18		On military space programs for 1964-69, including the R spaceplane	MO
June 19		On termination of work on the R-56 launch vehicle and on schedule of the testing for the N1	–
August 3	655-268	"On Work on the Exploration of the Moon and Mastery of Space"—piloted LK-1 circumlunar and L3 lunar landing projects and the Ye-6M lunar lander	TsK and SM
August 3		On formation of the State Commission for Voskhod	TsK and SM
October		On formation of TsUKOS in the Strategic Missile Forces	MO
October 28		On assignment of lunar programs to OKB-52 and OKB-1	VPK
1965			
During the year		On end of work on the Raketoplan at OKB-52	–
During the year		On plan of work on Spiral at OKB-155	–
During the year		On work on space stations at OKB-52	TsK and SM
January 12		"On Detailed Work on Ampulized R-36 and R-36-O Missiles"—design work on the R-36 and R-36-O missiles	GKOT
February 10		On approval of the L3 draft plan	MNTS-KI

TABLE IX

Date	Decree No.*	Title of Decree and Description	Issuing Body**
March 2	126-47	"On Improved Leadership of Defense Branch Industry"—creation of ministries for aviation, defense, ship building, radio-technology, electronics, and general machine building for managing the Soviet defense industry	TsK and SM
April 15		"On Cooperation of the USSR and Socialist Countries in the Sphere of Research and Use of Space"—international cooperation	TsK and SM
May 15		On establishment of the Institute of Space Research in the USSR Academy of Sciences on July 14. 1965	TsK and SM
June 22		"On Preparations in 1965-66 of 18 Small Unifunctional Earth Satellites for Carrying out Scientific Investigations"—creation of small scientific satellites at OKB-586	MOM
July 21		On adoption of the R-9A in shaft and surface variants into armaments	–
July 28	156	"On the Manufacture of 'Voskhod' Space Ship-Satellites"—manufacture of five more Voskhod spacecraft	VPK
August		On creation of military Voskhod and Soyuz spacecraft	VPK
August 18	180	"On the Order of Work on the Soyuz Complex"—approval of the schedule of work for Soyuz spacecraft	VPK
August 24		"On Creation of an R-36 Based Carrier Rocket for Launching the IS and US KA"—start of work on an R-36-based launch vehicle for the IS and US programs	–
August 24		On expansion of military space research and on 7K-VI Zvezda	TsK and SM
September 6		On delays in work on piloted lunar programs	MOM
October 20		On approval of work on the draft plan of the UR-700-LK-700 lunar complex	MOM
October 25		"On the Concentration of Forces of Industrial Design Organizations for the Creation of Rocket-Space Complex Means for Circling the Moon"—work on the UR-500K-L1 program	TsK and SM

Date	Decree No.*	Title of Decree and Description	Issuing Body**
November 13		On work on the UR-500K-LI program	MOM
1966			
February 22		"On Performing in March 1966 the Launch of 3KV n6 With Two Cosmonauts, for Solving Problems of Extended Space Flight (18-20 Days)"—course of Voskhod-3 preparations	TsK and SM
March 6		On renaming OKB-1 as TsKBEM and OKB-52 as TsKBM	MOM
March 30	145ss	On approval of the 7K-TK as transport for the Almaz station	MOM
April 27	101	"On Approving the Work Plan to Build the Piloted Spacecraft 7K-LI"—approving the plan for the UR-500K-LI and terminating the UR-500K-I	VPK
May 15	144	On assessing preparations for flights of the 7K-OK spacecraft	VPK
May 23	43 or 47	On creation of the civilian detachment of cosmonauts	OKB-I
June 15	144	On preparation of crews for the 7K-OK spacecraft and civilian cosmonauts	VPK
June 21		On long-range military use of space in 1966-70	TsK and SM
September		On approval of the N1-L3 mission profile	AN SSSR
September 14		On course of work on the N1-L3	VPK
September 17		On creation of a commission to compare the UR-700-LK-700 and the N1-L3	MOM
October		On renaming OKB-586 as KB Yuzhnoye	–
November		On lag of work on the N1-L3 and UR-500K-LI programs	VPK
December 28	304	On changes in the timeline for the Almaz program and suspension of the 7K-TK	VPK
December 28	305	On approval of work on the 7K-VI Zvezda and course of work on Almaz	VPK

TABLE IX

Date	Decree No.*	Title of Decree and Description	Issuing Body**
1967			
During the year	0015	On transfer of Zenit-2 from the 8A92 to the 11A57 launcher	MO
February 4	115-46	"On the Progress of the Work on the Development of the UR500K-L1"—confirmation of schedule for piloted lunar missions	TsK and SM
February 9		On approval of work on Almaz	MOM
February 14		On construction of the N1 payload fairing by the Khrunichev Plant	MOM
March 15	42	On search service for returning missions from the Moon	VPK
March 27	270-105	"On Preparation of Cosmonaut-Testers and Cosmonaut Researchers"—formation of group of research and test-cosmonauts to support future missions	TsK and SM
March 30		On formation of Anti-Space and Anti-Missile Forces of the Air Defense Forces (RKO) to operate Soviet ASAT systems	MO General Staff
April 22	145	"On the Preparation of Test-Cosmonauts and Research-Cosmonauts"—selection of the group of engineer-cosmonauts under the Ministry of General Machine Building	MOM
May		On adoption into armaments of the Raduga complex of DS-P1-Yu	–
June		On full approval of the Almaz and 7K-TK programs	TsK and SM
June 21		On approval of the Almaz draft plan	VPK
July		On use of the R-36-based launcher for the Kosmos and Meteor satellites	SM
July 21		On approval of the R-36 ICBM variant with means to overcome PRO and on adoption of the R-36 ICBM into armaments	–
July 21	715-240	"On the Creation of Space Systems for Naval Reconnaissance Comprising the US ISZ and the Rocket-Carrier on the Basis of the R-36"—further work on the US naval reconnaissance satellite, approval of work on the Yantar-2K, and course of work on the 7K-VI Zvezda and OIS	–

Date	Decree No.*	Title of Decree and Description	Issuing Body**
July 24	220	On approval of work on the Yantar-2K	MOM
August 14		On schedule of work on the Almaz space station	TsK and SM
November 14		On revision of the timetable for the N1-L3	-
November 17	1070-363	On approval of work on the UR-700 launch vehicle	TsK and SM
1968			
February 21		"On Introduction of Hydrogen in Rocket-Space Technology"—future of liquid hydrogen stages	TsK and SM
March 11		On formation of the Space Branch of NII-4	-
March 13		On approval of the training program for lunar cosmonauts	VVS
March 19	88	On use of liquid hydrogen in the space program	MOM
May 27	163	On formation of a new group of engineer-cosmonauts under MOM	MOM
October 24		On establishment of the Kristall communications system based on Molniya-2 satellites	TsK and SM
October 28		On renaming of Zelenyy as Zvezdnyy gorodok	Moscow Oblast Exec. Committee
November 19		On adoption of the R-36-O into armaments	-
1969			
January 8	19-10	"On Work on Research of the Moon, Venus and Mars by Automatic Stations"—work on automated lunar and interplanetary spacecraft	TsK and SM
June 30	232	On start of work on the UR-700M rocket	MOM

TABLE IX

Date	Decree No.*	Title of Decree and Description	Issuing Body**
1970			
January 2		"On the Creation of the Carrier-Rocket 11K68 on The Basis of 11K69 RN and SSM Stage for Launch of Space Apparatus 'Tselina' and 'Meteor'"—approval of work on the Tsiklon-3 RN	TsK and SM
February 9	105-41	On creation of the DOS using Almaz as a basis	TsK and SM
February 16	57ss	On creation of the DOS using Almaz as a basis	MOM
March		On formation of GUKOS on the basis of TsUKOS and subordinated to RVSN	MO
June 4		On standardized weather satellite system	VPK
June 16	437-160	On creation of the TKS and termination of the 7K-TK	TsK and SM
1971			
June 8		On work on nuclear rocket engines	–
September		On cooperation to build an Indian satellite	TsK and SM
October		"On the Development of a System of Global Television Reconnaissance (TGR)– 'Tayfun' ISZ"—approval of work on the Tayfun reconnaissance satellite	–
October		"On the Development of an Adjustment Complex with the 'Tayfun-2' ISZ"—approval of work on the Tayfun-2 system	–
December 21		"On Expansion of Work on Research of the Earth's Natural Resources by Space Systems Technology"—Meteor-Prioda system	TsK and SM
1972			
February 16		On approval of work on the draft plan for the N1-L3M two-launch lunar landing proposal	VPK
February 23		On work on the technical proposal for the creation of the MOK	VPK

Date	Decree No.*	Title of Decree and Description	Issuing Body**
March 26		On adoption of Tselina-O into armaments	TsK and SM
April 5		On use of Molniya and Gran for a unified satellite communications system	TsK and SM
May 15		On approval of the N1-L3M proposal	SGK
June 15		On schedule of work for the Almaz and TKS programs	MOM
June 26		"On the Creation of Automatic Universal Orbital Stations (AUOS)"—on approval of work on the AUOS satellite bus	MOM
September		On termination of production work on the L3	–
September 25		On formation of TsNII-50 on the basis of NII-4's Space Branch	–
November 19		"On Cooperation of USSR and India on Space Research"—USSR-India cooperation	TsK and SM
December 16		On establishment of the Planeta-S weather satellite system	VPK
1973			
February		On adoption of the Tayfun-1 into armaments	TsK and SM
March 26	182-63	On development of the Yantar-1KFT reconnaissance and cartographic satellite and the 11A511K launcher	TsK and SM
December		"On Carrying out Work on Reusable Space Systems"—response to NASA's Space Shuttle	VPK
1974			
May 19		On suspension of further launches of the N1	MO
May 22		On formation of NPO Energiya	–
June 21		On establishment of the State Commission for testing the Soyuz-T	VPK

TABLE IX

Date	Decree No.*	Title of Decree and Description	Issuing Body**
June 24		On suspension of work on the N1-L3	NPO Energiya
July 30		On separation of TsSKB from NPO Energiya and creation of the Volzhkiy Branch	–
December 31	314	On development of the Topaz-1 thermionic nuclear reactor for Plazma-A spacecraft	VPK
1975			
June 5	178	On development of the 11A511U2 launch vehicle	MOM
September 21		On USSR-France cooperation in space	TsK and SM
October		On adoption of US-A with Tsiklon-2 into armaments	–
1976			
January 19	46-13	On course of work on Almaz and the TKS	TsK and SM
February 3		On USSR-France cooperation in space	TsK and SM
February 17		On work on Energiya-Buran, DOS-7K nos. 7 and 8, Gamma, Geyzer, and Altair and cancellation of the N1	TsK and SM
February 27		On long-range military use of space up to 1990	TsK and SM
March 16		"On the Creation of a Universal Space Missile Complex 11K77 'Zenit'" — approval of work on the Zenit launcher	TsK and SM
May 31	409-147	"On Creation of Yantar-1KFT Space Complex for Solving Goals of Cartography" — development of the Yantar-1KFT reconnaissance satellite	TsK and SM
June 11		On selection of design layout for Buran	SGK

Date	Decree No.*	Title of Decree and Description	Issuing Body**
June 15		On course of work on nuclear rocket engines	–
November 8		On approval of a tactical-technical requirement for Buran	MO
December		"On Expansion of the Global Navigation Satellite System (GLONASS) Single Space Navigation System"—development of GLONASS	TsK and SM
December 10		On adoption of the Tselina-D into armaments	TsK and SM
December 10	342	On development of the Topaz-1 thermionic nuclear reactor for Plazma-A spacecraft	VPK
December 18		On course of work on Energiya-Buran	VPK

* All known decrees related to the Soviet space program in the period 1945–76 are shown above. All decrees related to ballistic missile development in the period 1945–57 are shown. In the period 1958–76, only the R-7A, R-9, and R-16 ICBMs are shown.

** Acronyms for the issuing bodies are as follows:

AN SSSR	USSR Academy of Sciences
GKO	State Committee for Defense
GKOT	State Committee for Defense Technology
MAP	Ministry of Aviation Industry
MNTS-KI	Interdepartmental Scientific-Technical Council on Space Research
MO	Ministry of Defense
MOM	Ministry of General Machine Building
MOP	Ministry of Defense Industries
MV	Ministry of Armaments
MVS	Ministry of Armed Forces
NKA	People's Commissariat of Armaments
NKAP	People's Commissariat of Aviation Industry
SGK	Council of Chief Designers
SM	Council of Ministers
TsK	Central Committee
VPK	Military-Industrial Commission
VVS	The Air Force
VS SSSR	USSR Supreme Soviet
VSNKh	All-Russian Council of the National Economy

APPENDICES

Appendix A
Soviet Piloted Space Projects, 1945–74

1. VR-190

Lead institutions	NII-1 (1945–46), NII-4 (1946–49)
Lead designer	M. K. Tikhonravov
Initiation of studies	1944–45
Project termination	1949
Spacecraft	VR-190
Launch vehicle	A-4 derivative
Objective	Launch of "stratonauts" on vertical flights to upper atmosphere

2. Antipodal Bomber

Lead institutions	NII-1, TsIAM
Lead scientists	M. V. Keldysh, V. F. Bolkhovitinov
Initiation of studies	1945–46
Preparation of design documentation	1947
Project termination	1950
Spacecraft	Sänger-Bredt winged bomber
Objective	Transatlantic upper atmospheric piloted flight

3. Vertical/Suborbital Program

Lead institutions	NII-88 OKB-1, OKB-1
Chief Designer	S. P. Korolev
Lead designer	N. P. Belov
Initiation of studies	April 1955
Preparation of design documentation	May 1956
Termination of studies	November 1958
Launch vehicles	R-1Ye, R-2A
Objective	Launch of humans on vertical and suborbital trajectories

4. Vostok

Lead institution	OKB-1
Chief Designer	S. P. Korolev
Deputy Chief Designer (for Vostok)	K. D. Bushuyev
Chief of Planning Department (for Vostok)	M. K. Tikhonravov
Group Chief (for Vostok)	K. P. Feoktistov
Lead designers	O. G. Ivanovskiy, Ye. A. Frolov
Initiation of studies	April 1957
Preparation of design documentation	August 18, 1958
Approval by Council of Chief Designers	November 1958
TsK KPSS/SM approval	January 5, 1959, May 22, 1959
Draft plan signed	April–May 1959 (for 1K), July 31, 1961 (for 3KA)
First orbital launch attempt	May 15, 1960 (Korabl-Sputnik)
Last orbital launch attempt	June 16, 1963 (Vostok 6)
Program termination	March–April 1964
Spacecraft	1K/11F61, 3KA/11F63
Launch vehicles	8K72/Luna, 8K72K/Vostok
Objective	Piloted orbital flight with a single cosmonaut

5. Gliding Cosmic Apparatus (PKA)

Lead institutions	OKB-256 (spacecraft), OKB-1 (launcher)
Chief Designer	P. V. Tsybin

Initiation of design studies	1957–58
Predraft plan signed	May 17, 1959
Termination of studies	October 1, 1959
Spacecraft	PKA/Lapotok
Launcher	8K72K/Vostok
Objective	Piloted military operations in Earth orbit with reusable spaceplane

6. M-48/VKA-23

Lead institutions	OKB-23 (spacecraft and launcher), OKB-1 (launcher)
General Designer	V. M. Myasishchev
Initiation of design studies	1957–58
TsK KPSS/SM approval	December 10, 1959
Project termination	October 3, 1960
Spacecraft	M-48
Launch vehicles	8K72K/Vostok, M-1
Objective	Piloted military operations in Earth orbit with reusable spaceplane

7. Sever/Space Station/1L Circumlunar Spacecraft

Lead institution	OKB-1
Chief Designer	S. P. Korolev
Initiation of studies	April 1959
Technical prospectus signed	March 10, 1962
Termination of studies	mid-1962
Spacecraft	5K/Sever, 5KA & 5KB/space station, 7K/Vostok, 9K/rocket stage, 1L/circumlunar vehicle
Launch vehicle	8K711
Objectives	All-purpose Earth-orbital operations with guided reentry, space station, piloted circumlunar flight

8. Heavy Interplanetary Ship (TMK)

Lead institution	OKB-1
Chief Designers	S. P. Korolev, M. P. Mishin
Chief of Planning Department	M. K. Tikhonravov
Group Chiefs	G. Yu. Maksimov, K. P. Feoktistov
Initiation of studies	1959
Predraft plan signed	May 1966
Experimental design signed	1969
Termination of studies	1969
Spacecraft	MEK
Launch vehicle	N1, N1M
Objective	Piloted spacecraft for orbiting and landing on Mars

9. Raketoplan

Lead institutions	OKB-52 (spacecraft and launchers); OKB-586, OKB-1, and OKB-52 Branch No. 1 (launchers)
General Designer	V. N. Chelomey
Initiation of studies	1959
TsK KPSS/SM approval	June 23, 1960
First launch attempt in program	December 22, 1961 (MP-1)
Draft plan signed	1963
Project termination	1965
Spacecraft	SR, MP-1, M-12, R-1, R-2
Launch vehicles	R-12, R-14, 8K81K, 8K82/Proton
Objective	Piloted reusable spaceplane for suborbital, orbital, and lunar missions

10. Kosmoplan

Lead institutions	OKB-52 (spacecraft and launchers), OKB-1 and OKB-52 Branch No. 1 (launchers)
General Designer	V. N. Chelomey
Initiation of studies	1959
TsK KPSS/SM approval	June 23, 1960
Predraft plan signed	1961
Project termination	May 22, 1964
Spacecraft	AK-1-7, AK-1-300, AK-3-300, AK-4
Launch vehicles	8K72K, A-300, 8K82/Proton
Objective	Automated and piloted reusable spacecraft to the Moon, Mars, and Venus

11. Soyuz Complex

Lead institutions	OKB-1 (7K and launcher), SKB-10 (11K), SKB-385 (9K)
Chief Designer	S. P. Korolev
Initiation of studies	January 26, 1962
Predraft plan signed	December 24, 1962
Technical prospectus signed	May 10, 1963
TsK KPSS/SM approval	December 3, 1963
Program termination	August 3, 1964
Spacecraft	7K/Soyuz-A, 9K/Soyuz-B, 11K/Soyuz-V
Launch vehicles	11A55, 11A56
Objective	Piloted circumlunar flight

12. R-56

Lead institution	OKB-586
Chief Designer	M. K. Yangel
TsK KPSS/SM approval	April 16, 1962
Termination of studies	June 19, 1964
Launch vehicles	R-56, SK-100
Objectives	Robotic lunar landing, piloted circumlunar missions

13. Zvezda/Heavy Orbital Station (TOS)

Lead institution	OKB-1
Chief Designers	S. P. Korolev, V. P. Mishin
Initiation of design studies	1960
Predraft plan signed	May 3, 1961
Termination of studies	1969
Spacecraft	TOS/Zvezda
Launch vehicle	N1 derivatives
Objective	Large piloted space station in Earth orbit

14. Soyuz-R

Lead institution	OKB-1 Branch No. 3
Chief Designer	D. I. Kozlov
Initiation of studies	1962-63
MO approval	June 18, 1964
Predraft plan signed	July 15, 1965
Program termination (11F71 station)	Early 1966
Program termination (7K-TK ferry)	June 21, 1967
Spacecraft	7K-TK/11F72/Soyuz-R, 11F71 station
Launch vehicle	Soyuz-type
Objective	Piloted reconnaissance platform in Earth orbit

15. Soyuz-P

Lead institution	OKB-1 Branch No. 3
Chief Designer	D. I. Kozlov
Initiation of studies	1962–63
Program termination	1965
Spacecraft	7K-PPK/Soyuz-P
Launch vehicle	11A514
Objective	Piloted anti-satellite spacecraft in Earth orbit

16. N1-L3

Lead institution	OKB-1
Chief Designers	S. P. Korolev, V. P. Mishin, B. A. Dorofeyev
Deputy Chief Designers	K. D. Bushuyev, S. S. Kryukov, S. O. Okhapkin
Initiation of studies	March 1963
TsK KPSS/SM approval	August 3, 1964
Predraft plan signed	December 30, 1964
Draft plan signed	November 11, 1965
First orbital launch attempt	February 21, 1969
Last orbital launch attempt	November 23, 1972
Project suspension	June 24, 1974
Program termination	February 18, 1976
Spacecraft	7K-L1S/11F92, 7K-LOK/11F93, LK/11F94, T1K, T2K, 7K-L1E, Blok D (originally included L1, L2, L3, L4, and L5)
Launch vehicles	N1, N1 derivatives, 8K82K/Proton-K
Objective	Landing of one cosmonaut on the Moon

17. Voskhod

Lead institutions	OKB-1 (spacecraft), OKB-1 Branch No. 3 (launcher)
Chief Designers	S. P. Korolev, V. P. Mishin
Lead designer	Ye. A. Frolov
Initiation of studies	December 1963
VPK approval	March 13, 1964
TsK KPSS/SM approval	April 13, 1964
Draft plan signed	August 1964
First orbital launch attempt	October 6, 1964 (Kosmos-47)
Last orbital launch attempt	February 22, 1966 (Kosmos-110)
Program termination	September–October 1966
Spacecraft	3KV/11F63, 3KD/11F63
Launch vehicle	11A57/Voskhod
Objective	Propaganda goals in Earth orbit (multicrews, EVA, long duration, tethers)

18. UR-500K/LK-1

Lead institutions	OKB-52 (spacecraft), OKB-52 Branch No. 1 (launcher)
General Designer	V. N. Chelomey
Initiation of studies	Late 1963
Predraft plan signed	August 3, 1964
TsK KPSS/SM approval	August 3, 1964
Draft plan signed	July 1965
Project termination	April 27, 1966
Spacecraft	LK-1, Blok A
Launch vehicle	8K82K/Proton-K
Objective	Piloted circumlunar flight

19. UR-700/LK-700

Lead institutions	OKB-52 (spacecraft), OKB-52 Branch No. 1 (launch vehicle)
General Designer	V. N. Chelomey
Initiation of studies	1964
Approval for work on draft plan	October 20, 1965
Predraft plan signed	August–September 1966
Program suspended	November 1966
TsK KPSS/SM approval	September 17, 1967
LK-700 draft plan signed	October 1968
Project termination	Early 1969
Spacecraft	LK-700
Launch vehicle	UR-700
Objective	Direct ascent piloted lunar landing

20. Soyuz

Lead institutions	OKB-1 (spacecraft), OKB-1 Branch No. 3 (launcher)
Chief Designers	S. P. Korolev, V. P. Mishin, V. P. Glushko
Lead designers	Ye. A. Frolov, A. F. Topol, Yu. P. Semenov, Ye. P. Vyatkin, V. P. Guzenko
Initiation of studies	Late 1964
VPK approval	August 18, 1965
Draft plan signed	October 23, 1965
First orbital launch attempt	November 28, 1966 (Kosmos-133)
Last orbital launch attempt	May 14, 1981 (Soyuz 40)
Program termination	May 1981
Spacecraft	7K-OK/11F615, 7K-T/11F615A8, 7K-TA/11F615A9, 7K-TM/11F615A12
Launch vehicles	11A511/Soyuz, 11A511U/Soyuz-U
Objectives	Master rendezvous and docking techniques in Earth orbit, station ferry

21. Almaz Orbital Piloted Station (OPS)/*Salyut*

Lead institutions	OKB-52 (spacecraft), OKB-52 Branch No. 1 (launch vehicle)
General Designer	V. N. Chelomey
Lead designer	V. A. Polyachenko
Initiation of studies	October 1964
Draft plan signed	June 23, 1967
TsK KPSS/SM approval	August 14, 1967
First orbital launch attempt	April 3, 1973 (*Salyut 2*)
Last orbital launch attempt	June 22, 1976 (*Salyut 5*)
Project termination	December 19, 1981
Spacecraft	OPS/11F71 station, 7K-TK/11F72 ferry, 7K-TA/11F6159 ferry (see also TKS)
Launch vehicle	8K82K/Proton-K
Objective	Piloted military station in Earth orbit

22. N11-Soyuz

Lead institution	OKB-1
Chief Designer	S. P. Korolev
Initiation of studies	Late 1964
Technical prospectus signed	February 5, 1965
Termination of studies	August 1965
Spacecraft	7K-PLK
Launch vehicle	N11
Objective	Piloted lunar orbital flight

23. 7K-VI Zvezda

Lead institution	OKB-1 Branch No. 3
Chief Designer	D. I. Kozlov
Initiation of studies	Late 1964
TsK KPSS/SM approval	August 24, 1965
Draft plan signed (first variant)	1965
Draft plan signed (second variant)	1967
MOM approval	July 7, 1966
Program termination	January–February 1968
Spacecraft	7K-VI/11F73/Zvezda
Launch vehicle	11A511M/Soyuz-M
Objective	Piloted military operations in Earth orbit

24. Spiral

Lead institutions	OKB-155 and Gromov LII (spaceplane), OKB-1 (conventional launcher), OKB-52 (booster), OKB-156 (GSR)
General Designer	A. I. Mikoyan
Chief Designer	G. Ye. Lozino-Lozinskiy
Initiation of studies	1964
Predraft plan signed	June 29, 1966
First launch attempt	July 15, 1969 (BOR-1)
Last launch attempt	1974 (BOR-3)
First airdrop	October 11, 1976 (105.11)
Last airdrop	September 1978 (105.11)
Project termination	September 1978
Spacecraft	Orbital Aircraft/50, EPOS, booster rocket, BOR-1, BOR-2, BOR-3, 105.11, 105.12, 105.13
Launch vehicle	GSR/50-50, 11A511/Soyuz, Tu-95K
Objective	Reusable military spaceplane for Earth-orbital operations

25. Zvezda Spaceplane

Lead institution	OKB-156
General Designer	A. N. Tupolev
Initiation of studies	Early 1960s
Termination of studies	1966
Spacecraft	Zvezda
Launch vehicle	Tu-95K
Objective	Air-launched reusable military spaceplane

26. Zond

Lead institutions	OKB-1 (spacecraft and upper stage), OKB-52 Branch No. 1 (launcher)
Chief Designers	S. P. Korolev, V. P. Mishin
Lead designers	B. V. Rublev, Yu. P. Semenov
Initiation of studies	August 1965
TsK KPSS/SM approval	October 25, 1965
MOM approval	November 13, 1965
Predraft plan signed	November 30, 1965
First orbital launch attempt	March 10, 1967 (Kosmos-146)
Last orbital launch attempt	October 20, 1970 (Zond 8)
Program termination	October 1970
Spacecraft	7K-L1/11F91/Zond, 7K-OK-T/Soyuz
Launch vehicle	8K82K/Proton-K
Objective	Piloted circumlunar flight

27. Multirole Orbital Complex (MOK)

Lead institution	TsKBEM
Chief Designers	S. P. Korolev, V. P. Mishin
Initiation of studies	September 30, 1963
VPK decree on issue of technical plan	February 23, 1972
Draft plan signed	Early 1973
Program termination	May 1974
Spacecraft	MBKS, 19K modules, TKS
Launch vehicles	N1, N1 derivatives
Objective	Massive piloted complex in Earth orbit

28. Long-Duration Lunar Base (DLB)

Lead institutions	KB OM (spacecraft), GSMZ Lavochkin (spacecraft), TsKBEM (spacecraft and launcher)
Chief Designers	V. P. Barmin, V. P. Mishin
Initiation of studies	1965-66
Termination of studies	Late 1970s
Spacecraft	Bolshoye koltso, Kolumb, Dal, Osvoyeniye
Launch vehicles	N1, N1 derivatives
Objective	Permanent piloted base on lunar surface

29. Transport-Supply Ship (TKS)

Lead institutions	OKB-52 and OKB-52 Branch No. 1 (spacecraft), OKB-52 Branch No. 1 (launcher)
Initiation of studies	1966-67
Draft plan signed	1969
First orbital launch attempt/TKS	July 17, 1977 (Kosmos-929)
Last orbital launch attempt/TKS	September 27, 1985 (Kosmos-1686)
First orbital launch attempt/TKS VA	December 15, 1976
Last orbital launch attempt/TKS VA	May 23, 1979 (Kosmos-1100/1101)
Project termination	1986
Spacecraft	TKS/11F72, TKS VA/11F74, TKS FGB/11F77
Launch vehicle	8K82K/Proton-K
Objective	Transport ship for Almaz and Salyut space stations

30. Soyuz-VI

Lead institutions	TsKBEM an TsKBEM Branch No. 3 (spacecraft), TsKBEM Branch No. 3 (launcher)
Chief Designers	V. P. Mishin, D. I. Kozlov
Initiation of studies	Late 1967
Project approval	January–February 1968
Draft plan signed	June 23, 1968
Project termination	February 1970
Spacecraft	OB-VI station/11F731, 7K-S/11F732, 7K-S-I/11F733, 7K-S-II/11F734, 7K-G/11F735
Launch vehicle	11A511/Soyuz
Objectives	Small military space station in Earth orbit with different ferry craft

31. UR-700M/MK-700

Lead institutions	TsKBM (spacecraft), TsKBM Branch No. 1 (launcher)
General Designer	V. N. Chelomey
Initiation of studies	Early 1969
MOM approval	June 30, 1969
Predraft plan signed (MK-700)	April 1970

Predraft plan signed (UR-700M)   October 1970
Project termination   Late 1970
Spacecraft   MK-700
Launch vehicle   UR-700M
Objective   Piloted landing on Mars

32. Long-Duration Orbital Station (DOS)/*Salyut*

Lead institutions   TsKBEM and TsKBM Branch No. 1 (spacecraft), TsKBM Branch
                                   No. 1 (launcher)
Chief/General Designers   V. P. Mishin, V. P. Glushko, Yu. P. Semenov, V. N. Bugayskiy
Lead designer   Yu. P. Semenov, V. V. Pallo
Initiation of studies   December 1969
TsK KPS/SM approval   February 9, 1970
First orbital launch attempt   April 19, 1971 (*Salyut*)
Last orbital launch attempt   February 19, 1986 (*Mir*)
Project termination   —
Spacecraft   17K/DOS, 17KS/*Mir*, 17KSM/ISS
Launch vehicle   8K82K/Proton-K
Objective   Small piloted station in Earth orbit with ferry craft

33. N1-L3M

Lead institution   TsKBEM
Chief Designers   V. P. Mishin, V. A. Borisov
Initiation of studies   1969–70
Draft plan signed   Late 1971
Approval by Council of Chief Designers   May 15, 1972
Project termination   May 1974
Spacecraft   L3M
Launch vehicle   N1F
Objective   Long-duration piloted landings on the Moon

Appendix B
Dramatis Personae, 1945–74

Full Name	Date of Birth/Death	Contribution to the Soviet Space Program
Designers and Scientists		
Abramov, Anatoliy Petrovich	1919–August 15, 1998	Deputy Chief Designer in 1966–80 at OKB Korolev worked on launch complexes.
Alekseyev, Semyon Mikhaylovich	1909–	Chief Designer in 1952–73 at OKB Zvezda worked on spacesuits and airlocks.
Avduyevskiy, Vsevolod Sergeyevich	July 28, 1920–	Scientist at NII-1 in 1953–73 and First Deputy Director at TsNIIMash in 1973–87.
Babakin, Georgiy Nikolayevich	November 14, 1914– August 3, 1971	Chief Designer in 1965–71 at OKB Lavochkin led work on lunar and interplanetary spacecraft.
Barmin, Vladimir Pavlovich	March 17, 1909– July 17, 1993	Chief Designer in 1941–93 at GSKB SpetsMash designed launch complexes.
Blagonravov, Anatoliy Arkadyevich	June 1, 1894– February 4, 1975	President of the Academy of Artillery Sciences in 1946–50 and public spokesperson.
Blokhin, Yuriy Dmitryevich	Unknown	Head of Mikoyan KB space branch worked on the Spiral spaceplane.
Bogomolov, Aleksey Fedorovich	June 2, 1913–	Chief Designer in 1954–88 at OKB MEI worked on telemetry and guidance systems.
Bogomolov, Vladislav Nikolayevich	September 14, 1919– February 9, 1997	Chief Designer in 1971–85 at OKB Isayev worked on rocket engines and succeeded Isayev.
Boguslavskiy, Yevgeniy Yakovlevich	1917–May 18, 1969	Deputy Chief Designer in 1950–69 at Ryazanskiy NII worked on spacecraft guidance systems.
Bolkhovitinov, Viktor Fedorovich	1989–1970	At NII-1, he worked on the Sänger-Bredt antipodal bomber.
Bondaryuk, Mikhail Makarovich	1908–1969	Chief Designer in 1950–69 at OKB-670 worked on ramjet engines for Burya and Buran.
Borodin, Sergey Aleksandrovich	1935–	Chief Designer from 1975 on at SOKB of Gromov LII designed simulators and cockpit consoles.
Budnik, Vasiliy Sergeyevich	June 24, 1913–	Deputy Chief Designer in 1954–72 at OKB Yangel worked on missiles and was a Korolev protégé.
Bugayskiy, Viktor Nikifirovich	Unknown	He headed OKB Chelomey Branch No. 1 in 1960–73 and worked on rockets and spacecraft.
Bushuyev, Konstantin Davidovich	May 23, 1914– October 26, 1978	Deputy Chief Designer in 1954–72 at OKB Korolev led piloted spacecraft projects.

Full Name	Date of Birth/Death	Contribution to the Soviet Space Program
Bykov, Yuriy Sergeyevich	1916–1970	Chief Designer in 1959–70 at NII-695 worked on communications systems for piloted spacecraft.
Chelomey, Vladimir Nikolayevich	June 30, 1914– December 8, 1984	Chief Designer/General Designer in 1955–84 at OKB-52 led work on cruise missiles, ICBMs, and spacecraft.
Chertok, Boris Yevseyevich	March 1, 1912–	Deputy Chief Designer in 1956–91 at OKB Korolev worked on guidance systems.
Darevskiy, Sergey Grigoryevich	Unknown	Chief Designer in 1965–75 at SOKB of Gromov LII designed simulators and cockpit consoles.
Dorofeyev, Boris Arkadyevich	November 25, 1927– July 9, 1999	Deputy Chief Designer at OKB Korolev, was Chief Designer for N1 rocket in 1972–74 (demoted in 1974).
Eidis, Arkadiy Ionovich	1913–	He headed OKB Chelomey Branch No. 3 in 1962–65 and was later Chelomey's First Deputy General Designer.
Feoktistov, Konstantin Petrovich	February 7, 1926–	Department Chief at OKB-Korolev worked on Vostok and other piloted spacecraft.
Gazenko, Oleg Georgiyevich	December 12, 1918–	Director of IMBP in 1969–88 performed early work on space medicine.
Glushko, Valentin Petrovich	September 2, 1908– January 10, 1989	Chief Designer/General Designer in 1946–89 at OKB-456 designed rocket engines for missiles and launchers.
Gubanov, Boris Ivanovich	March 14, 1930– March 18, 1999	He was First Deputy Chief Designer/ General Designer in 1972–82 at OKB Yangel and in 1982–93 at OKB Korolev.
Gubenko, Yevgeniy Stepanovich	Unknown–1959	Chief Designer in 1950–59 at SKB-567 worked on ground communications segment.
Gusev, Leonid Ivanovich	1922–	Director of NII-695 and from 1965 on Director of NII P led work on guidance systems.
Iosifyan, Andronik Gevondovich	1905–1993	Chief Designer in 1941–74 at NII-627 worked on power sources and remote-sensing craft.
Isayev, Aleksey Mikhailovich	October 24, 1908– June 10, 1971	Chief Designer in 1947–71 at OKB-2 led work on engines for piloted spacecraft.
Ishlinskiy, Aleksandr Yulevich	August 6, 1913–	Director of Institute of Mechanics in 1964–89 prepared space communiqués.
Ivanov, Ivan Ivanovich	1918–	Deputy Chief Designer at OKB Yangel, led work on LK lander engine.
Ivanovskiy, Oleg Genrikhovich	January 18, 1922–	He worked at OKB-1 on Sputnik and Vostok and was Deputy Chief Designer in 1971–83 at OKB Lavochkin.

Full Name	Date of Birth/Death	Contribution to the Soviet Space Program
Ivensen, Pavel Albertovich	1908–	At OKB-52, he worked on the early development of Proton and *Salyut.*
Izotov, Sergey Petrovich	June 30, 1917– May 6, 1983	Chief Designer/General Designer in 1960–83 at OKB-117 worked on Chelomey's lunar lander engines.
Kartukov, Ivan Ivanovich	Unknown	Chief Designer of KB-2 at Plant No. 81 worked on solid-propellant engines for spacecraft.
Keldysh, Mstislav Vsevolodovich	February 10, 1911– June 24, 1978	Director of NII-1 in 1946–55, Chief of IPM in 1953–78, and President of Academy of Sciences in 1961–75 led scientific work on missiles/spacecraft.
Kemurdzhian, Aleksandr Leonovich	Unknown	Chief Designer at VNII-100 worked on robotic lunar rovers.
Khomyakov, Mikhail Stepanovich	Unknown	At OKB-1, he was lead designer for Sputnik; later, he was Deputy General Designer at NPO Energiya.
Khristianovich, Sergey Aleksandrovich	November 9 1908–	He worked on ICBMs at TsAGI in 1942–53 and then at Institute of Theoretical and Applied Mechanics.
Khrushchev, Sergey Nikitich	1934–	Deputy Department Chief in 1958–68 at OKB Chelomey is son of Nikita Khrushchev.
Kisunko, Grigoriy Vasilyevich	July 20, 1918–1998	Chief Designer/General Designer in 1953–75 at KB-1 and later at OKB-30 led work on early anti-ballistic missile/ASAT.
Konopatov, Aleksandr Dmitriyevich	March 10, 1922–	Chief Designer in 1965–93 at OKB Kosberg led work on rocket engines.
Konoplev, Boris Mikhaylovich	1912–October 24, 1960	He worked on guidance at NII-885, NII-695, and OKB-692 and died in the R-16 accident.
Korolev, Sergey Pavlovich	January 12, 1907– January 14, 1966	Chief Designer in 1946 at OKB-1 and founder of the Soviet space program, his early prewar rocketry work was at GIRD and NII-3.
Kosberg, Semyon Ariyevich	December 14, 1903– January 3, 1965	Chief Designer in 1941–65 at OKB-154 led work on engines for ICBMs and launchers.
Kotelnikov, Vladimir Aleksandrovich	September 6, 1908–	He was at OKB MEI in 1947–54 and then at the Institute of Radio Technology and Electronics.
Kovtunenko, Vyecheslav Mikhaylovich	August 31, 1921– July 10, 1995	With early work at OKB Yangel, he later was Chief Designer/General Designer at NPO Lavochkin in 1977–95.
Kozlov, Dmitriy Ilich	October 1, 1919–	As head of OKB Korolev Branch No. 3/TsSKB from 1959 on, he worked on reconnaissance satellites.
Kryukov, Sergey Sergeyevich	1918–	He was Deputy Chief Designer in 1961–65 at OKB Korolev then Chief Designer in 1971–77 at OKB Lavochkin.

Full Name	Date of Birth/Death	Contribution to the Soviet Space Program
Kurchatov, Igor Vasilyevich	January 12, 1903–February 7, 1960	At KB-11, he worked on first hydrogen bomb-work coordinated with OKB-1.
Kuznetsov, Nikolay Dmitriyevich	June 23, 1911–July 30, 1995	Chief Designer/General Designer in 1949–94 at OKB-276 worked on rocket engines for the N1 and GR-1.
Kuznetsov, Viktor Ivanovich	April 27, 1913–March 22, 1991	Chief Designer in 1946–89 at NII-10 and NII-944 worked on missile and spaceship gyros.
Lapygin, Vladimir Lavrentyevich	February 4, 1925–	Deputy Chief Designer at Pilyugin NII worked on guidance and succeeded Pilyugin in 1982.
Lavochkin, Semyon Alekseyevich	September 11, 1900–June 9, 1960	Chief Designer in 1939–60 at OKB-301 worked on Burya cruise missile.
Lebedinskiy, Andrey Vladimirovich	1902–January 3, 1965	He was first Director of IMBP in 1963–65 and an early space medicine pioneer.
Lidorenko, Nikolay Stepanovich	April 15, 1916–	Chief Designer at NII IT worked on power sources for spacecraft, including Sputnik.
Likhushin, Valentin Yakovlevich	May 29, 1918–December 4, 1982	Director of NII-1 in 1955–88 worked on advanced engines.
Lobanov, Nikolay Aleksandrovich	1909–1978	Chief Designer in 1968–77 at NIEI PDS worked on parachutes and succeeded Tkachev.
Lozino-Lozinskiy, Gleb Yevgenyevich	December 25, 1909–	Chief Designer in 1966–76 at OKB Mikoyan led work on the Spiral spaceplane.
Lyulka, Arkhip Mikhaylovich	March 23, 1908–June 2, 1984	Chief Designer/General Designer in 1946–84 at OKB-165 worked on cryogenic engines for the N1.
Makeyev, Viktor Petrovich	October 25, 1924–October 25, 1985	This Chief Designer/General Designer in 1955–85 at SKB-385 was a Korolev protégé.
Melnikov, Mikhail Vasilyevich	September 22, 1919–1996	Deputy Chief Designer in 1960–74 at OKB Korolev worked on engines, including Blok D.
Mikoyan, Artem Ivanovich	August 5, 1905–December 9, 1970	Chief Designer/General Designer in 1942–69 at OKB-155 led work on the Spiral space-plane system.
Mishin, Vasiliy Pavlovich	January 18, 1917–	Chief Designer in 1966–74 at OKB Korolev led work on the N1–L3 lunar program, was fired in 1974, and was later at MAI.
Mnatsakanyan, Armen Sergeyevich	November 7, 1918–1992	Chief Designer in 1953–69 at NII-648 worked on spacecraft telemetry and radar systems.
Myasishchev, Vladimir Mikhaylovich	September 28, 1902–October 14, 1978	Chief Designer in 1951–60 at OKB-23 worked on a spaceplane and was later Director of TsAGI.
Nesmeyanov, Aleksandr Nikolayevich	September 9, 1899–1980	President of the Academy of Sciences in 1951–61 approved the first satellite project.

Full Name	Date of Birth/Death	Contribution to the Soviet Space Program
Nudelman, Aleksandr Emmanuilovich	1912–August 2, 1996	Chief Designer in 1965–87 at OKB-16 worked on a space cannon for Chelomey and Kozlov.
Okhapkin, Sergey Osipovich	1910–March 1980	Deputy Chief Designer in 1952–76 at OKB Korolev led work on the N1 and was Mishin's First Deputy.
Okhotsimskiy, Dmitriy Yevgenyevich	February 26, 1921–	This scientist at OPM MIAN did research work on an early ICBM.
Pallo, Vladimir Vladimirovich	Unknown	Deputy Chief Designer at OKB Chelomey Branch No. 1 led work on the DOS and *Salyut* stations.
Parin, Vasiliy Vasilyevich	March 18, 1903–June 15, 1971	Director of IMBP in 1965–69 was a premier space medicine specialist.
Paton, Boris Yevgenyevich	November 27, 1918–	Director of Institute of Electrical Welding from 1953 on worked on the N1 and the Vulkan unit.
Petrov, Boris Nikolayevich	March 11, 1913–August 23, 1980	Department Chief in 1951–1980 at Institute of Control Problems was a public spokesperson.
Petrov, Georgiy Ivanovich	May 31, 1912–May 17, 1987	After conducting aerodynamic research at NII-1, he was Director of Institute of Space Research in 1965–73.
Pilyugin, Nikolay Alekseyevich	May 18, 1908–August 2, 1982	Chief Designer in 1948–82 at NII-885 and NII AP worked on missile and spaceship guidance.
Pobedonostsev, Yuriy Aleksandrovich	February 7, 1907–October 1973	He was Chief Engineer in 1946–49 at NII-88 and was later at NII-125.
Polukhin, Dmitriy Alekseyevich	March 12, 1927–September 7, 1993	Chief Designer in 1973–93 at OKB Chelomey Branch No. 1 led the development of Proton.
Radovskiy, Viktor Petrovich	May 11, 1920–	Deputy Chief Designer at OKB-456 worked on rocket engines and succeeded Glushko in 1989.
Raspletin, Aleksandr Andreyevich	August 25, 1908–1967	Chief Designer in 1953–67 at KB-1 worked on the RORSAT, EORSAT, and ASAT programs.
Raushenbakh, Boris Viktorovich	January 18, 1915–	Department Chief in 1960–73 at OKB Korolev worked on guidance systems.
Reshetnev, Mikhail Fedorovich	November 10, 1924–January 26, 1996	Chief Designer/General Designer in 1961–96 at OKB-10 led work on communications satellites and was a Korolev protégé.
Rosselevich, Igor Aleksandrovich	1918–1991	Chief Designer in 1954–83 at NII-380 worked on TV systems for spacecraft.
Ryazanskiy, Mikhail Sergeyevich	April 5, 1909–August 7, 1987	Chief Designer in 1946–51 and 1955–87 at NII-885 worked on missile and spacecraft radio guidance.
Savin, Anatoliy Ivanovich	April 6, 1920–	General Designer from 1962 on at KB-1 and TsNII Kometa worked on the RORSAT, EORSAT, and ASAT programs.

Full Name	Date of Birth/Death	Contribution to the Soviet Space Program
Semenov, Yuriy Pavlovich	April 20, 1935–	He was the lead designer of Soyuz and Zond at OKB Korolev and then General Designer at RKK Energiya from 1989 on.
Sedov, Leonid Ivanovich	November 14, 1907–	He chaired the Commission for Promotion of Interplanetary Flights and was a public spokesperson.
Severin, Gay Ilich	July 24, 1926–	Chief Designer/General Designer from 1964 on at OKB Zvezda worked on spacesuits and EVA airlocks.
Shabarov, Yevgeniy Vasilyevich	1922–	Deputy Chief Designer at OKB Korolev led the flight testing of piloted spacecraft.
Sheremetyevskiy, Nikolay Nikolayevich	November 5, 1916–	Chief Designer in 1974–91 at NII Iosifyan in 1974–91 worked on power sources and Earth survey satellites.
Sisakyan, Norair Martirosovich	January 25, 1907–March 10, 1966	At the Second Division of Biological Sciences under the Academy of Sciences, he was an early medicine specialist.
Solovyev, Vsevolod Nikolayevich	Unknown	Chief Designer in 1963–92 at KB TransMash designed space launch complexes.
Stechkin, Boris Sergeyevich	1891–April 2, 1969	Chief Designer in 1955–69 at OKB Fakel under the Academy of Sciences performed attitude control engine work.
Stroyev, Nikolay Sergeyevich	1912–1997	Director of Gromov LII in 1954–66 worked on spacecraft testing and later was at VPK.
Struminskiy, Vladimir Vasilyevich	April 29, 1914–	Director of Institute of Theoretical and Applied Mechanics in 1966–71 worked on liquid hydrogen.
Tikhonravov, Mikhail Klavdiyevich	July 29, 1900–March 4, 1974	Designer at NII-4 and OKB Korolev worked on Sputnik and Vostok and performed early ICBM work and early work at GIRD and NII-3.
Tkachev, Fedor Dmitriyevich	Unknown	Chief Designer at NIEI PDS worked on parachutes and was fired in 1968 after Soyuz 1.
Tregub, Yakov Isayevich	Unknown	Deputy Chief Designer in 1964–73 at OKB Korolev led flight control for piloted flights.
Tritko, Karl Ivanovich	Unknown	Chief of SKB at NII-88 in 1946–49 led work on early missiles.
Trufanov, Yuriy Nikolayevich	Unknown	He was Deputy Chief Designer at OKB Chelomey Branch No. 1 and was then at NPO Energiya and NPO Lavochkin.
Tsybin, Pavel Vladimirovich	December 23, 1905–February 4, 1992	Deputy Chief Designer in 1960s at OKB Korolev performed early spaceplane work at OKB-256.
Tumanskiy, Sergey Konstantinovich	May 21, 1901–September 9, 1973	Chief Designer/General Designer in 1955–73 at OKB-300 worked on spacecraft attitude engines.

Full Name	Date of Birth/Death	Contribution to the Soviet Space Program
Tupolev, Andrey Nikolayevich	November 10, 1888– December 23, 1972	Chief Designer/General Designer in 1943–72 at OKB-156 worked on spaceplane carrier aircraft.
Tyurin, Petr Aleksandrovich	June 25, 1917– February 26, 2000	Chief Designer in 1953–81 of KB Arsenal worked on L3 components and later performed EORSAT work.
Utkin, Ivan Ivanovich	April 23, 1910– August 29, 1985	Chief Designer in 1960–70 at NII IT worked on spacecraft memory data recorders.
Utkin, Vladimir Fedorovich	October 17, 1923– February 15, 2000	He was Deputy Chief Designer in 1961–71 at OKB Yangel and succeeded Yangel in 1971.
Vernov, Sergey Nikolayevich	July 11, 1910– September 26, 1982	Director of NII-YaF of Moscow State University in 1960–82 worked on science experiments.
Vinogradov, Aleksandr Pavlovich	August 21, 1895–1975	Director of Institute of Geochemical and Analytical Chemistry worked on lunar samples.
Vitka, Vladimir Andreyevich	November 19, 1901– January 10, 1989	First Deputy Chief Designer in 1954–61 at OKB Glushko worked on rocket engines.
Voronin, Grigoriy Ivanovich	December 21, 1906– 1987	Chief Designer in 1939–85 at OKB-124 worked on life support systems for spacecraft.
Voskresenskiy, Leonid Aleksandrovich	July 14, 1913– December 15, 1965	Deputy Chief Designer in 1953–64 at OKB Korolev led flight testing of missiles.
Yangel, Mikhail Kuzmich	October 25, 1911– October 25, 1971	Chief Designer in 1954–71 at OKB-586 led work on missiles and robotic spacecraft.
Yefremov, Gerbert Aleksandrovich	March 15, 1933–	Deputy General Designer in 1971–84 at OKB Chelomey succeeded Chelomey in 1984.
Zaslavskiy, Mark Efimovich	1920–1995	He was Chief Designer in the 1960s at TsNII-108.

Military Officers

Agadzhanov, Pavel Artemyevich	May 21, 1923–	Department Chief in 1957–71 at TsKIK led flight control for piloted missions.
Agaltsov, Fillip Aleksandrovich	January 8, 1900–1980	Air Force Deputy Commander-in-Chief in 1958–62 prepared the selection of cosmonauts.
Alekseyev, Nikolay Nikolayevich	1914–November 12, 1980	He chaired the Science and Technical Committee, General Staff, Ministry of Defense, in 1960–70.
Anokhin, Sergey Nikolayevich	March 19, 1910– April 15, 1986	He was a test pilot for Gromov LII in 1941–64 and then worked at OKB Korolev.
Babiychuk, Aleksandr Nikolayevich	Unknown	Chief, Biomedical Service, at Air Force oversaw the early Vostok missions.
Beregovoy, Georgiy Timofeyevich	April 15, 1921– June 30, 1995	He was a cosmonaut who later became Director of the Cosmonaut Training Center in 1972–86.
Bibikov, Yakov Lvovich	Unknown	He was Director of NII-1 during German recovery operations in 1945–46.

Full Name	Date of Birth/Death	Contribution to the Soviet Space Program
Biryuzov, Sergey Semenovich	August 21, 1904–October 19, 1964	He was Commander-in-Chief of RVSN in 1962–63 and later Chief of General Staff, Ministry of Defense.
Bolshoy, Amos Aleksandrovich	Unknown	Department Chief at TsKIK led flight control teams for early missions.
Bulychev, Ivan Timofeyevich	Unknown	He was Chief of Communications Directorate, Ministry of Defense, in 1956–58.
Chechulin, Petr Petrovich	September 10, 1906–September 16, 1971	He was Director of NII-4 in 1951–55 during early research on satellites.
Fedorov, Petr Ivanovich	1898–February 7, 1945	First Director of NII-1 in 1944–45 oversaw the early search for the A-4.
Gagarin, Yuriy Alekseyevich	March 9, 1934–March 27, 1968	First human in space later became Deputy Director of the Cosmonaut Training Center in 1963–67 and then died in a plane crash.
Gallay, Mark Lazarevich	1914–1998	Test pilot at Gromov LII led training at the Cosmonaut Training Center.
Gaydukov, Lev Mikhaylovich	January 14, 1911–	Chief of the Interdepartmental Technical Commission in Germany in 1945–46.
Genin, Abram Moiseyevich	May 12, 1922–	He was Directorate Chief at Institute of Aviation and Space Medicine in 1964–75.
Gerchik, Konstantin Vasilyevich	September 27, 1918–	He was Commander of Tyura-Tam during the R-16 disaster in 1958–61.
Goreglyad, Leonid Ivanovich	1915–1986	He was General Staff representative at the Cosmonaut Training Center and an aide to Kamanin.
Grechko, Andrey Antonovich	October 17, 1903–April 26, 1976	Deputy Minister of Defense in 1967–76 was against piloted space programs.
Grigoryev, Mikhail Grigoryevich	October 23, 1917–November 12, 1981	First Commander of Mirnyy site in 1957–62 later chaired the State Commission for Almaz.
Gurovskiy, Nikolay Nikolayevich	Unknown	He was a doctor at Institute of Aviation and Space Medicine and later Deputy Director at IMBP.
Kamanin, Nikolay Petrovich	1909–March 13, 1982	Deputy Chief of General Staff in 1958–66 and then Aide to Air Force Commander in 1966–71 oversaw cosmonaut training.
Karas, Andrey Grigoryevich	September 27, 1918–January 2, 1979	He was Chief of TsKIK in 1959–65 and later Commander of TsUKOS/GUKOS in 1965–79.
Karpov, Yevgeniy Anatolyevich	1921–May 1990	He was first Director of the Cosmonaut Training Center in 1960–63.
Kerimov, Kerim Aliyevich	November 14, 1917–	First Commander of TsUKOS in 1964–65 and Directorate Chief at Ministry of General Machine Building in 1965–74 chaired the State Commission for Soyuz in 1966–91.

Full Name	Date of Birth/Death	Contribution to the Soviet Space Program
Kirillov, Anatoliy Semenovich	December 31, 1924–March 30, 1987	Chief, First Directorate, at Tyura-Tam in 1960–67 oversaw launch teams.
Krylov, Nikolay Ivanovich	April 29, 1903–February 9, 1972	Commander-in-Chief, Strategic Missile Forces, in 1963–72 was against piloted space programs.
Kurushin, Aleksandr Aleksandrovich	March 14, 1922–	He was Commander of Tyura-Tam in 1965–73 during the N1 launches.
Kutakhov, Pavel Stepanovich	August 6, 1914–	Commander-in-Chief of Air Force in 1969–84 succeeded Vershinin.
Kuznetsov, Nikolay Fedorovich	December 26, 1916–March 5, 2000	He was Commander of the Cosmonaut Training Center in 1963–72 during the Voskhod and Soyuz programs.
Kuznetsov, Nikolay Nikolayevich	1903–1983	He was first Chief of the Interdepartmental Technical Commission in Germany in 1945.
Maksimov, Aleksandr Aleksandrovich	August 29, 1923–October 12, 1990	Deputy Chief of TsKIK was later Commander of GUKOS/UNKS in 1979–89.
Malinovskiy, Rodion Yakovlevich	November 23, 1898–March 31, 1967	Minister of Defense in 1957–67 was against piloted space programs.
Morozov, Viktor Pavlovich	November 1, 1918–July 4, 1981	He chaired the Scientific-Technical Committee of the Strategic Missile Forces in 1962–67.
Moskalenko, Kirill Semenovich	May 11, 1902–June 17, 1985	He was Commander-in-Chief of the Strategic Missile Forces in 1960–62 and succeeded Nedelin.
Mozzhorin, Yuriy Aleksandrovich	December 28, 1920–May 15, 1998	Director of NII-88 in 1961–90 oversaw Soviet space policy.
Mrykin, Aleksandr Grigoryevich	August 15, 1905–October 6, 1972	He was First Deputy Commander of GURVO in 1955–65 and Strategic Missile Forces liaison with space.
Nedelin, Mitrofan Ivanovich	November 9, 1902–October 24, 1960	Deputy Minister of Defense in 1955–59 was first Commander-in-Chief of the Strategic Missile Forces and died in the R-16 accident.
Nesterenko, Aleksey Ivanovich	March 30, 1908–July 18, 1995	He was first Director of NII-4 in 1946–50 and then first Commander of Tyura-Tam in 1955–58.
Nitochkin, Aleksey Alekseyevich	Unknown	Engineer at TsPI-31 designed Tyura-Tam launch range.
Nosov, Aleksandr Ivanovich	March 27, 1913–October 24, 1960	Chief of launch command at Tyura-Tam in 1955–58 died in the R-16 disaster.
Odintsov, Mikhail Petrovich	1921–	Director of the Cosmonaut Training Center in 1963 was fired.
Ostashev, Yevgeniy Ilich	March 22, 1924–October 24, 1960	First Directorate Chief at Tyura-Tam in 1956–60 died in the R-16 disaster.

Full Name	Date of Birth/Death	Contribution to the Soviet Space Program
Pokrovskiy, Aleksey Vasilyevich	1903–1988	He was Director of Institute of Aviation and Space Medicine from the 1940s to 1959.
Rudenko, Sergey Ignatyevich	October 7, 1904–1990	First Deputy Commander-in-Chief of Air Force in 1958–68 oversaw cosmonaut training.
Semenov, Anatoliy Ivanovich	November 12, 1908–April 16, 1973	He was Commander of GURVO during Sputnik and Vostok in 1954–64.
Shatalov, Vladimir Aleksandrovich	December 18, 1927–	Commander-in-Chief's Aide of Air Force in 1971–87 succeeded Kamanin.
Shubnikov, Georgiy Maksimovich	May 1, 1903–July 31, 1965	He was Chief of Construction Directorate at Tyura-Tam in 1955–65.
Smirnitskiy, Nikolay Nikolayevich	August 9, 1918–April 15, 1993	Commander of GURVO in 1967–75 later moved to Ministry of General Machine Building.
Sokolov, Andrey Illarionovich	October 30, 1910–February 5, 1976	He was Director of NII-4 during the early space program in 1955–70.
Spiridinov, Aleksey Sergeyevich	Unknown	He was in Seventh Chief Directorate of Ministry of Armaments and Director of NII-88 in 1953–59.
Spitsa, Ivan Ivanovich	1919–1992	He was Commander of TsKIK during the N1 launches in 1965–73.
Titov, German Stepanovich	September 11, 1935–	Second human in orbit was later First Deputy Commander of GUKOS in 1979–91.
Tolubko, Vladimir Fedorovich	November 25, 1914–June 17, 1989	First Deputy Commander-in-Chief of the Strategic Missile Forces in 1960–68 was later Commander-in-Chief in 1972–85.
Tveretskiy, Aleksandr Fedorovich	November 17, 1904–December 31, 1992	He was first Commander of Special Purpose Brigade, precursor to the Strategic Missile Forces, in 1946–49.
Tyulin, Georgiy Aleksandrovich	October 9, 1914–April 22, 1990	First Deputy Chairman of GKOT in 1961–65 and First Deputy Minister of General Machine Building in 1965–76 oversaw many State Commissions.
Vasilyev, Anatoliy Alekeseyevich	November 28, 1921–November 12, 1973	Commander of GURVO in 1964–67 later chaired the Scientific-Technical Committee of the Strategic Missile Forces in 1967–69.
Vershinin, Konstantin Andreyevich	June 3, 1900–December 30, 1973	He was Commander of Air Force during the Vostok, Voskhod, and early Soyuz missions in 1957–69.
Vitruk, Andrey Avksentyevich	1906–	He was first Commander of TsKIK during the Sputnik and Luna missions in 1957–59.
Volynkin, Yuvenaliy Mikhaylovich	February 7, 1907–	He was Director of Institute of Aviation and Space Medicine in 1960–69 during Vostok.
Voronov, Nikolay Nikolayevich	1899–February 28, 1968	Commander of Artillery Forces in 1941–1950 later became President of Academy of Artillery Sciences.

Full Name	Date of Birth/Death	Contribution to the Soviet Space Program
Votintsev, Yuriy Vsevolodovich	1919–	First Commander of PRO/PKO forces in 1967–85 was in charge of ASAT forces.
Voznyuk, Vasiliy Ivanovich	January 1, 1907–September 12, 1976	First Commander of Kapustin Yar range in 1946–73 selected the Tyura-Tam site.
Yakovlev, Nikolay Dmitryevich	1898–May 10, 1972	He was Chief of Chief Artillery Directorate in 1941–48.
Yazdovskiy, Vladimir Ivanovich	1913–	Deputy Director of Institute of Aviation and Space Medicine was a space medicine pioneer.
Zakharov, Aleksandr Grigoryevich	February 20, 1921–	He was Commander of Tyura-Tam range during Vostok and Voskhod in 1961–65.
Zakharov, Matvey Vasilyevich	1898–January 31, 1972	He was Chief of Ministry of Defense General Staff in 1960–73.
Zhukov, Georgiy Konstantinovich	December 1, 1896–June 18, 1974	He was Minister of Defense in 1955–57 during the selection of Tyura-Tam.

Party and Government Officials

Afanasyev, Sergey Aleksandrovich	August 30, 1918–	First Minister of General Machine Building in 1965–83 oversaw N1 project.
Balmont, Boris Vladimirovich	October 6 1927–	He was Chief of Chief Directorate at Ministry of General Machine Building in 1965–73 and First Deputy Minister of General Machine Building in 1976–81.
Beriya, Lavrentiy Pavlovich	March 29, 1899–December 23, 1953	He was Soviet security apparatus chief through 1953.
Brezhnev, Leonid Illich	December 19, 1912–November 10, 1982	He was Secretary of Central Committee for defense and space in 1957–60 and 1963–65.
Brezhnev, Mikhail Aleksandrovich	Unknown	Deputy Minister of General Machine Building was responsible for guidance systems.
Bulganin, Nikolay Aleksandrovich	June 11, 1895–February 24, 1975	Minister of Defense in 1947–49 and 1953–55 chaired Special Committee No. 2 in 1947–49.
Burnazyan, Avetik Ignatyevich	1906–	Deputy Minister of Health from 1947 was involved in Voskhod crew selection.
Butoma, Boris Yevstafyevich	May 1, 1907–July 11, 1976	He was Minister of Ship Building Industry in 1957–76.
Dementyev, Petr Vasilyevich	January 24, 1907–May 14, 1977	Minister of Aviation Industry in 1953–77 was a supporter of Chelomey.
Dmitryev, Igor Fedorovich	1909–	First Deputy Chief, Central Committee Defense Industries Department, in 1965–81 succeeded Serbin.
Domrachev, Aleksandr Vasiliyevich	October 1906–January 26, 1961	First Chairman of GKOT in 1957–58 participated in Tyura-Tam's selection.

Full Name	Date of Birth/Death	Contribution to the Soviet Space Program
Gonor, Lev Robertovich	1906–November 13, 1969	First Director of NII-88 in 1946–50 was dismissed in 1950.
Grishin, Lev Arkhipovich	1920–October 24, 1960	Deputy Chairman of GKOT in 1958–60 died in the R-16 disaster.
Ivashutin, Petr Ivanovich	1903–	First Deputy Chairman of KGB during Vostok in 1959–63 was later GRU Chief in 1963–88.
Kalmykov, Valeriy Dmitriyevich	August 28, 1908–March 22, 1974	He was Minister of Radio-Technical Industry in 1954–74.
Khokhlov, Nikolay Dmitriyevich	Unknown	Deputy Minister of General Machine Building in 1965–83 was responsible for quality control.
Khrunichev, Mikhail Vasilyevich	April 4, 1901–June 2, 1961	Minister of Aviation Industries in 1946–53 was later in Gosplan.
Khrushchev, Nikita Sergeyevich	April 5, 1894–September 11, 1971	First Secretary of Central Committee in 1953–64 during the early space era chaired Council of Ministers in 1958–64.
Kozlov, Frol Romanovich	August 18, 1908–January 30, 1965	He was Secretary of Central Committee for defense and space during Vostok in 1960–63.
Leshchenko, Sergey Mikhaylovich	Unknown	He was First Deputy Minister of Aviation Industries in 1957–64.
Litvinov, Valentin Yakovlevich	1910–1983	Director of Progress Plant in 1944–62 later was Deputy Minister of General Machine Building in 1965–73.
Malenkov, Georgiy Maksimiliyanovich	January 2, 1902–January 23, 1988	First Chairman of Special Committee No. 2 in 1946–47 oversaw missile program.
Malyshev, Vyecheslav Aleksandrovich	February 16, 1902–February 20, 1957	Minister of Medium Machine Building in 1953–55 was first manager of Soviet defense industry.
Mazur, Yevgeniy Vasilyevich	Unknown-1982	He was Deputy Minister of General Machine Building in 1965–82.
Pashkov, Georgiy Nikolayevich	1911–	He was at Gosplan Second Department in 1946–51 and Deputy Chairman of Military-Industrial Commission in 1957–70.
Petrovskiy, Boris Vasilyevich	June 27, 1908–	Minister of Health from 1965 to 1980 operated on Korolev.
Pleshakov, Petr Stepanovich	July 13, 1922–September 11, 1987	Director of TsNII-108 in 1958–64 was then Minister of Radio Industry 1974–87.
Pravetskiy, Vladimir Nikolayevich	Unknown	He was Chief of Third Chief Directorate in Ministry of Health.
Rudnev, Konstantin Nikolayevich	June 22, 1911–August 13, 1980	Director of NII-88 in 1950–52 later chaired GKOT during Vostok in 1958–61.

Full Name	Date of Birth/Death	Contribution to the Soviet Space Program
Ryabikov, Vasiliy Mikhailovich	January 14, 1907–July 19, 1974	Chief of Third Chief Directorate of Council of Ministers in 1951–53 chaired Military-Industrial Commission in 1955–57 and Sputnik State Commission.
Serbin, Ivan Dmitryevich	1910–February 16, 1981	He was Chief of Defense Industries Department in 1958–81.
Serov, Ivan Aleksandrovich	September 29, 1905–July 1, 1990	First Deputy Minister of Internal Affairs later chaired KGB in 1954–58.
Shakhurin, Aleksey Ivanovich	February 25, 1904–July 3, 1975	He was People's Commissar for Aviation Industries in 1940–46.
Smirnov, Leonid Vasilyevich	April 16, 1916–	Director of YuzhMash Plant in 1952–61 chaired Military-Industrial Commission in 1963–85.
Stalin, Iosif Vissarionovich	December 21, 1879–March 5, 1953	General Secretary of Central Committee in 1924–53 was Chairman of Council of Ministers in 1941–53.
Stroganov, Boris Aleksandrovich	Unknown	He was Sector Chief, Central Committee Defense Industries Department.
Tabakov, Gleb Mikhaylovich	1912–1993	Director of NII-229 in 1958–63 was later Deputy Minister of General Machine Building in 1965–81.
Udarov, Grigoriy Rafailovich	1904–1991	Deputy Minister of General Machine Building in 1965–79 was responsible for ground complexes.
Ustinov, Dmitriy Fedorovich	October 30, 1908–December 20, 1984	Chairman of Military-Industrial Commission during Sputnik and Vostok in 1957–63 was later Secretary of Central Committee for defense and space in 1965–76.
Vetoshkin, Sergey Ivanovich	September 25, 1905–July 19, 1991	Directorate Chief in Ministry of Armaments was then First Deputy Chairman of Military-Industrial Commission in 1958–65.
Vladimirskiy, Sergey Mikhaylovich	Unknown	Deputy Minister of Radio-Technical Industries in 1954–79 was earlier at KB-1.
Zubovich, Ivan Gerasimovich	1901–July 18, 1956	Deputy Chairman of Special Committee No. 2 was then Deputy Minister of Armaments in 1949–51.
Zverev, Sergey Alekseyevich	October 18, 1912–December 17, 1978	He was Chairman of GKOT during the Voshkod program in 1963–65.

INDEX

Note: The following index serves both this book and
its companion volume, *Sputnik and the Soviet Space Challenge.*

* Note that page numbers in italics indicate photographs.

A

INDEX

All-Union Scientific-Research Institute for Digital Computer Technology, 594
All-Union Scientific-Research Institute for Electromechanics, 594, 683; see also NII-627
Almaz program, 597, 598, 607, 633, 635, 729, 766, 768, 770, 787, 800, 804, 814, 840; and conversion to DOS, 720–21; and coordination of schedules with DOS, 808–09; and delays in late 1960s, 716–17; and description of spaceship, 592–96; and first crews for, 808; and origins of, 590–92; and preparations for launch of Almaz 1/Salyut 2, 810–11; and Salyut 2 mission, 811–12; and work in 1970–72, 806–07; and work in 1970s and 1980s, 841–43
Amak-3 experiment, 778
Ambartsumyan, Viktor A., 770
An-2 aircraft, 248
An-12 aircraft, 198, 263, 422, 473, 569, 630, 656
Analytical Instrument Building Design Bureau, 472
Anders (a German), 58
Anders, William A., 667, 674
Andropov, Yuriy, 432, 846
Anikeyev, Ivan N., 246, 247, 374–75
Anna-3 telescope, 767, 778
Anokhin, Sergey N., 566, 567, 588
Anti-Party Group affair, 161, 177
antipodal bomber, see Sänger-Bredt bomber
Antonov, Oleg K., 198, 218
Appazov, Refat F., 44
Apollo Applications Program (NASA), 714; see also Skylab
Apollo program/spaceship (NASA), 383, 384, 396, 397, 398, 399, 402, 405, 406, 408, 409, 444, 446, 471, 475, 482, 483, 497, 499, 502, 539, 544, 550, 553, 554, 562, 595, 607, 614, 641, 646, 655, 661, 662, 668, 712, 736, 739, 741, 742, 750, 757, 772, 779, 827, 853, 857, 859; and financial comparison with N1-L3, 838; and Soviet comparison with Luna sample return missions, 740
Apollo 1 fire, 554, 562, 629, 651, 658
Apollo 4 mission, 643, 644
Apollo 5 mission, 734
Apollo 7 mission, 658, 660, 663
Apollo 8 mission, 667, 674, 693, 701, 713, 716; and decision on, 662–63; and Soviet response to, 665–68, 674–678, 687, 746, 754
Apollo 9 mission, 684, 734
Apollo 10 mission, 684, 686, 687
Apollo 11 mission, 687, 688, 693, 694–96, 699, 714, 729, 733, 737, 740, 750, 765, 788, 856; and Soviet response to, 696–97, 703
Apollo 12 mission, 729, 740
Apollo 15 mission, 763
Apollo 16 mission, 793, 804
Apollo 17 mission, 793, 806
Apollo-Soyuz Experimental Flight (EPAS), see Apollo-Soyuz Test Project
Apollo-Soyuz Test Project (ASTP), 43, 814, 848; and crews for, 814–15; and origins of, 793–94; and test flights for, 826
Argon-11 computer, 558, 614
Argon-12A computer, 594
Armenian Academy of Sciences, 778, 817

Armstrong, Neil A., 694, 695, 696, 699, 713, 724
Arsenal Machine Building Plant, 648, 733, 754
artificial gravity experiments, see IT project
Artyukhin, Yuriy, 811
Arzamas-16, 120
AS-203 Apollo mission, 734
ASA-34R camera, 593
Atlas booster, 846
Atlas ICBM, 80
Atlet spacesuit, 767
atomic bomb development, 36, 51, 86
Avduyevskiy, Vsevolod S., 189, 313
Aviation Week & Space Technology journal, 552
Azov Machine Building Plant, 769

B

B-29 bomber, 36
B-52A bomber, 600
Babakin, Georgiy N., 437–38, 528, 533, 534, 547, 548, 640, 668, 675, 679, 687, 688, 737, 738, 753, 782, 848; and background of, 530; and death of, 795–96; and development of Ye-8-5 lunar sample return spaceship, 641–43; and Luna 15 mission, 694–96; and Mars sample return, 753–54
Babiychuk, Aleksandr N., 425
Bakurin, 28
Balanina, Mariya N., 11
Barani training device, 248
Baranov Central Institute of Aviation Engine Building, see TsIAM
Bardin, Ivan P., 145
Barmin, Vladimir P., 29, 35, 46, 47, 132, 134, 136, 155, 156, 159, 170, 177, 192, 201, 254, 272, 289, 293, 330, 331, 356, 393, 429, 459, 480, 504, 538, 545, 550, 560, 563, 611, 619, 647, 692, 704, 755, 851; and background of, 549; and becomes Academician, 519–20; and death of, 846; and gets first Hero of Socialist Labor, 121; and gets second Hero of Socialist Labor, 284; and Long-Duration Lunar Base, 764–66
Baryshev, Vladimir M., 300
Bashkin, Yevgeniy V., 813
Batitskiy, Pavel F., 790
Bauman Moscow Higher Technical School, 3, 85, 127, 183, 189
Baykonur Cosmodrome, 512, 574, 579, 621, 625, 629, 655, 658, 659, 665, 680, 681, 688, 699, 700, 703, 705, 723, 772, 773, 805, 810, 851; and creation of name, 284; see also Tyura-Tam
Bazhinov, Igor K., 85, 139, 140
Belka (dog), 253
Belov, Nikolay P., 183, 186
Belyayev, Pavel I., 246, 247, 421, 461, 506, 512, 551, 609, 620, 717; and background of, 451–52; and Voskhod 2 mission, 454–60
Benderov, Vladimir N., 415
Beregovoy, Georgiy T., 522, 524, 622, 629, 630, 674, 693, 724; and Soyuz 2/3 mission, 657–62
Bereznyak, Aleksandr Ya., 17, 19, 29, 31

H

I

O

S

W

Asif A. Siddiqi is a Ph.D. candidate in history at Carnegie Mellon University. He has contributed articles on the space program and the Cold War to *Spaceflight*, *Quest: The Journal of Spaceflight*, and *Countdown*.